Springer Asia Pacific Mathematics Series

Volume 1

The Springer Asia Pacific Mathematics Series promotes high-quality scientific research connected to Asia-Pacific countries. Monographs, edited volumes, lecture notes, and textbooks will cover topics related to pure and applied mathematics as well as interdisciplinary research.

All material will be rigorously peer-reviewed and should be related to mathematical research conducted at or with institutions in the APAC region. This series is open to subseries, which may be topical or exclusively supplied by a single research institution.

Kenichiro Ishii • Naonori Ueda • Eisaku Maeda •
Hiroshi Murase

Pattern Recognition and Machine Learning for Self-Study I

Supervised Learning

 Springer

Kenichiro Ishii
Nagoya University
Nagoya, Japan

Naonori Ueda
RIKEN Center for Advanced Intelligence
Tokyo, Japan

Eisaku Maeda
Graduate School of System Design and
Technology
Tokyo Denki University
Tokyo, Japan

Hiroshi Murase
Nagoya University
Nagoya, Japan

ISSN 3091-2555   ISSN 3091-2563   (electronic)
Springer Asia Pacific Mathematics Series
ISBN 978-981-95-1477-9   ISBN 978-981-95-1478-6   (eBook)
https://doi.org/10.1007/978-981-95-1478-6

Mathematics Subject Classification: 68T07, 68T10, 68T05

This Springer imprint is published by the registered company Springer Nature Singapore Pte Ltd.
The registered company address is: 152 Beach Road, #21-01/04 Gateway East, Singapore 189721,
Singapore

If disposing of this product, please recycle the paper.

Preface

This book explains the fundamental principles of pattern recognition and machine learning in a beginner-friendly manner, targeting those who are trying to learn these subjects on their own. Within this field, the primary focus of the book is on supervised learning, while unsupervised learning will be covered in a sequel. The expected prerequisite knowledge for readers is limited to foundational understanding of linear algebra, probability, and statistics.

The expectations for pattern recognition and machine learning, considered the core technologies of artificial intelligence, have been steadily increasing in recent years. Research in this field began with the advent of computers, and numerous technologies have been proposed to date. One writing style involves addressing as many of these fascinating topics as possible, organizing them systematically, and presenting them to readers. However, when faced with such a book, beginners must extract fundamental topics from a vast array of subjects and further decide in what order to learn them. For beginners with no prior knowledge of the field, executing this task independently is not easy. While having a suitable guide allows individuals to select topics tailored to their respective levels, this is not something one can expect when studying independently.

Therefore, in writing this book, consideration has been given to the stance of "for beginners' self-study," focusing on the following points:

1. Rather than comprehensively introducing a diverse range of technologies, the book carefully selects the essential topics necessary to understand the field. These chosen topics are then explained in a focused and detailed manner, arranged in an order that is easy for beginners to learn.
2. Areas that may be challenging for beginners to grasp or prone to misunderstanding are extensively explained over multiple pages. Additionally, the derivation of formulas is presented meticulously to ensure that readers can easily follow along.
3. To enhance readers' intuitive understanding, the book incorporates as many concrete examples and experimental cases as possible.
4. The book is self-contained, eliminating the need for readers to refer to other books or literature.

5. The book explores where the sources of new techniques and methods originated, describing the historical process in which new technologies were created as a result of the combination of classical ideas.
6. A number of "coffee break" sections are included for relaxation, incorporating insights and expertise derived from the authors' experiences.
7. To facilitate a deeper understanding of the book's content, each chapter concludes with exercise problems, and detailed solutions are made available on the web for anyone to access.

This book is composed of the following three parts:

Part I: Linear Classification
Part II: Nonlinear Classification
Part III: Bayesian Unified Framework

Part I covers linear classification, introducing fundamental concepts in pattern recognition and machine learning. These include classical concepts and methods such as perceptron, Bayes error, Karhunen–Loève expansion linear discriminant methods, subspace methods, and others. To understand various subsequently developed technologies, it is crucial to firmly grasp these classical and fundamental principles that have been handed down over time.

Part II expands the scope from linear to nonlinear classification, explaining topics such as generalized linear discriminant functions, potential function methods, support vector machines, kernel methods, and convolutional neural networks. Among these, generalized linear discriminant functions and potential function methods have played an extremely important role in the subsequent development of support vector machines and kernel methods. Therefore, the book explores how these methods are interconnected and have led to new ideas. A crucial aspect not to be overlooked in nonlinear classification is deep learning. While deep learning holds promising prospects for the future, it is still in its infancy and highly dynamic. Given the limitations of available pages, the book focuses on convolutional neural networks, a topic of interest within deep learning. Understanding convolutional neural networks requires knowledge of regular neural networks, which is explained in a separate chapter.

Part III discusses how various technologies introduced in Parts I and II can be unified under the framework of the Bayes decision rule. The inclusion of this explanation at the end of the book aims to invite readers to revisit the basic techniques introduced in Parts I and II from the perspective of the Bayes decision rule. By adopting this viewpoint, readers can recognize the importance of the Bayes decision rule that permeates this field and can gain a macroscopic view of the entire technology.

Exercise problems are placed at the end of each chapter, and detailed solutions, not just abbreviated ones, are provided on the Springer website. Since these exercises are all valuable for understanding the main text, readers are encouraged to solve them on their own. However, due to page constraints, some topics that should be explained in the main text are covered in the form of exercise problems and their solutions, marked with *. These solutions should be treated as supplementary material. Additionally, exercises requiring programming are distinguished with†.

Readers who pick up this book may feel that the number of topics covered is relatively small compared to the page count, and that classical themes are more abundant than the latest technologies. Instead of introducing each latest technology in pattern recognition and machine learning individually, the book extensively explains essential topics, even if classical, that one should master. We believe that acquiring the foundational knowledge covered in this book will significantly enhance the efficiency of learning the latest technologies.

As mentioned earlier, we have endeavored to write this book in a variety of ways with beginners in mind. However, we acknowledge that some points may not have been fully conveyed through the text alone. Therefore, to achieve a deeper understanding, we strongly encourage readers to work with real-world data. Nowadays, various data sets for classification and learning are publicly available, and the environment for engaging with real data is much richer than before. Programming and applying the techniques introduced in this book to specific problems can yield invaluable insights and skills.

Finally, we express our heartfelt thanks to Dr. Hiroshi Kaneko, Professor Emeritus at Toho University, who strongly encouraged the publication of this book. We also extend our gratitude to Dr. Makoto Tsukada, Professor Emeritus at Toho University, and Dr. Kiyoshi Shirayanagi, Former Professor at Toho University, who provided valuable comments during the writing process, and to Masayuki Nakamura from Springer Japan for his continuous support throughout the publication process.

Nagoya, Japan   Kenichiro Ishii
Tokyo, Japan   Naonori Ueda
Tokyo, Japan   Eisaku Maeda
Nagoya, Japan   Hiroshi Murase
July 2025

Declarations

Competing Interests The authors have no competing interests to declare that are relevant to the content of this manuscript.

Contents

List of Coffee Breaks

Notation

c	number of classes
ω_i	i-th class $(i = 1, \ldots, c)$
n_i	number of patterns in class ω_i
$n = \sum_{i=1}^{c} n_i$	total number of patterns
d	dimensionality of the feature space
x_j	j-th feature $(j = 1, \ldots, d)$
$\boldsymbol{x} = (x_1, \ldots, x_j, \ldots, x_d)^t$	feature vector
$\boldsymbol{w} = (w_1, \ldots, w_d)^t$	weight vector
$\mathbf{x} = (1, x_1, \ldots, x_d)^t$	augmented feature vector
$\mathbf{w} = (w_0, w_1, \ldots, w_d)^t$	augmented weight vector
$\mathcal{X}_i$	set of patterns belonging to class ω_i
$\mathcal{X} = \cup_{i=1}^{c} \mathcal{X}_i$	set of all patterns
$g(\boldsymbol{x})$	discriminant function
$g_i(\boldsymbol{x})$	discriminant function of class ω_i
$g(\boldsymbol{x}) = \mathbf{w}^t \mathbf{x}$	linear discriminant function
$\boldsymbol{x}_s$	support vector
$P(\omega_i)$	prior probability of class ω_i
$P(\omega_i|\boldsymbol{x})$	posterior probability of class ω_i
$p(\boldsymbol{x})$	probability density function of $\boldsymbol{x}$
$p(\boldsymbol{x}|\omega_i)$	probability density function of $\boldsymbol{x}$ belonging to class ω_i
m	number of prototypes, number of component vectors
$\mathbf{p}_k$	k-th prototype $(k = 1, \ldots, m)$
ρ	learning rate $(\rho > 0)$
$b_k = b(\boldsymbol{x}_k)$	teaching signal for pattern $\boldsymbol{x}_k$
$\mathbf{b} = (b_1, \ldots, b_k, \ldots, b_n)^t$	vector of teaching signals for all patterns
$\mathbf{t} = (b_1, \ldots, b_i, \ldots, b_c)^t$	teaching vector (teaching signal for pattern of class ω_i is b_i)
$\mathbf{A} = (a_{ij})$	matrix whose (i, j) components are a_{ij}
$\mathbf{m}_i$	mean vector of class ω_i

$\mathbf{m}$	mean vector of all patterns
$\boldsymbol{\Sigma}_i$	covariance matrix of class ω_i
$\boldsymbol{\Sigma}$	covariance matrix of all patterns
$\boldsymbol{\Sigma}_W$	within-class covariance matrix
$\boldsymbol{\Sigma}_B$	between-class covariance matrix
$\boldsymbol{\Sigma}_T$	total covariance matrix
$\mathbf{S}_W$	within-class scatter matrix
$\mathbf{S}_B$	between-class scatter matrix
$\mathbf{S}_T$	total scatter matrix
$\mathbf{R}$	autocorrelation matrix
r_k	distance from pattern $\boldsymbol{x}_k$ to decision boundary $g(\boldsymbol{x}) = 0$
r^*	minimum value of r_k
R	margin
$\alpha_k\ (\geq 0)$	number of times $\boldsymbol{x}_k$ is not correctly classified during the learning
$\boldsymbol{\alpha} = (\alpha_1, \alpha_2, \ldots, \alpha_n)^t$	error counter vector
$\mathbf{X} = (\mathbf{x}_1, \ldots, \mathbf{x}_n)^t$	pattern matrix
D	dimensionality of Φ space
$\phi_j(\boldsymbol{x})$	scalar function with $\boldsymbol{x}$ as variable ($j = 1, \ldots, D$)
$\boldsymbol{\phi}(\boldsymbol{x}) = (\phi_1(\boldsymbol{x}), \ldots, \phi_D(\boldsymbol{x}))^t$	vector mapping $\boldsymbol{x}$ onto Φ space
$\Phi(\boldsymbol{x}) = \sum_{j=1}^{D} w_j \phi_j(\boldsymbol{x})$	generalized linear discriminant function (Φ function)
$\mathbb{N}$	set of natural numbers
$\mathbb{R}$	set of real numbers
$\mathbb{R}^d$	set of d-dimensional column vectors whose elements are real numbers
$\mathbb{R}^{l \times m}$	set of $l \times m$ matrices whose elements are real numbers
$\mathbf{I}_l$	l-dimensional identity matrix
$\mathbf{1}_l = (1, \ldots, 1)^t$	l-dimensional column vector with all elements 1
$\mathbf{1}_{lm} = \mathbf{1}_l \mathbf{1}_m^t$	matrix of $l \times m$ with all elements 1
$\mathbf{0}_l = (0, \ldots, 0)^t$	l-dimensional column vector with all elements 0
$\mathbf{0}_{lm}$	matrix of $l \times m$ with all elements 0
	(In the above, the orders l and m can be omitted if they do not cause confusion)
$K(\boldsymbol{x}, \boldsymbol{y})$	potential function
$k(\boldsymbol{x}_i, \boldsymbol{y}_j) = \boldsymbol{\phi}(\boldsymbol{x}_i)^t \boldsymbol{\phi}(\boldsymbol{y}_j)$	kernel function ($i = 1, \ldots, l, \quad j = 1, \ldots m$)
$\mathbf{K} = (k_{ij})$	kernel matrix ($k_{ij} = k(\boldsymbol{x}_i, \boldsymbol{y}_j)$)
$\mathbf{F} = (f_{ij})$	Gram matrix ($f_{ij} = \boldsymbol{\phi}(\boldsymbol{x}_i)^t \boldsymbol{\phi}(\boldsymbol{x}_j)$)

Part I
Linear Classification

Chapter 1
Basic Concepts of Pattern Recognition

Abstract Research on pattern recognition began with the advent of computers. The goal of pattern recognition technology is to realize an intelligent machine that can recognize various media such as characters, images, and speech with high accuracy. In other words, it is to realize human's excellent recognition functions on a computer. The pattern recognition functions of humans are considered to be acquired through learning. Therefore, the research to realize the learning functions by machine, that is, machine learning research, is inseparable from pattern recognition. Although they used to refer to almost the same research field, machine learning has developed on its own and expanded its scope of activities since around 2000. Therefore, machine learning has come to refer to a broader research area, including pattern recognition. However, there is little difference between the two fields if we limit ourselves to the fundamentals that beginners should learn. This chapter describes the basic processes of pattern recognition, namely feature extraction and nearest neighbor rule, and prepares for understanding the learning algorithms introduced in the next chapter.

1.1 Structure of Pattern Recognition System

Pattern recognition is the process of mapping an observed *pattern* to one of several predefined *classes*.[1] For example, in the case of alphabetical *character recognition*, the process maps the input pattern to one of 26 classes. The word "pattern" may bring to mind a two-dimensional pattern that is visible to the human eye, but pattern recognition deals with a much wider range of objects. For example, speech recognition, which tanslates human speech into text, is a typical pattern recognition. Another example of pattern recognition is to analyze the waveform

---

Supplementary Information The online version contains supplementary material available at https://doi.org/10.1007/978-981-95-1478-6_1.

---

[1] Sometimes called *category* instead of class.

of an electrocardiogram to determine whether or not there is an abnormality. Pattern recognition is also used to judge situations using not only visual and auditory senses, but also olfactory, tactile, and other sensors. Research on *machine learning* is indispensable for the efficient design of pattern recognition systems. As mentioned at the beginning of this chapter, pattern recognition and machine learning are inseparable.

Humans are capable of advanced pattern recognition, and attempts to realize this intellectual functionality in machines have been one of the central issues for researchers since the advent of computers. However, as research progressed, it became increasingly clear that, contrary to initial expectations, the problem was not as simple as it appeared. Although there was a time when the research on pattern recognition itself was questioned, research in this field has remained active not only because of the genuine intellectual interest in realizing human intellectual functions with machines, but also because of the high potential practical value of pattern recognition. Recognition devices for text, speech, and images have already reached a practical level and are used in various fields. As the demand and expectation for more advanced recognition systems are increasing, research on pattern recognition is expected to become more and more active in the future.

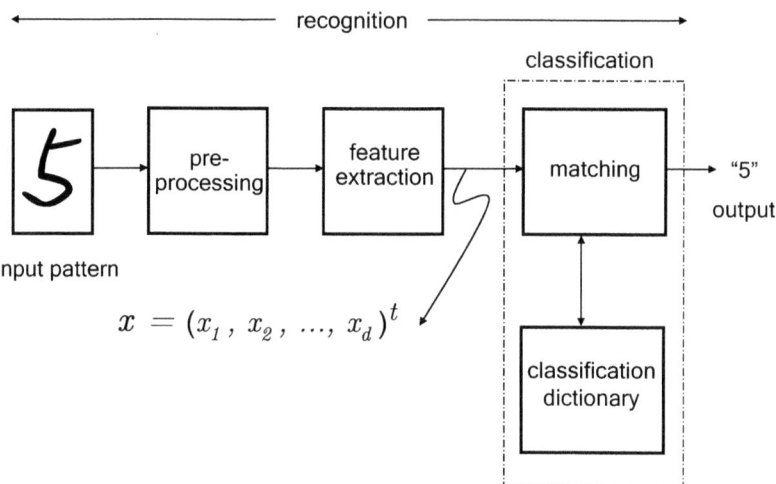

Fig. 1.1 Configuration of a recognition system

A pattern recognition system generally takes the structure of Fig. 1.1. As shown in the figure, a pattern recognition system consists of three parts. When a pattern is input, the *preprocessing* part first performs noise removal, normalization, and so on. The *feature extraction* part then extracts only the essential *features* necessary for classification from the original pattern, which has a vast amount of information. Using these features, the *classification* part assigns one of the classes to the input pattern.

For this purpose, a *classification dictionary* is prepared in advance, and by matching the extracted features with this dictionary, the class to which the input pattern belongs is output as the recognition result.

In this book, this part of dictionary matching is called classification, and the process from pattern input to output, that is, preprocessing, feature extraction, and classification, are collectively referred to as *recognition*.

1.2 Feature Vectors and Feature Space

As shown in Fig. 1.1, it is first necessary for recognition to extract essential features from the original pattern. Feature extraction is an extremely important process that determines the recognition performance. Traditionally, feature extraction has relied on heuristic methods based on human intuition. With the advent of deep learning, it is becoming possible to automate feature extraction for some recognition targets. However, it is difficult for humans to judge from the results of deep learning what features have been extracted. We proceed on the assumption that such features are explicitly obtained, whether by human intuition or as a result of deep learning.

1.2.1 Feature Vectors

Various types of features are possible. For example, in the case of character recognition, the slope, width, curvature, area, and number of loops of a character have been conventionally used. Each feature is represented as a scalar value, and a vector of these scalar values as components is usually used. Assuming that d features are used, the pattern is represented as a d-dimensional vector x as shown in the following equation:

$$x = (x_1, x_2, \ldots, x_d)^t, \tag{1.1}$$

where x_j ($j = 1, \ldots, d$) is the j-th feature and t is the transpose to convert the row vector to a column vector. In this book, all vectors are assumed to be column vectors unless otherwise noted. In the above equation, x is called a *feature vector*, and a d-dimensional space spanned by feature vectors is called a *feature space*. Therefore, a pattern is represented as a single point in the feature space as shown in Fig. 1.2. Let c be the total number of classes, and each class is denoted by $\omega_1, \omega_2, \ldots, \omega_c$. In general, the case of $c \geq 3$ is called a *multi-class problem*, and in particular the case of $c = 2$ is called a *two-class problem*. Since patterns belonging to the same class are similar to each other, patterns in the feature space should be observed as *clusters* of classes, as shown in Fig. 1.2.

Fig. 1.2 Distribution of
patterns in feature space

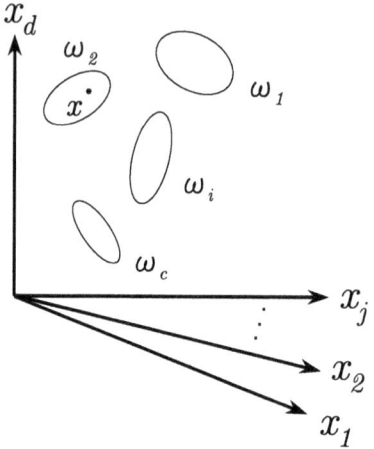

Here, we consider the recognition of patterns that have a two-dimensional extent. This corresponds to the recognition of characters or the determination of lesions from radiographs. For example, Fig. 1.3(a) is a grayscale image with density information. If the number of gray levels is represented by q, this image has a gray level $q = 256$. For such a pattern, we restrict the gray level to a finite number, and replace the actual value with the closest level among them. In Fig. 1.3(b), $q = 4$. This process is called *quantization*. The pattern is further divided into a grid and each element is represented by a certain gray level, as shown in Fig. 1.3(c). This process is called *sampling*. The quantization and sampling operations together are called *digitization*. Each element with a gray level is called a *picture element* or *pixel*. In Fig. 1.3(c), the lattice is set to 32×32, so the number of pixels is 1024. Let x_j be the gray level of the j-th pixel obtained by digitization. The image can be described by the vector shown in Eq. (1.1), where the dimension d is equal to the total number of pixels. Let q be the number of gray levels, then the total number of patterns that can be described by Eq. (1.1) is q^d. Figure 1.3(c) shows the pattern obtained as one of the q^d possibilities.

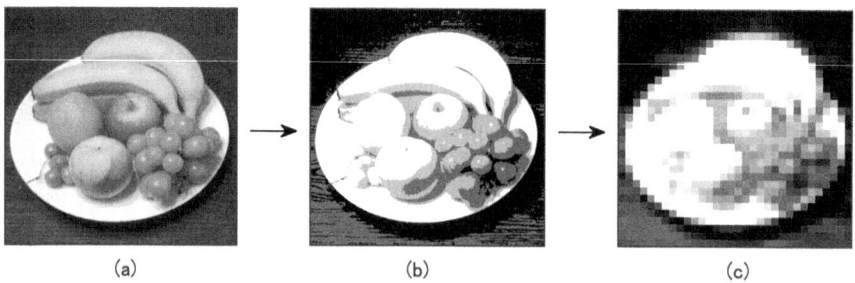

      (a)   (b)   (c)

Fig. 1.3 Digitization of a grayscale pattern. (**a**) Original image. (**b**) Quantization (q = 4). (**c**) Sampling (d = 32 × 32)

The process described above is merely a digitization. However, this process is regarded as feature extraction in the sense that the feature vector of Eq. (1.1) is obtained from the original pattern. In this book, feature extraction is interpreted in this broad sense.

1.2.2 Diversity of Feature Vectors

In the following, we apply these features to handwritten digit recognition. The number of classes is 10. Here, the input pattern is converted to a 5×5 image ($d = 25$), by sampling. Since the character is basically a black-and-white binary pattern, the elements of the feature vector can be considered as the following two values:

$$\begin{cases} x_j = 1 & \text{(black: character)} \\ x_j = 0 & \text{(white: background)} \quad (j = 1, \ldots, d). \end{cases} \tag{1.2}$$

Under this condition, since $q = 2$, the number of patterns that can be represented by 25 pixels is $2^{25} = 33, 354, 432$. Some of these 2^{25} patterns are shown in Fig. 1.4 as examples. Various patterns can be expressed, from (a) to (y) in the figure. The figure shows that 5×5 pixels is a rather coarse sampling to represent numeral patterns.

The simplest way to construct a classification system is to store all 33,554,432 patterns with their class names in a classification dictionary. This is equivalent to creating a reference table, where a class name is assigned to each of the 25-bit data. In this example, the classification dictionary in Fig. 1.1 corresponds to the reference table, and the classification procedure corresponds to the matching process of the reference table. The patterns that have gone through the sampling process in the feature extraction part always match one of the patterns in the classification dictionary. The class of the matched pattern is output as the classification result. However, as is clear from Fig. 1.4, the dictionary includes many patterns that are inappropriate as numerals. For such patterns, we can assign the *reject*[2] as the 11th class. Figure 1.5 shows a feature space consisting of two regions: the region where patterns can exist as numerals and the *reject region*. One point in this space corresponds to one of the 33,554,432 possibilities.[3]

---

[2] There are two types of rejects. One is a reject when the pattern is judged not to belong to any class, which corresponds to the example described here. The other is a case where multiple classes are listed as candidates and it is difficult to determine any of them. For example, Fig. 1.4(u) is rejected, since it is difficult to judge it as either "4" or "9" (See Problems 1.1 and 1.2).

[3] Since the features we are dealing with here are binary, a pattern will occupy a hypercubic grid point in the feature space. Although the figure does not show a discrete expression, we will continue to use a continuous representation in the figures, considering the extension to the general case where the features have continuous values.

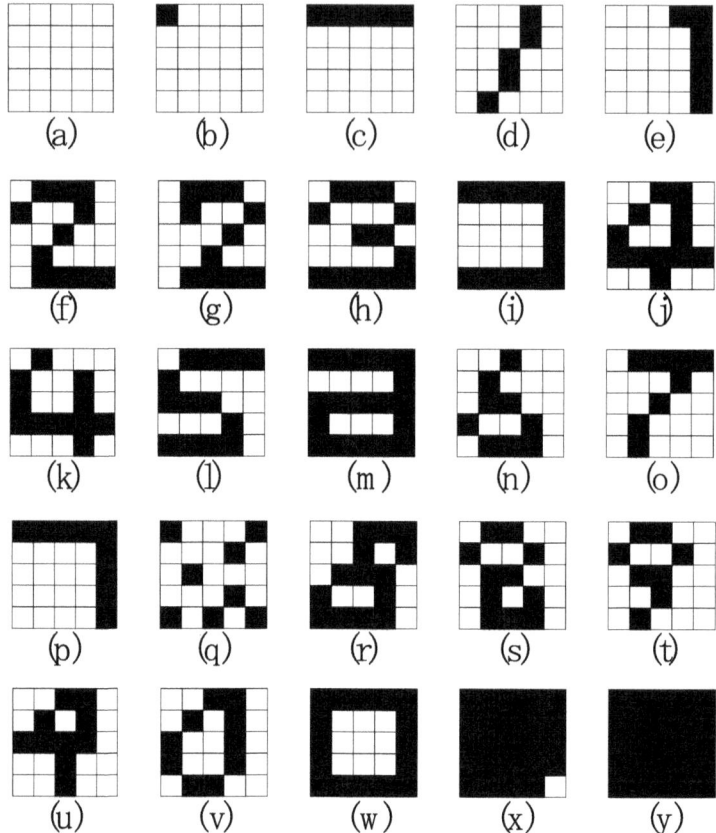

Fig. 1.4 Binary patterns with 5 × 5 pixels

Now, how much effort would it take to create this classification dictionary? It is not possible to automate the creation of the dictionary itself, because the process of assigning a class name to each pattern itself is nothing but the classification. In the end, the creation of a classification dictionary can only be done manually by humans. Let us assume that it takes one person one second to look at a pattern and enter a class name, and that he/she continues this process for eight hours a day. Then, according to a simple calculation, it would take 3 years and 2 months to finish entering all 33,554,432 class names. This is an estimate based on a coarse sampling of 5 × 5. To represent a numeral pattern, at least about 50 × 50 pixels are needed. If the number of pixels increases or a pattern is grayscale, the time required becomes astronomical, and it is no longer possible to create a classification dictionary in a realistic amount of time. Here, we have considered a simple feature, the black and white of a pixel. The situation would be the same even if other features were used.

Fig. 1.5 Feature space and reject region

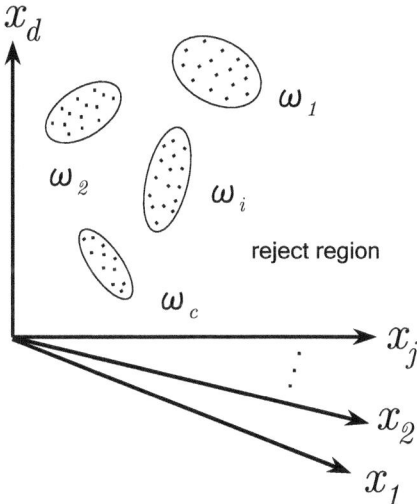

1.3 Prototypes and Nearest Neighbor Rule

The classification dictionary design described above is a method that covers and stores all the possibilities that can occur as feature vectors. This method, although theoretically possible, is impractical in terms of storage capacity and classification time.

As the next best solution, instead of covering all possibilities, only representative patterns can be stored. Such patterns are called *prototypes*. The input pattern is compared to these prototypes in the feature space, and the class to which the prototype with the closest distance to the input pattern belongs, is output as the classification result. The prototype closest to the input pattern is called a *nearest neighbor*. This approach is based on the assumption that patterns in close proximity in the feature space are similar to each other in terms of their properties. The distance is usually defined as the *Euclidean distance*. The classification method mentioned above is called the *nearest neighbor rule*. In the following, we describe the classification procedure using the prototypes and the nearest neighbor rule.

1.3.1 Prototypes

Let m be the number of prototypes. Denote the lth prototype by the d-dimensional vector $\mathbf{p}_l$ and its belonging class by θ_l. This can be written as follows:

$$\mathbf{p}_l \in \theta_l, \quad \theta_l \in \{\omega_1, \ldots, \omega_c\} \quad (l = 1, \ldots, m). \tag{1.3}$$

Denoting the distance between the pattern x of the unknown class and the prototype $\mathbf{p}_l$ as $D(x, \mathbf{p}_l)$, the nearest neighbor rule is expressed as follows: [4]

$$\min_{l=1,\ldots,m} \{D(x, \mathbf{p}_l)\} = D(x, \mathbf{p}_j) \quad \Longrightarrow \quad x \in \theta_j. \tag{1.4}$$

The nearest neighbor rule is a decision rule that finds the closest prototype $\mathbf{p}_j \in \theta_j$ to the unknown pattern x and classifies x as the class θ_j. A more general method is to take the k prototypes that are closest to x and classify x as the class with the highest number of prototypes. This is called the *k-nearest neighbor rule*.[5] The nearest neighbor rule shown in Eq. (1.4) corresponds to the 1-nearest neighbor rule.

Fig. 1.6 Prototypes and the nearest neighbor rule

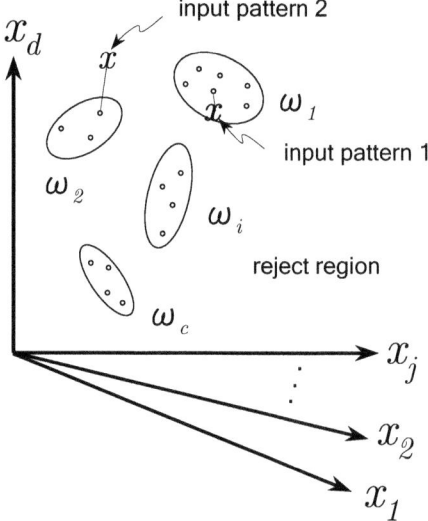

Figure 1.6 shows the processing of the nearest neighbor rule. The prototypes are indicated by white circles. For example, the input pattern 1 is determined to be of class ω_1 because its nearest neighbor belongs to class ω_1. If the distance from the nearest neighbor is too large, the pattern may be meaningless as a character, so it is rejected. For example, the input pattern 2 would be judged as class ω_2 if the nearest neighbor rule is unconditionally applied. However, it is rejected because the distance exceeds a predefined threshold. By applying this procedure, all points in the feature space are assigned a class name. For prototypes, typical patterns representing the class should be selected. Figure 1.7 shows an example of such prototypes for 5×5 pixel patterns.

---

[4] We write $x \in \theta_j$ to denote that the pattern x belongs to the class θ_j.

[5] The nearest neighbor rule and the k-nearest neighbor rule are sometimes abbreviated as the *NN rule* and the *k-NN rule*, respectively. However, in this book, we avoid using these notations as much as possible. This is because a neural network is sometimes abbreviated as an NN, which can easily lead to confusion.

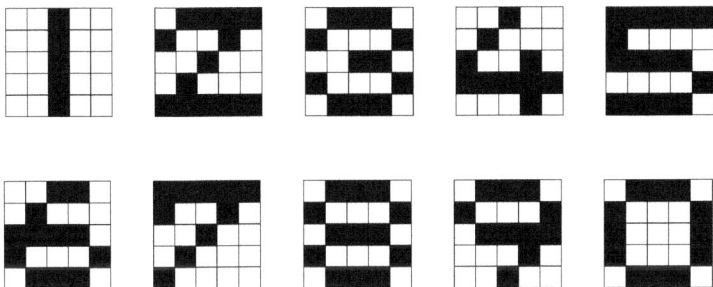

Fig. 1.7   Prototype examples

Since this method only works with prototypes, it drastically reduces the effort required to create a classification dictionary and eliminates the problems of storage capacity and classification time. It is much more efficient than the previous method.

Coffee Break

Nearest Neighbor Rule as a Baseline

Although the nearest neighbor rule is a simple decision rule, it is known to have statistically very interesting properties (see Sect. 5.4.1). For this reason, it has been a favorite topic of researchers for a long time and has been the subject of a large number of papers. (See the references Dasarathy et al. (2000) and Bhatia and Vandana (2010).) At that time, when computer power and storage capacity were poor, the nearest neighbor rule was regarded only as a subject for pursuing the theoretical aspects of pattern recognition. On the contrary, it was precisely because of such a situation that many researchers have been working on the nearest neighbor rule in search of a better classification method in terms of computational and storage capacity. However, with today's dramatic improvements in computer performance, the nearest neighbor rule can now be performed in a realistic amount of time. As a result, the nearest neighbor rule can now at least serve as a baseline for various classification methods.

1.3.2   Partitioning of Feature Space

How should prototypes be set up in order to provide a machine with pattern recognition functions? Let us take the example of digit recognition mentioned earlier. First, it is necessary to know how real handwritten numerals are distributed in the feature space. Since it is not possible to directly determine the exact distribution, we collect actual handwritten characters and consider them as data reflecting the real-world distribution. Since individual characteristics appear as various deformations in handwritten characters, we need to collect a sufficient number of patterns to cover these deformations. The collected patterns are then considered as typical of real patterns, and the number of prototypes and their positions in the feature space are determined so that these patterns can be correctly classified.[6]

The simplest and most reliable way to achieve this goal is to use all the collected patterns as prototypes. This approach is called a *complete storage scheme*. To realize this approach, the throughput and storage capacity of the computer must be sufficient. Note, however, that even if the complete storage scheme is used, it can only cover a small fraction of real patterns that may occur.

Figure 1.8 plots the patterns on a two-dimensional feature space ($d = 2$), assuming that patterns belonging to one of the three classes ω_1, ω_2, and ω_3 were collected. There are 30 patterns in each class, for a total of 90 patterns. The classes ω_1, ω_2, and ω_3 are indicated by the symbols $\bigcirc$, $\triangle$, and $\square$ in the figure, respectively. The elliptical contours representing the shape of the distribution are indicated by thin lines.

Fig. 1.8 Collected patterns

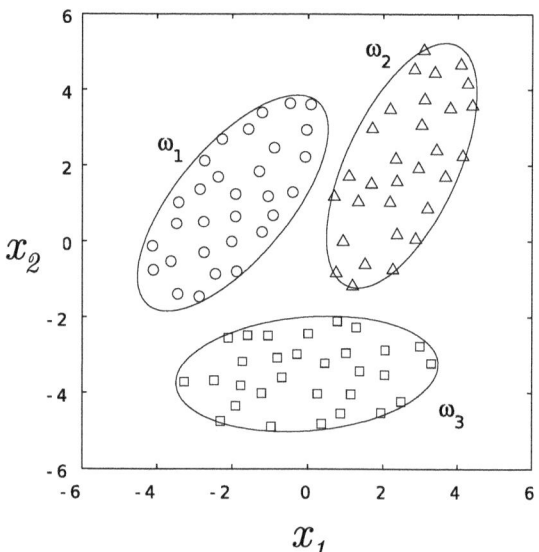

Figure 1.9 shows an example of the complete storage scheme using all of these patterns as prototypes. As shown by the thin lines in the figure, the feature space is finely divided by the closed region containing each prototype. If a pattern x exists in a closed region defined by a certain prototype, then the nearest neighbor of the pattern x is that prototype. As a result, the pattern x is judged to belong to the same class as this prototype. In other words, this closed region indicates the range of patterns with the prototype as its nearest neighbor. Such a diagram is called the *Voronoi diagram*.

If the prototypes of two adjacent regions belong to different classes, the boundary between the regions is a *decision boundary* that separates the classes. The decision

---

[6] Although we should select suitable patterns as prototypes from the collected patterns, it is better to move the prototypes in the feature space for a more sophisticated design.

boundary for the data in Fig. 1.9 is shown in Fig. 1.10, where the decision boundaries are indicated by thick lines.

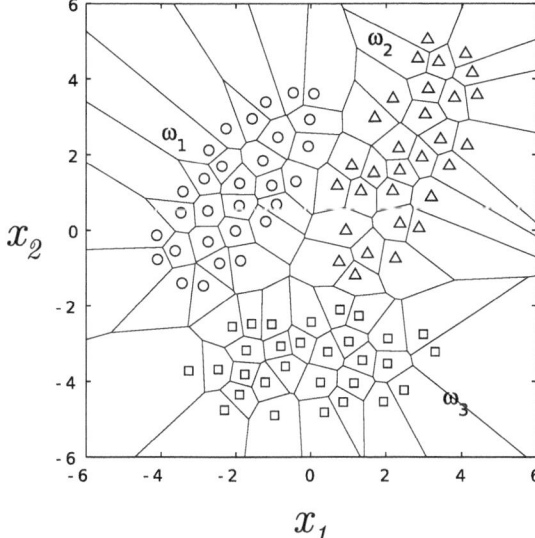

Fig. 1.9 Voronoi diagram with all learning patterns as prototypes

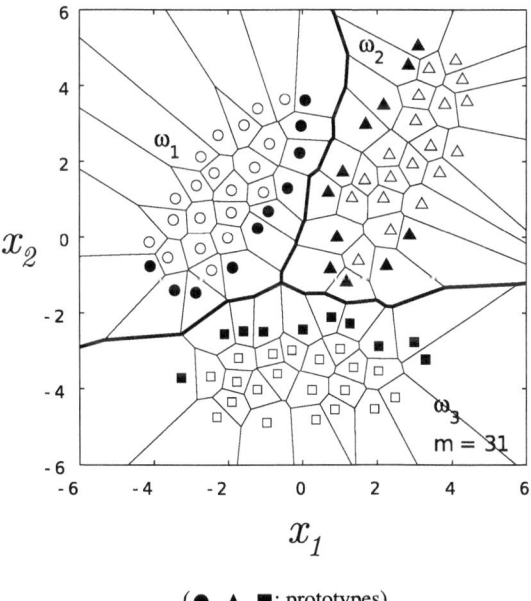

Fig. 1.10 Decision boundaries determined by 31 prototypes

($\bullet$, $\blacktriangle$, $\blacksquare$: prototypes)

As can be seen in the figure, these decision boundaries completely separate the three classes. The figure shows that the decision boundaries are determined by a small number of patterns located near the boundaries. In Fig. 1.10, these patterns are indicated by ●, ▲, and ■ for each class, respectively. The number of such patterns is 31 out of the total 90 patterns. This number m is shown in the lower right corner of the figure. Therefore, when applying the nearest neighbor rule, the distance to the input pattern is calculated only for these m prototypes, not for all of them. That is, since these prototypes contribute to the determination of the decision boundaries, we can say that they play the same role as component vectors discussed in Sect. 2.5.

In this example, the decision boundaries are set by not a few number of prototypes ($m = 31$), resulting in a complex shape. For a simple distribution such as the one discussed here, it is more efficient to further reduce the number of prototypes and set simpler decision boundaries. The extreme case is to have each class represented by a single prototype.

In the following, we will consider the case where the nearest neighbor rule is applied using one prototype per class. The decision boundary for separating two classes on a two-dimensional feature space ($d = 2$) is a straight line equidistant from the prototypes of both classes, i.e. the perpendicular bisector of the line segment connecting the two prototypes. In a three-dimensional feature space ($d = 3$), the decision boundary is a plane. In general, the decision boundary on a d-dimensional feature space is a $(d - 1)$-dimensional *hyperplane*, i.e. a $(d - 1)$-dimensional *subspace* of a d-dimensional space. A hyperplane is a generalization of the usual two-dimensional plane to other dimensions. A subspace is a linear space spanned by vectors in the original space. In general, its dimensionality is smaller than that of the original space. For a rigorous definition, refer to a textbook of linear algebra (Strang 2023).

In the example shown in Fig. 1.8, it is reasonable to choose the center of gravity of the distribution as the prototype representing the class. Figure 1.11(a) shows

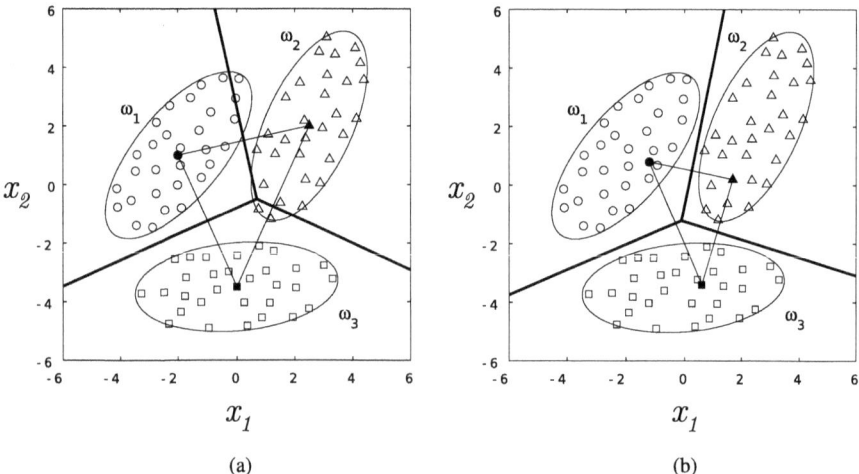

(a)   (b)

Fig. 1.11 Feature space partitioning by one prototype per class. (**a**) Prototypes = centers of gravity. (**b**) Prototypes ≠ centers of gravity

the decision boundaries with the centers of gravity as the prototypes. The decision boundaries are represented by thick lines.

This figure shows that these decision boundaries do not correctly separate the three classes. To correctly separate the classes, it is rather necessary to shift the prototypes from the center of gravity and select them as shown in (b) of the same figure. This allows the feature space to be correctly divided into three regions corresponding to each class.

Since this example is two-dimensional, the correct location of the prototypes could be determined visually. However, on a higher dimensional space, it is not possible to determine the position of the prototype in this way. Is it possible to automatically determine the correct location of the prototypes for class separation even in a high-dimensional space? In fact, this is made possible by the learning procedure described in the next chapter.

─────────────── **Coffee Break** ───────────────

Feature Extraction Now and Then

Around 1970, when research on character recognition was active, various feature extraction methods were proposed. However, each method had its own merits and demerits, and none of them could become the definitive method. At that time, feature extraction should be devised by humans through trial and error based on intuition and experience.

The emergence of deep learning has required a modification of this view. This is because results are being obtained suggesting that even feature extraction can be automated by computer if a large number of patterns are provided. However, deep learning cannot always explicitly explain what kind of features it is trying to extract and what causes its high performance. On the other hand, the classical feature extraction methods, for example, as shown in Appendix C, the meaning and aim of the features are clear, which is in contrast to the features obtained by deep learning.

Until now, artificial intelligence has gone through booms and winters. Although deep learning has shown great promise, looking back at the history of artificial intelligence, we may need more time before we can say with confidence that feature extraction can be fully automated.

Problems

1.1 Classify 5×5 pixel patterns x_1, x_2, x_3 and x_4 shown in Fig. 1.12 using the nearest neighbor rule. For classification, use one prototype per class as shown in Fig. 1.7. Both the input patterns and the prototypes are converted to 25-dimensional binary feature vectors according to Eq. (1.2). In classification, appropriate reject conditions shall be set and the classification should also include the reject process.

1.2 The following features are considered for classifying patterns. First, project the binarized pattern along the vertical and horizontal axes to obtain a histogram of black pixels. Figure 1.13 is the result of the histogram calculation using the numeral "2", which is one of the prototypes shown in Fig. 1.7. The histogram of Fig. 1.13 is represented by the following 10-dimensional vector:

Fig. 1.12 Input patterns to
be classified

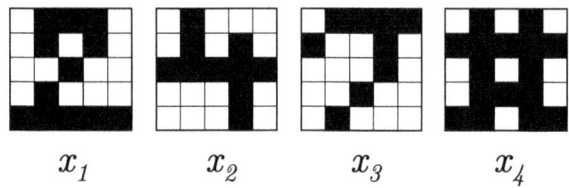

$$x = (x_1, x_2, \ldots, x_{10})^t$$
$$= (4, 2, 1, 1, 5, 2, 3, 3, 3, 2)^t.$$

Using this 10-dimensional vector as a feature, classify the four patterns of Fig. 1.12 under the same conditions as Problem 1.1.

Fig. 1.13 Features based on
histogram

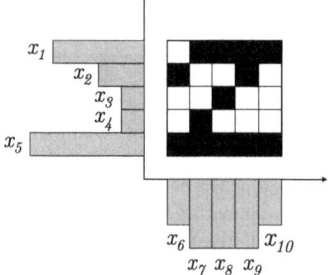

Chapter 2
Linear Discriminant Functions and Their Learning

Abstract In the previous chapter, we described a method for separating classes by means of a hyperplane designed with one prototype per class. In this case, it is important to determine the position of the prototype in the feature space. As shown in the previous chapter, a simple method that aligns the prototype with the center of gravity of the distribution does not always yield correct results. The learning procedure described in this chapter automatically finds the correct position of the prototypes in the high-dimensional feature space and divides the feature space into classes. This chapter first describes the relationship between the nearest neighbor rule and the linear discriminant function. Then, we explain the perceptron learning rule, a method for learning linear discriminant functions, with experiments. The piecewise linear discriminant function, which is an advanced form of the linear discriminant function, is also discussed. In addition, we discuss the concepts of margin, component vectors, and how they relate to support vector machines, which will be covered in Chap. 10.

2.1 What Is Learning?

In the previous chapter, we discussed the need to collect a large number of patterns that reflect the tendencies in real-world patterns for classification part design. Such patterns are called *learning pattern*s. *Learning* is the process of finding the optimal positions of the prototypes using the learning patterns and finding decision boundaries that correctly classify all of the learning patterns.[1] A complete storage scheme can be considered as a special case of learning according to this definition. On the other hand, the classification performance is evaluated by *test pattern*s that are prepared independently of the learning patterns. Test patterns are sometimes

---

Supplementary Information The online version contains supplementary material available at https://doi.org/10.1007/978-981-95-1478-6_2.

---

[1] Learning is sometimes referred to as *training* and learning patterns as *training pattern*s.

© The Author(s), under exclusive license to Springer Nature Singapore Pte Ltd. 2026
K. Ishii et al., *Pattern Recognition and Machine Learning for Self-Study I*, Springer Asia Pacific Mathematics Series 1, https://doi.org/10.1007/978-981-95-1478-6_2

called *unknown patterns* because they are not included in the learning patterns and the classes to which they belong are considered to be unknown.

In this book, learning patterns are assumed to be given together with information indicating the class to which they belong. Such a learning method is called *supervised learning*. Conversely, learning using learning patterns without information about the class they belong to is called *unsupervised learning*. Learning patterns used in supervised learning are called *labeled patterns*. The reason is that these patterns can be regarded as having labels assigned to them that indicate the class to which they belong. For the same reason, learning patterns used in unsupervised learning are called *unlabeled patterns*. For more information on unsupervised learning, see Ishii and Ueda (2026).

2.2 Nearest Neighbor Rule and Linear Discriminant Functions

Before proceeding with learning, let us formulate the nearest neighbor rule with one prototype per class.[2]

Suppose we choose vectors $\mathbf{p}_1, \mathbf{p}_2, \ldots, \mathbf{p}_c$ as prototypes for c classes $\omega_1, \omega_2, \ldots, \omega_c$, respectively. Let x be the input pattern[3] and use Euclidean distance $D(x, \mathbf{p}_i)$ as the distance between x and $\mathbf{p}_i$. Then, the nearest neighbor rule is to find i that minimizes the following equation:

$$D(x, \mathbf{p}_i)^2 = \|x - \mathbf{p}_i\|^2$$
$$= \|x\|^2 - 2\mathbf{p}_i^t x + \|\mathbf{p}_i\|^2 \qquad (i = 1, \ldots, c). \qquad (2.1)$$

Noting that $\|x\|^2$ does not depend on i in the above equation, minimizing $D(x, \mathbf{p}_i)^2$ is equivalent to maximizing $g_i(x)$ in the following equation:

$$g_i(x) \overset{\text{def}}{=} -\frac{1}{2}\|\mathbf{p}_i\|^2 + \mathbf{p}_i^t x. \qquad (2.2)$$

In the end, the classification is performed according to the following formula, which is called the *classification rule*:[4]

---

[2] A classification method that calculates the distance with one prototype per class is called a *minimum distance method*, which may be distinguished from the nearest neighbor rule, but is considered here as a form of the nearest neighbor rule.

[3] Hereafter x will be sometimes be referred to as a pattern instead of a feature vector.

[4] The discriminant function $g_i(x)$ and the classification rule together are called a *classifier*.

$$\max_{i=1,\ldots,c} \{g_i(\boldsymbol{x})\} = g_k(\boldsymbol{x}) \implies \boldsymbol{x} \in \omega_k. \tag{2.3}$$

In this way, a function $g_i(\boldsymbol{x})$ $(i = 1, \ldots, c)$ is assigned to each class, and the class to which the pattern $\boldsymbol{x}$ belongs is determined by the value of $g_i(\boldsymbol{x})$. This method is called the *discriminant function method* and the function used in this method is called the *discriminant function*. The most typical classification method is the one that outputs the class with the largest function value $g_i(\boldsymbol{x})$, as shown in Eq. (2.3). We deal with this method hereafter. Figure 2.1 shows a block diagram of the discriminant function method.

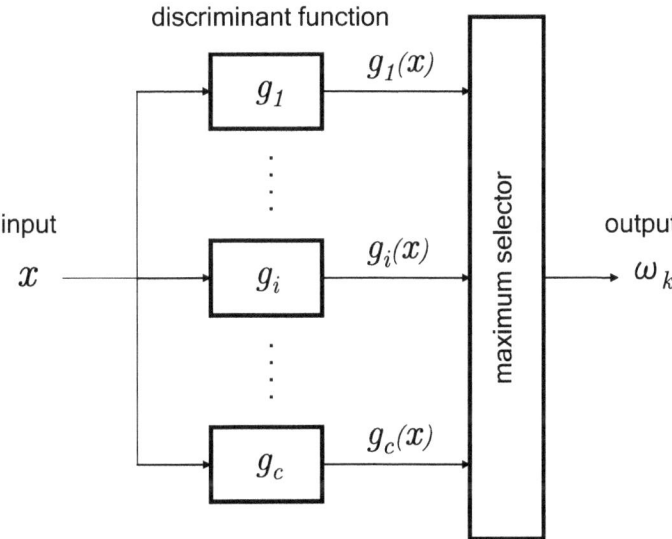

Fig. 2.1 Classification by discriminant function method

In particular, a discriminant function that is linear with respect to $\boldsymbol{x}$, as in Eq. (2.2), is called the *linear discriminant function*. The perceptron described later is a typical example. On the other hand, the discriminant function obtained by applying nonlinear processing to $\boldsymbol{x}$ is called the *nonlinear discriminant function*, which includes neural networks, etc., as described later. The nonlinear discriminant functions will be discussed in Part II.

Omitting the subscript i and writing Eq. (2.2) in a more general form, the linear discriminant function is expressed as

$$g(\boldsymbol{x}) = w_0 + \sum_{j=1}^{d} w_j x_j, \tag{2.4}$$

where $w_0, w_1, \ldots, w_d$ are *weight coefficients*. If there is no confusion, we will refer to the weight coefficients simply as *weights*. Using vector notation, $g(\boldsymbol{x})$ can be written as

$$g(x) = w_0 + \boldsymbol{w}^t \boldsymbol{x}. \tag{2.5}$$

In the above equation,

$$\boldsymbol{w} = (w_1, w_2, \ldots, w_d)^t, \tag{2.6}$$

where $\boldsymbol{w}$ is called the *weight vector*, and the weight coefficient w_0 is called the *bias*. Let us define $(d + 1)$ dimensional vectors $\mathbf{x}$ and $\mathbf{w}$ as follows:

$$\mathbf{x} = (x_0, x_1, \ldots, x_d)^t = \begin{pmatrix} x_0 \\ \boldsymbol{x} \end{pmatrix}, \qquad x_0 \equiv 1 \tag{2.7}$$

$$\mathbf{w} = (w_0, w_1, \ldots, w_d)^t = \begin{pmatrix} w_0 \\ \boldsymbol{w} \end{pmatrix}. \tag{2.8}$$

Note that, as shown in Eq. (2.7), the element x_0 of the vector $\mathbf{x}$ is identically 1. Using these vectors, Eq. (2.5) can be expressed more concisely as

$$g(x) = \mathbf{w}^t \mathbf{x}. \tag{2.9}$$

The newly defined $\mathbf{x}$ and $\mathbf{w}$ are called an *augmented feature vector* and an *augmented weight vector*, respectively. The reason for defining $\mathbf{x}$ and $\mathbf{w}$ in this way is to ensure that all weight coefficients, including bias, are learned consistently with the same learning rule, as described later. Hereafter, if there is no confusion, we will refer to the augmented feature vector and the augmented weight vector simply as the feature vector and the weight vector, respectively.

Let $g_i(\boldsymbol{x})$ $(1, \ldots, c)$ be the linear discriminant function of class ω_i, which can be written as

$$g_i(\boldsymbol{x}) = \sum_{j=0}^{d} w_{ij} x_j \tag{2.10}$$

$$= w_{i0} + \boldsymbol{w}_i^t \boldsymbol{x} \tag{2.11}$$

$$= \mathbf{w}_i^t \mathbf{x}, \tag{2.12}$$

where $\boldsymbol{w}_i$ and $\mathbf{w}_i$ $(i = 1, \ldots, c)$ are the weight vector and the augmented weight vector of class ω_i, respectively. They are written as

$$\boldsymbol{w}_i = (w_{i1}, \ldots, w_{id})^t, \tag{2.13}$$

$$\mathbf{w}_i = (w_{i0}, w_{i1}, \ldots, w_{id})^t. \tag{2.14}$$

A block diagram of the linear discriminant function is shown in Fig. 2.2.

In Eq. (2.11), if we set

$$w_i = \mathbf{p}_i, \tag{2.15}$$

$$w_{i0} = -\frac{1}{2}\|\mathbf{p}_i\|^2, \tag{2.16}$$

then, Eq. (2.2) is obtained, so we can see that the nearest neighbor rule with one prototype per class is a classification method based on a linear discriminant function (Problem 2.1). Determining the position of the prototypes is equivalent to determining the weights of the linear discriminant function. In the recognition system using the linear discriminant function, the weight coefficients are stored in the classification dictionary shown in Fig. 1.1, and the classification part performs the calculation of Eq. (2.10) and the maximum value selection.

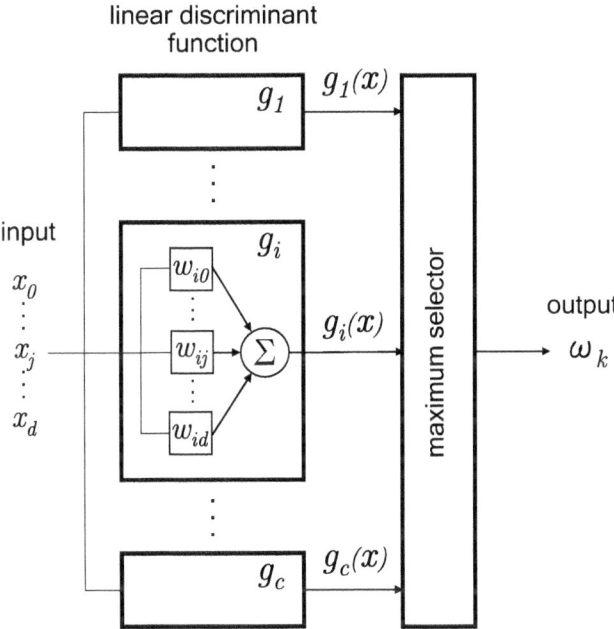

Fig. 2.2 Perceptron composed of linear discriminant functions

A classification system like Fig. 2.2, which consists of linear summation of inputs and maximum selector, is called the *perceptron*, and was proposed by Frank Rosenblatt in 1957 as a model of the brain with a learning function. The following perceptron learning rule is known as a method of determining the weights to correctly classify the learning patterns.

2.3 Perceptron Learning Rule

In the previous section, we mentioned that the nearest neighbor rule with one prototype per class is a classification method with a linear discriminant function. As a result, it is clear that determining the location of the prototypes is equivalent to determining the weights of the linear discriminant function. Therefore, learning is the process of automatically determining these weights using the learning patterns. This is made possible by the perceptron learning rule described below.

2.3.1 Weight Space and Solution Region

Denoting by the set $\mathcal{X}$ the entire set of learning patterns and by $\mathcal{X}_i$ ($i = 1, \ldots, c$) the set of learning patterns belonging to class ω_i, learning of a linear discriminant function is, from Eq. (2.3), to determine weight vectors $\mathbf{w}_i$ ($i = 1, \ldots, c$) such that

$$g_i(\mathbf{x}) > g_j(\mathbf{x}) \qquad (j = 1, \ldots, c \quad j \neq i) \tag{2.17}$$

for all $\mathbf{x}$ belonging to $\mathcal{X}_i$. Here, $g_i(\mathbf{x}) = g_j(\mathbf{x})$ represents the decision boundary between classes ω_i and ω_j.

In general, for c-class classification problems, we define *linearly separable* as the case where every class pair can be separated by a linear discriminant function, and *linearly nonseparable* otherwise. If there exists at least one set of weights satisfying Eq. (2.17), every class pair can be separated by a linear discriminant function. Therefore, the condition Eq. (2.17) is one example of a linearly separable case. For more details refer to Sect. 4.3.2.

First, for simplicity, we consider the case with two classes ($c = 2$), ω_1 and ω_2. In this case, instead of comparing the values of the discriminant functions $g_1(\mathbf{x})$ and $g_2(\mathbf{x})$, we can simply examine the sign of one discriminant function

$$g(\mathbf{x}) = g_1(\mathbf{x}) - g_2(\mathbf{x}) = (\mathbf{w}_1 - \mathbf{w}_2)^t \mathbf{x} \tag{2.18}$$

$$= \mathbf{w}^t \mathbf{x}, \tag{2.19}$$

where

$$\mathbf{w} \overset{\text{def}}{=} \mathbf{w}_1 - \mathbf{w}_2. \tag{2.20}$$

That is, only one weight vector, $\mathbf{w}$, is required. The classification rule by this discriminant function is expressed as

$$\begin{cases} g(\mathbf{x}) = \mathbf{w}^t \mathbf{x} > 0 & \Longrightarrow \quad \mathbf{x} \in \omega_1 \\ g(\mathbf{x}) = \mathbf{w}^t \mathbf{x} < 0 & \Longrightarrow \quad \mathbf{x} \in \omega_2, \end{cases} \tag{2.21}$$

and the decision boundary[5] between the two classes is

$$g(x) = \mathbf{w}^t \mathbf{x} = 0. \tag{2.22}$$

Therefore, in learning, we are looking for $\mathbf{w}$ such that

$$\begin{cases} g(x) = \mathbf{w}^t \mathbf{x} > 0 & \text{(for all } x \text{ belonging to } X_1) \\ g(x) = \mathbf{w}^t \mathbf{x} < 0 & \text{(for all } x \text{ belonging to } X_2). \end{cases} \tag{2.23}$$

To satisfy the above equation, it is necessary that there exists a weight vector $\mathbf{w}$ satisfying Eq. (2.23), i.e. X must be linearly separable. Here, the $(d+1)$-dimensional space spanned by $\mathbf{w}$ is called the *weight space*. In the weight space, $\mathbf{w}$ is represented by a single point whose coordinates are the weight coefficients. For any pattern[6] $\mathbf{x}$, $\mathbf{w}^t \mathbf{x} = 0$ determines one hyperplane through the origin in the weight space. One side of the weight space divided by this hyperplane is the region (positive side) where $g(x)$ is positive and the other side is the region (negative side) where $g(x)$ is negative.[7] If the total number of learning patterns is n, there are n hyperplanes in the weight space corresponding to each learning pattern. Equation (2.23) indicates in which side of the hyperplane $\mathbf{w}$ must exist for the corresponding learning pattern. Thus, Eq. (2.23) specifies by n hyperplanes the region where $\mathbf{w}$ must be located in the weight space. This region is called the *solution region*. Linear separability implies the existence of a solution region.

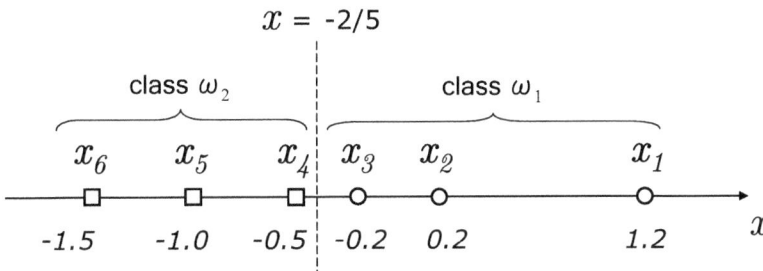

Fig. 2.3 Linearly separable learning patterns in one-dimensional feature space

Here is an example. Figure 2.3 is a one-dimensional ($d = 1$) example, and the feature space is a number line. Therefore, the features are scalars, not vectors. However, to avoid confusion, the pattern itself is denoted by a vector x, and the features, i.e., the coordinate values on the number line, are denoted by a scalar x. Suppose that the learning patterns of two classes ($c = 2$) are distributed on this

---

[5] If $g(x) = \mathbf{w}^t \mathbf{x} = 0$ at classification, then x is undecidable.

[6] Hereafter $\mathbf{x}$ may be referred to as a pattern as well as x.

[7] Note that this hyperplane does not define a decision boundary, because it is set in the weight space, not in the feature space. This differs from the hyperplane described in Sect. 1.3.2.

number line as shown in the figure. The total number of patterns is $6(n = 6)$, where x_1, x_2, x_3 belong to class ω_1, x_4, x_5, x_6 belong to class ω_2.

As is clear from the figure, they are linearly separable. The discriminant function is represented by two weight coefficients w_0 and w_1, so the weight space is two-dimensional. Figure 2.4 represents the weight space, showing the six hyperplanes (straight lines in the figure) defined by each learning pattern and the solution region determined by them (shaded areas). For convenience, the horizontal axis is w_1 and the vertical axis is w_0.

Fig. 2.4 Weight space and the solution region

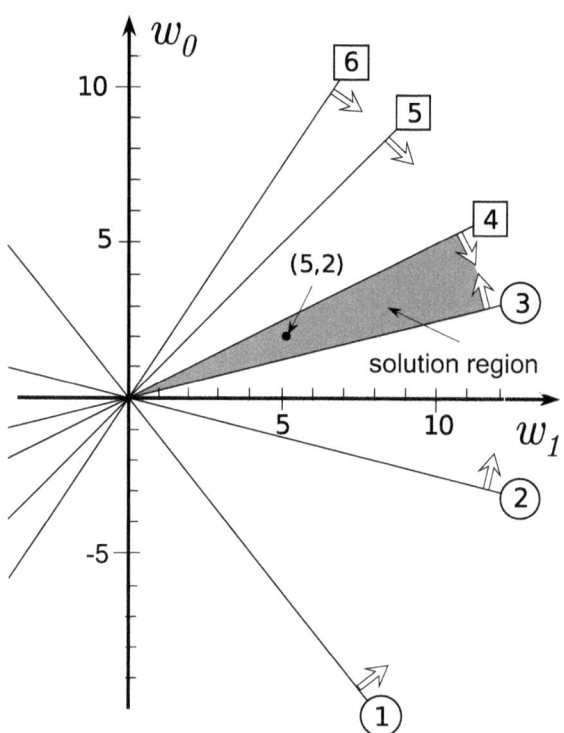

In the figure, ◯ and □ indicate that the pattern belongs to the class ω_1 and ω_2, respectively, and the numbers in them correspond to the pattern numbers in Fig. 2.3. The arrows attached to the hyperplanes indicate the side of the weight vector that correctly classifies the pattern. Any weight vector **w** in the solution domain gives the discriminant function that correctly separates the learning patterns. For example, a point $(w_1, w_0) = (5, 2)$ in the solution domain corresponds to a discriminant function $g(x) = w_0 + w_1 x = 2 + 5x$. Applying Eq. (2.21) gives the classification rule

$$\begin{cases} x > -\dfrac{2}{5} & \Longrightarrow \quad x \in \omega_1 \\[3mm] x < -\dfrac{2}{5} & \Longrightarrow \quad x \in \omega_2, \end{cases} \tag{2.24}$$

which is verified to be the correct weight vector by comparing with Fig. 2.3.

2.3.2   Perceptron Convergence Theorem

Let us describe how to determine the weight vector of a linear discriminant function by learning. The *perceptron learning rule* is a well-known method for determining the weight vector, and the procedure is shown below for the two-class case.

Perceptron Learning Rule (for Two-Class Problem)

Step 1   Prepare n learning patterns $x_1, \ldots, x_n$ whose classes are known.

Step 2   Set initial values of weight vector $\mathbf{w}$.

Step 3   Select one pattern x_k $(k = 1, \ldots, n)$ from the learning patterns, and calculate $g(x_k)$ in the following equation:

$$g(x_k) = \mathbf{w}^t \mathbf{x}_k \qquad (k = 1, \ldots, n). \tag{2.25}$$

Step 4   Modify $\mathbf{w}$ according to the classification result and replace it with a new weight vector $\mathbf{w}'$ as follows:[8]

$$\mathbf{w}' = \mathbf{w} + \rho\, \mathbf{x}_k \qquad (g(x_k) \le 0 \text{ for } x_k \in \omega_1), \tag{2.26}$$

$$\mathbf{w}' = \mathbf{w} - \rho\, \mathbf{x}_k \qquad (g(x_k) \ge 0 \text{ for } x_k \in \omega_2), \tag{2.27}$$

$$\mathbf{w}' = \mathbf{w} \qquad\qquad (\text{otherwise}). \tag{2.28}$$

Step 5   If all the learning patterns are correctly classified, the program terminates. Otherwise, return to Step 3 and repeat the above process with another pattern.

Equations (2.26) and (2.27) are the weight modification process performed when the pattern $\mathbf{x}_k$ is not correctly classified, and ρ is a positive constant indicating the learning step size, called the *learning rate*. When the pattern $\mathbf{x}_k$ is correctly classified, the weight vector is not modified, as shown in Eq. (2.28).

---

[8] Note that the weight vector needs to be modified even when $g(x_k) = 0$.

Each learning pattern x_k $(k = 1, \ldots, n)$ is given with the information about its class. We denote this information by the label $b(x_k)$ and call it the *teaching signal*. The teaching signal is given by the following equation:

$$b_k \stackrel{\text{def}}{=} b(x_k) = \begin{cases} 1 & (x_k \in \omega_1) \\ -1 & (x_k \in \omega_2) \end{cases} \qquad (k = 1, \ldots, n). \qquad (2.29)$$

Hereafter, unless otherwise noted, $b(x_k)$ will be abbreviated as b_k as shown in the above equation. With this teaching signal, Eqs. (2.26), (2.27), and (2.28) can be combined, and the weight vector modification in the two-class case is written as follows:

$$\begin{cases} \mathbf{w}' = \mathbf{w} + \rho\, b_k \mathbf{x}_k & (b_k\, g(x_k) \le 0) \\ \mathbf{w}' = \mathbf{w} & (\text{otherwise}). \end{cases} \qquad (2.30)$$

As can be seen from the above procedure, in the learning process, n learning patterns are used over and over again to modify the weight vector. In this book, the learning of one pattern is considered to be one *iteration*. When all the learning patterns have been learned, i.e., when n iterations have been completed, we say that the learning of one *epoch* has been completed. The above algorithm is a method of iterating modifications until the correct weight vector is obtained, and the algorithm is called the *iterative method*.

Fig. 2.5 Modification of a
weight vector

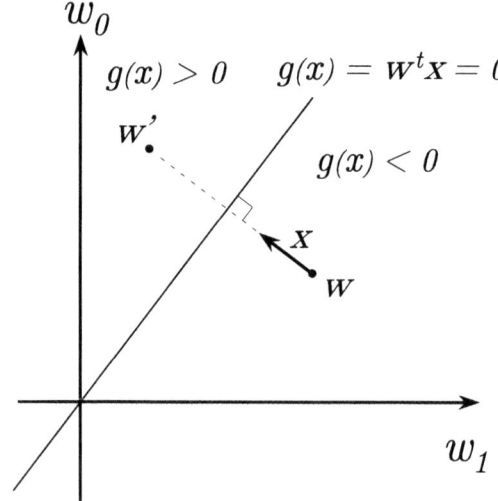

As shown in Fig. 2.5, since the vector $\mathbf{x}_k$ is orthogonal to the hyperplane $\mathbf{w}^t \mathbf{x} = 0$, Eqs. (2.26) and (2.27) indicate that the weight vector $\mathbf{w}$ is moved in the direction orthogonal to the hyperplane. That is, Eq. (2.26) represents a perpendicular shift from the negative to the positive side of the hyperplane and Eq. (2.27) represents a shift from the positive to the negative side. If the learning rate ρ is large enough, the sign of $\mathbf{w}^t \mathbf{x}_k$ can be reversed in a single modification. It is easy to verify that each modification according to Eqs. (2.26), (2.27) will always update the discriminant function in the direction of improvement (Problem 2.2).

If the learning patterns are linearly separable, then the above algorithm reaches the weight vector in the solution domain in a finite number of iterations. This is known as the *perceptron convergence theorem*. See Appendix A for the proof.

The weight learning method described here can be extended to more general functions. That is, a function represented by a linear combination of arbitrary functions $\phi_1(\mathbf{x}), \phi_2(\mathbf{x}), \ldots, \phi_d(\mathbf{x})$ of $\mathbf{x}$ is called a generalized linear discriminant function or Φ function. It is known that the perceptron convergence theorem can also be applied to it. The generalized linear discriminant function is described in Sect. 8.5.

Using the previous example, Fig. 2.6 shows how the weight vector moves in the weight space.

Fig. 2.6 Movement of weight vectors by learning

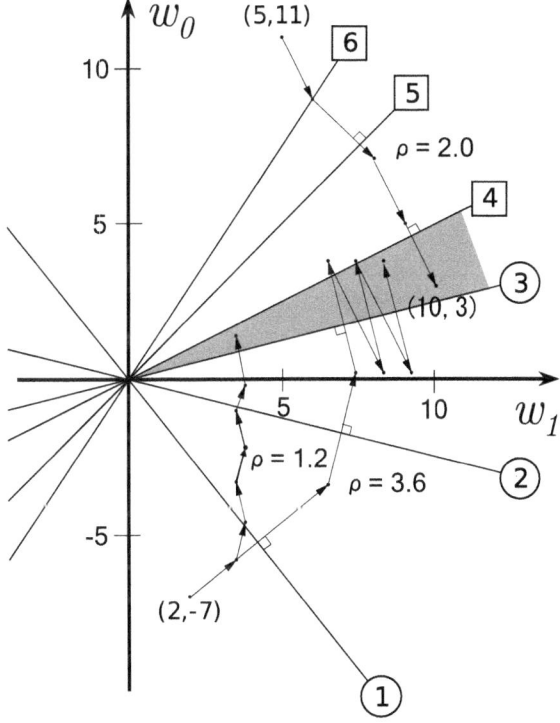

In learning, patterns are given in the order $\mathbf{x}_1, \mathbf{x}_2, \ldots, \mathbf{x}_6$. Regardless of the initial value and the value of ρ, we see that the weight vector eventually reaches the solution region. For example, when the initial weights are set to $(w_1, w_0) = (5, 11)$ and $\rho = 2.0$, the solution reaches $(10, 3)$ as shown in the figure. The decision boundary in this case is $x = -w_0/w_1 = -3/10$, and it can be confirmed from Fig. 2.3 that the correct weights are obtained. If the learning rate ρ is too small, it is inefficient because of the repetition of small increments of modification, and if it is too large, it is undesirable because of the oscillating convergence. The learning method described above is a method of fixing the value of ρ during the learning, and is therefore called the *fixed increment rule*.

The perceptron learning rule can be extended to multi-class as follows.

Perceptron Learning Rule (for Multi-Class Problem)

Step 1   Prepare n learning patterns $x_1, \ldots, x_n$ whose belonging classes are known.

Step 2   Set initial values of weight vectors $\mathbf{w}_1, \ldots, \mathbf{w}_c$.

Step 3   Select one pattern x_k ($k = 1, \ldots, n$) from the learning patterns and classify it by c discriminant functions to obtain $g_i(x_k) = \mathbf{w}_i^t x_k$ ($i = 1, \ldots, c$).

Step 4   If the class of the pattern x_k is ω_i, the following process is performed:

(1) If there exists j ($\neq i$) such that $g_i(x_k) \leq g_j(x_k)$[9] replace $\mathbf{w}_i$ and $\mathbf{w}_j$ with $\mathbf{w}_i'$ and $\mathbf{w}_j'$, respectively according to the following formula:

$$\begin{cases} \mathbf{w}_i' = \mathbf{w}_i + \rho\, \mathbf{x}_k \\ \mathbf{w}_j' = \mathbf{w}_j - \rho\, \mathbf{x}_k. \end{cases} \tag{2.31}$$

(2) If $g_i(x_k) > g_j(x_k)$ for all $j (\neq i)$, no weight modification is performed.

Step 5   If all learning patterns are correctly classified, the program terminates. Otherwise, return to Step 3 and repeat the above process with another pattern.

Step 4 (1) above is the case when a pattern of ω_i is misclassified as ω_j, or when both ω_i and ω_j are candidates ($i \neq j$). In this case, not only $\mathbf{w}_i$ but also $\mathbf{w}_j$ are modified. In the case of Step 4 (2), when the pattern is correctly classified, the weights are not modified.

Since the perceptron learning rule modifies the weights only when the pattern is not correctly classified, we call this learning method the *error-correction method*.

---

[9] Note that the weight vector must also be modified when $g_i(x_k) = g_j(x_k)$. In this case, it is not a misclassification, but a reject. (See footnote 2 in Chap. 1.)

2.3.3   Linear Discriminant Function and Projection Axis

The linear discriminant function can be viewed as a projection from a d-dimensional feature space to a 1-dimensional space (a straight line). For simplicity, we will deal with the two-class problem. In the two-class case, as shown in Eq. (2.21), the classification is performed based on $g(x) = w_0 + w^t x \gtrless 0$, and $g(x) = 0$, i.e.

$$w_0 + w^t x = 0 \tag{2.32}$$

is the decision boundary separating the two classes.

Equation (2.32) is a $(d - 1)$-dimensional hyperplane in a d-dimensional feature space   and w is the normal vector of this hyperplane. Here, w is assumed to be normalized so that $\|w\| = 1$.

If we set the axis y passing through the origin O in the direction normal to the hyperplane, $w^t x$ is the projection value of the vector x onto the y axis as shown in Fig. 2.7. There are innumerable hyperplanes with vector w as a normal, among which Eq. (2.32) is a hyperplane that intersects the y-axis at $- w_0$ ($w^t x = -w_0$) and is indicated by H in the figure. Given the above geometric structure, the learning process of the perceptron can be described as follows.

Fig. 2.7 Projection axis y and hyperplane H in the feature space   ($\|w\| = 1$)

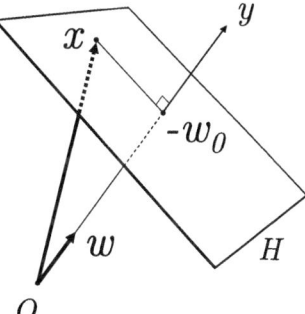

That is, after obtaining the projection value $w^t x$ for each learning pattern and projecting the pattern on the y axis, the modification of the weight vector w and w_0 is repeated until all learning patterns are correctly classified by $w^t x \gtrless -w_0$.

As mentioned above, determining a linear discriminant function is equivalent to defining a 1-dimensional projection axis on a d-dimensional feature space. The projection axis must be optimized to separate classes on the axis as efficiently as possible. The perceptron learning rule uses the number of misclassified patterns,[10] and the weights are modified repeatedly until the number of misclassifications is reduced to zero. The Widrow–Hoff learning rule and Fisher's method, which will

---

[10] Strictly speaking, "the number of patterns not correctly classified".

be discussed in subsequent chapters, are similar to the perceptron learning rule in that they seek the optimal projection axis, but they use different evaluation measures.

2.4 Experiments on Perceptron Learning Rule (1)

In this section, we perform an experiment with the perceptron using a concrete example, and confirm the functions of the perceptron described so far. The data used are the learning patterns of the following two classes distributed in a three-dimensional feature space ($d = 3$). The total number of patterns is 8 ($n = 8$).

$$\begin{aligned}
&x_1 = (1, 1, 1)^t, \quad x_2 = (0, 1, 1)^t, \quad x_3 = (1, 0, 1)^t, \quad x_4 = (1, 1, 0)^t, \\
&x_5 = (0, 0, 0)^t, \quad x_6 = (1, 0, 0)^t, \quad x_7 = (0, 1, 0)^t, \quad x_8 = (0, 0, 1)^t.
\end{aligned} \tag{2.33}$$

Among them, x_1, x_2, x_3, x_4 belong to class ω_1 and x_5, x_6, x_7, x_8 to class ω_2, respectively. In Fig. 2.8, patterns of class ω_1 and class ω_2 are indicated by ● and ○, respectively.

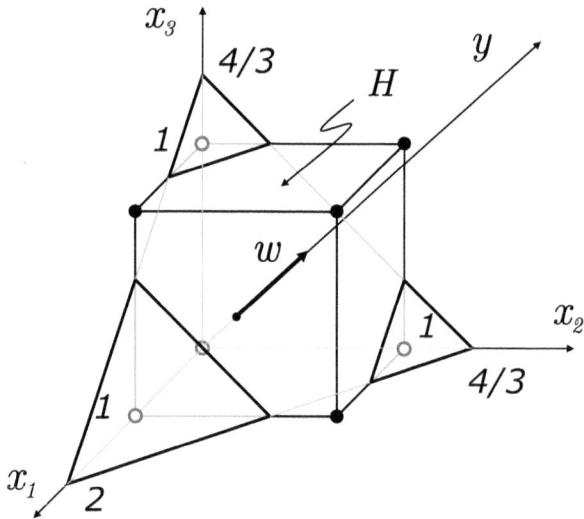

Fig. 2.8 Learning patterns and a decision boundary in three-dimensional feature space

As is clear from the figure, the patterns exist at the vertices of the unit cubic lattice, and they are linearly separable.

Now, we want to set a linear discriminant function

$$g(x) = w_0 + w_1 x_1 + w_2 x_2 + w_3 x_3 \tag{2.34}$$

and determine the weights w_0, w_1, w_2, w_3 so that

$$\begin{cases} g(\boldsymbol{x}_k) > 0 & (\boldsymbol{x}_k \in \omega_1) \\ g(\boldsymbol{x}_k) < 0 & (\boldsymbol{x}_k \in \omega_2) \end{cases} \quad (k = 1, \ldots, 8), \quad (2.35)$$

using the perceptron learning rule. Here, the initial values of the weights are set to $(w_0, w_1, w_2, w_3) = (-5, 1, 1, 1)$ and the learning rate is set to $\rho = 1$. The learning patterns are given repeatedly from $\boldsymbol{x}_1$ to $\boldsymbol{x}_8$ in this order.

The results of the experiment are shown in Table 2.1. In this experiment, the number of iterations is shown in the second column of the table. Since there are eight learning patterns, eight iterations correspond to one epoch, which is shown in the first column.

The third and fourth columns of the table show the input patterns and the classes to which they belong, respectively. Next, the contents of the augmented feature vectors and augmented weight vectors are shown. The former is fixed, while

Table 2.1 Learning process of weights by perceptron

ep: epoch   iter. iteration   adj.: adjustment (- : not modified)   M: modified

				Feature				Weight					New weight			
ep.	iter.			x_0	x_1	x_2	x_3	w_0	w_1	w_2	w_3	$g(x)$	adj.	w_0'	w_1'	w_2'	w_3'
1	1	$\boldsymbol{x}_1$	ω_1	1	1	1	1	−5	1	1	1	−2	M	−4	2	2	2
	2	$\boldsymbol{x}_2$	ω_1	1	0	1	1	−4	2	2	2	0	M	−3	2	3	3
	3	$\boldsymbol{x}_3$	ω_1	1	1	0	1	−3	2	3	3	2	−				
	4	$\boldsymbol{x}_4$	ω_1	1	1	1	0	−3	2	3	3	2	−				
	5	$\boldsymbol{x}_5$	ω_2	1	0	0	0	−3	2	3	3	−3	−				
	6	$\boldsymbol{x}_6$	ω_2	1	1	0	0	−3	2	3	3	−1	−				
	7	$\boldsymbol{x}_7$	ω_2	1	0	1	0	−3	2	3	3	0	M	−4	2	2	3
	8	$\boldsymbol{x}_8$	ω_2	1	0	0	1	−4	2	2	3	−1	−				
2	9	$\boldsymbol{x}_1$	ω_1	1	1	1	1	−4	2	2	3	3	−				
	10	$\boldsymbol{x}_2$	ω_1	1	0	1	1	−4	2	2	3	1	−				
	11	$\boldsymbol{x}_3$	ω_1	1	1	0	1	−4	2	2	3	1	−				
	12	$\boldsymbol{x}_4$	ω_1	1	1	1	0	−4	2	2	3	0	M	−3	3	3	3
	13	$\boldsymbol{x}_5$	ω_2	1	0	0	0	−3	3	3	3	−3	−				
	14	$\boldsymbol{x}_6$	ω_2	1	1	0	0	−3	3	3	3	0	M	−4	2	3	3
	15	$\boldsymbol{x}_7$	ω_2	1	0	1	0	−4	2	3	3	−1	−				
	16	$\boldsymbol{x}_8$	ω_2	1	0	0	1	−4	2	3	3	−1	−				
3	17	$\boldsymbol{x}_1$	ω_1	1	1	1	1	−4	2	3	3	4	−				
	18	$\boldsymbol{x}_2$	ω_1	1	0	1	1	−4	2	3	3	2	−				
	19	$\boldsymbol{x}_3$	ω_1	1	1	0	1	−4	2	3	3	1	−				
	20	$\boldsymbol{x}_4$	ω_1	1	1	1	0	−4	2	3	3	1	−				
	21	$\boldsymbol{x}_5$	ω_2	1	0	0	0	−4	2	3	3	−4	−				
	22	$\boldsymbol{x}_6$	ω_2	1	1	0	0	−4	2	3	3	−2	−				

the latter is updated as needed during the learning process. The element of the augmented feature vector x_0 is identically 1.

The $g(x)$ in the table are the values of the linear discriminant function computed for each iteration. The next column indicates whether or not weight vector modification is required. That is, if the pattern is correctly classified and no modification is necessary, it is marked with "-". On the other hand, if the pattern is not correctly classified and the weight vector needs to be modified, it is marked with "M" and the modified weight vectors are shown in the last four columns of the table as w'_0, w'_1, w'_2, w'_3.

Repeating this process, 8 patterns are correctly classified consecutively from iteration 15 to iteration 22, at which point convergence is achieved. In other words, convergence occurs at iteration 22 in the middle of the third epoch.[11] The final weight vector $\mathbf{w}$ obtained is

$$\mathbf{w} = (w_0,\ w_1,\ w_2,\ w_3)^t = (-4,\ 2,\ 3,\ 3)^t. \tag{2.36}$$

Therefore, from Eq. (2.32), the decision boundary is a plane in the 3D feature space

$$-4 + 2x_1 + 3x_2 + 3x_3 = 0 \tag{2.37}$$

and it is shown as the plane H of Fig. 2.8. In the figure, $\mathbf{w}$ is the normal vector of the plane H and y is the projection axis in the normal direction. It is confirmed that the plane H is the decision boundary that correctly separates the two classes.

As shown in Table 2.1, the weights were modified a total of five times, i.e., iterations 1, 2, 7, 12, and 14, with new weights obtained each time. The learning process is shown in Fig. 2.9. The figure shows how each pattern is projected onto the y-axis in the initial state and just after the weight modification. In order to make the figure easier to read, the patterns are shifted by w_0 after projection and are classified by the positive and negative values on the y-axis. Each position of the pattern on the projection axis is indicated by ● or ○ with the pattern name x_k. Along with the iterations, we can observe the process of separating the two classes on the y-axis with zero as the threshold. The reason why the sum of both marks does not equal the number of patterns is that some patterns have the same projection value and overlap in the figure. See also Problem 2.3 for other experimental examples.

---

[11] The convergence decision could be made after each epoch, in which case the convergence decision is made at the end of epoch 3, i.e., at iteration 24 in this experiment. However, iterations 23 and 24 are unnecessary because the patterns x_7, x_8 are already correctly classified in iterations 15 and 16.

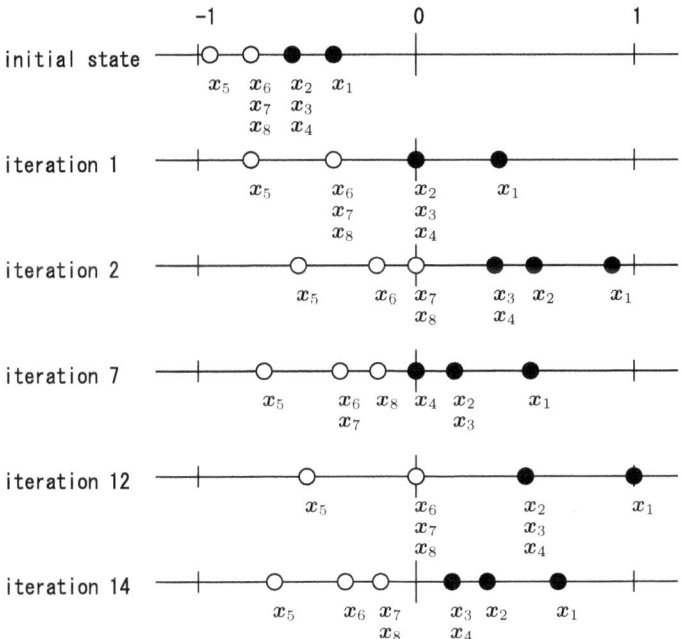

Fig. 2.9 Learning process observed on the projection axis

2.5 Perceptron and Component Vector

Let us see what kind of discriminant function is finally obtained by applying the perceptron learning rule. Since the positive constant ρ in the perceptron learning rule is arbitrary, let $\rho = 1$, then the modified weight vector shown in the first equation of Eq. (2.30) is

$$\mathbf{w}' = \mathbf{w} + b_k \mathbf{x}_k. \tag{2.38}$$

The modification of the above equation is applied when the pattern is not correctly classified. Since $b_k = \pm 1$ from Eq. (2.29), the operation of adding or subtracting $\mathbf{x}_k$ to or from the current weight vector is repeated each time the pattern $\mathbf{x}_k$ that is not correctly classified occurs. Since the initial value of the weight vector $\mathbf{w}$ can be set arbitrarily, if $\mathbf{w} = \mathbf{0}$, the final weight vector obtained after convergence can be written in the form of

$$\mathbf{w} = \sum_{k=1}^{n} \alpha_k b_k \mathbf{x}_k. \tag{2.39}$$

The α_k in the above equation represents the number of times the pattern $\boldsymbol{x}_k$ is not correctly classified during the learning process. Therefore, $\alpha_k = 0$ for $\boldsymbol{x}_k$ that are always correctly classified in the learning. Naturally, α_k is a non-negative integer. Using Eq. (2.39), the final discriminant function obtained after convergence is expressed by the following equation:

$$g(\boldsymbol{x}) = \mathbf{w}^t \mathbf{x} \tag{2.40}$$

$$= \sum_{k=1}^{n} \alpha_k b_k \mathbf{x}_k{}^t \mathbf{x}. \tag{2.41}$$

As is clear from the above, only $\alpha_k \neq 0$ contributes to the construction of the discriminant function $g(\boldsymbol{x})$. We call such $\mathbf{x}_k$ as a *component vector*.

Denote α_k $(k = 1, \ldots, n)$ in Eq. (2.41) by a vector as

$$\boldsymbol{\alpha} = (\alpha_1, \alpha_2, \ldots, \alpha_n)^t \tag{2.42}$$

and call this vector an *error counter vector*. In expressing the discriminant function $g(\boldsymbol{x})$ here, Eq. (2.40) using $\mathbf{w}$ and Eq. (2.41) using $\boldsymbol{\alpha}$ are shown. The fact that it can be expressed in these two ways is called duality, and will be introduced again in Chap. 8 and Chap. 9.

There are some points to be noted in the linear discriminant function obtained by the perceptron learning rule. They will be cited several times later. They are particularly important for support vector machine. Let m $(\leq n)$ be the number of component vectors.

Point 1   As shown in Eq. (2.39), the weight vector $\mathbf{w}$ is represented as a linear combination of the m $(\leq n)$ component vectors of the n learning patterns that are $\alpha_k \neq 0$.

Point 2   As a result, the linear discriminant function $g(\boldsymbol{x})$ is represented as a linear combination of m of the inner product $\mathbf{x}_k{}^t \mathbf{x}$ as shown in Eq. (2.41).

Point 3   On the other hand, when using perceptron learning rules, it is not uniquely determined which learning pattern is chosen as the component vector and how many m of component vectors are chosen. Consequently, the decision boundary is not uniquely determined. These depend on the order of the learning patterns given in the iterative learning process.

2.6   Experiments on Perceptron Learning Rule (2)

In the following, let us confirm the points discussed in the previous section through experiments. The learning patterns used in the experiment are shown in Fig. 2.10. The learning patterns consist of two classes $(c = 2)$, ω_1 and ω_2, distributed in the two-dimensional feature space ($d = 2$), and they are marked with $\bigcirc$ and $\triangle$, respectively. The number of patterns is 100 for each class, for a total of 200 patterns. As is clear from the figure, these are linearly separable.

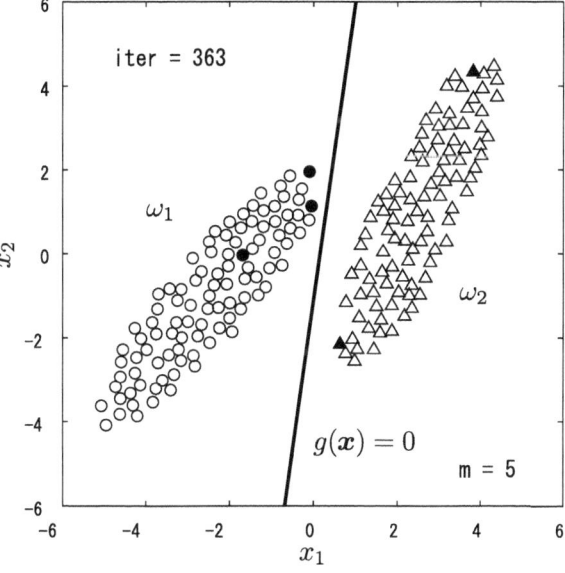

Fig. 2.10 Decision boundary and component vectors for linearly separable learning patterns

($\bullet$, $\blacktriangle$: component vectors)

The decision boundary obtained by applying the perceptron learning rule (for two-class problem) shown in Sect. 2.3.2 to this learning pattern is shown in the figure with a thick line. In the experiment, $\rho = 1$, and the initial weight vector was set to $\mathbf{w} = \mathbf{0}$ as shown in the previous section. The learning patterns were selected from two classes alternately and randomly. The number of iterations required for convergence is shown in the upper left corner of the figure, which was 363 (the number of epochs was 2) for this data.

Let us now review Points 1 and 2 described in the previous section. As mentioned earlier, the patterns that were not correctly classified during the learning process contribute to the construction of the discriminant function as component vectors. In the figure, component vectors are indicated by $\bullet$ and $\blacktriangle$ for each of ω_1 and ω_2, respectively (the same notation is used hereafter). As shown in the figure, component vectors have been selected from the classes ω_1 and ω_2 with 3 and 2 respectively. The total number of component vectors indicated as $m = 5$ in the lower right corner of the figure. That is, the number of the learning patterns n is 200, but the number of component vectors m is only 5, confirming the phenomenon described in Points 1 and 2. Also, it can be confirmed that the decision boundary $g(\mathbf{x}) = 0$ determined from these 5 component vectors correctly separates the two classes.

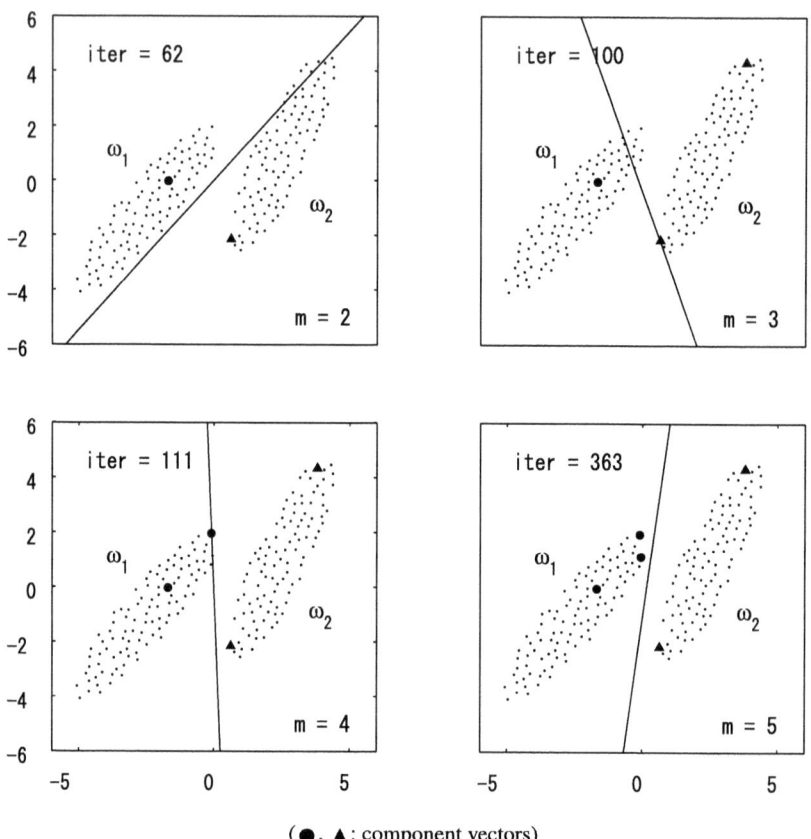

Fig. 2.11 Perceptron convergence process and component vectors

Figure 2.11 shows the progress up to Fig. 2.10, which is the convergence state, along with the number of iterations "iter" and the number of component vectors m. It can be seen that each time a pattern that cannot be correctly classified occurs, the pattern is added as a component vector. To make the figure easier to read, all patterns in both classes except component vectors are marked with ".".

Next, to confirm Point 3, the following experiment was conducted. In the learning process, four types of learning were prepared by changing the order in which the learning patterns were given, and examined component vectors and decision boundaries after convergence. The results are shown in Fig. 2.12.

Let us find the pattern closest to the decision boundary and its distance. Suppose the discriminant function for classifying two classes ω_1 and ω_2 is defined as

$$g(x) = w_0 + w^t x. \tag{2.43}$$

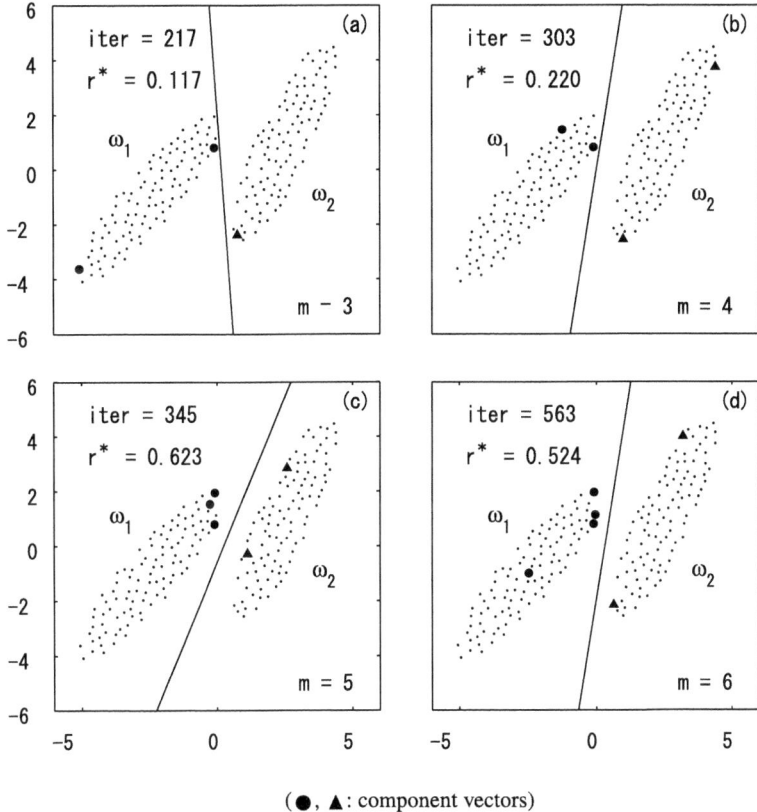

($\bullet$, $\blacktriangle$: component vectors)

Fig. 2.12 Decision boundaries and component vectors that vary with learning conditions

Let r_k (> 0) be the distance from the vector $\boldsymbol{x}_k$ in the feature space to the hyperplane (in this example, a straight line) $g(\boldsymbol{x}) = 0$, which is the decision boundary. Then, r_k is expressed as

$$r_k = \frac{|g(\boldsymbol{x}_k)|}{\|\boldsymbol{w}\|}. \tag{2.44}$$

For the derivation of the above equation, see Problem 2.4. In Fig. 2.12, Eq. (2.44) was calculated for each pattern and its minimum value r^* was obtained.

The figure shows the number of iterations to convergence (iter), r^* and the number of component vectors m. As can be seen from the figure, the number of component vectors is small, ranging from 3 to 6, and the decision boundaries are diverse. As a result, the value of r^* also varies. This phenomenon is described in Ppoint 3.

2.7 Linear Discriminant Functions with Margin

In the perceptron learning rule, the number of incorrectly classified patterns is used as the criteria, and the weight vector is repeatedly modified until it reaches zero. If the learning patterns are linearly separable, as described in Point 3, there are an infinite number of decision boundaries that can correctly classify all learning patterns, and they are not uniquely determined. There is also no guarantee that the decision boundary obtained at the point of convergence is optimal. In other words, there may be decision boundaries that achieve class separation with a sufficient margin, and vice versa. For example, (a) and (b) in Fig. 2.12 are $r^* = 0.117, 0.220$. The decision boundaries are quite close to learning patterns, and can be said to be examples of no margin. With such decision boundaries, even a slight disturbance in the pattern distribution cause misclassification, and high classification accuracy cannot be expected for unknown patterns.

The solution to this problem is described below. In the learning rule described so far, the weights are modified when the learning pattern crosses the decision boundary and enters the region of misclassification.[12] As an alternative, if we modify the weights when the pattern approaches the decision boundary beyond a certain distance $\delta(> 0)$, we can obtain a decision boundary with more leeway. For this purpose, by using Eq. (2.44), we can modify the weights when the following equation holds:

$$\frac{b_k g(\boldsymbol{x}_k)}{\|\boldsymbol{w}\|} < \delta. \tag{2.45}$$

From the above, the weight modification for the two-class case shown in Eq. (2.30) can be rewritten as the following equation:

$$\begin{cases} \boldsymbol{w}' = \boldsymbol{w} + \rho\, b_k \boldsymbol{x}_k & (b_k\, g(\boldsymbol{x}_k) < \delta\|\boldsymbol{w}\|) \\ \boldsymbol{w}' = \boldsymbol{w} & \text{(otherwise)}. \end{cases} \tag{2.46}$$

Naturally, if $\delta = 0$, the above equation is identical to the conventional weight modification method.[13]

Using the weight vector $\boldsymbol{w}$ obtained by the above learning method, the equation

$$\frac{b_k g(\boldsymbol{x}_k)}{\|\boldsymbol{w}\|} = \frac{b_k(w_0 + \boldsymbol{w}^t \boldsymbol{x}_k)}{\|\boldsymbol{w}\|} \geq \delta \qquad (k = 1, \ldots, n) \tag{2.47}$$

---

[12] Strictly speaking, the weights are also modified when the learning pattern is on the decision boundary.

[13] To make Eq. (2.30) a special case of Eq. (2.46), the modification of the weights should be "when $b_k\, g(\boldsymbol{x}_k) \leq \delta\|\boldsymbol{w}\|$", but for consistency with the support vector machine discussed later, the equal sign is omitted.

holds, and all learning patterns can be separated from the decision boundary with a distance of δ or more. This constant δ plays a role in controlling the margin of the decision boundary. That is, the larger δ is, the more the margin increases, and the classification performance for unknown patterns improves.

Here, we conduct an experiment using the weight modification of Eq. (2.46) to confirm the effectiveness of the learning method described in this section. We use the same learning patterns as shown in Fig. 2.10. As in the previous section, $\rho = 1$, and the initial weight vector is set to $\mathbf{w} = \mathbf{0}$. Four types of learning are prepared by changing the order in which the learning patterns are given.

The problem here is how to set the value of δ. Looking at Fig. 2.12, the largest value of r^* is $r^* = 0.623$ in (c), and the margin of the decision boundary is also relatively large. Therefore, in the experiment, a value slightly larger than this r^* was chosen, and $\delta = 0.670$ was used. The result is shown in Fig. 2.13, which shows the component vectors and the decision boundaries after convergence. The interpretation of the figure is the same as that of Fig. 2.12.

Comparing Fig. 2.13 with Fig. 2.12 confirms the following. First, in all four experiments, the values of r^* are within the range $0.670 < r^* < 0.680$, and as a result, there is no large variation in the decision boundaries as in Fig. 2.12. On the other hand, both the number of component vectors and the number of iterations increase. This result is reasonable because $\delta = 0$ in Fig. 2.12, whereas $\delta = 0.670$ in Fig. 2.13, which imposes a stricter condition on the decision boundary.

Let us take (a) of Fig. 2.13 as an example and examine the results in more detail below. Fig. 2.14 is an enlarged figure, and the interpretation of the figure is the same as that of Fig. 2.10.

The obtained decision boundary is indicated by a thick line as a hyperplane H_0 (here, a straight line). The closest patterns to the decision boundary H_0 were determined for each class ω_1 and ω_2. Letting the distances between H_0 and these patterns be R_1 and R_2, respectively, they were

$$\begin{cases} R_1 = 0.673 \\ R_2 = 0.676. \end{cases} \tag{2.48}$$

Both of these two proximity patterns were found to be component vectors.[14] The perpendicular lines from these patterns down to H_0 are indicated by the thin lines in the figure.

The value 0.673 of r^* noted in Fig. 2.13(a) refers to R_1 in the above equation, and the closest pattern to H_0 among all patterns is the pattern of class ω_1.

Let H_1 and H_2 be the hyperplanes that include the above proximity patterns and are parallel to H_0. These are indicated by the thin lines in the figure. The distance R between both hyperplanes is obtained as

---

[14] The pattern closest to the decision boundary is not necessarily the component vector.

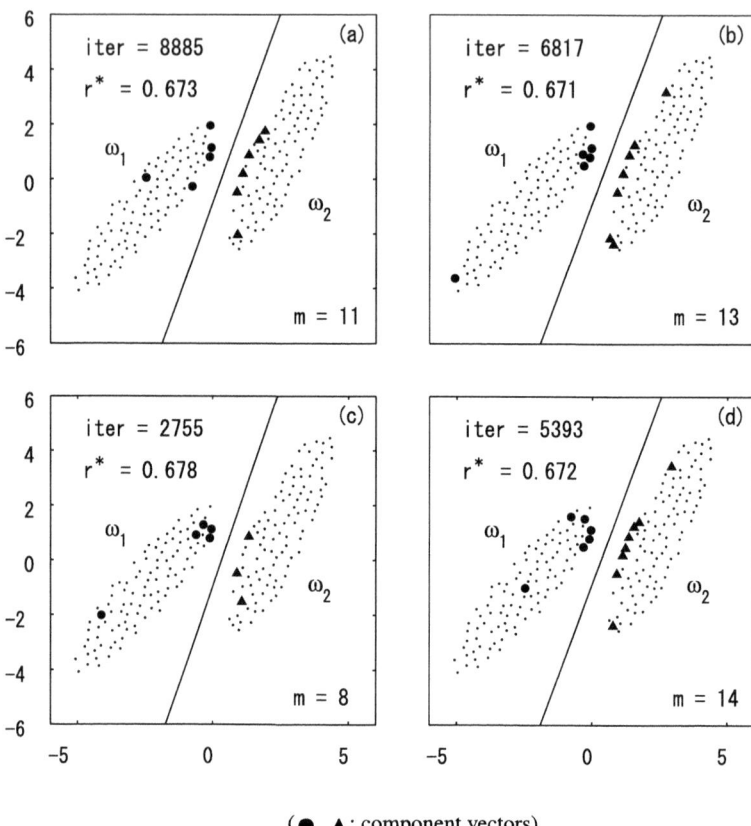

($\bullet$, $\blacktriangle$: component vectors)

Fig. 2.13 Decision boundaries with margins and component vectors

$$R = R_1 + R_2 = 0.673 + 0.676 = 1.349. \tag{2.49}$$

Naturally, there are no learning patterns in the feature space between the hyperplanes H_1 and H_2, which are separated from each other by R. The R in the above equation represents the *margin* of the decision boundary. It can be expected that the larger the margin, the higher the classification accuracy for unknown patterns.

Figure 2.15 is an enlarged version of (a) in Fig. 2.12. As in Fig. 2.14, by calculating the distances to the pattern closest to H_0 for each ω_1 and ω_2, $R_1 = 0.117$ and $R_2 = 0.303$ are obtained. Unlike Fig. 2.14, neither of the two closest patterns is a component vector. From this, the margin R is

$$R = R_1 + R_2 = 0.117 + 0.303 = 0.420. \tag{2.50}$$

The margin is smaller than in Eq. (2.49), and the difference is obvious when comparing both figures. See Problems 2.5 and 2.6 for other concrete examples.

Fig. 2.14 Decision boundaries with large margins (Fig. 2.13(a) is enlarged)

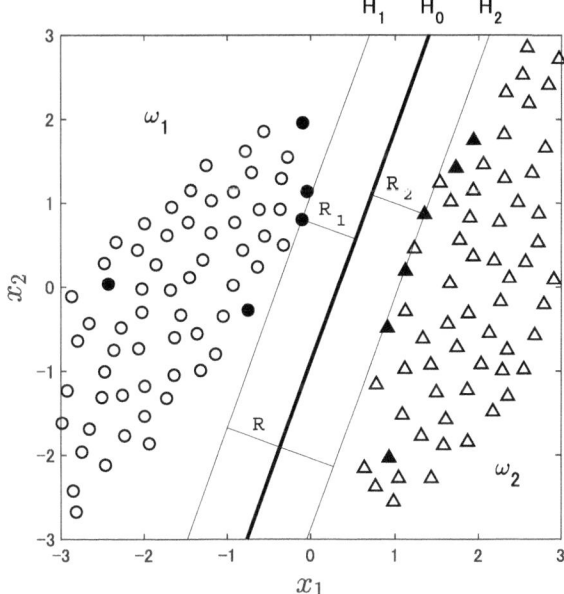

($\bullet$, $\blacktriangle$: component vectors)

As described above, it was experimentally confirmed that the weight modification method of Eq. (2.46) provides a decision boundary with a margin. However, the following problems still remain in this method.

The optimal decision boundary is the hyperplane with the largest margin R. The δ controls the margin R, and the larger the value of δ is set, the larger the margin R obtained as a result of learning. However, if δ is too large, there is no $\mathbf{w}$ that satisfies Eq. (2.47), and the learning rule will not converge. Conversely, if δ is too small, the decision boundary may vary widely and only a decision boundary with a small margin may be obtained. Thus, although δ can control the margin R, the final R is not known until the learning is performed. Therefore, to obtain the largest possible margin R, one must rely on trial and error. In any case, the decision boundary cannot be uniquely determined, and the obtained decision boundary is not necessarily the optimal one that maximizes R.

The support vector machine can solve the above problems. The support vector machine can uniquely determine the decision boundary with the largest margin. Support vector machines and margins are discussed in detail in Chap. 10.

Fig. 2.15 Decision boundaries that ended up with small margins (enlarge Fig. 2.12(a))

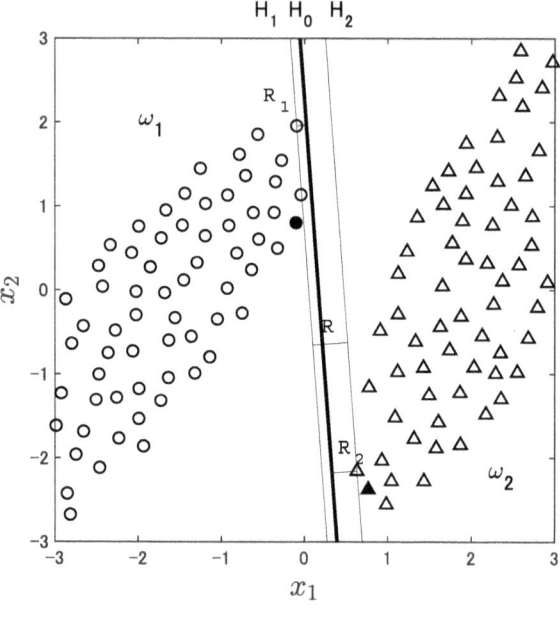

($\bullet$, $\blacktriangle$: component vectors)

2.8 Piecewise Linear Discriminant Functions

The linear discriminant function is realized with one prototype per class and the computational complexity is small due to its simple structure. However, this method cannot separate between classes for linearly nonseparable distributions. The piecewise linear discriminant function described below was devised to overcome this drawback. The piecewise linear discriminant function is equivalent to the nearest neighbor rule with multiple prototypes per class and is an advanced form of the linear discriminant function.

2.8.1 Properties of Piecewise Linear Discriminant Functions

In Fig. 1.10, complex decision boundaries were set with a large number of prototypes for a 2-dimensional feature vector, 3-class data ($d = 2$, $c = 3$). However, as already shown, this data is linearly separable, so the nearest neighbor rule based on one prototype per class, i.e., linear discriminant functions, can be used to separate the classes.

On the other hand, Fig. 2.16, which is also data for $d = 2$, $c = 3$, is an example of a linearly nonseparable distribution.

Fig. 2.16 Linearly
nonseparable distribution

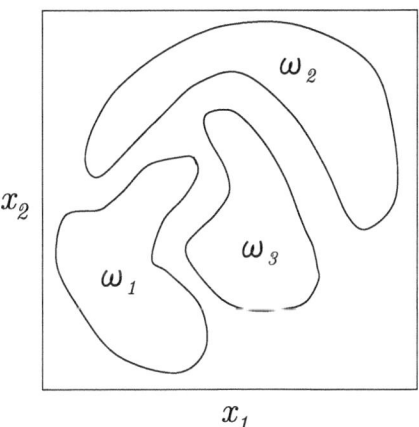

In such cases, the linear discriminant function cannot separate the classes. In other words, with one prototype per class, the nearest neighbor rule cannot correctly classify the learning pattern no matter where the prototypes are placed. To solve this problem, the nearest neighbor rule with multiple prototypes per class should be used.

Figure 2.17(a) shows an example of the nearest neighbor rule using multiple prototypes per class to separate classes.

Let m_i denote the number of prototypes of class ω_i and m the total number of prototypes, then $m_1 = m_2 = m_3 = 7$ and $m = 21$. The figure is represented by a Voronoi diagram, and the decision boundaries are indicated by thick lines.

If the purpose is limited to the separation between classes, prototypes that do not contribute to the determination of the decision boundaries can be omitted, even if they represent the distribution of patterns. Figure 2.17(b) is an example of reducing the number of prototypes to $m = 12$ with $m_1 = 3$, $m_2 = 4$, $m_3 = 5$. Although reducing the number of prototypes results in coarser decision boundaries, the classes are correctly separated in both cases.

As is clear from the example above, the decision boundaries consist of a combination of hyperplanes (polylines in this example). Based on the discussion of Sect. 2.2 on prototypes and linear discriminant functions, it is clear that the discriminant function used here is represented by a combination of linear discriminant functions. That is, the discriminant function $g_i(x)$ of class ω_i is represented by m_i linear discriminant functions $g_i^{(l)}(x)$ $(l = 1, \ldots, m_i)$ as follows:

$$g_i(x) = \max_{l=1,\ldots,m_i} \{g_i^{(l)}(x)\}, \tag{2.51}$$

$$g_i^{(l)}(x) = w_{i0}^{(l)} + \sum_{j=1}^{d} w_{ij}^{(l)} x_j \qquad (i = 1, 2, \ldots, c). \tag{2.52}$$

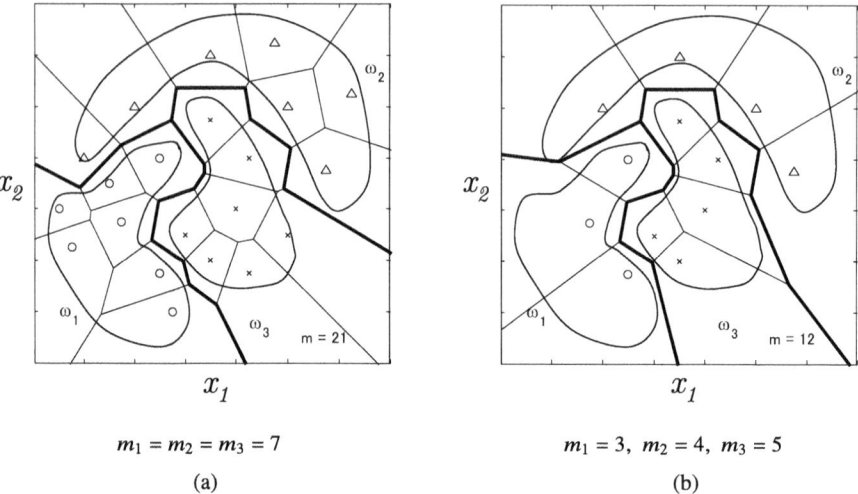

$$m_1 = m_2 = m_3 = 7 \qquad\qquad\qquad m_1 = 3, \ m_2 = 4, \ m_3 = 5$$

$$\text{(a)} \qquad\qquad\qquad\qquad\qquad \text{(b)}$$

Fig. 2.17 Class separation by piecewise linear discriminant function (**a**) Total number of proto-types $m = 21$ (**b**) Total number of prototypes $m = 12$

The m_i in the above equation is the number of prototypes of class ω_i. Such a discriminant function $g_i(x)$ is called a *piecewise linear discriminant function*. The linear discriminant function $g_i^{(l)}(x)$ in the above equation is called a *subsidiary discriminant function* (Problem 2.7).

Thus, the nearest neighbor rule realizes a piecewise linear discriminant function, and the special case of one prototype ($m_i = 1$) per class becomes a linear discriminant function. A block diagram of the piecewise linear discriminant function is shown in Fig. 2.18. The piecewise linear discriminant function outputs, as the classification result, the class having the subsidiary discriminant function with the maximum output value for the input pattern.

The piecewise linear discriminant function is extremely effective and can approximate any complex decision boundaries with arbitrary accuracy. Thus, a finite number of learning patterns can be completely separated into classes by the piecewise linear discriminant function. Unfortunately, however, the piecewise linear discriminant function does not have an effective algorithm equivalent to the perceptron learning rule. In order to obtain a piecewise linear discriminant function by learning, both the number of subsidiary discriminant functions m_i and their weights must be determined by learning. The former problem can be avoided by setting m_i large in advance, but this is wasteful. If the total number of subsidiary discriminant functions is limited, the procedure of moving subsidiary discriminant functions from one class to another should be included in the learning. A learning method for the piecewise linear discriminant function is presented in Nilsson (1965), but the conditions for convergence to the optimal solution are not known.

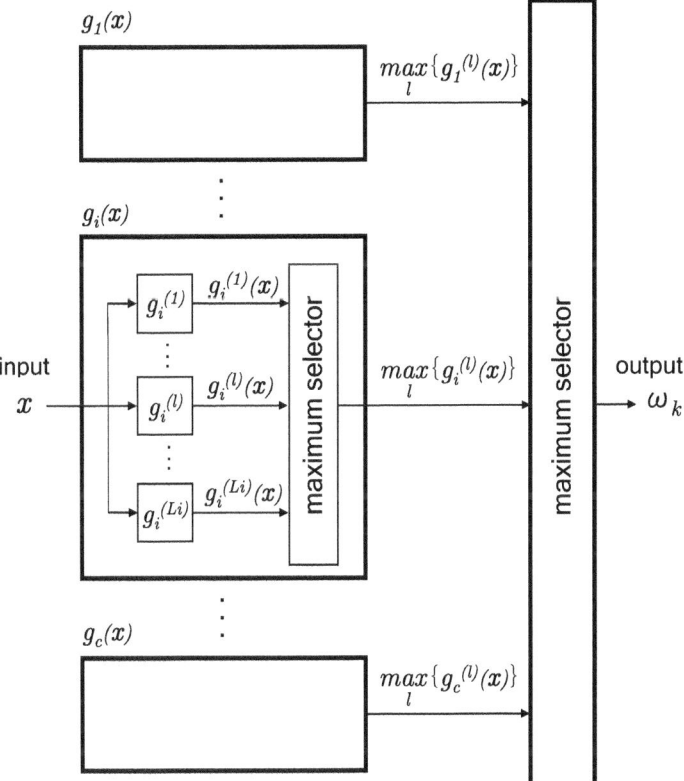

Fig. 2.18 Block diagram of a piecewise linear discriminant function

2.8.2   Relationship to Neural Networks

Neural networks will be introduced in detail later, but here is a brief description of the relationship between the previous discussion and neural networks. There are various forms of neural networks, but the one discussed here is a feedforward and multi-layer neural network.

Neural networks have been proved to be equivalent in the limit to piecewise linear discriminant functions (Nilsson, 1965).[15] Considering that a neural network realizes a nonlinear discriminant function and that a piecewise linear discriminant function can approximate a nonlinear discriminant function with arbitrary precision, it is

---

[15] However, the neural network treated in Nilsson (1965) is a classical neural network using the threshold function as a nonlinear element, which is different from a neural network using the sigmoid function. However, since the sigmoid function is equivalent to the threshold function in the limit, this conclusion can be considered to apply to neural networks as well.

intuitively clear that the two are equivalent. In fact, a comparison of the two methods shows that they have several points in common. For example, when using a neural network, the number of internal layers and the number of units must be determined prior to learning. This is the same as having to determine the number of subsidiary discriminant functions (the number of prototypes) in a piecewise linear discriminant function. It is also known that the classification ability of a neural network increases as the number of intermediate layers and units is increased, which corresponds to the fact that the more the number of subsidiary discriminant functions is increased, the more complex the decision boundaries can be made.

If neural networks are equivalent to piecewise linear discriminant functions, it is also clear that there is no difference in their classification ability. Both neural networks and piecewise linear discriminant functions are extremely powerful in the sense that they can approximate decision boundaries with arbitrary precision. However, whether such classification performance can be acquired by learning is another matter. As already mentioned, there is no powerful learning algorithm for piecewise linear discriminant functions. On the other hand, the backpropagation method, which was developed for neural networks, has been experimentally confirmed to be an effective learning algorithm. In other words, it should be noted that the difference between piecewise linear discriminant functions and neural networks is not their classification ability but their learning ability.

Coffee Break

Perceptron is the Basis of Learning: Misunderstood Minsky

The Perceptron, proposed by Rosenblatt in 1957, attracted much attention for its ability to realize the human intellectual activity of learning on a computer, and was the start of the first neural boom. However, the limitations of the perceptron became apparent, and artificial intelligence research entered a long period of winter. It is commonly believed that this was caused by Marvin Minsky's criticism of the perceptron in his book (Minsky and Papert 1969). However, he himself refutes this as follows Minsky and Papert (1988):

(1) In his book (Minsky and Papert 1969), he claims that some problems can be easily solved with the perceptron, but the computational cost of solving large-scale problems with the perceptron is extremely high. He never meant to imply that the perceptron is unimportant.

(2) In a nutshell, he clarified the computational cost of the perceptron as a function of the problem size.

(3) The perceptron is highly suggestive in terms of functionality and architecture, and many of the ideas learned from the perceptron should continue to be valid in the future.

Needless to say, the perceptron is the foundation of the learning model and the basic for anyone studying pattern recognition and machine learning. Even the multi-layer neural networks used in deep learning are composed of a final layer and a previous layer, which is the structure of a perceptron itself. The structure of the support vector machine is also common to the perceptron. In fact, researchers will be keenly aware of the above point (3) in various situations.

Problems

2.1 The learning patterns of three classes ω_1, ω_2, and ω_3 are distributed on the two-dimensional feature space. Here, we wish to apply the minimum distance method, where the mean vector of each class is used as a prototype. The prototypes $\mathbf{p}_1$, $\mathbf{p}_2$, and $\mathbf{p}_3$ of each class are as follows:

$$\mathbf{p}_1 = (4, 16)^t, \quad \mathbf{p}_2 = (12, 6)^t, \quad \mathbf{p}_3 = (12, 18)^t.$$

(1) Find a linear discriminant function $g_i(\mathbf{x})$ $(i = 1, 2, 3)$ that implements the minimum distance method using these prototypes.
(2) Find all decision boundaries $g_{ij}(\mathbf{x}) \overset{\text{def}}{=} g_i(\mathbf{x}) - g_j(\mathbf{x}) = 0$ $(i < j)$, and plot them on a figure. Then, show each region that is determined to be of class ω_1, ω_2, and ω_3.
(3) To the above learning patterns, a further learning pattern of class ω_4 was added, whose prototype $\mathbf{p}_4$ was $\mathbf{p}_4 = (2, 4)^t$. In the same way as above, find a linear discriminant function $g_4(\mathbf{x})$ that implements the minimum distance method based on the prototypes.
(4) Add the decision boundaries $g_{i4}(\mathbf{x}) = 0$ $(i = 1, 2, 3)$ to the figure, and also show each region determined to be of class ω_1-ω_4.
(5) Show the results of classifying patterns $\mathbf{x}_1 = (2, 9)^t$ and $\mathbf{x}_2 = (2, 11)^t$ with the discriminant functions $g_1(\mathbf{x})$ to $g_4(\mathbf{x})$. Confirm that the results are correct by plotting the patterns on the figure.

2.2 Show that the discriminant function $g(\mathbf{x})$ is always updated in the direction of improvement each time the weight vector is modified according to Eq. (2.30).

2.3 Suppose that six learning patterns $\mathbf{x}_1, \mathbf{x}_2, \ldots, \mathbf{x}_6$ are given in the two-dimensional feature space as follows:

$$\mathbf{x}_1 = (11, 8)^t, \ \mathbf{x}_2 = (10, 10)^t, \ \mathbf{x}_3 = (6, 3)^t,$$
$$\mathbf{x}_4 = (6, 5)^t, \quad \mathbf{x}_5 = (2, 8)^t, \quad \mathbf{x}_6 = (1, 2)^t.$$

Among them, $\mathbf{x}_1, \mathbf{x}_2, \mathbf{x}_3$ belong to class ω_1 and $\mathbf{x}_4, \mathbf{x}_5, \mathbf{x}_6$ belong to class ω_2. Now, we want to set a linear discriminant function

$$g(\mathbf{x}) = w_0 + w_1 x_1 + w_2 x_2$$

and determine weights w_0, w_1, w_2 so that

$$g(\mathbf{x}_k) > 0 \quad (\text{when } \mathbf{x}_k \text{ belongs to class } \omega_1),$$

$$g(\mathbf{x}_k) < 0 \quad (\text{when } \mathbf{x}_k \text{ belongs to class } \omega_2),$$

for the learning patterns $\mathbf{x}_k$ $(k = 1, \ldots, 6)$.

(1) Find the weights w_0, w_1, w_2 using the perceptron learning rule. The initial values of the weights are $(w_0, w_1, w_2) = (-54, 13, -15)$ and the learning rate is $\rho = 1$. The learning patterns are given repeatedly from x_1 to x_6 in this order.

(2) Plot the learning patterns on the two-dimensional feature space and also illustrate the decision boundary determined by the weights set as initial values.

(3) Illustrate the decision boundary determined by the weights obtained by the perceptron learning rule.

2.4 Derive Eq. (2.44).

2.5 † The following 8 patterns $x_1, \ldots, x_8$ are distributed in the 2-dimensional feature space, of which x_1, x_2, x_3, x_4 belong to class ω_1 and x_5, x_6, x_7, x_8 belong to class ω_2 respectively:

$$\left. \begin{array}{l} x_1 = (2, 5)^t, \ x_2 = (3, 3)^t, \ x_3 = (1, 2)^t, \ x_4 = (2, 1)^t \quad \in \omega_1 \\ x_5 = (8, 9)^t, \ x_6 = (9, 8)^t, \ x_7 = (7, 7)^t, \ x_8 = (9, 6)^t \quad \in \omega_2 \end{array} \right\}$$

We now set up a linear discriminant function

$$g(x) = w_0 + w_1 x_1 + w_2 x_2$$

and want to determine the weights w_0, w_1, w_2 so that we can correctly classify all patterns using the classification rule shown in Eq. (2.21).

(1) Find the weights w_0, w_1, w_2 using the perceptron learning rule (for two-class problem) in Sect. 2.3.2, where the initial weight vector is $\mathbf{w} = (w_0, w_1, w_2)^t = (0, 0, 0)^t$ and the learning rate is $\rho = 1$. The learning patterns are given repeatedly from x_1 to x_8 in this order.

(2) Illustrate the decision boundary H_0 determined by the weights obtained. Also, find and illustrate the hyperplanes H_1 and H_2 that determine the margin as shown in Fig. 2.14, and calculate the value of the margin R.

2.6 † Using the same 8 patterns as in the previous problem, we want to determine the weights of the linear discriminant function by the perceptron learning rule. In order to obtain a decision boundary with a margin, we apply the weight modification method shown in Eq. (2.46). Set δ of Eq. (2.46) to $\delta = 1$, and with the other conditions being the same as in the previous problem, show the final weights obtained. Also, as (2) in the previous problem, illustrate the decision boundary H_0, the hyperplanes H_1 and H_2 that determine the margin, and calculate the value of the margin R. Furthermore, compare these results with those of the previous problem.

2.7 Eight learning patterns $x_1, x_2, \ldots, x_8$ are given on the two-dimensional feature space as follows:

$$x_1 = (3, 0)^t, \ x_2 = (4, 3)^t, \ x_3 = (6, 4)^t, \ x_4 = (7, 1)^t,$$
$$x_5 = (1, 2)^t, \ x_6 = (3, 5)^t, \ x_7 = (4, 6)^t, \ x_8 = (0, 3)^t.$$

Among them, $x_1, \ldots, x_4$ belong to the class ω_1, and $x_5, \ldots, x_8$ belong to the class ω_2, respectively.

(1) As a classification method, adopt all 8 learning patterns as prototypes and apply the nearest neighbor rule (method 1). Illustrate the decision boundary of the piecewise linear discriminant function determined by this method.

(2) Illustrate the decision boundary when the minimum distance method (method 2) is applied, using the mean of the learning patterns of each class as the prototype.

(3) Show the results of classifying the test pattern $x_9 = (3, 3)^t$ by methods 1 and 2, respectively.

Chapter 3
Learning Based on Minimum Square Error Criterion

Abstract The drawback of the perceptron learning rule introduced in the previous chapter is that it requires the learning pattern to be linearly separable, i.e., there must be a linear discriminant function that makes misclassification zero. For learning patterns that are linearly nonseparable, the weight modification procedure is repeated infinitely, and no solution can be reached. Even if we forcefully terminate the process because there is no possibility of convergence, there is no guarantee that the weights obtained at that time will be optimal. In general, it is difficult to confirm in advance whether or not linear separation is possible. This chapter introduces a general learning algorithm that can be applied to the case where linear separation is not possible.

3.1 Minimum Square Error Learning

The perceptron learning rule aims at a linear discriminant function that makes the number of misclassified patterns[1] zero. In other words, the perceptron learning rule uses the number of misclassified patterns as its evaluation measure. However, as long as this evaluation measure is used, no solution can be obtained for learning patterns that are linearly nonseparable.

In the following, we propose an alternative evaluation measure to the number of misclassified patterns. The basic idea is to define an evaluation function that can be applied even when linear separability is not possible, and to minimize it. It is an extremely important fact to point out that the learning algorithms described below are all closely related to the Bayes decision rule. This will be discussed in detail in Chap. 15.

Supplementary Information The online version contains supplementary material available at https://doi.org/10.1007/978-981-95-1478-6_3.

---

[1] This number includes the number of reject patterns. See footnote 2 in Chap. 1.

3.1.1   Evaluation Functions for Learning

Here, we consider the following learning method. For each learning pattern, the desired output value of the discriminant function is predefined. Set the teaching signal described in Eq. (2.29) as the desired output value. As an evaluation function for learning, we consider the error between the output value of the discriminant function and the teaching signal. In learning, the weights are modified to minimize the error.

As a teaching signal, Eq. (2.29) is defined for the two-class ($c = 2$) case. Here we generalize it to the multi-class ($c > 2$) case. Now, when the kth pattern x_k ($k = 1, 2, \ldots, n$) is input, the desired output value of the ith discriminant function $g_i(x_k)$ is defined as the teaching signal b_{ik}. Furthermore, if the output values of the c discriminant functions are represented by the vector $(g_1(x_k), g_2(x_k), \ldots, g_c(x_k))^t$, the corresponding teaching signal is also represented by the vector $(b_{1k}, b_{2k}, \ldots, b_{ck})^t$. This vector representation of the teaching signal is called a *teaching vector*. Due to the nature of the discriminant function, each component of the teaching vector must be set so that

$$b_{ik} > b_{jk} \quad (j \neq i) \qquad \text{if} x_k \in \omega_i. \tag{3.1}$$

If we assign the same teaching vector $\mathbf{t}_i$ to all patterns belonging to class ω_i, then

$$\mathbf{t}_i = (b_1, \ldots, b_i, \ldots, b_c)^t \qquad (b_i > b_j, \quad j \neq i) \quad (i = 1, \ldots, c). \tag{3.2}$$

In this case, we only need c teaching vectors $\mathbf{t}_1, \mathbf{t}_2, \ldots, \mathbf{t}_c$. As the teaching vector $\mathbf{t}_i$, one way is, for example, to choose the c-dimensional unit vector of the following equation, obtained by setting $b_i = 1, \; b_j = 0 \; (j \neq i)$:

$$\mathbf{t}_i = (\overset{1}{0}, \ldots, 0, \overset{i}{1}, 0, \ldots, \overset{c}{0})^t \qquad (i = 1, \ldots, c). \tag{3.3}$$

That is, the following settings are used:

$$\text{if} x_k \in \omega_i \qquad (b_{1k}, b_{2k}, \ldots, b_{ck})^t = (\overset{1}{0}, \ldots, 0, \overset{i}{1}, 0, \ldots, \overset{c}{0})^t. \tag{3.4}$$

A vector as shown in Eq. (3.3), where only one specific element is 1 and all other elements are 0, is called the *one-hot vector*. The teaching vectors will be discussed again in Chaps. 14 and 15.

The error ε_{ik} between the actual output and the teaching signal for the input pattern x_k is

$$\varepsilon_{ik} = g_i(x_k) - b_{ik} \tag{3.5}$$

and defining the sum of squares of ε_{ik} as the evaluation function J_k, J_k can be written as a function of the weight vector $\mathbf{w}_i$ as follows:[2]

$$J_k(\mathbf{w}_1, \mathbf{w}_2, \ldots, \mathbf{w}_c) = \frac{1}{2} \sum_{i=1}^{c} \varepsilon_{ik}^2 \tag{3.6}$$

$$= \frac{1}{2} \sum_{i=1}^{c} (g_i(\mathbf{x}_k) - b_{ik})^2 \tag{3.7}$$

$$= \frac{1}{2} \sum_{i=1}^{c} (\mathbf{w}_i^t \mathbf{x}_k - b_{ik})^2, \tag{3.8}$$

where $\mathbf{x}_k$ is the augmented feature vector of $\mathbf{x}_k$.

The square error J for all patterns can be expressed as

$$J(\mathbf{w}_1, \mathbf{w}_2, \ldots, \mathbf{w}_c) = \sum_{k=1}^{n} J_k(\mathbf{w}_1, \mathbf{w}_2, \ldots, \mathbf{w}_c) \tag{3.9}$$

$$= \frac{1}{2} \sum_{k=1}^{n} \sum_{i=1}^{c} (g_i(\mathbf{x}_k) - b_{ik})^2 \tag{3.10}$$

$$= \frac{1}{2} \sum_{k=1}^{n} \sum_{i=1}^{c} (\mathbf{w}_i^t \mathbf{x}_k - b_{ik})^2. \tag{3.11}$$

Therefore, the optimal weight vector can be found as the solution that minimizes Eq. (3.11). Such an optimization method for the weight vector is called the *minimum square error learning*.

3.1.2 Closed-Form Solution

For a function $J(\mathbf{w})$ of weight vector $\mathbf{w} = (w_0, w_1, \ldots, w_d)^t$, define the *gradient vector* as follows:

$$\nabla J = \frac{\partial J}{\partial \mathbf{w}} = \left(\frac{\partial J}{\partial w_0}, \frac{\partial J}{\partial w_1}, \ldots, \frac{\partial J}{\partial w_d} \right)^t. \tag{3.12}$$

The gradient vector corresponding to the weight vector $\mathbf{w}_i$ of class ω_i is denoted by $\nabla_i J$ or $\partial J / \partial \mathbf{w}_i$. For vector differentiation, see Appendix B.

---

[2] The reason for multiplying by a factor $1/2$ is to simplify the notation of Eq. (3.15) below.

The direct way to find the minimum solution of $J(\mathbf{w}_1, \mathbf{w}_2, \ldots, \mathbf{w}_c)$ is to find the solution of the following equation:

$$\frac{\partial J}{\partial \mathbf{w}_i} = \nabla_i J = \mathbf{0} \qquad (i = 1, 2, \ldots, c). \tag{3.13}$$

That is, from Eq. (3.11), we can solve

$$\frac{\partial J}{\partial \mathbf{w}_i} = \sum_{k=1}^{n} \frac{\partial J_k}{\partial \mathbf{w}_i} \tag{3.14}$$

$$= \sum_{k=1}^{n} (\mathbf{w}_i^t \mathbf{x}_k - b_{ik}) \mathbf{x}_k = \mathbf{0} \qquad (i = 1, 2, \ldots, c). \tag{3.15}$$

If we now define an $n \times (d + 1)$ matrix $\mathbf{X}$ and an n-dimensional vector $\mathbf{b}_i$ [3] with

$$\mathbf{X} \overset{\text{def}}{=} (\mathbf{x}_1, \mathbf{x}_2, \ldots, \mathbf{x}_n)^t \tag{3.16}$$

$$\mathbf{b}_i \overset{\text{def}}{=} (b_{i1}, b_{i2}, \ldots, b_{in})^t \qquad (i = 1, 2, \ldots, c), \tag{3.17}$$

then Eqs. (3.11) and (3.15) can be easily expressed as follows:

$$J(\mathbf{w}_1, \mathbf{w}_2, \ldots, \mathbf{w}_c) = \frac{1}{2} \sum_{i=1}^{c} \|\mathbf{X}\mathbf{w}_i - \mathbf{b}_i\|^2 \tag{3.18}$$

$$\frac{\partial J}{\partial \mathbf{w}_i} = \mathbf{X}^t (\mathbf{X}\mathbf{w}_i - \mathbf{b}_i) = \mathbf{0} \qquad (i = 1, 2, \ldots, c). \tag{3.19}$$

The matrix $\mathbf{X}$ is called the *pattern matrix*. From Eq. (3.19), we obtain

$$\mathbf{X}^t \mathbf{X} \mathbf{w}_i = \mathbf{X}^t \mathbf{b}_i \qquad (i = 1, 2, \ldots, c). \tag{3.20}$$

Assuming that the $(d + 1) \times (d + 1)$ matrix $\mathbf{X}^t \mathbf{X}$ is *non-singular*, we have

$$\mathbf{w}_i = (\mathbf{X}^t \mathbf{X})^{-1} \mathbf{X}^t \mathbf{b}_i \qquad (i = 1, 2, \ldots, c). \tag{3.21}$$

The process of obtaining $\mathbf{w}_i = (\mathbf{X}^t \mathbf{X})^{-1} \mathbf{X}^t \mathbf{b}_i$ as the minimum solution of $\|\mathbf{X}\mathbf{w}_i - \mathbf{b}_i\|^2$, as shown above, is the same method as *multiple regression analysis*, where $\mathbf{x}_k$ is an *explanatory variable* and b_{ik} is an *objective variable*. The $\mathbf{w}_i$

---

[3] Be careful not to confuse the vector $\mathbf{b}_i$ of Eq. (3.17) with the teaching vector $(b_{1k}, b_{2k}, \ldots, b_{ck})^t$ of Eq. (3.4). The elements in the vectors are all teaching signals b_{ik}, but the former is an n-dimensional vector created by varying k from 1 to n and the latter is a c-dimensional vector created by varying i from 1 to c.

thus obtained is the *globally optimal solution* and the unique minimum point. See Problem 3.1 for details. Equation (3.3) indicates that different classes should correspond to different teaching vectors that are easily distinguishable from each other. Minimizing Eq. (3.11) also indicates that patterns belonging to the same class should be concentrated in the neighborhood of the same teaching vector. Therefore, the above process corresponds to minimizing the within-class variance under the constant between-class variance, and can be interpreted as a special case of the linear discriminant method. This will be discussed again in Chap. 15.

3.1.3 Solution by the Steepest Descent Method (Widrow–Hoff Learning Rule)

The method described above is not applicable when $\mathbf{X}^t\mathbf{X}$ is singular. Also, when d is large, the method is not practical due to the large amount of computation required to compute the inverse matrix. As an alternative, we describe here a method to determine weights by successive approximation. The most commonly used method of this type is the *steepest descent method*. That is, the weight vector is successively updated by

$$\mathbf{w}_i' = \mathbf{w}_i - \rho \cdot \frac{1}{n} \cdot \frac{\partial J}{\partial \mathbf{w}_i} \tag{3.22}$$

$$= \mathbf{w}_i - \rho \cdot \frac{1}{n} \sum_{k=1}^{n} \frac{\partial J_k}{\partial \mathbf{w}_i} \qquad (i = 1, 2, \ldots, c), \tag{3.23}$$

and finally, the minimum solution of J is reached. In this equation, a positive constant ρ is the learning rate already introduced in Eqs. (2.26) and (2.27).

The above equation indicates that after all learning patterns are presented, the weights are modified in a batch. Such a learning method is called *batch learning*. In the above equation, ρ is divided by n so that ρ does not need to be adjusted depending on the number of patterns.

On the other hand, the modification can be made each time a pattern is presented. Such a learning method is called *online learning*. In this case, the modification of the weights is performed by

$$\mathbf{w}_i' = \mathbf{w}_i - \rho \frac{\partial J_k}{\partial \mathbf{w}_i} \qquad (i = 1, 2, \ldots, c). \tag{3.24}$$

An intermediate learning method between batch learning and online learning is *minibatch learning*. In this learning method, n learning patterns are divided into m ($1 \le m \le n$) groups, and weights are modified when all patterns in each group are presented. The case $m = 1$ corresponds to batch learning, and the case $m = n$

corresponds to online learning. The maximum number of weight modifications per epoch is 1 for batch learning, n for online learning, and m for mini-batch learning.

In the following, we will take online learning as an example. For simplicity, $g_i(x_k)$ is abbreviated to g_{ik}, then

$$\frac{\partial J_k}{\partial \mathbf{w}_i} = \frac{\partial J_k}{\partial g_{ik}} \cdot \frac{\partial g_{ik}}{\partial \mathbf{w}_i}. \tag{3.25}$$

From Eq. (3.7), the first term on the right-hand side of the above equation is

$$\frac{\partial J_k}{\partial g_{ik}} = g_{ik} - b_{ik} = \varepsilon_{ik}, \tag{3.26}$$

and from $g_{ik} = \mathbf{w}_i^t \mathbf{x}_k$, the second term is

$$\frac{\partial g_{ik}}{\partial \mathbf{w}_i} = \mathbf{x}_k. \tag{3.27}$$

Therefore, Eq. (3.25) can be written as

$$\frac{\partial J_k}{\partial \mathbf{w}_i} = (g_{ik} - b_{ik})\mathbf{x}_k \tag{3.28}$$

$$= \varepsilon_{ik}\, \mathbf{x}_k. \tag{3.29}$$

Substituting Eq. (3.29) into Eq. (3.24) yields

$$\mathbf{w}_i' = \mathbf{w}_i - \rho\, \varepsilon_{ik}\, \mathbf{x}_k \tag{3.30}$$

$$= \mathbf{w}_i - \rho(g_{ik} - b_{ik})\mathbf{x}_k \tag{3.31}$$

$$= \mathbf{w}_i - \rho(\mathbf{w}_i^t \mathbf{x}_k - b_{ik})\mathbf{x}_k \qquad (i = 1, 2, \ldots, c), \tag{3.32}$$

as a modified weight vector. In the case of batch learning,

$$\mathbf{w}_i' = \mathbf{w}_i - \rho \cdot \frac{1}{n} \sum_{k=1}^{n} \varepsilon_{ik}\, \mathbf{x}_k \tag{3.33}$$

$$= \mathbf{w}_i - \rho \cdot \frac{1}{n} \sum_{k=1}^{n} (\mathbf{w}_i^t \mathbf{x}_k - b_{ik})\, \mathbf{x}_k \qquad (i = 1, 2, \ldots, c) \tag{3.34}$$

can be obtained by substituting Eq. (3.29) into Eq. (3.23). The weight modification for mini-batch learning can be performed in the same way. This is called the *Widrow–Hoff learning rule*. This is sometimes called the *delta rule* (Rumelhart and McClelland 1986).

This method can be applied to both linearly separable and non-separable learning patterns. In the non-separable case, zero misclassification cannot be achieved, and there is no guarantee that the number of misclassified patterns can be minimized. Even if the patterns are linearly separable, the method does not necessarily achieve zero misclassification. These points should be taken into account when using this method.

3.1.4 Two-Class Case

The above discussion has been based on the assumption of multi-class. If the classification target is a two-class case, only one weight vector is required from Eq. (2.19), and the following formula can be used instead of Eq. (3.8):

$$J_k(\mathbf{w}) = \frac{1}{2}(g(\mathbf{x}_k) - b_k)^2 = \frac{1}{2}(\mathbf{w}^t \mathbf{x}_k - b_k)^2. \tag{3.35}$$

As shown in Eq. (2.29), teaching signal b_k is set here as follows:

$$b_k = \begin{cases} 1 & (\mathbf{x}_k \in \omega_1) \\ -1 & (\mathbf{x}_k \in \omega_2) \end{cases} \qquad (k = 1, \dots, n). \tag{3.36}$$

From Eqs. (3.9) and (3.35), the square error $J(\mathbf{w})$ is

$$J(\mathbf{w}) = \sum_{k=1}^{n} J_k(\mathbf{w}) \tag{3.37}$$

$$= \frac{1}{2}\sum_{k=1}^{n}(g(\mathbf{x}_k) - b_k)^2 = \frac{1}{2}\sum_{k=1}^{n}(\mathbf{w}^t \mathbf{x}_k - b_k)^2. \tag{3.38}$$

Let $\mathbf{b}$ be defined as

$$\mathbf{b} \overset{\text{def}}{=} (b_1, b_2, \dots, b_n)^t. \tag{3.39}$$

Each component b_k ($k = 1, \dots, n$) of the vector $\mathbf{b}$ is defined according to Eq. (3.36). To obtain a closed-form solution, first transform $J(\mathbf{w})$ as follows:

$$J(\mathbf{w}) = \frac{1}{2}\|\mathbf{Xw} - \mathbf{b}\|^2. \tag{3.40}$$

Then, by setting $\partial J/\partial \mathbf{w} = \mathbf{0}$ as in Eq. (3.19) to obtain

$$\mathbf{w} = (\mathbf{X}^t \mathbf{X})^{-1} \mathbf{X}^t \mathbf{b}. \tag{3.41}$$

The Widrow–Hoff learning rule as a successive approximation is as follows:

$$\mathbf{w}' = \mathbf{w} - \rho (\mathbf{w}^t \mathbf{x}_k - b_k) \mathbf{x}_k \qquad \text{(online learning)}, \qquad (3.42)$$

$$\mathbf{w}' = \mathbf{w} - \rho \cdot \frac{1}{n} \sum_{k=1}^{n} (\mathbf{w}^t \mathbf{x}_k - b_k) \mathbf{x}_k \qquad \text{(batch learning)}. \qquad (3.43)$$

3.1.5   Experiments on Widrow–Hoff Learning Rule

Let us confirm the operation and effectiveness of the learning method described above by simple experiments. The data used are the same as in Fig. 2.3, i.e., six learning patterns $x_1, \dots, x_6$ distributed on a number line in a one-dimensional feature space. However, although the arrangement of each pattern is the same as in Fig. 2.3, the classes to which the patterns x_3 and x_4 belong are reversed. That is, x_1, x_2, x_4 belong to class ω_1 and x_3, x_5, x_6 belong to class ω_2.

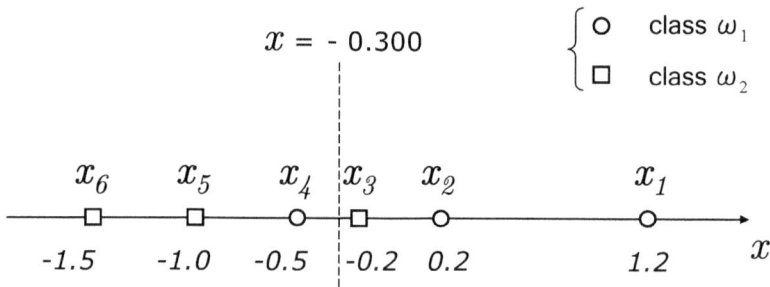

Fig. 3.1 Linearly nonseparable learning patterns in a one-dimensional feature space

The locations of the learning patterns and their classes are shown in Fig. 3.1, and they are clearly linearly nonseparable. Therefore, the perceptron learning rule of the previous chapter is not applicable. Since we are dealing with a two-class case, we will follow the procedure described in Sect. 3.1.4.

Let us first find a closed-form solution according to Eq. (3.41). The pattern matrix $\mathbf{X}$ is

$$\mathbf{X}^t = \begin{pmatrix} 1.0 & 1.0 & 1.0 & 1.0 & 1.0 & 1.0 \\ 1.2 & 0.2 & -0.2 & -0.5 & -1.0 & -1.5 \end{pmatrix} \qquad (3.44)$$

and from Eqs. (3.36) and (3.39), $\mathbf{b}$ is

$$\mathbf{b} = (1,\ 1,\ -1,\ 1,\ -1,\ -1)^t. \qquad (3.45)$$

Using Eq. (3.44), we obtain

$$(\mathbf{X}^t\mathbf{X})^{-1}\mathbf{X}^t$$
$$= \begin{pmatrix} 0.267 & 0.200 & 0.173 & 0.153 & 0.120 & 0.086 \\ 0.335 & 0.112 & 0.022 & -0.045 & -0.156 & -0.268 \end{pmatrix}. \tag{3.46}$$

Then, from Eq. (3.41), $\mathbf{w}$ is

$$\mathbf{w} = (w_0,\ w_1)^t = (\mathbf{X}^t\mathbf{X})^{-1}\mathbf{X}^t\mathbf{b} \tag{3.47}$$
$$= (0.241,\ 0.804)^t. \tag{3.48}$$

Since the decision boundary for separating the two classes is

$$g(x) = \mathbf{w}^t\mathbf{x} = w_0 + w_1 x = 0, \tag{3.49}$$

we obtain

$$x = -w_0/w_1 = -0.241/0.804 = -0.300. \tag{3.50}$$

This decision boundary is indicated by the dotted line in Fig. 3.1. The result shows that the patterns x_3 and x_4 are misclassified. However, this decision boundary is optimal in terms of minimizing the square error J among the decision boundaries obtained by the linear discriminant function. See Problem 3.2 for an example of using Eq. (3.21) instead of Eq. (3.41).

Next, let us find the solution $\mathbf{w}$ by successive approximation, i.e., by the Widrow–Hoff learning rule. Given learning patterns, $J(\mathbf{w})$ is determined by Eq. (3.40). The $J(\mathbf{w})$ obtained by substituting the learning patterns $x_1, \ldots, x_6$ and the teaching signal $\mathbf{b}$ into Eq. (3.40) is shown in Fig. 3.2. In the figure, the contour lines of $J(\mathbf{w})$ are shown as thin lines on the two-dimensional plane with (w_1, w_0) as the coordinate. The optimal value of $(w_1, w_0) = (0.804, 0.241)$ for $\mathbf{w}$ obtained as a closed-form solution in Eq. (3.48) is indicated by a small black circle.

As already mentioned, this optimal value is a globally optimal and unique solution, so the result obtained does not depend on the initial value. This is evident from the shape of $J(\mathbf{w})$ shown in Fig. 3.2.

In the experiments, the initial value of the weight vector $\mathbf{w}$ was set to $(w_1, w_0) = (5, 11)$. This initial value was also used in Fig. 2.6. The initial position of the weight vector $\mathbf{w}$ is marked with "×" in Fig. 3.2. Experiments were conducted for both batch and online learning, and the learning rate was set to $\rho = 0.1$ in Eqs. (3.42) and (3.43). Convergence was considered to occur when the change in $J(\mathbf{w})$ became smaller than a predetermined threshold value of 0.01, during the iteration. The trajectory of the weight vector $\mathbf{w}$ up to convergence is shown in Fig. 3.2 with a thick line, and the change in the value of $J(\mathbf{w})$ up to convergence is shown in Fig. 3.3.

The number of weight modifications required for convergence is 69 for batch learning and 90 for online learning. The weights (w_1, w_0) obtained by batch learning and online learning are $(0.907, 0.321)$ and $(0.833, 0.251)$, respectively. It can be also confirmed from Fig. 3.2 that both are close to the optimal value $(0.804, 0.241)$ obtained as a closed-form solution.

Fig. 3.2 Widrow–Hoff
learning rule (linearly
nonseparable data)

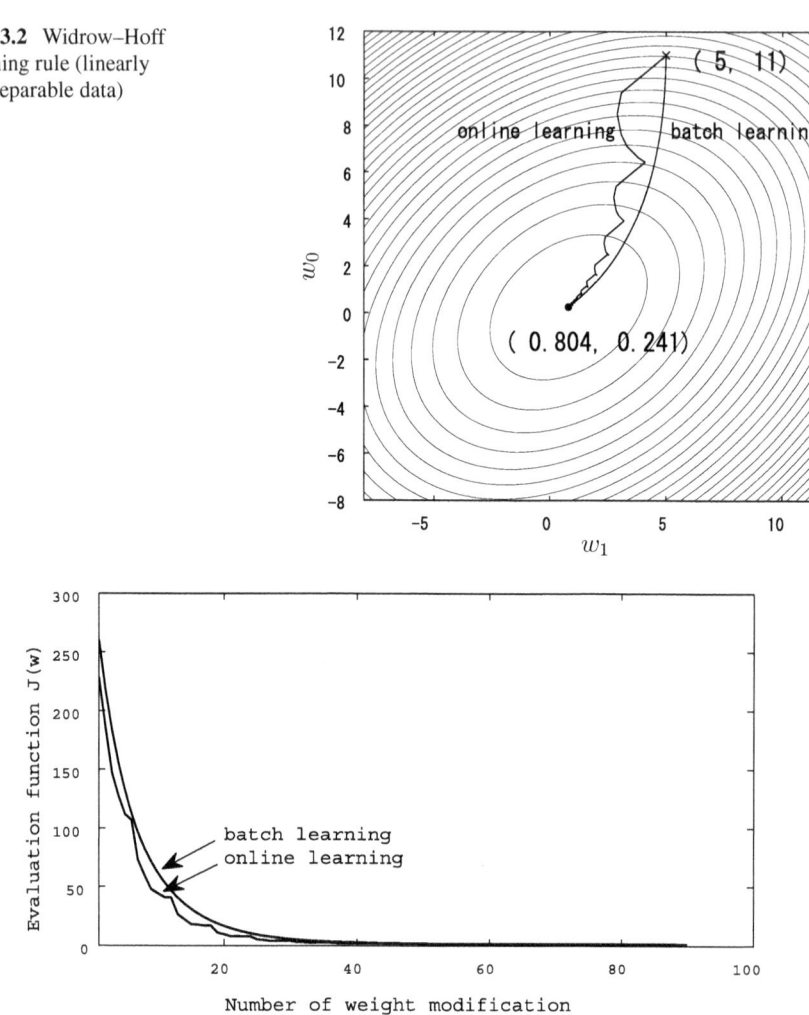

Fig. 3.3 Convergence process of the Widrow–Hoff learning rule (linearly nonseparable data)

The figure shows that in batch learning, the modification is almost in the direction perpendicular to the contour, i.e., the direction of the steepest descent of $J(\mathbf{w})$. On the other hand, in online learning, the direction of each modification does not necessarily coincide with the direction of the steepest descent of $J(\mathbf{w})$,[4] but the optimal solution is almost reached in the end. In batch learning, $J(\mathbf{w})$ can be reduced most efficiently with a single modification, and thus convergence is achieved with fewer modifications than in online learning.

---

[4] It coincides with the direction of the steepest descent of $J_k(\mathbf{w})$.

The difference between the two also appears in Fig. 3.3. That is, $J(\mathbf{w})$ decreases monotonically and smoothly in the repeated modifications in batch learning, whereas the decreasing trend of $J(\mathbf{w})$ is not smooth in online learning.

As described above, the experiments on learning based on square error used linearly nonseparable data. When the same method is applied to the linearly separable data shown in Fig. 2.3, there is no significant difference in the results obtained (Problem 3.3).

3.2 Minimum Square Error Learning and Perceptron

Although the minimum square error learning described in the previous section is a method designed to overcome the shortcomings of the perceptron, it is closely related to the perceptron learning rule.

In the following, we first show that the Widrow–Hoff learning rule includes the perceptron learning rule as a special case. Then, we show that the perceptron learning rule can also be interpreted as an algorithm for minimizing the evaluation function by the steepest descent method.

3.2.1 Binary Error Evaluation

Let us compare the Widrow–Hoff learning rule with the perceptron learning rule. The ε_{ik} term in Eq. (3.30) is the difference between the teaching signal and the actual output, and the amount of weight modification is proportional to it.

Here, after obtaining $g_i(\mathbf{x})$ ($= \mathbf{w}_i{}^t\mathbf{x}$) in Fig. 2.2, we apply to $g_i(\mathbf{x})$ the operation by *threshold function* T_i and put the result again as $g_i(\mathbf{x})$. The threshold function T_i is defined by

$$T_i(u) = \begin{cases} 1 \ (u > 0) \\ 0 \ (u < 0). \end{cases} \qquad (i = 1, 2, \ldots, c) \qquad (3.51)$$

Then, $g_i(\mathbf{x})$ will be the binary value of 1 or 0. The new classification system obtained in this way is shown in Fig. 3.4. Such a structure, consisting of weights, summing operator, and threshold function is called a *threshold logic unit*, and is considered the basic building block for perceptrons and other multi-layer networks with learning abilities.

If we set the weight vector to be

$$\begin{cases} \mathbf{w}_i^t\mathbf{x} > 0 \ (\mathbf{x} \in \omega_i) \\ \mathbf{w}_i^t\mathbf{x} < 0 \ (\mathbf{x} \notin \omega_i), \end{cases} \qquad (i = 1, 2, \ldots, c) \qquad (3.52)$$

Fig. 3.4 Classification
system including threshold
functions

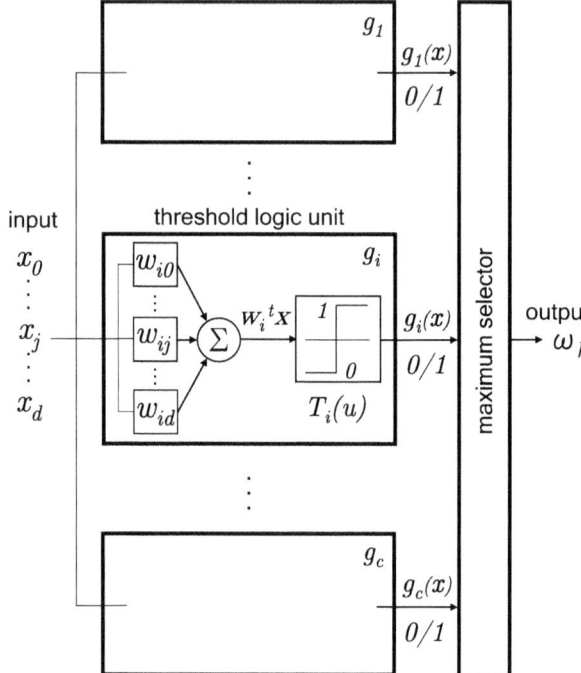

the discriminant functions for $x \in \omega_i$ will be

$$\begin{cases} g_i(x) = 1 \\ g_j(x) = 0 \quad (j \neq i), \end{cases} \qquad (i, j = 1, 2, \ldots, c) \qquad (3.53)$$

and thus the maximum selector of Fig. 3.4 can correctly classify the pattern. If we
set the teaching signal b_{ik} to

$$b_{ik} = \begin{cases} 1 \ (x_k \in \omega_i) \\ 0 \ (x_k \notin \omega_i), \end{cases} \qquad (i = 1, 2, \ldots, c) \qquad (3.54)$$

the pattern $x_k \in \omega_i$ is misclassified as ω_j, when

$$\begin{cases} g_i(x_k) = 0, \quad b_{ik} = 1 \\ g_j(x_k) = 1, \quad b_{jk} = 0. \end{cases} \qquad (j \neq i)$$

As a result, Eq. (3.31) is

$$\begin{cases} \mathbf{w}'_i = \mathbf{w}_i + \rho \, \mathbf{x}_k \\ \mathbf{w}'_i = \mathbf{w}_j - \rho \, \mathbf{x}_k. \end{cases} \qquad (3.55)$$

When the classification is correct, $g_i(x_k) - b_{ik} = 0$, no modification occurs. The Eq. (3.55) is the same as Eq. (2.31), indicating that the Widrow–Hoff learning rule includes the perceptron learning rule as a special case.[5]

3.2.2   Evaluation by Distance from Hyperplane

As shown in Eqs. (3.53) and (3.54), the perceptron learning rule is that both the discriminant function $g_i(x_k)$ and the teaching signal b_{ik} are binary, and the weights are repeatedly modified until the output and the teaching signal match for all learning patterns. If learning patterns are linearly separable, the procedure always reaches weights of misclassification zero, but if patterns are linearly nonseparable, it does not converge.

On the other hand, the Widrow–Hoff learning rule sets the output of the discriminant function as a continuous value and aims at minimizing the sum of the square errors with the teaching signal. Therefore, for individual learning patterns, the difference between the output and the teaching signal is not necessarily small. In other words, while the method guarantees convergence regardless of linearly separable or non-separable, the weights obtained in the linearly separable case do not necessarily achieve misclassification zero. This is where it differs from the perceptron learning rule.

Next, let us derive the perceptron learning rule as an algorithm for minimizing the evaluation function. For simplicity, we will take a two-class problem. The discriminant function is $g(x) = w^t x$ from Eq. (2.19), and the classification rule is defined by Eq. (2.21).

The evaluation function of the perceptron is considered as follows. First, define the function $J_k(w)$ for the pattern x_k $(k = 1, \ldots, n)$ as shown in the following equation (Tou and Gonzalez 1974):

$$J_k(w) = \frac{1}{2}\left(|w^t x_k| - b_k w^t x_k \right). \tag{3.56}$$

In the above equation, $|\cdot|$ is the absolute value and b_k is the teaching signal defined by Eq. (3.36). It is easy to verify that the following equation holds for the pattern x_k[6]:

$$J_k(w) = \begin{cases} 0 & (x_k \text{ is correctly classified}) \\ -b_k w^t x_k \quad (> 0) & (x_k \text{ is misclassified}). \end{cases} \tag{3.57}$$

---

[5] $g_i(x_k)$ defined in this section is non-differentiable because it involves a threshold operation, so strictly speaking, Eq. (3.27) does not hold.

[6] $J(w) = 0$ also holds when the pattern x_k is on the decision boundary, but Eq. (3.57) excludes this case.

If we define the evaluation function $J(\mathbf{w})$ as

$$J(\mathbf{w}) = \sum_{k=1}^{n} J_k(\mathbf{w}), \tag{3.58}$$

as is clear from Eq. (3.57), when all the learning patterns $x_1, \ldots, x_n$ are correctly classified, $J(\mathbf{w}) = 0$, which is the minimum.[7]

Denoting the set of misclassified patterns by $\mathcal{E}$, as is clear from Eq. (3.57), we can write

$$J(\mathbf{w}) = - \sum_{x_k \in \mathcal{E}} b_k \mathbf{w}^t \mathbf{x}_k \quad (> 0) \tag{3.59}$$

$$= \sum_{x_k \in \mathcal{E}} |\mathbf{w}^t \mathbf{x}_k|. \tag{3.60}$$

If $\mathcal{E}$ does not contain any misclassified patterns, we define $J(\mathbf{w}) = 0$.

Now consider a hyperplane $g(\mathbf{x}) = \mathbf{w}^t \mathbf{x} = 0$ in the weight space (see Fig. 2.5). The distance r between the weight vector $\mathbf{w}$ and the hyperplane can be obtained by a simple calculation as (Problem 3.4)

$$r = \frac{|\mathbf{w}^t \mathbf{x}|}{\|\mathbf{x}\|}. \tag{3.61}$$

If a pattern is misclassified, the weight vector is shifted from the hyperplane to the wrong side by r. The value of r indicates the degree of deviation of the weight vector from the correct position. From Eq. (3.61), $|\mathbf{w}^t \mathbf{x}| \propto r$, so $J(\mathbf{w})$ in Eq. (3.60) is a reasonable evaluation function for the perceptron, and the optimal $J(\mathbf{w})$ can be found by minimizing $J(\mathbf{w})$.

Therefore, $J(\mathbf{w})$ defined by Eq. (3.58) is minimized by the steepest descent method. Again, online learning is used here. Partial differentiation of $J_k(\mathbf{w})$ in Eq. (3.56) with respect to $\mathbf{w}$ yields

$$\frac{\partial J_k(\mathbf{w})}{\partial \mathbf{w}} = \frac{1}{2} \left(\mathbf{x}_k \cdot \mathrm{sgn}(\mathbf{w}^t \mathbf{x}_k) - b_k \mathbf{x}_k \right). \tag{3.62}$$

The function $\mathrm{sgn}(\cdot)$ in the above equation is defined as follows:

$$\mathrm{sgn}(u) = \begin{cases} 1 & (u > 0) \\ -1 & (u < 0). \end{cases} \tag{3.63}$$

---

[7] However, as is clear from Eq. (3.56), if $\mathbf{w} = \mathbf{0}$, then $J(\mathbf{w}) = 0$ always holds, but this solution is meaningless and must be excluded.

Substituting Eq. (3.62) into Eq. (3.24) yields the following equation, where the weight vector $\mathbf{w}$ is successively modified to a new weight vector $\mathbf{w}'$:[8]

$$\mathbf{w}' = \mathbf{w} - \rho \, \frac{\partial J_k(\mathbf{w})}{\partial \mathbf{w}}$$

$$= \mathbf{w} - \frac{1}{2} \rho \left(\mathbf{x}_k \cdot \mathrm{sgn}(\mathbf{w}'\mathbf{x}_k) - b_k \mathbf{x}_k \right) \tag{3.64}$$

$$= \begin{cases} \mathbf{w} + \rho \mathbf{x}_k & (\text{if } \mathbf{w}'\mathbf{x}_k \leq 0 \text{ for } \mathbf{x}_k \in \omega_1) \\ \mathbf{w} - \rho \mathbf{x}_k & (\text{if } \mathbf{w}'\mathbf{x}_k \geq 0 \text{ for } \mathbf{x}_k \in \omega_2) \\ \mathbf{w} & (\text{otherwise}). \end{cases} \tag{3.65}$$

It is clear that Eq. (3.65) is equivalent to Eqs. (2.26)–(2.28). In other words, the perceptron learning rule is equivalent to the procedure for minimizing the evaluation function $J(\mathbf{w})$ by the steepest descent method.

Fig. 3.5 Perceptron learning using the evaluation function $J(\mathbf{w})$

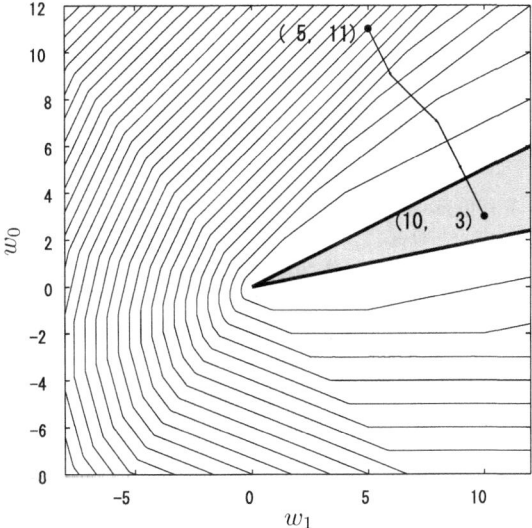

We show the results of applying the above process to the learning patterns in the one-dimensional feature space of Fig. 2.3. The contour plot of the evaluation function $J(\mathbf{w})$ of Eq. (3.58) is shown in Fig. 3.5. The gray area surrounded by thick lines in the figure is the solution region where $J(\mathbf{w})$ takes the minimum value zero, which coincides with the solution region shown in Fig. 2.4. The figure also shows the trajectory until the minimum solution of $J(\mathbf{w})$ is obtained by the steepest descent method with initial values $(w_1, w_0) = (5, 11)$ and $\rho = 2.0$. From the figure, $\mathbf{w}$ converges to $(w_1, w_0) = (10, 3)$ and reaches the solution region. It can

---

[8] If $\mathbf{w}'\mathbf{x}_k = 0$, it is not a misclassification but the weight vector needs to be corrected.

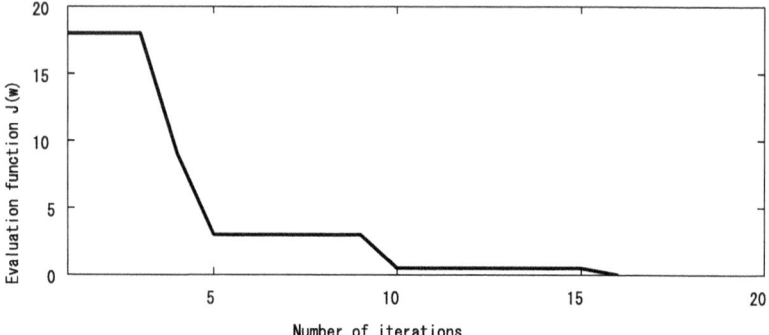

Fig. 3.6 Convergence process of the perceptron learning rule

be seen that all these results are the same as those shown in Fig. 2.6. The number of iterations required for convergence is 16, and the process up to convergence is shown in Fig. 3.6 as the value of $J(\mathbf{w})$, which changes with iteration.

3.2.3   Comparison of Evaluation Functions

In this section, we have shown that the perceptron learning rule is the same as the Widrow–Hoff learning rule in terms of evaluation function minimization. In the following, we will discuss the difference between the two evaluation functions. Since both methods were tested on 6 learning patterns distributed in a one-dimensional feature space, we will use the results of this experiment as a concrete example for comparison.

First, consider the case where the learning patterns are linearly separable, as shown in Fig. 2.3. The solution region of the perceptron learning rule is shown in Fig. 3.5. As can be seen in the figure, since $J(\mathbf{w})$ is a *convex function*, a point in the solution region, i.e., $J(\mathbf{w}) = 0$, is the globally optimal solution. However, since $J(\mathbf{w})$ is not a *strictly convex function*, the solution is not uniquely determined.

On the other hand, the diagram corresponding to Fig. 3.5 in the Widrow–Hoff learning rule is that shown in the solution to Problem 3.3, which is not much different from Fig. 3.2. In this case, since $J(\mathbf{w})$ is a strictly convex function, the solution is globally optimal and uniquely determined. However, the optimal solution is not necessarily $J(\mathbf{w}) = 0$.

Next, consider the case where the learning patterns are linearly nonseparable, as shown in Fig. 3.1. In this case, according to the perceptron learning rule, there is no solution region in $J(\mathbf{w})$, i.e., no region where $J(\mathbf{w}) = 0$. Therefore, even if the steepest descent method is applied, it will not converge and a solution cannot be obtained.

On the other hand, the Widrow–Hoff learning rule gives the contours of $J(\mathbf{w})$ as shown in Fig. 3.2. That is, as in the linearly separable case, $J(\mathbf{w})$ is a strictly convex function, and the solution is globally optimal and uniquely determined.

In summary, the above can be stated as follows. The evaluation function $J(\mathbf{w})$ of the perceptron learning rule only works when patterns are linearly separable. The solution obtained is globally optimal at $J(\mathbf{w}) = 0$, but not uniquely determined.

In contrast, the evaluation function $J(\mathbf{w})$ of the Widrow–Hoff learning rule is a strictly convex function regardless of whether it is linearly separable or not. The solution is globally optimal and uniquely determined. The optimal solution $J(\mathbf{w})$, however, is not necessarily equal to zero.

Problems

3.1 Show that $\mathbf{w}_i$ of Eq. (3.21) is the globally optimal solution and the unique minimum point.

3.2 The following six learning patterns $x_1, x_2, \ldots, x_6$ are given in the two-dimensional feature space:

$$\left.\begin{array}{l} x_1 = (0, 5)^t, \ x_2 = (1, 1)^t, \ x_3 = (5, 0)^t \\ x_4 = (6, 2)^t, \ x_5 = (2, 6)^t, \ x_6 = (2, 2)^t \end{array}\right\}. \tag{3.66}$$

Among them, x_1, x_2, x_3 belong to class ω_1 and x_4, x_5, x_6 belong to class ω_2. Find $\mathbf{w}_1$ and $\mathbf{w}_2$ by the minimum square error learning using Eq. (3.21), and illustrate the decision boundary determined by the obtained linear discriminant function.

3.3 Find the optimal weights for the linearly separable learning patterns shown in Fig. 2.3 by the minimum square error learning.

3.4 Derive Eq. (3.61).

Chapter 4
Design of Classifiers

Abstract We have introduced methods for obtaining discriminant functions by learning. The clues used so far were learning patterns themselves, and the information about the probability distribution of learning patterns was assumed to be unknown. In this chapter, we first describe a method for obtaining discriminant functions assuming that the information about the probability distribution of learning patterns is known. Next, guidelines for designing discriminant functions are introduced from a different perspective. Next, we explain that it is extremely important to understand the relationship between the dimensionality of the feature space and the number of learning patterns in order to design a classifier, along with the reasons for this. Finally, we describe how to evaluate the designed classifier based on the error probability. Specific methods for setting up learning and test patterns will be introduced.

4.1 Parametric Learning and Nonparametric Learning

The patterns being classified are considered to be generated based on *probability density function* $p(x|\omega_i)$ $(i = 1, \ldots, c)$.[1] We have described a method for obtaining discriminant functions by learning, What we used as a clue was not the probability density function itself, but learning patterns that are thought to have been generated based on the probability density function. Although we cannot directly know the probability density function, let us assume that we know it.

Supplementary Information The online version contains supplementary material available at https://doi.org/10.1007/978-981-95-1478-6_4.

---

[1] The function $p(x|\omega_i)$ denotes the probability density function of the pattern x belonging to the class ω_i.

We denote the probability of occurrence of class ω_i by $P(\omega_i)$ and the probability density of occurrence of x by $p(x)$.[2]

This $P(\omega_i)$ is called *prior probability*. When x occurs, the probability that its belonging class is ω_i is expressed as $P(\omega_i|x)$. This is called *posterior probability*.[3] The prior probability $P(\omega_i)$ is the probability of occurrence of ω_i obtained only with *a priori* knowledge before observing x. On the other hand, the posterior probability $P(\omega_i|x)$ is the probability of occurrence of ω_i after observing x. The terms "prior" and "posterior" indicate before or after the observation.

The following equations hold between the terms mentioned above:

$$\sum_{i=1}^{c} P(\omega_i) = 1, \tag{4.1}$$

$$\sum_{i=1}^{c} P(\omega_i|x) = 1, \tag{4.2}$$

$$p(x) = \sum_{i=1}^{c} P(\omega_i)\, p(x|\omega_i). \tag{4.3}$$

Furthermore, the following equation, known as *Bayes' theorem*, holds:

$$P(\omega_i|x) = \frac{p(x|\omega_i)}{p(x)} P(\omega_i) \qquad (i = 1, \ldots, c). \tag{4.4}$$

Bayes' theorem can be viewed as a transformation equation that changes the prior probability $P(\omega_i)$ to the posterior probability $P(\omega_i|x)$ by obtaining the observed value x.

As mentioned above, $P(\omega_i|x)$ indicates the certainty that the class to which the input unknown pattern x belongs is ω_i. Therefore, the most natural way to classify a pattern x is to output as the classification result ω_i whose posterior probability $P(\omega_i|x)$ $(i = 1, \ldots, c)$ is the largest. That is, the classification rule is

$$\max_{i=1,\ldots,c} \{P(\omega_i|x)\} = P(\omega_k|x) \quad \Longrightarrow \quad x \in \omega_k. \tag{4.5}$$

Such a classification rule is called the *Bayes decision rule* and will be discussed again in Sect. 5.3. Equation (4.5) is equivalent to setting $g_i(x) = P(\omega_i|x)$ in Eq. (2.3). Note that $p(x)$ in Eq. (4.4) is a factor common to each class, so the

---

[2] In this book, we use the capital letter $P(\cdot)$ for *probability mass function* defined for discrete events, and the lower case letter $p(\cdot)$ for probability density functions defined for continuous variables.

[3] The prior probability and the posterior probability are also called *a priori* probability and *a posteriori* probability, respectively.

discriminant function $g_i(x)$ is

$$g_i(x) = p(x|\omega_i) P(\omega_i) \qquad (i = 1, \ldots, c). \tag{4.6}$$

Alternatively, the logarithm of the right side can be taken and the following equation can be used:

$$g_i(x) = \log p(x|\omega_i) + \log P(\omega_i) \qquad (i = 1, \ldots, c). \tag{4.7}$$

Let us consider the case where the probability density function $p(x|\omega_i)$ is expressed as a *multivariate normal distribution* as shown in the following equation:

$$p(x|\omega_i) = \frac{1}{(2\pi)^{d/2} |\Sigma_i|^{1/2}} \exp\left\{ -\frac{1}{2}(x - \mu_i)^t \Sigma_i^{-1}(x - \mu_i) \right\}$$

$$(i = 1, \ldots, c). \tag{4.8}$$

Here, μ_i and Σ_i are a *mean vector*[4] and a *covariance matrix* of class ω_i, respectively, and are defined by

$$\mu_i = \frac{1}{n_i} \sum_{x \in X_i} x, \tag{4.9}$$

$$\Sigma_i = \frac{1}{n_i} \sum_{x \in X_i} (x - \mu_i)(x - \mu_i)^t, \tag{4.10}$$

where n_i is the number of patterns of class ω_i, and X_i is the pattern set of class ω_i. The term $|\Sigma_i|$ of Eq. (4.8) is the determinant[5] of Σ_i. Substituting this into Eq. (4.7), we obtain

$$g_i(x) = -\frac{1}{2}(x - \mu_i)^t \Sigma_i^{-1}(x - \mu_i)$$

$$\qquad - \frac{1}{2} \log |\Sigma_i| - \frac{d}{2} \log 2\pi + \log P(\omega_i) \tag{4.11}$$

$$= -\frac{1}{2} x^t \Sigma_i^{-1} x + x^t \Sigma_i^{-1} \mu_i - \frac{1}{2} \mu_i^t \Sigma_i^{-1} \mu_i$$

$$\qquad - \frac{1}{2} \log |\Sigma_i| - \frac{d}{2} \log 2\pi + \log P(\omega_i). \tag{4.12}$$

In other words, for a multivariate normal distribution, the discriminant function is a quadratic function of x. If putting

---

[4] In this book, the general mean vector is represented by **m**, and the mean vector of a normal distribution is represented by μ.

[5] See Appendix B.

$$D_M^2(x, \mu_i) \overset{\text{def}}{=} (x - \mu_i)^t \Sigma_i^{-1}(x - \mu_i),$$ (4.13)

in Eq. (4.11), $D_M(x, \mu_i)$ is called the *Mahalanobis distance*[6] of x and μ_i.

If the covariance matrix is equal for all classes as

$$\Sigma_i = \Sigma_0 \quad (i = 1, \dots, c),$$ (4.14)

we can write

$$g_i(x) = x^t \Sigma_0^{-1} \mu_i - \frac{1}{2} \mu_i^t \Sigma_0^{-1} \mu_i + \log P(\omega_i),$$ (4.15)

by omitting the term independent of i in Eq. (4.12). This is clearly a linear discriminant function. Furthermore, in Eq. (4.15), let Σ_0 be the identity matrix. In other words, there is no correlation between the features, and their variances are assumed to be equal $(= 1)$. Then the above equation becomes

$$g_i(x) = \mu_i^t x - \frac{1}{2} \|\mu_i\|^2 + \log P(\omega_i).$$ (4.16)

If the prior probabilities are equal for each class, that is, $P(\omega_i) = 1/c$ $(i = 1, \dots, c)$, then we can set

$$g_i(x) = \mu_i^t x - \frac{1}{2} \|\mu_i\|^2.$$ (4.17)

This is nothing but the minimum distance method introduced in Eq. (2.2).

Taking the two-class case as an example, if the covariance matrices of each class are equal $(\Sigma_1 = \Sigma_2 = \Sigma_0)$, then the optimal discriminant function is a linear discriminant function from Eq. (4.15), and the decision boundary is represented as a hyperplane perpendicular to $\Sigma_0^{-1}(\mu_1 - \mu_2)$. Then, when $P(\omega_1) = P(\omega_2)$, the optimal decision boundary passes through the midpoint of μ_1 and μ_2 (Problem 4.1). When the covariance matrices of each class are different, the optimal discriminant function is represented by a quadratic function and the decision boundary is a quadratic surface.

Intuitively, one would think that the quadratic discriminant function would have a practical advantage over the linear discriminant function, since the quadric surface can describe more complex boundaries than the hyperplane, and the covariance matrices for each class are generally different. However, this is not necessarily true. It is known that methods using linear discriminant functions often yield rather better results, and this is called the *robustness* of the linear discriminant function. The reasons for this are as follows.

---

[6] It is formally called the *Mahalanobis generalized distance* in the sense that it generalizes the concept of ordinary distance (see also Fig. 6.7).

That is, the quadratic discriminant function estimated from patterns asymptotically approaches the optimal discriminant function as the number of patterns is increased, but the quadratic discriminant function requires more patterns than the linear discriminant function to achieve a certain level of accuracy. This is because the quadratic discriminant function has more parameters (of order d^2). For example, when the covariance matrices are equal for each class, the optimal decision boundary is the same hyperplane regardless of whether the linear or quadratic discriminant function is used. In this case, it is known that the asymptotic property to the optimal decision boundary is worse for the quadratic discriminant function when the number of patterns n is increased. Refer also to the coffee break in Sect. 5.5.1.

Let us consider the following case. That is, the probability density function is a function with a finite number of parameters, and the form of the function is known, but the parameters are unknown. For example, we know that the probability density function is a multivariate normal distribution, but we do not know the mean vector or covariance matrix. In this case, the parameters are estimated from the given learning patterns. The estimated parameters are then regarded as true values, and the discriminant function can be constructed by Eqs. (4.6) or (4.7). This method of constructing the classifier by estimating the parameters of the probability density function using the learning patterns is called the *parametric learning*. In contrast, the learning algorithms described in Chaps. 2 and 3 were methods that obtained the discriminant function directly from the learning patterns without assuming the form of the probability density function. Such a method is called the *nonparametric learning*.

4.2 Parameter Estimation

This section describes a method for estimating the parameters of the probability density function. Since both the learning patterns and parameters can be considered independent across classes, we omit the symbols that distinguish the classes to avoid complications. In actual application, the following procedures should be performed for each class.

Let $X = \{x_1, x_2, \ldots, x_n\}$ be the learning pattern set containing n patterns, and denote the probability density function to be estimated by $p(x; \theta)$,[7] where θ is a vector representing a set of parameters, called a *parameter vector*. Here, we consider which θ is the most likely among the various possible candidates for θ that yielded the learning pattern set X. Since each pattern in the pattern set X is considered to have arisen independently according to the probability $p(x; \theta)$, the probability $p(X; \theta)$ of obtaining such a pattern set can be expressed as follows:

---

[7] A set of learning patterns is prepared for each class as $X_1, \ldots, X_c$. The probability density function $p(x|\omega_i; \theta_i)$ of class ω_i is estimated using the learning patterns X_i, but as mentioned above, the subscripts indicating the class are omitted because they are complicated.

$$p(\mathcal{X}; \boldsymbol{\theta}) = \prod_{k=1}^{n} p(\boldsymbol{x}_k; \boldsymbol{\theta}). \tag{4.18}$$

Therefore, it is natural to assume that the most likely $\boldsymbol{\theta}$ is the $\boldsymbol{\theta}$ that maximizes Eq. (4.18). If we put such $\boldsymbol{\theta}$ as $\hat{\boldsymbol{\theta}}$ and use it as an estimate, then

$$\hat{\boldsymbol{\theta}} = \underset{\boldsymbol{\theta}}{\mathrm{argmax}} \ \{p(\mathcal{X}; \boldsymbol{\theta})\}, \tag{4.19}$$

which can be obtained by solving the following equation:

$$\nabla p(\mathcal{X}; \boldsymbol{\theta}) = \frac{\partial}{\partial \boldsymbol{\theta}} \ p(\mathcal{X}; \boldsymbol{\theta}) = \mathbf{0}, \tag{4.20}$$

or taking the logarithm

$$\frac{\partial}{\partial \boldsymbol{\theta}} \ \log p(\mathcal{X}; \boldsymbol{\theta}) = \sum_{k=1}^{n} \frac{\partial}{\partial \boldsymbol{\theta}} \ \log p(\boldsymbol{x}_k; \boldsymbol{\theta}) = \mathbf{0}. \tag{4.21}$$

For the vector and matrix derivative operations necessary for calculating Eqs. (4.20) or (4.21), see Appendix B.

In Eq. (4.18), when $\mathcal{X}$ is fixed and $p(\mathcal{X}; \boldsymbol{\theta})$ is regarded as a function of $\boldsymbol{\theta}$, then $p(\mathcal{X}; \boldsymbol{\theta})$ is called the *likelihood function* or simply *likelihood*. An estimation method such as Eq. (4.19) is called the *maximum likelihood method*.

As an example of the application of the maximum likelihood method, let us take the case where we know that patterns are distributed according to a multidimensional normal distribution, but the mean vector and covariance matrix are unknown. The parameters $\boldsymbol{\theta}$ in this case are $\boldsymbol{\mu}$ and $\boldsymbol{\Sigma}$. By applying Eq. (4.21) to Eq. (4.8), we obtain the following estimators $\hat{\boldsymbol{\mu}}$ and $\hat{\boldsymbol{\Sigma}}$ for $\boldsymbol{\mu}$ and $\boldsymbol{\Sigma}$ respectively (see Problem 4.2). These estimators are intuitively natural:[8]

$$\hat{\boldsymbol{\mu}} = \frac{1}{n} \sum_{k=1}^{n} \boldsymbol{x}_k, \tag{4.22}$$

$$\hat{\boldsymbol{\Sigma}} = \frac{1}{n} \sum_{k=1}^{n} (\boldsymbol{x}_k - \hat{\boldsymbol{\mu}})(\boldsymbol{x}_k - \hat{\boldsymbol{\mu}})^t. \tag{4.23}$$

In most cases, probability density functions with a single maximum point, i.e., *unimodal* functions, are taken as an example. In reality, however, we must also deal with probability density functions with multiple maxima, i.e., *multimodal* functions. This is the case, for example, when multiple normal distributions overlap to form a

---

[8] Note that the covariance matrix defined by Eq. (4.23) is not an unbiased estimator.

single probability density function. More generally, we need to deal with the case where the probability density function $p(x; \theta)$ is expressed as a linear combination of r probability density functions, as in the following equation:

$$p(x; \theta) = \sum_{i=1}^{r} \pi_i \, p_i(x; \theta_i), \tag{4.24}$$

where $p_i(x; \theta_i)$ $(i = 1, \ldots, r)$ is a probability density function whose functional form is known in advance, θ_i is its parameter vector and π_i is the mixture ratio of r probability density functions. A probability density function such as Eq. (4.24) is called a *mixture density*.[9] When we try to obtain such a probability density function by the maximum likelihood method, not only θ_i $(i = 1, \ldots, r)$ but also the mixture ratio of each distribution π_i $(i = 1, \ldots, r)$ must be estimated as parameters. That is, the parameter vector to be estimated is

$$\theta^t = \left(\theta_1^t, \ldots, \theta_r^t, \pi_1, \ldots, \pi_r \right). \tag{4.25}$$

The probability density functions described so far have been based on the assumption that they can be estimated independently for each class. This is because we have assumed that the learning patterns are labeled patterns described in Sect. 2.1. Therefore, the parameter estimation is supervised learning.

On the other hand, in the case of a mixture distribution, learning patterns are unlabeled patterns, and the parameter estimation is unsupervised learning. In this case, since patterns are presented with all classes mixed, the probability density function can no longer be estimated independently for each class. In other words, the only information we can rely on is that patterns are distributed according to $p(x)$ of Eq. (4.3). However, the number of classes c is assumed to be known in advance.[10] The problem of estimating $P(\omega_i)$ and $p(x|\omega_i)$ under such conditions corresponds to the parameter estimation problem for a mixture distribution, where π_i is set to $P(\omega_i)$ in Eq. (4.24) $(i = 1, \ldots, c)$. In this case, instead of solving the problem analytically using Eqs. (4.20) or (4.21), it is common to solve it by iterative operations using the *hill-climbing method*.[11] (Ishii and Ueda 2026; Duda et al. 2001)

As already mentioned in Sect. 4.1, the probability density function is rarely known in the real world. Actual patterns are considered to have complex distributions, and a simple probability density function such as the normal distribution

---

[9] Equation (4.24) can be called a *mixture distribution*, but when the random variable is continuous as in this example, it is called a mixture density.

[10] If the number of classes is not known, the maximum likelihood method can no longer be used. Instead, *clustering* will be used. This method finds clusters of distributions in the feature space. Clustering is one of the methods of unsupervised learning, but is not covered in this book. For unsupervised learning including clustering, see Ishii and Ueda (2026).

[11] If we add a negative sign to the function to be optimized by the hill-climbing method, we obtain the steepest descent method.

should be regarded as rather unfeasible. Of course, increasing the number of parameters can approximate complex probability density functions with arbitrary precision, but this is not realistic. The nonparametric learning is easier to handle and has more practical value because it does not assume a probability density function.

──────────────────────── Coffee Break ────────────────────────

Maximum Likelihood Method: Most Likely

In the maximum likelihood method, it is important to note that we are not making the claim that "the most likely θ to occur is $\hat{\theta}$." Since θ is not a random variable, but a constant, it does not fluctuate in the probability of being likely or unlikely to happen. This is why the word "likely" is used. The maximum likelihood method is an estimation method based on the assumption that "if something happened, it is most likely that something happened."

In general, the preferred properties of an estimator are, first, when the number of patterns n is sufficiently large, the expected value of the estimate agrees with the true value, i.e. *asymptotic unbiasedness*. Secondly, as the number of patterns n grows, the probability that the absolute value of the error between the estimate and the true value exceeds any positive value becomes infinitesimally small, i.e. *asymptotic consistency*. Thirdly, if the number of patterns n is sufficiently large, the variance of the estimate is minimized, i.e. *asymptotic efficiency*. It has been theoretically shown that the maximum likelihood method has all of these properties and is therefore widely used in many fields. As mentioned in footnote 8 in this Chapter, $\hat{\Sigma}$ of Eq. (4.23) is not unbiased, but it is unbiased as a property when n grows, i.e., asymptotically unbiased.

As mentioned earlier, the maximum likelihood method treated the parameter θ as an unknown constant. In contrast, *Bayesian estimation* regards the parameter as a random variable. When prior knowledge of the parameters (prior distribution $p(\theta)$) is available, Bayesian estimation generally gives better estimates than the maximum likelihood method. However, when the prior knowledge is unknown, the superiority of the two methods is indeterminable. Bayesian estimation is not covered in this book, so interested readers should refer to Ishii and Ueda (2026) and Duda and Hart (1973) or other sources.

──

4.3 Linear Discriminant Function Design

In this section, we first introduce an evaluation function to find the optimal linear discriminant function for the two-class problem. By maximizing this function, the weights of the linear discriminant function can be determined. Next, we introduce how to set a linear discriminant function for multi-class problems, and discuss the problems specific to multi-class and points to be considered. The generalized linear discriminant function can be regarded as an advanced form of the linear discriminant function, but this will be discussed in Sect. 8.5.

4.3.1 Two-Class Classification

The discriminant function is a function of the feature vector $\boldsymbol{x}$, and is used to describe the classification rule that determines the class to which the feature vector belongs. For a two-class classification problem, $g(\boldsymbol{x})$ satisfying

$$\begin{cases} g(\boldsymbol{x}) > 0 \Longrightarrow \boldsymbol{x} \in \omega_1 \\ g(\boldsymbol{x}) < 0 \Longrightarrow \boldsymbol{x} \in \omega_2 \end{cases} \tag{4.26}$$

is an example of a discriminant function that classifies two classes, ω_1 and ω_2. The discriminant functions consist of the linear discriminant functions and the nonlinear discriminant functions. The linear discriminant functions mentioned so far have been well studied and most frequently used.

As already mentioned in Sect. 2.3.3, classifying patterns of two classes is to perform the following process. That is, a one-dimensional space is defined as a subspace in the d-dimensional feature space by the following linear discriminant function:

$$g(\boldsymbol{x}) = w_0 + \boldsymbol{w}^t \boldsymbol{x} = \mathbf{w}^t \mathbf{x}. \tag{4.27}$$

Then, patterns are projected onto this one-dimensional space, and the decision boundary is determined on this space. The method for determining $\mathbf{w}$ of Eq. (4.27) from the viewpoint of minimum square error or expected loss is described in Chaps. 3 and 15, respectively. Also, the method for determining $\mathbf{w}$ from the viewpoint of the transformation of feature space is described in Chap. 6. Here, let us look at the design of discriminant function $g(\boldsymbol{x})$ from a different point of view.

First, let J denote the function that represents the evaluation for the discriminant function $g(\boldsymbol{x})$. As mentioned above, defining a hyperplane as a decision boundary is equivalent to defining an axis represented by its normal vector and a boundary point on the axis. Therefore, let $\tilde{m}_i$ and $\tilde{\sigma}_i^2$ $(i = 1, 2)$ denote the mean and variance of each class ω_i in the one-dimensional space represented by the axis, respectively.

Suppose that the evaluation function J is given as a function of $\tilde{m}_i$ and $\tilde{\sigma}_i^2$ as shown in the following equation:

$$J \stackrel{\text{def}}{=} J\left(\tilde{m}_1, \tilde{m}_2, \tilde{\sigma}_1^2, \tilde{\sigma}_2^2\right). \tag{4.28}$$

If we express the discriminant function as Eq. (4.27), it represents a projection onto a one-dimensional subspace, so the following equations hold:

$$\tilde{m}_i = \frac{1}{n_i} \sum_{\boldsymbol{x} \in \mathcal{X}_i} g(\boldsymbol{x}) \tag{4.29}$$

$$= \boldsymbol{w}^t \mathbf{m}_i + w_0 \qquad (i = 1, 2), \tag{4.30}$$

$$\tilde{\sigma}_i^2 = \frac{1}{n_i} \sum_{x \in X_i} (g(x) - \tilde{m}_i)^2 \tag{4.31}$$

$$= w^t \frac{1}{n_i} \sum_{x \in X_i} (x - m_i)(x - m_i)^t w \tag{4.32}$$

$$= w^t \Sigma_i w \qquad (i = 1, 2), \tag{4.33}$$

where m_i and Σ_i are the mean vector and covariance matrix of class ω_i. Now let us find w and w_0 that maximize J.

First, partial differentiation of Eqs. (4.30), (4.33) by w and w_0 yields

$$\frac{\partial \tilde{m}_i}{\partial w} = m_i, \quad \frac{\partial \tilde{m}_i}{\partial w_0} = 1, \quad \frac{\partial \tilde{\sigma}_i^2}{\partial w} = 2 \Sigma_i w, \quad \frac{\partial \tilde{\sigma}_i^2}{\partial w_0} = 0, \tag{4.34}$$

so partial differentiation of J by w and w_0 setting $\mathbf{0}$ and 0 respectively yields[12]

$$\frac{\partial J}{\partial w} = \frac{\partial J}{\partial \tilde{\sigma}_1^2} \cdot \frac{\partial \tilde{\sigma}_1^2}{\partial w} + \frac{\partial J}{\partial \tilde{\sigma}_2^2} \cdot \frac{\partial \tilde{\sigma}_2^2}{\partial w} + \frac{\partial J}{\partial \tilde{m}_1} \cdot \frac{\partial \tilde{m}_1}{\partial w} + \frac{\partial J}{\partial \tilde{m}_2} \cdot \frac{\partial \tilde{m}_2}{\partial w}$$

$$= 2 \left(\frac{\partial J}{\partial \tilde{\sigma}_1^2} \Sigma_1 + \frac{\partial J}{\partial \tilde{\sigma}_2^2} \Sigma_2 \right) w + \left(\frac{\partial J}{\partial \tilde{m}_1} m_1 + \frac{\partial J}{\partial \tilde{m}_2} m_2 \right) = 0, \tag{4.35}$$

$$\frac{\partial J}{\partial w_0} = \frac{\partial J}{\partial \tilde{\sigma}_1^2} \cdot \frac{\partial \tilde{\sigma}_1^2}{\partial w_0} + \frac{\partial J}{\partial \tilde{\sigma}_2^2} \cdot \frac{\partial \tilde{\sigma}_2^2}{\partial w_0} + \frac{\partial J}{\partial \tilde{m}_1} \cdot \frac{\partial \tilde{m}_1}{\partial w_0} + \frac{\partial J}{\partial \tilde{m}_2} \cdot \frac{\partial \tilde{m}_2}{\partial w_0}$$

$$= \frac{\partial J}{\partial \tilde{m}_1} + \frac{\partial J}{\partial \tilde{m}_2} = 0. \tag{4.36}$$

Substituting Eq. (4.36) into Eq. (4.35), we obtain the relation

$$w = \frac{1}{2} \cdot \frac{\partial J}{\partial \tilde{m}_1} \left(\frac{\partial J}{\partial \tilde{\sigma}_1^2} \Sigma_1 + \frac{\partial J}{\partial \tilde{\sigma}_2^2} \Sigma_2 \right)^{-1} (m_2 - m_1) \tag{4.37}$$

$$\propto (s \Sigma_1 + (1 - s) \Sigma_2)^{-1} (m_1 - m_2), \tag{4.38}$$

where

$$s \overset{\text{def}}{=} \frac{\partial J / \partial \tilde{\sigma}_1^2}{\partial J / \partial \tilde{\sigma}_1^2 + \partial J / \partial \tilde{\sigma}_2^2}. \tag{4.39}$$

---

[12] Note the distinction between 0 for a scalar and $\mathbf{0}$ for a vector.

Since w is a vector representing the normal direction of the hyperplane, only the direction is required and the constant multiplication can be ignored. The term w_0 can be obtained using Eq. (4.36).

From the above results, we can generally find w and w_0 that maximize any J defined as a function of $\tilde{m}_1$, $\tilde{m}_2$, $\tilde{\sigma}_1^2$, and $\tilde{\sigma}_2^2$.

As an example,[13] let us consider the case where J is defined as

$$J \stackrel{\text{def}}{=} \frac{(\tilde{m}_1 - \tilde{m}_2)^2}{k_1 \tilde{\sigma}_1^2 + k_2 \tilde{\sigma}_2^2} \tag{4.40}$$

with arbitrary positive constants k_1 and k_2. Maximizing J means finding w such that the difference between the class means of the projected patterns onto the one-dimensional space is as large as possible and the variance of each class is as small as possible. Here, the following equation holds:

$$\frac{\partial J}{\partial \tilde{\sigma}_i^2} = -k_i \frac{(\tilde{m}_1 - \tilde{m}_2)^2}{(k_1 \tilde{\sigma}_1^2 + k_2 \tilde{\sigma}_2^2)^2} \qquad (i = 1, 2). \tag{4.41}$$

Therefore, from Eqs. (4.38) and (4.39), we obtain

$$w \propto (k_1 \Sigma_1 + k_2 \Sigma_2)^{-1} (m_1 - m_2). \tag{4.42}$$

The linear discriminant method described in Sect. 6.4 can be viewed as a method to maximize J with $k_1 = P(\omega_1)$ and $k_2 = P(\omega_2)$ when the prior probability of class ω_i is $P(\omega_i)$. For example, compare Eqs. (4.42) and (6.127).

On the other hand, if $\partial J / \partial \tilde{m}_i$ is computed and substituted into Eq. (4.36), the term w_0 disappears and w_0 becomes indeterminate. In other words, when J is defined as Eq. (4.40), w is obtained but w_0 is not uniquely determined. The linear discriminant method described in Sect. 6.4 corresponds to this example. For an example applied to another evaluation expression J, refer to Problem 4.3.

As in the example above, if the position of the boundary cannot be determined even though the normal vector of the decision boundary can be determined, w_0 must be determined using another method. Since the projection result can be observed as a distribution in a one-dimensional space, the decision boundary can be set visually. If it is to be set automatically, for example, the following methods are possible:

(1) A method that uses the midpoint of the transformed class averages as the boundary. That is,

$$w_0 = -\frac{\tilde{m}_1 + \tilde{m}_2}{2}. \tag{4.43}$$

---

[13] For several other examples of $J(\tilde{m}_1, \tilde{m}_2, \tilde{\sigma}_1^2, \tilde{\sigma}_2^2)$, refer to Chapter 4 of Fukunaga (1990).

(2) A method of internal division based on the variance of each class after transformation. That is,

$$w_0 = -\frac{\tilde{\sigma}_2^2 \tilde{m}_1 + \tilde{\sigma}_1^2 \tilde{m}_2}{\tilde{\sigma}_1^2 + \tilde{\sigma}_2^2}, \tag{4.44}$$

or, a method of internal division based on the standard deviation of each class after transformation. That is,

$$w_0 = -\frac{\tilde{\sigma}_2 \tilde{m}_1 + \tilde{\sigma}_1 \tilde{m}_2}{\tilde{\sigma}_1 + \tilde{\sigma}_2}. \tag{4.45}$$

(3) A method of internal division that also takes prior probabilities into account. That is,

$$w_0 = -\frac{P(\omega_2)\tilde{\sigma}_2^2 \tilde{m}_1 + P(\omega_1)\tilde{\sigma}_1^2 \tilde{m}_2}{P(\omega_1)\tilde{\sigma}_1^2 + P(\omega_2)\tilde{\sigma}_2^2}. \tag{4.46}$$

In applying the method using Eq. (4.40), it should be noted that for any J, w is always expressed in the form of Eq. (4.38), and differences in the evaluation function J are reflected only in s of Eq. (4.39). For example, this method can also be applied to the evaluation function $J(\mathbf{w})$ of the minimum square error learning shown in Eq. (3.38). It is confirmed that the optimal w can be obtained as the form of Eq. (4.38). However, since the evaluation function $J(\mathbf{w})$ of the minimum square error learning is a function of the augmented weight vector $\mathbf{w} = (w_0, \mathbf{w}^t)^t$, not only the optimal w but also w_0 is obtained (Problem 4.4).

4.3.2 Multi-Class Classification

In this subsection, we attempt to extend the idea of linear discriminant functions for two-class problems to multi-class ($c > 2$) problems. When obtaining a multi-class decision boundary, cases arise in which more than c linear discriminant functions are needed. As described below, methods are proposed that use such linear discriminant functions to create classification rules for multi-class classification. Assuming that c classes are linearly separable, that is, any pair of c classes is linearly separable (see Sect. 2.3.1), the following three cases (a), (b), and (c) can be considered for the distribution of patterns. The classification methods for these cases are described in order. In the figures used, we take the case $d = 2$ and $c = 3$ as examples.

(a) Any Class ω_i Is Linearly Separable from All Classes Other than ω_i

This is the case where the class distribution is as shown in Fig. 4.1(a) and can be considered as a combination of two-class problems: class ω_i and other classes

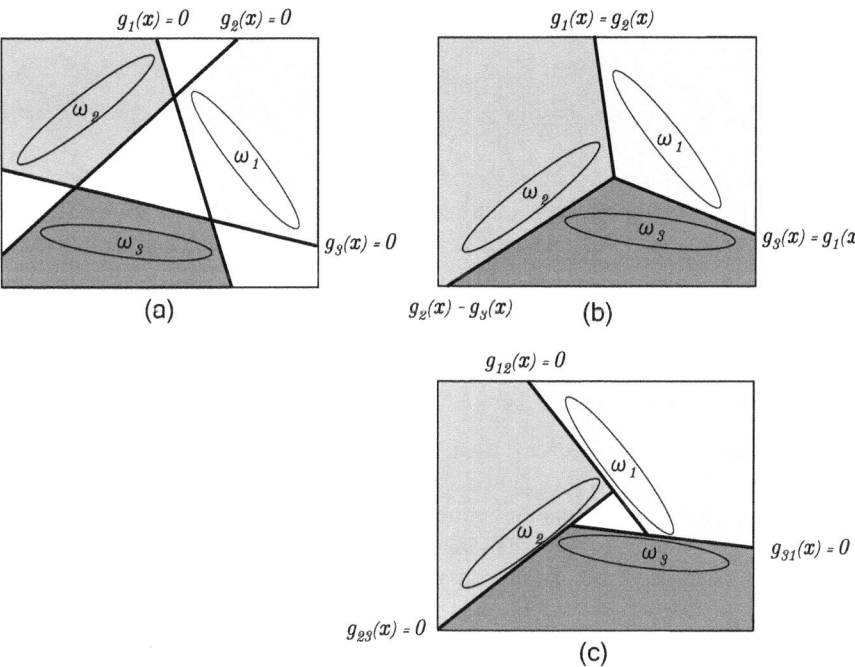

Fig. 4.1 Multi-class classification by linear discriminant functions

$(i = 1, \ldots, c)$. In this case, there exist linear discriminant functions $g_i(x)$ $(i = 1, \ldots, c)$ that separate class ω_i from classes other than ω_i, satisfying

$$\begin{cases} x \in \omega_i \implies g_i(x) > 0 \\ x \notin \omega_i \implies g_i(x) < 0 \end{cases} \quad (i = 1, \ldots, c). \tag{4.47}$$

In this way, a linear discriminant function can be defined for each class. Therefore, the number of linear discriminant functions to be prepared is c. The classification rule to classify multiple classes is expressed as follows:

$$g_i(x) > 0 \quad \text{and} \quad \forall j \neq i, \quad g_j(x) < 0 \quad \implies \quad x \in \omega_i. \tag{4.48}$$

In this case, there is a possibility that a region (the white area in the figure) may occur that cannot be determined to be any class. This corresponds to the reject region described in Sect. 1.2.

(b) Belonging Class Can Be Determined by Comparing $g_i(x)$ Values

Next, consider the case where the class distribution is as shown in Fig. 4.1(b). For such a distribution, the classification method described in (a) cannot be applied. In this case, the linear discriminant function shown in Eq. (2.17) can be applied.

That is, a linear discriminant function $g_i(x)$ is prepared for the class ω_i, and the classification rule is

$$\forall j \neq i, \quad g_i(x) > g_j(x) \quad \Longrightarrow \quad x \in \omega_i \quad (i = 1, \ldots, c). \qquad (4.49)$$

The number of required linear discriminant functions is c as in (a). Since the magnitude relationship between multiple $g_i(x)$ values can always be determined, one of the classes ω_i is assigned to every region except the boundary, and there are no reject regions as in (a). If the classification method shown in (a) is applicable to the distribution, then the classification method described here is also applicable. However, the converse is not necessarily true.

(c) Any Two Classes ω_i and ω_j Are Linearly Separable

Finally, consider the case of the distribution shown in Fig. 4.1(c). For such a distribution, the classification method described in (b) cannot be applied. That is, it is not possible to set a decision boundary such that every region corresponds to one of the classes. However, as mentioned at the beginning of this subsection, any class pair is assumed to be linearly separable, so there exists a linear discriminant function $g_{ij}(x)$ that separates classes ω_i and ω_j, and satisfies

$$\begin{cases} x \in \omega_i \Longrightarrow g_{ij}(x) > 0 \\ x \in \omega_j \Longrightarrow g_{ij}(x) < 0. \end{cases} \qquad (4.50)$$

In the same way, $c(c-1)/2$ linear discriminant functions can be defined. Therefore, if we define $g_{ij}(x) = -g_{ji}(x)$, the classification rule to classify multiple classes is

$$\forall j \neq i, \quad g_{ij}(x) > 0 \quad \Longrightarrow \quad x \in \omega_i \quad (i, j = 1, \ldots, c). \qquad (4.51)$$

In this case, a reject region is generated as shown in Fig. 4.1(c).

   As a variation of this method, the following *majority voting* scheme is also often used. First, the number of j $(= 1, \ldots, c)$ for which the following conditions hold for all i $(= 1, \ldots, c)$ is found:

$$g_{ij}(x) > 0 \qquad (j \neq i). \qquad (4.52)$$

Then, this number is regarded as the number of votes that ω_i received, and is set as $N(i)$. The classification rule is set to (see Problem 4.6 (2))

$$\forall j \neq i, \quad N(i) > N(j) \quad \Longrightarrow \quad x \in \omega_i \quad (i, j = 1, \ldots, c). \qquad (4.53)$$

This majority voting scheme can also be used in the classification method (a) introduced at the beginning (Problem 4.6 (3)).

   Even though it is difficult to set up the linear discriminant functions $g_i(x)$ $(i = 1, \ldots, c)$ described in (b) above, this classification rule is effective when $g_{ij}(x)$ can be easily obtained by Fisher's method, etc.. The disadvantage is that the number of

$g_{ij}(x)$ to be prepared increases in proportion to $c(c-1)/2$. However, the number of $g_{ij}(x)$ actually used as decision boundaries is usually less than $c(c-1)/2$ (see Problem 4.6 (1)).

If the classification method described in (b) is applicable to the distribution, the classification method described here is necessarily applicable. However, the converse is not necessarily true.

What has been explained so far regarding multi-class classification can be summarized as (1), (2), and (3) below, and the important points to be pointed out as (4) and (5), respectively. The classification methods applicable to the three cases (a), (b), and (c) shown above will be referred to as method (a), method (b), and method (c).

(1) There are three methods of multi-class classification, (a), (b) and (c) mentioned above, depending on the distribution of the classes in the feature space.

(2) If method (a) is applicable to the distribution, then method (b) is also applicable, but the converse is not necessarily true. If method (b) is applicable to the distribution, then method (c) is also applicable, but the converse is not necessarily true. That is, the above is expressed as follows (see Problem 4.5):

$$(a) \Longrightarrow (b) \Longrightarrow (c) \tag{4.54}$$

In other words, the distribution in (a) can be viewed as a special case of (b) and the distribution in (b) as a special case of (c).

(3) When using method (b), every region in the feature space corresponds to one of the classes, but when using methods (a) and (c), there is a possibility that regions which cannot be determined as any class, i.e., reject regions, will be generated.

(4) The occurrence of reject regions is not necessarily a drawback. For example, in Fig. 4.1(a) and (c), the regions where each class exists are narrower than in (b), which may help stabilize the classification. In many cases, it is appropriate to reject the input pattern rather than to classify it as one of the classes if it exists in a region where the classes are close to each other or in a region far from the distribution of the classes. Reject judgment is effective in cases where misclassification is to be avoided as much as possible.

(5) The aforementioned (c) is the very definition of linear separability for the c class, as described in Sect. 2.3.1 of Chap. 2. On the other hand, Eq. (2.17) is the classification method described in (b) above. As already mentioned, even if the classes cannot be separated by method (b), there is still a possibility that they can be separated by method (c).

The above is the case assuming linear separability. On the other hand, in the linearly nonseparable case, we must first define some evaluation criteria and then find $g_i(x)$, $g_{ij}(x)$, etc. to minimize or maximize the criteria, depending on the method. For this purpose, we can use methods such as the minimum square error learning described in Sect. 3.1. For example, if g_i is to be used as the solution that minimizes Eq. (3.7), method (a) can be used without modification.

4.4   Dimensionality of Feature Vector and Number of Learning Patterns

One of the practical problems faced when designing a classifier is how to determine the number of learning patterns. The number of learning patterns n should be considered in relation to the dimensionality of feature vectors (i.e., the dimensionality of the feature space) d, and cannot be discussed independently. To understand this intuitively, assume the case where the dimensionality of the feature vectors is increased while keeping the number of patterns constant. In this case, it is clear that the distribution of patterns becomes sparse in the feature space and the statistical reliability decreases. Therefore, in order to realize an advanced classifier, it is necessary to prepare a sufficient number of learning patterns commensurate with the dimensionality. Then, how many learning patterns should be prepared? Unfortunately, there is no general answer to this question. Instead, we will give some examples that suggest the relationship between the dimensionality and the number of learning patterns.

First, let us assume that the number of learning patterns is less than or equal to the dimensionality of the feature space ($n \leq d$). In this case, although a d-dimensional feature space is prepared, only ($n - 1$) dimensions of the space are actually used, which means that ($d - n + 1$) dimensions are wasted. This is evident in the following example.

Consider a three-dimensional ($d = 3$) space as a feature space, and assume that there are only three learning patterns ($n = 3$) in it. These patterns determine a two-dimensional plane in the three-dimensional space. In other words, even though the space is three-dimensional, the distribution remains on a two-dimensional plane due to the small number of patterns. In order to have a three-dimensional spread, the number of patterns must be at least greater than the dimensionality, i.e., four or more. However, this is a minimum requirement, and even if the number of patterns is 4, the situation is the same if they happen to be on the same plane. Therefore, the relation

$$n \gg d \tag{4.55}$$

is required for the distribution of patterns to have a spread in the feature space that is commensurate with the dimensionality. Although this seems quite obvious considering the above example, classifier designers often overlook this condition. This is because designers are in such a hurry to improve classification performance that they forget that the number of learning patterns is finite and resort to the easy way of adding features. Normally, if features are added, the number of learning patterns must be increased accordingly, but most designers do not take the time and effort to collect patterns and simply add more features.

Equation (4.55) is also true when applying the KL expansion of Chap. 6 or Fisher's method. That is, KL expansion requires the calculation of the covariance matrix, and the obtained covariance matrix will not be non-singular unless there are at least $(d + 1)$ patterns. In addition, when Fisher's method is applied, problems arise such as the inverse matrix not being obtained. Here, $n = d + 1$ is a minimum condition, and a much larger n is needed to obtain statistically reliable results.

However, as described in Sect. 6.5.2, when there is a strong correlation between features, even if the dimensionality is large, it is only apparent, and in reality, it is often possible to represent patterns with a smaller dimensionality (intrinsic dimensionality). The d in Eq. (4.55) should be interpreted as the intrinsic dimensionality, not the apparent dimensionality.

Next, let us consider the case where $n > d$. Assume that n patterns are in *general position*.[14] If each pattern belongs to one of the two classes ω_1 or ω_2, then there are a total of 2^n ways to assign classes. If we choose any one of them, what is the probability $p(n, d)$ that these patterns are linearly separable?

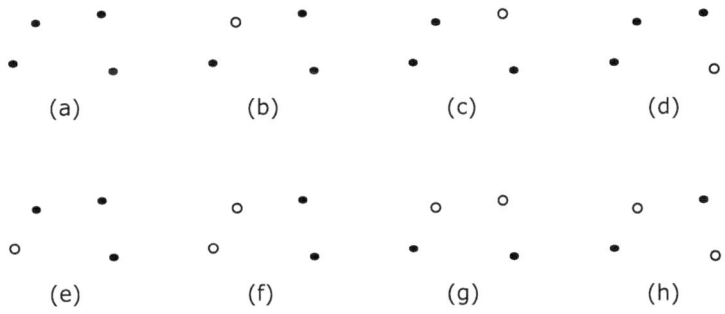

Fig. 4.2 Four patterns distributed in a two-dimensional feature space

For example, let us take the case $d = 2$, $n = 4$, i.e., four patterns are distributed in a two-dimensional feature space. Since each pattern belongs to one of the classes ω_1 and ω_2, the number of combinations is $2^4 = 16$. The distribution is shown in Fig. 4.2. In this figure, the two classes are distinguished by ● and ○, and eight types of distributions from (a) to (h) are shown. By replacing ● and ○ in this figure, eight more distributions are added, giving a total of 16 distributions.

It is clear from the figure that these four patterns are in general position, since three of them do not lie on a straight line. Of the eight distributions in the figure, only (h) is linearly nonseparable, and the other seven are all linearly separable, so the next equation holds:

$$p(n, d) = p(4, 2) = 14/16 = 0.875. \tag{4.56}$$

In general, $p(n, d)$ for d, n is expressed by the following equation:

---

[14] Patterns are said to be in a general position in the following cases.

For $n > d$, it is when no $(d + 1)$ patterns are on a $(d - 1)$-dimensional hyperplane, and for $n \leq d$, it is when an $(n - 2)$-dimensional hyperplane doesn't contain any n patterns. For example, when four patterns are not on the same plane in a three-dimensional feature space $(d = 3)$.

$$p(n, d) = \begin{cases} 2^{1-n} \cdot \sum_{j=0}^{d} {}_{n-1}C_j & (n > d) \\ \\ 1 & (n \leq d). \end{cases}$$
(4.57)

See Problems 4.7 and 4.8 for the derivation.

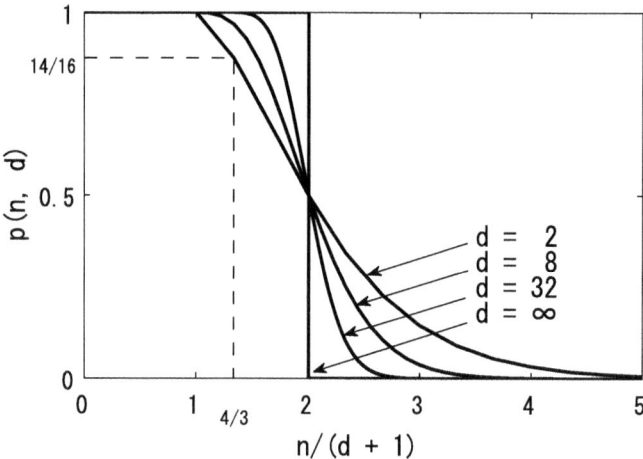

Fig. 4.3 Linearly separable probabilities

Figure 4.3 is a plot of $p(n, d)$ of Eq. (4.57) with $n/(d+1)$ as the horizontal axis. The figure is plotted for the cases $d = 2$, 8 and 32. It can be seen that $p(n, d) = 1/2$ when $n/(d+1) = 2$ (Problem 4.9). As d increases, the threshold effect becomes more pronounced around $n/(d+1) = 2$, i.e., $n = 2(d+1)$, and in the limit of $d \to \infty$, the following can be confirmed:

$$\begin{cases} p(n, d) \approx 1 \ (n < 2(d+1)) \\ \\ p(n, d) \approx 0 \ (n > 2(d+1)). \end{cases}$$
(4.58)

The result obtained by Eq. (4.56) is shown by the dashed line in the figure.

This $2(d+1)$ is called the *capacity* of the hyperplane. For more information on capacity, refer to Nilsson (1965).

–––––––––––––––––––––––––––– Coffee Break ––––––––––––––––

Additional Features Degrade Classification Performance?

In the above discussion, we assumed that the dimensionality of feature vectors d is fixed and that the number of learning patterns n can be freely set. However, in reality, the number of learning patterns is often limited due to the high cost of data collection.

Therefore, if the number of learning patterns is not sufficient relative to the dimensionality, the dimensionality may have to be reduced by removing or merging features instead of increasing the number of patterns.

In order to improve classification performance, designers naturally add new features that are as independent of existing features as possible. It is natural to expect that the addition of new features will improve classification performance, or at worst, maintain it at the current level. However, it often happens that the addition of new features leads to a degradation of classification performance. How can this be explained? The cause lies in adding too many features, forgetting that the number of learning patterns is finite, and is ultimately related to the curse of dimensionality, which will be discussed later. That is, while the dimensionality d is small relative to the number of patterns n, the addition of features leads to improvement in classification performance. However, when d becomes large enough relative to n, the statistical reliability decreases, and contrary to expectations, the classification performance deteriorates. This is known as the Hughes phenomenon and is the result of "bias" caused by the finite number of learning patterns. The details will be described in Sect. 5.5.1.

In any case, when designing the classifier, we must always keep in mind the relationship between the number of learning patterns and the dimensionality.

Now consider the case of finding a hyperplane that linearly separates two classes by learning. According to Eq. (4.58), when d is large, no matter how learning patterns are distributed, a hyperplane can almost certainly be found for $n < 2(d+1)$. Conversely, if $n > 2(d+1)$, the probability of finding such a hyperplane is infinitely close to 0. Therefore, if the desired hyperplane can be found under the condition $n > 2(d + 1)$, the reliability is extremely high. What we expect from learning is the effect that a desired hyperplane (decision boundary) can be found "inevitably" by strong constraints from a sufficient number of learning patterns. This "inevitability" is important, and the hyperplanes obtained when $n < 2(d + 1)$ are likely to be accidental. In other words, the number of learning patterns $n < 2(d + 1)$ is too weak as a constraint to determine the hyperplane.

From the above examples, it can be seen that when designing the classifier, a sufficient number of learning patterns compared to the dimensionality must be prepared. Unfortunately, however, even if we try to prepare a large number of learning patterns commensurate with the dimensionality, it is often impossible in reality. This is because the number of learning patterns required increases exponentially with the dimensionality. This extremely troublesome problem, known as the *curse of dimensionality*, has plagued researchers working in pattern recognition and machine learning.

The dimensionality d can be interpreted as the number of parameters required to describe the hyperplane, namely the weight coefficients $w_0, w_1, \ldots, w_d$. Therefore, the above statement can be rephrased to mean that a sufficient number of patterns should be used for learning compared to the number of parameters that make up the classifier. Over-fitting, which is often discussed in neural networks, is also caused by using too few learning patterns relative to the number of parameters. Learning can be thought of as finding the relationship between the inputs and outputs of a classifier by means of learning patterns. Over-fitting is the approximation of this relationship with high precision by a complex function so that it applies only to a small number of learning patterns. It brings the risk of not being able to produce accurate outputs for test patterns.[15] For a solution to the over-fitting problem, see the coffee break in Sect. 14.2.2.

4.5 Optimization of Classifier

In the previous sections, we have discussed methods for learning classifiers. In general, a classifier has a hyperparameter, i.e., a parameter for the unknown parameters. In this section, we describe three representative methods on how to determine this hyperparameter.

4.5.1 Parameters Determining Classifier

So far, we have introduced nonparametric classification methods such as linear discriminant functions and the k-nearest neighbor rule. All of them are representative methods that are often used in practical applications. However, there are parameters to be determined prior to learning, such as the number of dimensions d for the linear discriminant function and the value of k for the k-nearest neighbor rule. The same applies to the number of internal layers and the number of units in neural networks introduced in Chap. 12. Since these parameters can be interpreted as "parameters for the original parameters (e.g. weight parameters in the case of linear discriminant functions)" of the classifier, they are often called *hyperparameters*.

Because the setting of hyperparameters has a significant impact on classification performance, it is extremely important in practice. The misclassification rate evaluates the quality of the hyperparameters. In this section, we describe a typical method for determining hyperparameters from given learning patterns. Let

$$\mathcal{X} = \{x_1, x_2, \ldots, x_n\} \tag{4.59}$$

be given *labeled patterns*. Hyperparameters are determined by evaluating the classification performance for unknown patterns. Let λ be the hyperparameter of a certain classification method. For example, in the k-nearest neighbor rule, since $\lambda = k$, it becomes a natural number of $\lambda = 1, 2, 3, 4, \ldots$. Here, let us fix the value of λ and write e_λ as the average misclassification rate of the learned classifier for all possible unknown patterns that follow the same distribution as the given learning patterns. The hyperparameter decision problem is the problem of determining $\lambda \in \Lambda$ which minimizes e_λ. Here Λ is the set of all λ.

---

[15] The method of fitting individual patterns while taking into account the number of parameters is known as *AIC:Akaike Information Criterion*.

Needless to say, however, since the distribution is unknown, $\mathbf{e}_\lambda$ cannot be simply computed. Therefore, the problem here is that it reduces to the problem of estimating $\mathbf{e}_\lambda$ only from a given set of n patterns $X = \{x_1, x_2, \ldots, x_n\}$.

4.5.2 Hold-Out Method

First, we describe the simplest method. First, X is divided into a learning pattern set X_1 and a test pattern set X_2. Then a classifier is designed for each value of $\lambda \in \Lambda$ using X_1. A possible method for estimating $\mathbf{e}_\lambda$ is to evaluate the classification performance of the learned classifier by using X_2. In other words, a part of the given pattern set is regarded as a test pattern set (unknown pattern set). This method is called the *hold-out method*. Hereafter this is abbreviated as *H method*.

However, since this method uses a portion of the given pattern set as a test pattern set, the number of patterns used in actual learning is reduced, resulting in a degradation of classification performance. Conversely, if the number of learning patterns is divided to be as large as possible, the number of test patterns is reduced, and the reliability of the performance evaluation is decreased. Therefore, this method may be effective when the number of given patterns is sufficiently large, but when the number of patterns is small, the H method is not an accurate method for estimating $\mathbf{e}_\lambda$.

4.5.3 Cross-Validation Method

In the H method, the given pattern set was used for either learning or testing, In the *cross-validation method* (hereafter abbreviated as *CV method*),[16] the estimation of $\mathbf{e}_\lambda$ ensures that all elements of X are used for learning and testing. Specifically, X is first divided into m groups $X_1, X_2, \ldots, X_m$. The number of patterns in each group becomes n/m. Then, after learning with $(m-1)$ groups of patterns excluding X_i, the misclassification rate is calculated by X_i. This procedure is performed for all $i = 1, 2, \ldots, m$, and the average value over the obtained m misclassification rates is set to the estimate of $\mathbf{e}_\lambda$.

The simplest and most commonly used division is the one with the number of elements equal to 1. That is, the procedure of learning with $(X - x_i)$ and testing with x_i is performed for $i = 1, 2, \ldots, n$, and the average of the misclassification rates obtained from n tests is the estimate of $\mathbf{e}_\lambda$. This method is called the *leave-one-*

---

[16] The CV method was originally proposed as an estimation method for arbitrary statistics. It has been theoretically proved that estimators based on the CV method have asymptotic consistency (when $n \to \infty$, the estimator asymptotically approaches the true value).

out method.[17] Hereafter this is abbreviated as the *L method*. Obviously, in the case of the L method, since all patterns are used for learning and testing, the estimate of $\mathbf{e}_\lambda$ is more accurate than the H method. However, it requires n iterations of learning, which is a huge amount of processing.

4.5.4   Bootstrap Method

Similar to the CV method, the *bootstrap method* can be used to estimate arbitrary statistics.[18] Hereafter this is abbreviated as *BS method*. Compared to the CV method, the BS method has the advantage that the variance of the estimated value is smaller, i.e., the estimated value is more stable with respect to the variation of X. The BS method uses various estimation methods depending on the statistic, but its basic method is based on *sampling with replacement*, In other words, it is an extraction method that takes out and restores. In the following, we explain how the BS method is applied to estimate $\mathbf{e}_\lambda$. However, the following is only an intuitive explanation with emphasis on practical applications. For details, refer to the literature article (Efron and Tibshirani 1993).

Now, when estimating $\mathbf{e}_\lambda$, suppose that X is used for both learning and testing and that the estimated value $\widehat{\mathbf{e}}_\lambda$ is obtained. Since the learning pattern set was also used for testing, obviously the obtained estimate of the misclassification rate should be smaller than the true value. In this case, the deviation is

$$R = \mathbf{e}_\lambda - \widehat{\mathbf{e}}_\lambda \tag{4.60}$$

and if the value of R can be estimated in some way, then using the R value, we have

$$\mathbf{e}_\lambda = \widehat{\mathbf{e}}_\lambda + R. \tag{4.61}$$

In the BS method, the pseudo-pattern set $X^* = \{x_1^*, x_2^*, \ldots, x_n^*\}$ is generated from X by n times sampling with replacement from X. Using this pseudo-pattern set, this method finds the estimated value of R in Eq. (4.60).

That is, in the BS method, considering the pseudo-pattern set X^* as a learning pattern set and the original pattern set X as the test pattern set, the relationship in Eq. (4.60) is rewritten as

---

[17] It is called the *jackknife method*. The CV method is usually referred to as the jackknife method, which was originally designed for estimating arbitrary statistics. In other words, the CV method is applied here to estimate the misclassification rate. Since the jackknife is known as a universal knife, we call the CV method used for estimating arbitrary statistics the jackknife method with the meaning of "universal".

[18] The word bootstrap comes from the phrase "to pull oneself up by one's bootstrap", which means to get out of a tight spot by one's own efforts. That is, the intention is to somehow infer the true value only from a given set of patterns.

$$R^* = \mathbf{e}^*_\lambda - \widehat{\mathbf{e}}^*_\lambda. \tag{4.62}$$

Here, $\mathbf{e}^*_\lambda$ denotes the estimate of $\mathbf{e}_\lambda$ obtained by using X^* as the learning pattern set and X as the test pattern set. On the other hand, $\widehat{\mathbf{e}}^*_\lambda$ denotes the estimate of $\mathbf{e}_\lambda$ obtained by using X^* as both the learning and the test pattern sets. However, in order to eliminate the influence of specific sampling R^*, we generate B [19] pseudo-pattern sets $X^{*1}, X^{*2}, \ldots, X^{*B}$, compute $R^{*1}, R^{*2}, \ldots, R^{*B}$ for each of them, and set their average value to R^*. The BS method, like the L method, has a solid theoretical foundation. For more in-depth study, Efron and Tibshirani (1993) is a good textbook.

The estimation method of $\mathbf{e}_\lambda$ by the BS method is summarized as follows.

Step 1 After designing a classifier for the model λ by using the original pattern set X, calculate the misclassification rate for the same X, and set the value to $\widehat{\mathbf{e}}_\lambda$.

Step 2 Do the following for each of $b = 1, \ldots, B$.

Generate X^{*b} from the original pattern set X by n sampling with replacement and design a classifier for the model λ using X^{*b} as the pattern set. After designing the classifier for the model λ using the pattern set X^{*b}, compute the misclassification rate for X and denote the result as $\mathbf{e}^{*b}_\lambda$. Then, compute the misclassification rate for X^{*b} and denote the result as $\widehat{\mathbf{e}}^{*b}_\lambda$. Using these values, compute

$$R^{*b} = \mathbf{e}^{*b}_\lambda - \widehat{\mathbf{e}}^{*b}_\lambda.$$

Step 3 Let R^* be the mean value of $R^{*1}, \ldots, R^{*B}$ obtained in Step 2. The estimate to be obtained is $\widehat{\mathbf{e}}_\lambda + R^*$.

The model selection methods that do not require maximum likelihood estimates have been described above. For practical purposes, we recommend using the L or BS methods. Both of these methods require a huge amount of computation time[20] compared to the H method, but this disadvantage is well compensated for by the accuracy.

Problems

4.1 Show that the decision boundary determined by the linear discriminant function of Eq. (4.15) is a hyperplane perpendicular to $\Sigma_0^{-1}(\mu_1 - \mu_2)$. Also, show that if $P(\omega_1) = P(\omega_2)$ holds, then this decision boundary passes through the midpoint of μ_1 and μ_2.

---

[19] B is usually said to be about 50.

[20] When the number of patterns is n, for a single λ value, the CV (BS) method requires n (B) times computations. However, for the L method, a fast method for obtaining the covariance matrix and its inverse is known (see Chapter 5 in Fukunaga (1990)).

4.2 Derive Eqs. (4.22) and (4.23) (use the formula in Appendix B).

4.3 In the projected one-dimensional feature space, we seek w such that the spread between the two classes around the origin is as large as possible and the variance of each class is as small as possible. We can consider

$$J \overset{\text{def}}{=} \frac{k_1 \tilde{m}_1^2 + k_2 \tilde{m}_2^2}{k_1 \tilde{\sigma}_1^2 + k_2 \tilde{\sigma}_2^2}, \tag{4.63}$$

as the evaluation function J for this purpose. Using this evaluation function, find the optimal w and w_0 by the method described in Sect. 4.3.1.

4.4 * As the evaluation function J of Eq. (4.28), let us use the following J, which is defined for the minimum square error learning for two classes:

$$J(w) = \frac{1}{2} \sum_{k=1}^{n} (g(x_k) - b_k)^2 = \frac{1}{2} \sum_{k=1}^{n} (w^t x_k - b_k)^2. \tag{4.64}$$

Show that the optimal w and w_0 obtained by the method described in Sect. 4.3.1 are

$$w = a \Sigma_W^{-1} (m_1 - m_2) \tag{4.65}$$

$$w_0 = -m^t w + P(\omega_1) - P(\omega_2),$$

$$= -a\, m^t \Sigma_W^{-1} (m_1 - m_2) + P(\omega_1) - P(\omega_2), \tag{4.66}$$

where a is an arbitrary constant, Σ_W is the within-class covariance matrix defined by Eq. (6.113) and b_k is the teaching signal defined by Eq. (3.36). (see also Problem 15.2).

4.5 Show that Eq. (4.54) holds.

4.6 As shown in Fig. 4.4, patterns of four classes ω_1, ω_2, ω_3 and ω_4 are distributed in a two-dimensional feature space.

Suppose that these classes can be separated by the classification method (a) introduced in Sect. 4.3.2, and that $g_i(x)$ ($i = 1, \ldots, 4$) are expressed by the following equations:

$$\begin{cases} g_1(x) = -92 + 3x_1 + 4x_2 \\ g_2(x) = -10 - 9x_1 + 5x_2 \\ g_3(x) = 24 - x_1 - 2x_2 \\ g_4(x) = -28 + 7x_1 - 8x_2, \end{cases} \tag{4.67}$$

(1) As shown in Eq. (4.54), if the classification method (a) is applicable, then the method (b) is applicable, and the method (c) is also applicable. Therefore, for the distribution of Fig. 4.4, it is possible to separate classes by the method (c).

Fig. 4.4 Distribution of four classes

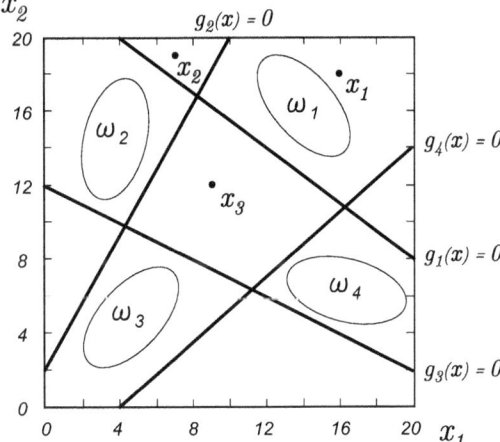

Then, find $g_{ij}(x)$ of Eq. (4.50) for the four classes with reference to the solution method of Problem 4.5, and illustrate the decision boundaries set by them.

(2) As shown in Fig. 4.4, there exist three patterns $x_1 = (16, 18)^t$, $x_2 = (7, 19)^t$ and $x_3 = (9, 12)^t$ in this feature space. Classify these patterns by the majority voting scheme using $g_{ij}(x)$ obtained in (1) above.

(3) In Sect. 4.3.2, it is stated that the majority voting scheme can be applied to the classification method (a) using Eq. (4.67). Give a concrete example of the majority voting scheme and show the results of classifying x_1, x_2, and x_3 by the scheme.

4.7 * Assume that a total of n patterns are distributed in a d-dimensional feature space and that they are in a general position. Let $L(n, d)$ be the number of ways to bisect these patterns into two groups ω_1 and ω_2 by a $(d-1)$-dimensional hyperplane. Such a bisection is called the *linear dichotomy*. Show that the next equation holds:

$$L(n, d) - L(n-1, d) + L(n-1, d-1).\tag{4.68}$$

Note that $L(n, d)$ should include the case where the number of patterns in the group is zero.

4.8 From Eq. (4.68), prove that the following equation holds using mathematical induction:

$$L(n, d) = \begin{cases} 2\sum_{j=0}^{d} {}_{n-1}C_j & (n > d) \\ \\ 2^n & (n \le d). \end{cases}\tag{4.69}$$

4.9 Show that $p(n, d)$ of Eq. (4.57) is $p(n, d) = 1/2$ when $n/(d+1) = 2$.

Chapter 5
Feature Evaluation and Bayes Error

Abstract In constructing a recognition system, it is important to evaluate the features to be used in advance. In this chapter, we describe the feature evaluation. First, the meaning and necessity of feature evaluation are described, and then the ratio of between-class variance to within-class variance which is known as a simple feature evaluation method, is discussed. Then, the Bayes error, which plays an important role in feature evaluation, is introduced. The Bayes error is an extremely important concept that forms the basis of pattern recognition and machine learning, and is explained in detail in Sect. 5.3, 5.4 and 5.5. It is impossible to obtain the Bayes error directly, and its estimation methods have been studied for a long time. After introducing them, we introduce experimental results of feature evaluation using concrete data.

5.1 Feature Evaluation

As described in Chap. 1, a recognition system consists of three parts: a preprocessing part, a feature extraction part, and a classification part. Suppose that the recognition system does not perform as expected. Since the recognition performance is an evaluation measure of the entire recognition system including the three parts, in order to improve the recognition performance, it is necessary to clarify which part of the recognition system is responsible for the poor performance.

As an example, consider the case where two classes are distributed in a two-dimensional feature space. Suppose that the distribution of the classes is observed in the feature space as shown in Fig. 5.1(a). In this case, since the two classes are completely separated, it should be possible to realize a recognition system that does not cause misrecognition if the classification part is appropriately designed. If, in spite of this, the recognition performance remains low, the cause lies not in the feature extraction part but in the classification part.

Supplementary Information The online version contains supplementary material available at https://doi.org/10.1007/978-981-95-1478-6_5.

On the other hand, suppose that distributions such as Fig. 5.1(b) or (c) are obtained. In either case, there is an overlap between the distributions of the classes, so no matter how the classification part is designed, misrecognition is inevitable. In this case, the problem lies not in the classification part but in the feature extraction part.

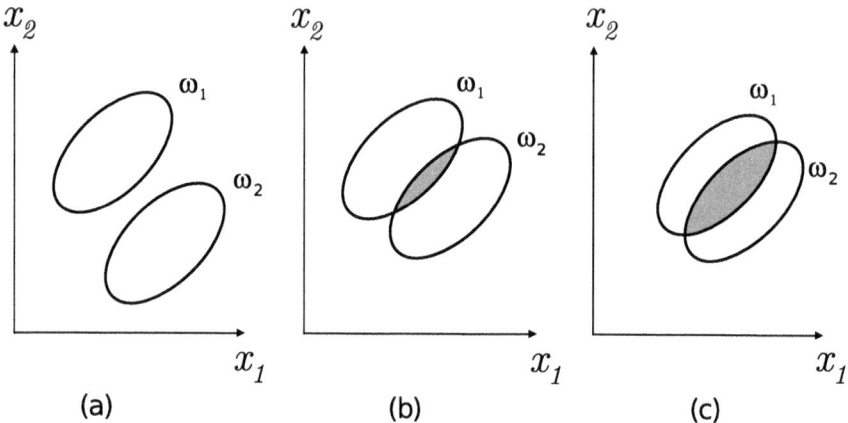

Fig. 5.1 Distributional overlap and Bayes error

When designing a feature extraction part, it is extremely important to evaluate the features[1] in advance. As shown in the above example, if the features are not appropriate, a highly accurate recognition system cannot be achieved no matter how much effort is put into the design of the classification part. The quality of the features can be evaluated based on their ability to separate classes. In the above example, (b) is better than (c), and (a) is better than (b). In the following, we describe how to evaluate features by the degree of separation between classes.

5.2 Ratio of Between-Class Variance to Within-Class Variance

In order to separate classes with high accuracy, it is desirable for patterns of the same class to be distributed as close as possible in the feature space, while patterns of different classes should be distributed as far apart as possible. In the following, we introduce a method for evaluating the degree of separation between classes from this perspective.

---

[1] Since we are discussing class separation in a multidimensional feature space defined by multiple features, it should be more accurate to call it an evaluation of feature vectors rather than an evaluation of individual features.

Let $\mathcal{X}_i$ denote the set of patterns belonging to class $\mathcal{X}_i$, n_i be the number of patterns included in $\mathcal{X}_i$, and $\mathbf{m}_i$ be the mean vector. Also, let n be the number of all patterns and $\mathbf{m}$ be the mean vector of all patterns. Let *within-class variance* be σ_W^2 and *between-class variance* be σ_B^2, we can write as follows:

$$\sigma_W^2 = \frac{1}{n} \sum_{i=1}^{c} \sum_{x \in \mathcal{X}_i} (x - \mathbf{m}_i)^t (x - \mathbf{m}_i), \tag{5.1}$$

$$\sigma_B^2 = \frac{1}{n} \sum_{i=1}^{c} n_i (\mathbf{m}_i - \mathbf{m})^t (\mathbf{m}_i - \mathbf{m}). \tag{5.2}$$

That is, the within-class variance represents the average of the spread of each class, and the between-class variance represents the spread among the classes. Therefore, by defining their ratio J_σ as

$$J_\sigma = \frac{\sigma_B^2}{\sigma_W^2}, \tag{5.3}$$

we can judge that the larger J_σ is, the better the feature is. The above J_σ is called the *ratio of between-class variance to within-class variance*. This can also be viewed as the interclass distance normalized by the intraclass distance (Problem 5.1).

Although the ratio of between-class variance to within-class variance is a simple evaluation method, it has the following drawbacks.

For multi-class problems, the values evaluated by this ratio do not necessarily reflect the actual separation of the distribution. For example, when considering a distribution consisting of four classes in a two-dimensional feature space such as Fig. 5.2, the values of J_σ for distributions (a) and (b) are equal. However, as a

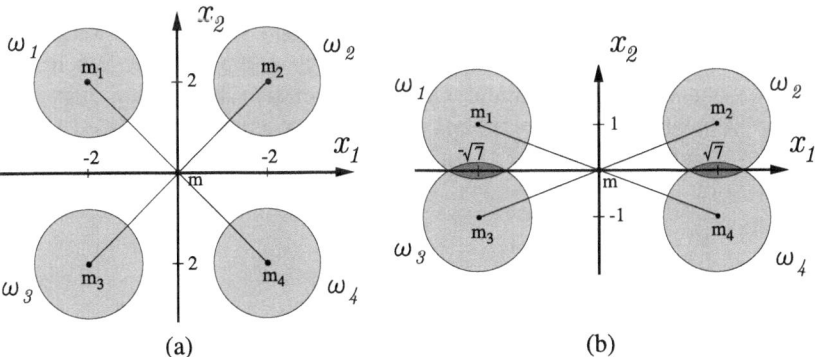

(a)   (b)

Fig. 5.2 Example of evaluation by ratio of between-class variance to within-class variance not working

feature vector (x_1, x_2), (a) is clearly superior to (b). This is because in (a), the four distributions are equally separated and do not overlap, while in (b), the distributions of ω_1, ω_3 and distributions of ω_2, ω_4 overlap. This phenomenon occurs because J_σ only considers the distance between classes and does not evaluate the overlap of the distributions. Moreover, since this method evaluates the average separation of the entire distribution, it does not reflect the separation of class pairs. To avoid this problem, calculate Eq. (5.3) for all class pairs and use the average as the evaluation value. However, this method is computationally expensive when the number of classes is large.

Then, how can we examine the degree of overlap of the distributions? In fact, this is closely related to the Bayes error described in the next section.

5.3 What Is the Bayes Error?

As an example, let us consider a two-class problem of determining whether or not a patient is infected with a certain disease. The test method is as follows:

- The test is performed using a certain reagent.
- The test result is expressed as a numerical value, and if the value exceeds a predetermined threshold, the test is positive and the patient is diagnosed as infected; if the value is below the threshold, the test is negative and the patient is not infected.
- However, the accuracy of this reagent is not perfect, and a positive result may be obtained even if the patient is not infected, or a negative result may be obtained even if the patient is infected. The former is called *false positive* and the latter is called *false negative*.

Let us formulate this two-class problem. Denote the test value by x and consider x as a feature. Let ω_1 (infected) and ω_2 (not infected) be two classes, and their prior probabilities be $P(\omega_1)$ and $P(\omega_2)$, respectively. Here, $P(\omega_1)$ represents the infection rate in the absence of any information such as test results, and can be considered as the so-called incidence rate of the disease. When a test result x is obtained, the probabilities that it belongs to ω_1 and ω_2, that is, the posterior probabilities, are $P(\omega_1|x)$ and $P(\omega_2|x)$, respectively. Let $p(x)$ denote the probability density function of x. As already mentioned in Sect. 4.1, the following equations hold:

$$P(\omega_1) + P(\omega_2) = 1, \tag{5.4}$$

$$P(\omega_1|x) + P(\omega_2|x) = 1, \tag{5.5}$$

$$p(x) = P(\omega_1)p(x|\omega_1) + P(\omega_2)p(x|\omega_2). \tag{5.6}$$

In addition, from Bayes' theorem of Eq. (4.4),

$$P(\omega_i|x) = \frac{p(x|\omega_i)}{p(x)}P(\omega_i) \qquad (i = 1, 2) \tag{5.7}$$

is obtained.

Suppose we get a graph like Fig. 5.3(a) when we plot $P(\omega_1|x)$ and $P(\omega_2|x)$ against x.

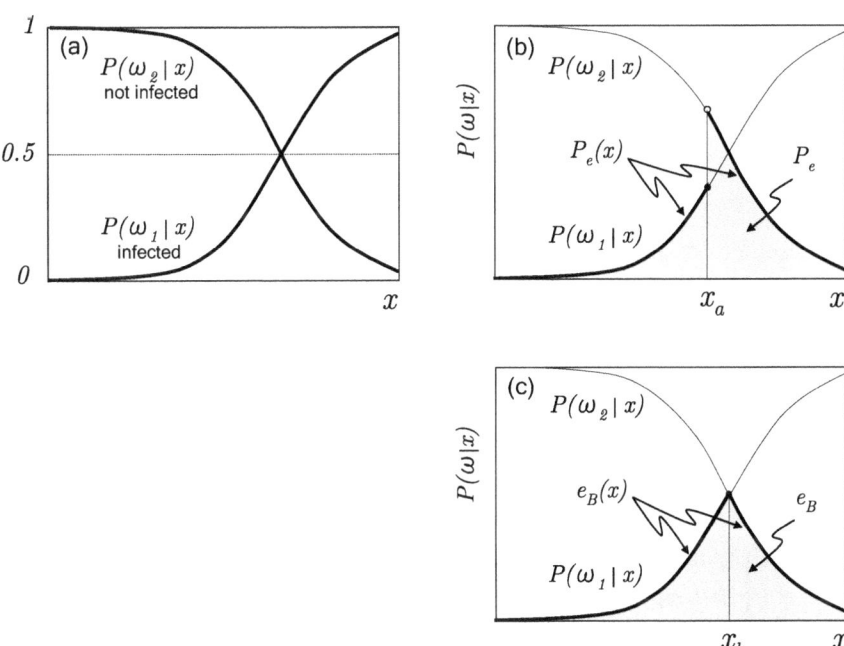

Fig. 5.3 Bayes error

We can see that the probability of infection $P(\omega_1|x)$ increases as the test value x increases, while the probability of non-infection $P(\omega_2|x)$ decreases. The two intersect at $P(\omega_1|x) = P(\omega_2|x) = 1/2$ from Eq. (5.5).

The x of the obtained test value is judged as ω_1 (infected) or ω_2 (not infected), but it is not possible to determine unambiguously from the x value whether it is ω_1 or ω_2. Because, as can be seen from the shape of the graph, for any x, neither $p(x|\omega_1)$ nor $p(x|\omega_2)$ is 0 or 1, but $0 < p(x|\omega_1),\ p(x|\omega_2) < 1$, so there is a possibility of both ω_1 and ω_2.

Thus, as long as the distributions overlap, there are always misclassifications. The *error probability* $P_e(x)$ for some x is represented by

$$P_e(x) = \begin{cases} P(\omega_2|x) \text{ (when judged as } \omega_1 \text{ (infected))} \\ P(\omega_1|x) \text{ (when judged as } \omega_2 \text{ (not infected)).} \end{cases} \tag{5.8}$$

Suppose that the threshold used to detect infection is set to x_a. In other words, the detection is performed as follows:

$$\begin{cases} \text{if } x > x_a, \text{ judged as } \omega_1 \text{ (infected)} \\ \text{if } x \leq x_a, \text{ judged as } \omega_2 \text{ (not infected).} \end{cases} \tag{5.9}$$

Here, using Eq. (5.8), $P_e(x)$ can be written as

$$P_e(x) = \begin{cases} P(\omega_2|x) & (x > x_a) \\ P(\omega_1|x) & (x \leq x_a). \end{cases} \tag{5.10}$$

The $P_e(x)$ obtained in this way is shown by thick lines in Fig. 5.3(b). The upper and lower expressions in Eq. (5.10) represent error probabilities as false positive and false negative, respectively. The error probability P_e for all possible x can be written as

$$P_e = \int P_e(x) \, p(x) dx. \tag{5.11}$$

The above P_e corresponds to the gray region in Fig. 5.3(b).[2]

To minimize P_e in the above equation, we can use a decision rule such that the smaller of $P(\omega_1|x)$ and $P(\omega_2|x)$ is chosen as $P_e(x)$ in Eq. (5.11). That is,

$$\begin{cases} \text{if } P(\omega_1|x) > P(\omega_2|x). \text{ judged as } \omega_1 \text{ (infected)} \\ \text{if } P(\omega_1|x) < P(\omega_2|x), \text{ judges as } \omega_2 \text{ (not infected).} \end{cases} \tag{5.12}$$

To perform the above, instead of x_a in Eq. (5.9), we can use x_b, the value of x where $P(\omega_1|x)$ and $P(\omega_2|x)$ intersect and set as follows:

$$\begin{cases} \text{if } x > x_b, \text{ judged as } \omega_1 \text{ (infected)} \\ \text{if } x \leq x_b, \text{ judged as } \omega_2 \text{ (not infected).} \end{cases} \tag{5.13}$$

This is a *decision rule* that outputs the classification result ω_i that maximizes the posterior probability $P(\omega_i|x)$ ($i = 1, 2$). Such a decision rule is called the Bayes decision rule as already described in Eq. (4.5).[3] If $P_e(x)$ that minimizes P_e of

---

[2] If $p(x)$ is a uniform distribution, P_e represents the area of the gray region.

[3] Although we have considered minimizing the error probability here, it should be formulated in a more general way by defining the *loss function* and minimizing the loss associated with the decision. The problem of detecting infection discussed here is a typical example. Because the loss of a false negative is much greater for the patient than the loss of a false positive for the same error.

Eq. (5.11) is expressed by $e_B(x)$, then

$$e_B(x) = \min\{P(\omega_1|x), P(\omega_2|x)\} \leq \frac{1}{2}. \tag{5.14}$$

We call $e_B(x)$ in the above equation, the *conditional Bayes error*. Similarly, if the minimum value of P_e is expressed by e_B, then

$$e_B = \min P_e \tag{5.15}$$

$$= \int e_B(x)\, p(x)\, dx \tag{5.16}$$

$$= \int \min\{P(\omega_1|x), P(\omega_2|x)\}\, p(x)dx. \tag{5.17}$$

Figure 5.3(c) shows the process described above. In the figure, $e_B(x)$ is indicated by thick lines, and the gray region corresponds to e_B.[4] The e_B in the above equation is the limit where the error probability cannot be smaller than this, in other words, the "overlap of distributions," which is the "inevitable error" introduced by this feature extraction system. This is called the *Bayes error*[5] (Problem 5.2).

If we decide the possibility of being infected with the disease without any tests, the only available information is the prior probability $P(\omega_1)$, i.e., the incidence rate. Since $P(\omega_1) < P(\omega_2)$, unless the disease is an epidemic, the best way to minimize the error probability is to judge ω_2 (not infected). On the other hand, after obtaining the test value x, there is more information for judgment, so a more accurate judgment can be made. In this case, the judgment method is Eq. (5.12).

In the above, the feature (test value) x is treated as a scalar, and the case of two-class was described, but it can be easily extended to the more general case, where the feature is a vector or the multi-class case. That is, if d kinds of test values are expressed as a d-dimensional feature vector $\boldsymbol{x} = (x_1, \ldots, x_j, \ldots, x_d)^t$, the discussion so far still holds as it is by replacing x with $\boldsymbol{x}$, where x_j is the j-th test value.[6]

In the multi-class case, the Bayes decision rule is given by the following equation instead of Eq. (5.12) (hereafter, features are denoted by vectors):

$$\text{if} \ \max_{i=1,\ldots,c} \{P(\omega_i|\boldsymbol{x})\} = P(\omega_k|\boldsymbol{x}), \ \text{judged as } \omega_k. \tag{5.18}$$

The minimum loss is called the *Bayes risk*. The Bayes risk includes the Bayes error as a special case. This is discussed in Sect. 14.2.

[4] As with P_e, if $p(x)$ is a uniform distribution, e_B represents the area of the gray region.

[5] Since e_B represents the error probability, it should be called the Bayes error probability, but in this field, it is often called the Bayes error.

[6] The notation $x \geq x_a$ and $x \geq x_b$ should be changed to the notation in a d-dimensional space.

This decision rule has already been shown in Eq. (4.5). Also, the Bayes error e_B is given by the following equation instead of Eq. (5.17):

$$e_B = \int \min_i \{1 - P(\omega_i | x)\}\, p(x) dx. \tag{5.19}$$

As mentioned in Sect. 4.1, Eq. (5.18) shows that $P(\omega_i | x)$ can be used as a discriminant function. That is, the discriminant function $g_i(x)$ is

$$g_i(x) = P(\omega_i | x) \qquad (i = 1, \ldots, c). \tag{5.20}$$

Thus, a discriminant function that realizes the Bayes decision rule is called the *Bayes discriminant function*.

Coffee Break

Feature Evaluation by Amount of Information

Since we have mentioned information, let us discuss the relationship between the amount of information and feature evaluation. In the example discussed in this section, let us calculate the degree of uncertainty (ambiguity) about infection and non-infection, that is, *entropy*.
The entropy H_0 before obtaining the test value x is

$$H_0 = - \sum_{i=1}^{2} P(\omega_i) \log P(\omega_i), \tag{5.21}$$

and the entropy $H(x)$ after obtaining the test value is

$$H(x) = - \sum_{i=1}^{2} P(\omega_i | x) \log P(\omega_i | x). \tag{5.22}$$

According to information theory, the amount of information $I(x)$ brought about by obtaining x is the decrease in uncertainty, that is, the difference in entropy, so we can write

$$I(x) = H_0 - H(x). \tag{5.23}$$

Therefore, the average amount of information I obtained by using the feature x is

$$I = \int I(x)\, p(x)\, dx. \tag{5.24}$$

The larger this I is, the more effective the feature is, so it can be defined as a measure of feature evaluation. Here we have treated the case where the number of classes is 2, but it can be generalized by setting $\sum_{i=1}^{c}$ in Eqs. (5.21) and (5.22) even if the number of classes increases.
It has already been mentioned in Sect. 4.1 that $P(\omega_i)$ and $P(\omega_i | x)$ above are called prior probability and posterior probability, respectively. In general, by obtaining x, $P(\omega_i | x)$ tends to have a large value in a particular class compared to $P(\omega_i)$, and this will appear as a decrease in entropy, or information, as shown in Eq. (5.23).

5.4 Bayes Error and the Nearest Neighbor Rule

In the previous section, we introduced the basic concept of the Bayes error and clarified its importance. How then, can the Bayes error be obtained? The Bayes error calculation formula presented in the previous section is based on the probability density function. However, since the probability density function is unknown, it cannot be applied to real patterns. What we know about the Bayes error is an estimate or an approximation from actual patterns generated according to an unknown probability distribution. In the following, we describe a method for estimating the Bayes error.

5.4.1 Error Probability of the Nearest Neighbor Rule

If the probability density function is known in advance, the Bayes error can be calculated analytically using Eq. (5.17) or (5.19). In reality, however, the probability density function is not known. What we can observe is not the probability density function itself, but the realizations of the probability density function, i.e., the individual patterns generated based on the probability density function. In other words, the Bayes error is an idealized concept that we cannot know directly.

Therefore, methods to approximate the Bayes error have been studied for a long time. Among them, the most well-known is the approximation by the nearest neighbor rule, which has been introduced in Sect. 1.3. The relation is expressed as Cover and Hart (1967)

$$e_B \leq e_N \leq e_B \left(2 - \frac{c}{c-1} e_B \right) \leq 2e_B, \qquad (5.25)$$

where e_B is the Bayes error, e_N is the error probability of the nearest neighbor rule, and c is the number of classes. The equation shows the very interesting result that with a sufficiently large number of prototypes, the error probability of the nearest neighbor rule is larger than the Bayes error (which is natural), but not more than twice the Bayes error. In other words, although the nearest neighbor rule is a simple process, it is a relatively good approximation of the Bayes error when the number of prototypes is large. In the following, let us derive Eq. (5.25) for the case $c = 2$.

First, prepare n prototypes $x_1, x_2, \ldots, x_n$ whose belonging classes are known. If x' is the nearest neighbor to the input pattern x, the next formula holds since x' is chosen from the prototype:

$$x' \in \{x_1, x_2, \ldots, x_n\}. \qquad (5.26)$$

An error of the nearest neighbor rule occurs when the input pattern and its nearest neighbor belong to different classes. Therefore, the error probability $e_n(x)$ of classifying the pattern x by the nearest neighbor rule using n prototypes is

$$e_n(x) = P(\omega_1|x)\, P(\omega_2|x') + P(\omega_2|x)\, P(\omega_1|x'). \tag{5.27}$$

The error probability e_n for all possible x is as follows:[7]

$$e_n = \int e_n(x)\, p(x)\, dx. \tag{5.28}$$

Here, the following assumption is made:

$$\lim_{n\to\infty} x' = x. \tag{5.29}$$

That is, if the number of prototypes n approaches infinity, x' approaches x infinitely. Thus,

$$\lim_{n\to\infty} P(\omega_i|x') = P(\omega_i|x) \qquad (i = 1, 2), \tag{5.30}$$

and from these we obtain

$$\lim_{n\to\infty} e_n(x) = 2P(\omega_1|x)\, P(\omega_2|x) \tag{5.31}$$

$$= 2e_B(x)\,(1 - e_B(x)). \tag{5.32}$$

The transformation from Eqs. (5.31) to (5.32) was done using the relation Eq. (5.14).
From Eqs. (5.16), (5.28), and (5.32), the error probability e_N of the nearest neighbor rule is

$$e_N = \lim_{n\to\infty} e_n \tag{5.33}$$

$$= \int \left(\lim_{n\to\infty} e_n(x) \right) p(x)dx \tag{5.34}$$

$$= \int 2e_B(x)\,(1 - e_B(x))\, p(x)dx \tag{5.35}$$

$$= 2e_B\,(1 - e_B) - 2 \cdot \text{Var}\,(e_B(x)) \tag{5.36}$$

$$\leq 2e_B\,(1 - e_B), \tag{5.37}$$

---

[7] Strictly speaking, $e_n(x, x')$ should be defined instead of $e_n(x)$ and by defining a probability density function $q(x, x')$, e_n should be written as

$$e_n = \int\int e_n(x, x')\, p(x)q(x, x')\, dxdx'.$$

However, they are abbreviated for the sake of simplicity. For more details, refer to Sect. 5.4.2.

where Var $(e_B(x))$ denotes the variance of $e_B(x)$.[8]
On the other hand, from Eqs. (5.28), (5.32), and (5.33), e_N is expressed as

$$e_N = \int 2e_B(x)\,(1 - e_B(x))\,p(x)dx$$

$$= \int \{e_B(x) + e_B(x)\,(1 - 2e_B(x))\}\,p(x)dx \tag{5.38}$$

$$= e_B + \int e_B(x)\,(1 - 2e_B(x))\,p(x)dx \tag{5.39}$$

$$\geq e_B. \tag{5.40}$$

Here, $e_B(x) \leq 1/2$ of Eq. (5.14) is applied.
The above is summarized as

$$e_B \leq e_N \leq 2e_B\,(1 - e_B) \leq 2e_B, \tag{5.41}$$

and the case $c = 2$ was proved in Eq. (5.25). For the proof of the general case of $c > 2$, refer to the original paper (Cover and Hart 1967).

─────────────── **Coffee Break** ───────────────

Nearest Neighbor Rule and Prototype Distribution

The nearest neighbor rule is an excellent classification method because its error probability is close to the Bayes error when the prototypes are densely distributed. The decision boundary by the Bayes decision rule is defined as the points where the posterior probabilities are equal, i.e., the points satisfying the next equation:

$$P(\omega_1|x) - P(\omega_2|x) = \frac{P(\omega_1)p(x|\omega_1)}{p(x)} - \frac{P(\omega_2)p(x|\omega_2)}{p(x)}$$

$$= 0. \tag{5.42}$$

In general, misclassifications often occur near the decision boundary. In many cases, the probability density of the pattern is low near the decision boundary. For example, in the above equation, if $p(x|\omega_i)$ $(i = 1, 2)$ are represented by normal distributions with different

───────────────

[8] Let $f(x)$ be a function of x and

$$\overline{f} \stackrel{\text{def}}{=} \int f(x)\,p(x)\,dx,$$

then,

$$\int f(x)\,(1 - f(x))\,p(x)\,dx = \overline{f(1 - f)} = \overline{f} - \overline{f^2} = \overline{f}(1 - \overline{f}) - (\overline{f^2} - \overline{f}^2)$$

$$= \overline{f}(1 - \overline{f}) - \text{Var}(f).$$

mean vectors, the decision boundary is set near the tail of the normal distribution, where the probability density is low. This tendency is more pronounced as the distance between the mean vectors increases.

Now, suppose that prototypes are collected to perform the nearest neighbor rule. Since they are considered to reflect the original probability density, a large number of prototypes will inevitably be collected in areas with high probability density. However, this means that only a small number of prototypes are collected near the decision boundary, where misclassifications are likely to occur.

On the other hand, to achieve high classification performance with the nearest neighbor rule, a large number of prototypes must be placed near the decision boundary where misclassifications are most likely to occur. Conversely, only a small number of prototypes are needed for other areas. Therefore, only those prototypes that contribute to determining the decision boundary should be retained. In other words, when placing prototypes, there is a conflict between faithfully reflecting the distribution of patterns and achieving high classification performance.

The *editing algorithm* Dasarathy et al. (2000), which can achieve more efficient classification, is proposed as a way to create a new set of prototypes from the collected prototypes.

5.4.2 Example of Error Probability Calculation

Here, let's confirm that Eq. (5.41) holds by using a concrete example. The example below was introduced in the paper Cover and Hart (1967) by Thomas Cover and Peter Hart. It is taken up because it is good material for understanding the relationship between the Bayes error and the nearest neighbor rule.

Assume that patterns of the two classes ω_1 and ω_2 are distributed on the interval $[0, 1]$ in the one-dimensional feature space. The prior probabilities of both classes are equal to $1/2$ as follows:

$$P(\omega_1) = P(\omega_2) = \frac{1}{2}. \tag{5.43}$$

Also, the probability density functions $p(x|\omega_1)$ and $p(x|\omega_2)$ for both classes are set to

$$p(x|\omega_1) = 2x \tag{5.44}$$

$$p(x|\omega_2) = 2 - 2x. \tag{5.45}$$

where x denotes the one-dimensional feature value. Since from Eq. (5.6),

$$p(x) = P(\omega_1)\, p(x|\omega_1) + P(\omega_2)\, p(x|\omega_2) \tag{5.46}$$

$$= \frac{1}{2} \cdot 2x + \frac{1}{2}(2 - 2x) \tag{5.47}$$

$$= 1, \tag{5.48}$$

then, the distribution of n patterns is found to be uniform. Now, suppose that a total of n patterns $x_1, x_2, \ldots, x_n$ in both classes are distributed according to the probability distribution described above. We plot $p(x|\omega_1)$ and $p(x|\omega_2)$ in Fig. 5.4, and furthermore show how the patterns of both classes are distributed on the interval $[0, 1]$.

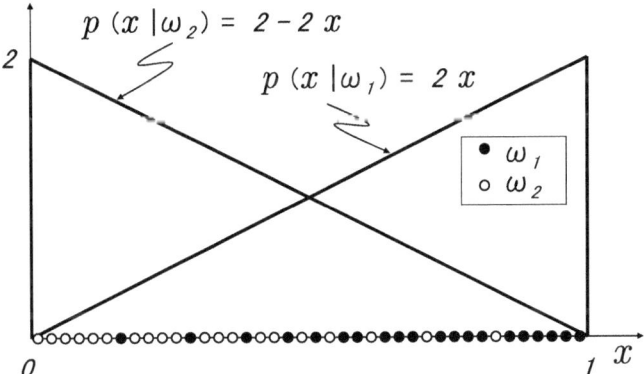

Fig. 5.4 Probability density functions for two classes

When these n patterns are used as prototypes and unknown patterns are classified by the nearest neighbor rule, the error probability e_n is obtained after some calculations as[9]

$$e_n = \frac{1}{3} + \frac{3n + 5}{2(n + 1)(n + 2)(n + 3)}. \tag{5.49}$$

For the derivation of the above equation, see Problem 5.3.

Let us examine the validity of the above equation. If the nearest neighbor rule is applied using only one prototype, all patterns in one of the two classes would be misclassified. In fact, in Eq. (5.49), if $n = 1$, then $e_n = 1/2$, which confirms this.

If $n \to \infty$ in Eq. (5.49), then the error probability e_N of the nearest neighbor rule is

$$e_N = \lim_{n \to \infty} e_n = \frac{1}{3}. \tag{5.50}$$

A plot of the relationship between n and e_n is shown in Fig. 5.5.

---

[9] This result differs from the equation

$$e_n = \frac{1}{3} + \frac{1}{(n + 1)(n + 2)}$$

given in the original paper, but Eq. (5.49) agrees with the result of the reference Peterson (1970). It is likely that there was an error in the derivation of the equation by Cover and Hart. However,

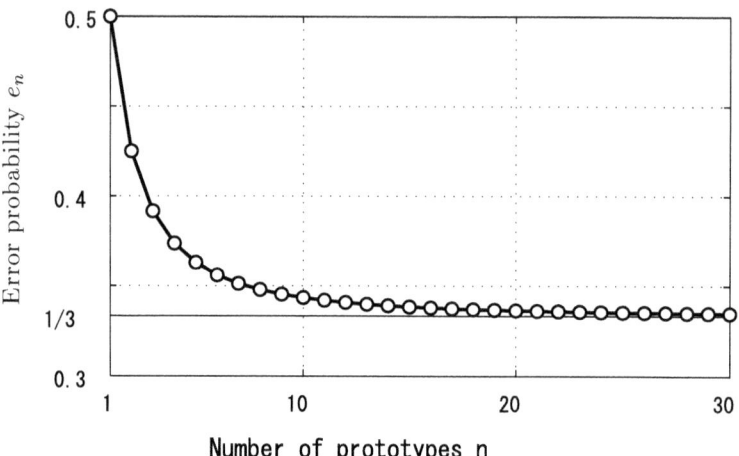

Fig. 5.5 Relationship between the error probability of the nearest neighbor rule and the number of prototypes

On the other hand, the Bayes error e_B is obtained from Eq. (5.17) as follows:

$$e_B = \int_0^1 \min\{P(\omega_1|x), P(\omega_2|x)\} \, p(x)dx$$

$$= \int_0^1 \min\{x, 1-x\} \, dx \tag{5.51}$$

$$= \frac{1}{4}. \tag{5.52}$$

Applying Eqs. (5.50) and (5.52) to Eq. (5.41) in the two-class case, we obtain

$$\frac{1}{4} < \frac{1}{3} < \frac{3}{8}, \tag{5.53}$$

indicating that the relationship is indeed valid.

─────────────────────── **Coffee Break** ───────────────────────

Error Probability of Nearest Neighbor Rule Exceeds Twice that of Bayes Error!

The paper by Cover and Hart (1967) that clarified the relationship between the Bayes error and the nearest neighbor rule is a landmark in the history of statistical pattern recognition, and has had a significant impact on subsequent pattern recognition and machine learning

─────────────────────

when $n = 1$ and $n \to \infty$, the values of the equations by Cover et al. are all consistent with the results of Eq. (5.49).

research. However, twenty years after Cover et al.'s paper, Keinosuke Fukunaga et al. overturned this established theory. This is discussed below.

The claim of Eq. (5.25) that "the error probability of the nearest neighbor rule is at most twice that of the Bayes error" supports the usefulness of the nearest neighbor rule not only as a pattern classifier but also as a means of estimating the Bayes error, and is the cornerstone of the nearest neighbor rule. Later, however, Fukunaga and others (Fukunaga 1987) questioned this claim. According to Fukunaga's theoretical analysis, when the dimensionality of the feature space (feature vector) is high, the asymptotic performance of the nearest neighbor rule in terms of error probability is not twice, but three or four times higher than that of the Bayes error. Then, what was wrong with the analysis by Cover?

In general, estimates are subject to bias (deviation from the true value) and variance (scatter from the mean value of the estimate) due to stochastic variation in patterns used for estimation. The bias and variance will be explained in Sect. 5.5.1. If the bias is large, the estimate deviates considerably from the true value, and if the variance is large, a single estimation result is unreliable. In fact, even in the estimation of error probability in the nearest neighbor rule, these nuisances, especially the bias in the estimate itself, have been working behind the scenes. However, the bias in the nearest neighbor rule, independent of the number of patterns, is highly related to the dimension of the feature space, the distance measure (the distance defining the nearest neighbor, generally the Euclidean distance), and the distribution of patterns. In particular, the effect of dimensionality is significant. The bias increases exponentially with increasing dimensionality, so if the dimensionality of the feature space is high, no matter how much the number of patterns is increased, the bias does not decrease, and as a result, a large bias is added to the error probability, which far exceeds twice the Bayes error.

In other words, Cover did not consider the effect of bias at all in the derivation of Eq. (5.25). Furthermore, there was also a problem in the derivation process of Eq. (5.25) by Cover. That is, as shown in Eq. (5.29), Cover uses the assumption that the distance between a pattern x and its nearest neighbor pattern x' is zero when a sufficiently large number of patterns are uniformly distributed. In the case of a two-dimensional feature space, the above assumption is plausible since a huge number of patterns are uniformly distributed in a two-dimensional plane. However, the story is not so simple in higher dimensions. In the high-dimensional case, shown in Fig. 5.6, consider a d-dimensional hypersphere of radius r centered at some x and assume that patterns are uniformly distributed within the sphere. Next, consider a d-dimensional hypersphere with the same center and radius $a \cdot r$ $(0 < a < 1)$. Let V_1 be the volume of the d-dimensional hypersphere of radius r and V_2 be the volume of the part outside the hypersphere of radius $a \cdot r$ and inside the hypersphere of radius r. Since the volume of the hypersphere is proportional to the dth power of its radius, the following equation holds:

$$\frac{V_2}{V_1} = \frac{r^d - (a \cdot r)^d}{r^d} = 1 - a^d.$$

For example, if $a = 0.8$, $d = 100$, then $V_2/V_1 \approx 1 - 2.03 \times 10^{-10}$. Interestingly, this indicates that when d is large, the volume of the hypersphere is almost entirely occupied by the volume near its surface because V_2/V_1 is almost one. Hence, the probability $P(\delta V)$ of a pattern existing in a small region of δV within the hypersphere is distribution $\times$ δV. However, in the vicinity of the center, $P(\delta V) \approx 0$, because $P(\delta V) \approx 0$ from the above calculation. In other words, the pattern in the neighborhood of x exists in the shaded area in Fig. 5.6 with probability ≈ 1. This is well known to those who have studied communication theory as the *spherical concentration phenomenon* in the proof of *Shannon's second fundamental theorem of communication*. Thus, the above assumption that the distance between x and its nearest neighbor pattern is zero is not appropriate in high-dimensional spaces.

Fig. 5.6 Spherical
concentration phenomenon in
higher-dimensional spaces

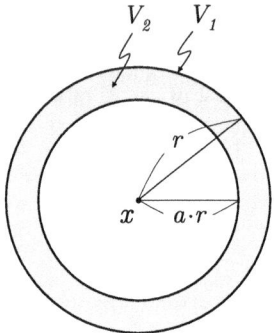

From the above, it should be recognized that Eq. (5.25) is not valid when the dimension of the feature space is high. To avoid misunderstanding, this is not to say that the nearest neighbor rule is not useful as a classification method. It is just that the dimension of the feature space, the number of patterns, and the distance scale must be set well in order to use the nearest neighbor rule as a desirable classifier. However, the setting method has not been fully established, and this is an important future research issue for the nearest neighbor rule. Some people take the pessimistic view that there are no more research problems in pattern recognition and machine learning, but this example shows that they are completely off the mark. Nevertheless, the curse of dimensionality is a troublesome one.

5.5 Estimation of Bayes Error

In the previous section, we discussed the importance of the Bayes error as a criterion for evaluating features. In this section, we describe specific estimation methods for the Bayes error.

5.5.1 Bias and Variance of Misclassification Rates

Since the Bayes error is defined using the true distribution, it is impossible to calculate the Bayes error directly in practical applications. Even if the distribution is known, in the case of a multidimensional feature space, calculating the Bayes error requires multidimensional integral calculations, which is difficult except in special cases, for example, when the functional form of the distribution is an exponential family. Therefore, it is not reasonable to attempt to directly estimate the Bayes error based on Eq. (5.19). In fact, the above approach is not used to estimate the Bayes error. In the following, we describe a practical method for indirectly estimating the Bayes error based on learning patterns, rather than from the formulas. In order to explain the estimation of the Bayes error, it is necessary to explain important issues concerning the bias and variance associated with the estimation of the misclassification rate. First, we explain what bias and variance are.

When some statistic $s(X)$ is estimated from a given pattern set X, the estimator is a random variable because it depends on the pattern set X with stochastic variation. We write this as $S(X)$.[10]

The bias of $S(X)$ is the difference between the mean (expected value) over all possible X of $S(X)$ and the true value s_0:

$$\text{Bias} = \underset{X}{E}\{S(X)\} - s_0. \tag{5.54}$$

Intuitively, the bias is a quantity that indicates how much the average of the estimated values over all possible patterns X that follow the same distribution is biased against the true value. And when the bias is zero, the estimator is called an *unbiased estimator*, or *unbiased*. Also, the variance of $S(X)$ is expressed as the variation among the estimates as

$$\text{Var} = \underset{X}{E}\{(S(X) - E\{S(X)\})^2\}. \tag{5.55}$$

The smaller the bias, the closer the estimator is to the true value, and the smaller the variance, the more reliable the estimator is. Therefore, bias and variance are used as measures of the goodness of the estimator.

The following important facts regarding bias and variance in the estimation of the misclassification rate have been revealed.[11] That is, the bias in the misclassification rate is due to the fact that learning patterns are finite, and the variance in the misclassification rate is due to the fact that test patterns are finite.

───────────────────────── **Coffee Break** ─────────────────────────

Bias and Variance Dilemma

The derivation is omitted, but the mean square error of the estimator, that is, the expected value of the squared error between the estimate and the true value can be decomposed into bias and variance as follows:[12]

$$\text{MSE} = \text{Bias}^2 + \text{Var}. \tag{5.56}$$

Equation (5.56) is an important relation that tells us the details of the MSE. For example, if we design a discriminant function by increasing the order of the discriminant function

───────────────────────────

[10] It should be noted that estimator and estimate are different concepts. Given an estimator, its realization is an estimate. In the case of the mean, for example, the function $M = 1/n \sum_{i=1}^{n} X_i$ of the random variables $X_1, \ldots, X_n$ is the estimator, and its realization $1/n \sum_{i=1}^{n} X_i$ by $X_1 = x_1, \ldots, X_n = x_n$ is the estimate. Thus, the estimate $s(X)$ is the realization of the estimators $S(X)$.

[11] See chapter 5 in Fukunaga (1990) for the details.

[12] Since MSE and Var are the second-order statistics and Bias is the first-order statistic, it is intuitively natural that Bias is squared. If the data contain noise, the variance of the noise is also added to the right-hand side of Eq. (5.56).

from the first order to the second order to the third order, and sequentially evaluate the classification performance using test patterns, the misclassification rate will decrease up to a certain order, but after that, the misclassification rate will increase. This is called the *Hughes phenomenon* (Hughes 1968). What does this mean? We can successfully explain this phenomenon by using Eq. (5.56). Bias and variance are generally closely related to the degrees of freedom of the estimator model. In other words, as shown in Fig. 5.7, the larger the degrees of freedom of the model, the smaller the bias (closer to the true value), but conversely, the larger the variance (greater dispersion). Increasing the order of the discriminant function increases its expressive power as a function, i.e., the degrees of freedom of the model. Therefore, as the order is increased, the bias is indeed reduced, but the variance is also increased. As a result, the MSE, the sum of the two, changes from decreasing to increasing after a certain point, as shown in Fig. 5.7.

Fig. 5.7 Relationship among MSE, Bias, and Var

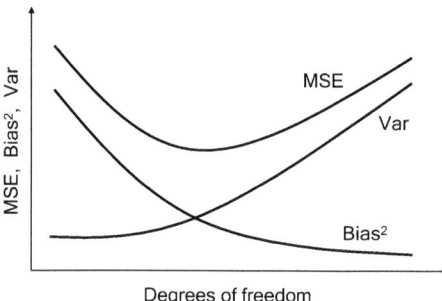

Degrees of freedom

The Bias–Var curve has a trade-off relationship in that decreasing one increases the other. Therefore, if we use an unnecessarily complex model, the classification rate may be almost 100% for learning patterns, but for test patterns, the classification performance is often considerably degraded. This is precisely due to the increase in variance.

When using a nonlinear model with a high number of degrees of freedom, such as a neural network, successful classification of learning patterns does not imply that the performance is directly applicable to test patterns. On the other hand, with a linear model, although the fit to the true value is not so good, the variance of the estimated value is small, and stable results can be obtained. In other words, there is little difference between the classification performance obtained with the learning patterns and that of the test patterns, which makes the evaluation easier. The determination of the degrees of freedom of the model is actually a search for the optimal balance of bias and variance.

5.5.2 Upper and Lower Bounds on Bayes Error

The misclassification rate is generally a function of the learning patterns and the test patterns. This is evident from the fact that learning patterns are used to design classifiers and test patterns are used to evaluate them. Therefore, we express the misclassification rate as a function of these two distributions. The true misclassification rate, or Bayes error, can be thought of as the error probability tested on the true distribution after learning on the true distribution using a classifier that can implement the Bayes decision rule. That is, let the set of true distributions be

$\mathcal{P}$,[13] then, the Bayes error is written as $\epsilon(\mathcal{P}, \mathcal{P})$. The first and second arguments of $\epsilon(\ )$ denote the distribution of the learning and test patterns, respectively.

On the other hand, if $\hat{\mathcal{P}}$ denotes the distribution estimated for a finite number of learning patterns, the following two inequalities hold:

$$\epsilon(\mathcal{P}, \mathcal{P}) \leq \epsilon(\hat{\mathcal{P}}, \mathcal{P}), \tag{5.57}$$

$$\epsilon(\hat{\mathcal{P}}, \hat{\mathcal{P}}) \leq \epsilon(\mathcal{P}, \hat{\mathcal{P}}). \tag{5.58}$$

The above inequalities are intuitively clear, since different distributions for both learning and testing will definitely increase the misclassification rate compared to the case where they are the same.

On the other hand, as described in Sect. 5.5.1, the bias in the misclassification rate is entirely due to the learning patterns. In other words, since the misclassification rate is unbiased with respect to the test patterns, the expected value for test patterns independent of the learning patterns is equal to the misclassification rate tested with the true distribution. Thus, if we write $\hat{\mathcal{P}}'$ for the distribution of test patterns which are independent of learning patterns, the following equation holds:

$$\underset{\hat{\mathcal{P}}'}{E}\{\epsilon(\hat{\mathcal{P}}, \hat{\mathcal{P}}')\} = \epsilon(\hat{\mathcal{P}}, \mathcal{P}). \tag{5.59}$$

From Eqs. (5.57) and (5.59), we obtain

$$\epsilon(\mathcal{P}, \mathcal{P}) \leq \underset{\hat{\mathcal{P}}'}{E}\{\epsilon(\hat{\mathcal{P}}, \hat{\mathcal{P}}')\}. \tag{5.60}$$

Also, from Eq. (5.58), we obtain

$$\underset{\hat{\mathcal{P}}}{E}\{\epsilon(\hat{\mathcal{P}}, \hat{\mathcal{P}})\} \leq \underset{\hat{\mathcal{P}}}{E}\{\epsilon(\mathcal{P}, \hat{\mathcal{P}})\}. \tag{5.61}$$

However, as above, since the expected value for the test patterns is unbiased,

$$\underset{\hat{\mathcal{P}}}{E}\{\epsilon(\mathcal{P}, \hat{\mathcal{P}})\} = \epsilon(\mathcal{P}, \mathcal{P}) \tag{5.62}$$

holds. From Eqs. (5.61) and (5.62), we obtain

$$\underset{\hat{\mathcal{P}}}{E}\{\epsilon(\hat{\mathcal{P}}, \hat{\mathcal{P}})\} \leq \epsilon(\mathcal{P}, \mathcal{P}). \tag{5.63}$$

Finally, from Eqs. (5.60) and (5.63), the following upper and lower bounds on the Bayes error are obtained:

$$\underset{\hat{\mathcal{P}}}{E}\{\epsilon(\hat{\mathcal{P}}, \hat{\mathcal{P}})\} \leq \epsilon(\mathcal{P}, \mathcal{P}) \leq \underset{\hat{\mathcal{P}}'}{E}\{\epsilon(\hat{\mathcal{P}}, \hat{\mathcal{P}}')\}. \tag{5.64}$$

---

[13] For example, in the case of c classes, $\mathcal{P} = \{p(\boldsymbol{x}|\omega_1), \ldots, p(\boldsymbol{x}|\omega_c)\}$.

From Eq. (5.64), the true misclassification rate $\epsilon(\mathcal{P}, \mathcal{P})$ can be estimated by pincer between the expected value of the misclassification rate estimated by using the learning patterns for testing as well and the expected value of the misclassification rate estimated by using the test patterns independent of the learning patterns. In other words, the Bayes error can be estimated indirectly by estimating the misclassification rate in the above two ways.

In actual applications where only one data set consisting of n learning patterns is given, it is not possible to calculate the expected value, so the lower and upper bounds of the misclassification rate can be approximated in the following way. The lower bound shown in Eq. (5.64) is approximated by simply designing a classifier with learning patterns, testing it with the same learning patterns, and calculating the misclassification rate. Since this method involves resubstitution of the learning patterns into the classifier, it is called the *resubstitution method*.[14] This is hereafter abbreviated as the *R method*.

On the other hand, taking the expected value of both sides of Eq. (5.64) with respect to $\hat{\mathcal{P}}$ does not change the inequality in any way, so the upper bound is $E_{\hat{\mathcal{P}}}\{E_{\hat{\mathcal{P}}'}\{\epsilon(\hat{\mathcal{P}}, \hat{\mathcal{P}}')\}\}$. And this is the H method described in Sect. 4.5.2. However, since only one data set is given now, the H method would have the disadvantage of reducing the number of patterns when designing and testing the classifier. Therefore, the L method is used instead of the H method. That is, by performing the procedure of learning with $\mathcal{X} - x_i$ and testing with x_i for $i = 1, 2, \ldots, n$, we obtain the misclassification rate and use the value as the estimate of the upper bound.

From the above, it seems that it is possible to indirectly estimate the Bayes error using the R and L methods with a given finite number of patterns, but the story is not simple. To actually perform the above procedure, the R and L methods are first executed using an appropriate classification method that can approximate the Bayes decision rule. The estimation accuracy depends greatly on the classification method, the number of given patterns, and the dimension of the features. For example, when the above process is performed, the estimated values of $\epsilon(\mathcal{P}, \mathcal{P})$ by a linear discriminant function and a neural network which will be explained in detail in Chap. 12 are clearly different. In addition, even with the same neural network, the estimates of $\epsilon(\mathcal{P}, \mathcal{P})$ are clearly different for learning patterns of 100 and 1,000. The reason for this is due to the bias of the estimated values. The bias is generally a function of the classification method, the number of feature dimensions, and the number of patterns.[15] When the classification method is fixed, the larger the ratio of the number of patterns to the dimensionality of the features, the smaller the bias. The degree of bias reduction depends on the classification method. Therefore, to estimate the Bayes error more accurately, it is necessary to estimate the above bias and correct it, but in general, the bias estimation is not easy.

---

[14] Resubstitution is only a conceptual explanation; in practice, the misclassification rate when learning with the learning patterns can be adopted as it is, and there is no need to recalculate the misclassification rate after learning.

[15] Let us recall that the coffee break in Sect. 5.4.2 was exactly describing this bias problem.

As can be seen from the above, the estimation of the Bayes error is a problem involving model selection and bias correction of the estimates. In particular, the estimation of the Bayes error in high-dimensional feature spaces with large bias is an extremely difficult problem due to the curse of dimensionality.[16] Unnecessarily increasing the number of feature dimensions makes not only the design of a classifier, but also the evaluation of features difficult. This is an important matter to remember.

――――――――――――――――――――― **Coffee Break** ―――――――――――

Difficulties Associated with Estimation of Bayes Error

The performance of a pattern recognition system depends on the feature selection method and the classifier design method. In particular, the former is a necessary condition for designing a highly accurate pattern recognition system. This is because a pattern recognition system with a high classification rate cannot be constructed for features with a large Bayes error, no matter how good the classifier is. This is the reason why, in the past, pattern recognition research such as character recognition focused more on feature selection rather than classifiers.

Therefore, as mentioned in Sect. 5.1, it is necessary to evaluate features separately from the evaluation of classifiers. As mentioned earlier, the features are evaluated in terms of the Bayes error, which can be estimated by the procedure described in Sect. 5.5.2. However, in order to obtain the upper and lower bounds of the Bayes error using Eq. (5.64), some classifier must be assumed. Therefore, it is not possible to evaluate features independently of the classifier. Therefore, the original intention to evaluate features independently of the classifiers cannot be achieved.

In the end, a realistic solution to this problem would be to estimate the Bayes error using a classifier that is as close to a Bayes discriminant function as possible. For low-dimensional feature spaces, a classifier based on the nearest neighbor rule is a good candidate, but for high-dimensional feature spaces, even the nearest neighbor rule is not sufficient, as described in the coffee break in Sect. 5.4.2. In fact, there is unfortunately no suitable classifier required for estimating the Bayes error in the case of high-dimensional feature spaces. Therefore, the method described in Sect. 5.5.2 cannot accurately estimate the Bayes error. As can be seen, estimating the Bayes error is one of the unsolved and important problems in statistical pattern recognition.

5.6 Experiments on Feature Evaluation

Two types of feature evaluation methods were introduced in this chapter: the ratio of between-class variance to within-class variance and the estimated Bayes error. In this section, we conduct evaluation experiments using these methods on real data to confirm the effectiveness of each method.

The features used are three types of Glucksman's features (Appendix C), GLK16, GLK81, and GLK256. Each of them consists of 1000 patterns per class, for a total

---

[16] The estimation of the Bayes error based on the nearest neighbor rule is discussed in detail in Chap. 7 in Fukunaga (1990).

of 10,000 patterns. Features GLK16, GLK81, and GLK256 are more advanced in this order, and the following experiments show that this can be evaluated quantitatively using the evaluation methods described so far. For the data used, refer to Appendix D.

5.6.1   Feature Evaluation by Ratio of Between-Class Variance to Within-Class Variance

In the following, we describe feature evaluation experiments based on the ratio of between-class variance to within-class variance. The results obtained by applying Eq. (5.3) to all classes are shown in light gray bars in Fig. 5.8 (evaluation method 1). The calculated evaluation values are shown above the bars. The evaluation values increase in order of 0.622, 0.691, and 0.694 for GLK16, GLK81, and GLK256, respectively, confirming that the features become more sophisticated in this order and that the separation between classes becomes easier.

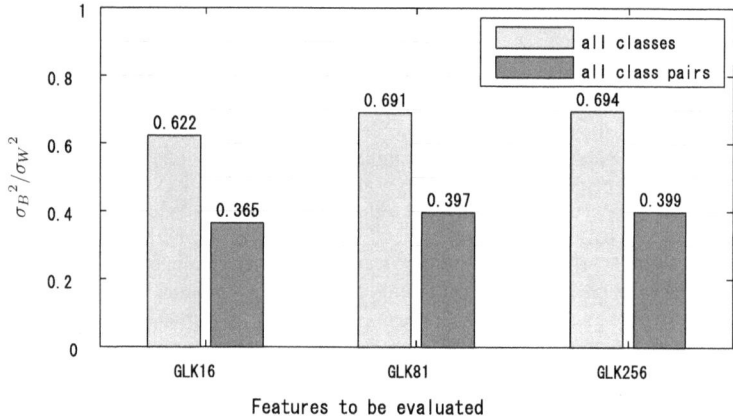

Fig. 5.8 Feature evaluation by the ratio of between-class variance to within-class variance

However, as described in Sect. 5.2, this evaluation method cannot evaluate the overlap of distributions when viewed in terms of class pairs, so we introduced a method (evaluation method 2) in which Eq. (5.3) is calculated for each class pair and averaged. Since the number of classes is 10 and the total number of class pairs is $_{10}C_2 = 45$ for numerals, the computational complexity is not a serious problem. The evaluation results obtained in this way are shown in the dark gray bars in the figure. As in evaluation method 1, the evaluation values are shown above the bars. The same trend as in evaluation method 1 is observed in evaluation method 2, confirming that it functions properly as an evaluation method. As far as this figure is concerned, there are no problems that are of concern in evaluation method 1, as pointed out in Sect. 5.2.

The feature evaluation based on the ratio of between-class variance to within-class variance described above is a simple and computationally inexpensive method, but it has the disadvantage that its evaluation values are not directly linked to the classification rate or misclassification rate.

5.6.2 Feature Evaluation by Estimated Bayes Error

The Bayes error is a measure of the degree of overlap of distributions in the feature space. Therefore, unlike the ratio of between-class variance to within-class variance, the Bayes error is an ideal feature evaluation method in that it allows evaluation based on the misclassification rate. However, unless the probability density function of each class is known, it is impossible to directly calculate the Bayes error.

As an alternative method, Eq. (5.64) shows that the Bayes error can be estimated by finding the upper and lower bounds of the Bayes error. It is also shown that the upper and lower bounds shown in Eq. (5.64) can be obtained indirectly by the L and R methods, respectively, using a finite number of patterns. Since the upper bound is more important than the lower bound as an estimated value,[17] the upper bound obtained by the L method will be regarded as the estimated value of the Bayes error in the following.

Since the Bayes error is the error probability when the Bayes decision rule is applied, a classifier that can realize the Bayes decision rule is necessary for using the L method. As shown in Eq. (5.25), the nearest neighbor rule has a clear relationship with the Bayes error and is therefore a promising classification method to use for estimation. However, there are several problems with this method. First, as mentioned in the coffee break in Sect. 5.4.2, the estimation of the Bayes error by the nearest neighbor rule is subject to a considerable bias in a high-dimensional feature space. Even if this bias problem can be avoided, the problem pointed out in the coffee break in Sect. 5.5.2 remains. That is, feature evaluation via a specific classification method, the nearest neighbor rule, violates the requirement that "feature evaluation must be independent of the classification method".

These problems are serious and not easy to solve. However, even with the various problems described above, the combination of the L method and nearest neighbor rule seems to be the best and most realistic solution currently available to meet the demand for estimating the Bayes error. This method seems to be sufficient to meet the requirement of comparing multiple feature extraction methods rather than strictly estimating the Bayes error.

Another reason for applying the nearest neighbor rule is as follows. Assuming that the number of patterns is n, in the L method, the operation of learning with $(n - 1)$ patterns and testing with the remaining 1 pattern must be repeated n times while changing patterns. In other words, learning with $(n - 1)$ patterns is repeated n

---

[17] For example, when learning is performed by the perceptron learning rule, the convergence condition is zero misclassification, so the lower bound estimated by the R method is always zero, which is meaningless. This is also the case when the nearest neighbor rule is used.

times, which requires a huge amount of processing, and this is considered to be the biggest drawback of the L method. However, this problem can be avoided when the nearest neighbor rule is used as a classification method. This is because the learning of the nearest neighbor rule is completed simply by registering $(n - 1)$ learning patterns. This method has already been introduced in Sect. 1.3.2 as the complete storage scheme.

A method for estimating the Bayes error by combining the L method and the nearest neighbor rule is described below. Here, the total number of patterns is assumed to be n.

Estimation Procedure of Bayes Error Using L Method and Nearest Neighbor Rule

(1) Register all patterns (n patterns).
(2) Among them, one pattern is selected as a test pattern, and the remaining $(n - 1)$ patterns are used as learning patterns.
(3) All learning patterns are regarded as prototypes, and the above test pattern is classified according to the nearest neighbor rule (complete storage scheme), and the result is recorded.
(4) Select another pattern as a test pattern, and execute (3) with the remaining $(n - 1)$ patterns as learning patterns.
(5) Repeat (4) above, and when all patterns are classified, calculate the misclassification rate, and use that value as an estimated Bayes error.

The results obtained by performing the above procedure on patterns with features GLK16, GLK81, and GLK256 are shown in Fig. 5.9. In this experiment, $n = 10,000$. The view of the figure is the same as that of Fig. 5.8. Values of the estimated Bayes error are shown as percentages above the bars. The Bayes error decreases in the order of GLK16, GLK81, and GLK256 to 19.19, 7.68, and 7.46%, confirming that the features are upgraded in this order as in Fig. 5.8.

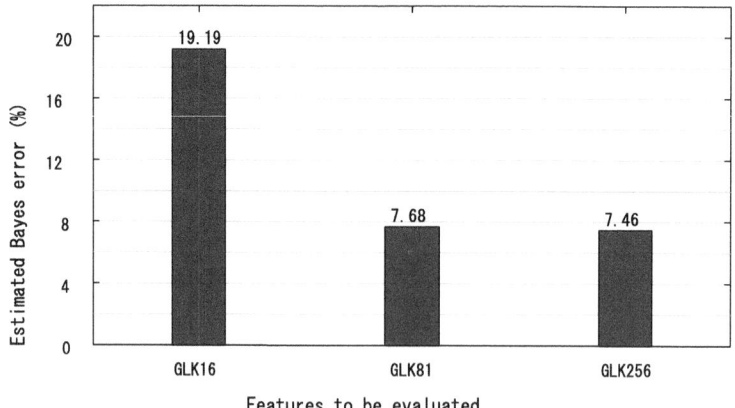

Fig. 5.9 Feature evaluation by estimated Bayes error

Coffee Break

Unbiasedness of the Least Squares Method

Some readers who have dabbled a bit in linear regression models may remember the important property that the "least squares estimator is an unbiased estimator in linear regression models". And, of course, they may wonder whether this contradicts the statement in the coffee break in Sect. 5.5.1 (linear models are highly biased). It is not good to be confused, so let us clear up this question.

In conclusion, there is no contradiction between the two. Both are correct. That is, unbiasedness in a linear regression model means that for $x \in \mathcal{R}^d$, the unbiasedness is a property of the least squares estimator of the parameter $\mathbf{w}$ in the regression equation $y = w_0 + w_1 x_1 + \ldots + w_d x_d = \mathbf{w}^t \mathbf{x}$, not of y. On the other hand, the previous coffee break was discussing y, the estimate of Y. In other words, the bias in the above linear regression model refers to the bias between the $\mathbf{w}$ estimator and the true value $\mathbf{w}_0$ and not the bias between the estimator Y and the true value y_0.

It is important to note that the estimator S can be written as $S(\theta)$ as a function of some random variable θ, and even if the estimator of θ is unbiased, there is no guarantee that the estimator S is unbiased in general. It should be clear from this that there is no contradiction between the unbiased estimator of $\mathbf{w}$ and the large bias of Y. Thus, one should pay attention to which estimator is at issue in discussing bias and variance.

Problems

5.1 There are two different feature extraction methods 1 and 2, both of which generate two-dimensional feature vectors. Suppose we apply these feature extraction methods to eight patterns and obtain two-dimensional feature vectors $x_1, x_2, \ldots, x_8$. Here, x_1, x_2, x_3, x_4 belong to class ω_1, and x_5, x_6, x_7, x_8 belong to class ω_2.

Now, suppose that feature extraction method 1 yields

$$x_1 = (1, 1)^t, \ x_2 = (1, 3)^t, \ x_3 = (2, 3)^t, \ x_4 = (4, 1)^t$$
$$x_5 = (5, 2)^t, \ x_6 = (6, 2)^t, \ x_7 = (7, 5)^t, \ x_8 = (6, 7)^t,$$

and feature extraction method 2 yields

$$x_1 = (0, 0)^t, \ x_2 = (0, 1)^t, \ x_3 = (1, 2)^t, \ x_4 = (3, 1)^t$$
$$x_5 = (5, 3)^t, \ x_6 = (6, 4)^t, \ x_7 = (4, 5)^t, \ x_8 = (5, 8)^t.$$

(1) Plot the eight feature vectors obtained by the feature extraction methods 1 and 2, respectively, on a two-dimensional feature space.
(2) Show which of the feature extraction methods is superior using the ratio of between-class variance to within-class variance.

(Four patterns per class is too few for actual feature evaluation. Since the purpose of this exercise is to learn the technique and calculation method, consideration was made to lighten the computational burden.)

5.2 Let $x = (x_1, x_2)^t$ denote a pattern in the two-dimensional feature space. In this space, patterns belonging to class ω_1 and class ω_2 are distributed as follows:

(1) Patterns of class ω_1: x_1, x_2 are independent and both are uniformly distributed in the interval $[0, 4]$.
(2) Patterns of class ω_2: x_1, x_2 are independent and both are uniformly distributed in the interval $[2, 5]$. The prior probabilities $P(\omega_1)$ and $P(\omega_2)$ are assumed to be

$$P(\omega_1) = 2/3,$$
$$P(\omega_2) = 1/3.$$

When the patterns distributed as above are classified by the Bayes decision rule, find the error probability, i.e., the Bayes error, Also, draw the decision boundary determined by the Bayes decision rule on the two-dimensional plane (x_1, x_2).

5.3 * Derive Eq. (5.49).

Chapter 6
Transformation of Feature Space

Abstract As described in Chaps. 1 and 2, the basic flow of pattern recognition is to convert patterns into feature vectors by feature extraction and then divide the learning patterns in the feature space into classes. However, the feature space and feature vectors, which should be called the original feature space and the original feature vectors, cause a number of problems in classification processing if they are used as they are. To solve these problems, scaling, feature normalization, and feature selection represented by dimensionality reduction, are implemented. In this chapter, these are treated as feature space transformations in a unified manner and described one by one.

6.1   Necessity of Feature Space Transformation

Before applying the obtained original feature vectors to the classification, it is necessary to solve several problems.

First, there is the problem of *scaling* between the components of the feature vectors. Usually, each component of the feature vector is measured in a different unit. The distribution of patterns in the feature space changes drastically by simply changing the unit of measurement, i.e., *scale*. To avoid this situation, a process called normalization of feature vectors is required.

Next, there is the issue of the dimensionality of the feature space. In general, when designing a classifier, there is a tendency to increase the number of features too much. This is due to the expectation that increasing the number of features will increase the amount of information and also increase the classification rate. The reasons why this is not always a good idea can be summarized in the following three points.

Supplementary Information The online version contains supplementary material available at https://doi.org/10.1007/978-981-95-1478-6_6.

First, the more the number of features is increased, the more likely it is that highly correlated feature pairs will be included, and the effect will not be as great as expected.

Second, the computational cost for statistical calculations is at least of the order of the power of the dimensionality, therefore, an increase in the dimensionality of the feature space leads to an explosion in computational complexity. This is a problem known as the so-called curse of dimensionality, and was discussed in Sect. 4.4.

Third, when designing a classifier from a finite number of learning patterns, the misclassification rate actually increases as the dimensionality is increased. This is known as the Hughes phenomenon, and was introduced in Sect. 5.5.1.

For this reason, *dimensionality reduction* of the feature space is one of the most important issues in pattern recognition. Dimensionality reduction of the feature space according to a certain criterion is called the *feature selection*.[1] A possible feature selection method is to simply construct a $\tilde{d}(< d)$-dimensional feature vector by extracting only useful components from a given d-dimensional feature vector. In this case, it is necessary to repeat the process of selecting $\tilde{d}$ components from d components and evaluating their usefulness, which is computationally expensive in the case of high dimensionality. On the other hand, sometimes a method is used to convert the original feature vector into a smaller dimensional feature vector based on specific criteria.

The above operations of transforming the original feature vector into a form suitable for subsequent processing, such as normalization or feature selection, are called *transformation of feature space*, and are expressed in most cases as linear transformations shown in the following equation:

$$y = A^t x, \tag{6.1}$$

where x is the original feature vector and y is the transformed vector, and the dimensions are d and $\tilde{d}$, respectively. Also, a matrix A is the *transformation matrix* for the linear transformation and has size $d \times \tilde{d}$.

In the normalization case, A is a diagonal matrix and $\tilde{d} = d$, as described later. If only $\tilde{d}$ components are extracted from d components to make a new feature vector, only the corresponding elements of $\tilde{d}$ column vectors of A should be set to 1 and the rest to 0.

The normalization process is described in Sect. 6.2, and the dimensionality reduction methods for feature selection are described in each subsequent section. That is, Sect. 6.3 describes the KL expansion and Sect. 6.4 the linear discriminant method. The last section, Sect. 6.5 summarizes the points to be noted when applying KL expansion.

---

[1] It should be noted that the use of the term feature selection and the aforementioned term feature extraction varies depending on textbooks. For example, Fukunaga (1990) applies the term feature extraction to feature selection used in this book.

―――――――――――――――――――――― **Coffee Break** ――――――――――――――――

Theorem of the Ugly Duckling: What is Feature Selection?

Our own brains are actually solving pattern recognition problems every day. What exactly does "feature selection" mean in human pattern recognition? The "Ugly Duckling Theorem" introduced here was created by Satoshi Watanabe. The gist of this is that if the similarity of two objects is measured by a certain criterion, the similarity of any two objects is equal.[2] Watanabe uses the term *predicate* instead of the feature. Therefore, following Watanabe, we will use the term predicate here.

Theorem of the Ugly Duckling An ugly duckling and a swan are just as similar to each other as are two swans (Watanabe 1969).

Proof In general, suppose that d predicates (features) are selected to classify birds. For example, S_1 (white feathers), S_2 (can fly), S_3 (large body), ..., S_d, etc. Since the predicates used here are binary, the bird species n that can be distinguished by d predicates is $n = 2^d$. Here, taking the case of $d = 3$ as an example, and considering the above S_1, S_2, and S_3, the Venn diagram of Fig. 6.1(a) is obtained. That is, $n = 2^d = 8$, and every bird belongs to one of the 8 sections $A_1, A_2, \ldots, A_8$ in the diagram. Watanabe called these n sections that do not overlap each other *atoms* . For example, A_2 shows $\bar{S}_1 \cap S_2 \cap \bar{S}_3$, i.e., the characteristics of "feathers not white, can fly, and body not large". In the figure, $x_2 \in A_2$ is such a bird, and a crow is an example of this. Similarly, $x_1 \in A_1$ corresponds to a duck and $x_7 \in A_7$ corresponds to a swan.

Watanabe generalized the predicate and defined it as follows. That is, any r $(1 \le r \le n)$ of n atoms are chosen and their union set is regarded as a predicate. Watanabe calls this r a dimension. Since there are two ways to choose or not to choose each atom as a predicate, the number of predicates N obtained in this way[3] is $N = 2^n - 1$.

Figure 6.1(b) shows $N(= 2^n - 1 = 255)$ predicates for $n = 8$. In this table, 1 and 0 indicate whether the atom is chosen or not as a predicate, respectively. For example, P_k in this table indicates the predicate $P_k = A_1 \cup A_4 \cup A_6 \cup A_7$ of dimension $r = 4$. It is clear from Fig. 6.1(a) that $P_k = S_1$, which is nothing but the "white feathers" predicate shown at the beginning.

Now let us consider the similarity between two arbitrary birds $x_i \in A_i$ and $x_j \in A_j$ $(i \ne j)$. It is natural to define this as

similarity = number of predicates they share out of N predicates.

In other words, the more predicates they share, the more similar they are. Let N_C be the number of shared predicates, then N_C is the number of predicates containing any r $(0 \le r \le n - 2)$ atoms other than A_i and A_j to which the two birds belong. Therefore, using the same calculation method[4] as when creating the table in Fig. 6.1(b), $N_C = 2^{n-2}$. In the example discussed here, $N_C = 2^6 = 64$.

―――――――――――――――

[2] For a rigorous formulation and proof of this theorem, see Watanabe (1969). This book is instructive for the study of pattern recognition and philosophy of recognition.

[3] We subtract 1 from 2^n since we exclude the case where none of the atoms is chosen (r = 0). A similar result is obtained as
$$N = \sum_{r=1}^{n} {}_nC_r = 2^n - 1.$$
That is, N can be obtained by adding ${}_nC_r$ with $r = 1, 2, \ldots, n$, where ${}_nC_r$ represents the number of ways to select r items from n items.

[4] In the same way as finding N, it can also be found as
$$N_C = \sum_{r=0}^{n-2} {}_{n-2}C_r = 2^{n-2}.$$
Here we include the case where none of the atoms is chosen, so $\sum$ is from $r = 0$ and $r = 0, 1, 2, \ldots, (n - 2)$.

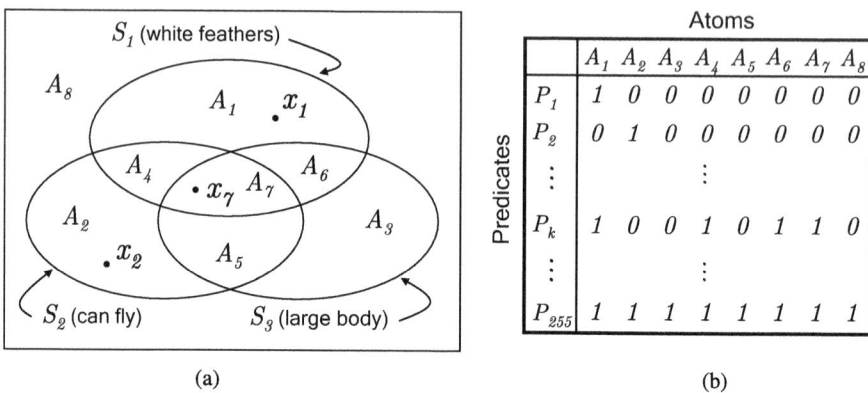

(a)   (b)

Fig. 6.1 Theorem of the ugly duckling. (**a**) Generation of atoms by S_1, S_2, S_3. (**b**) Configuring predicates by selecting atoms

This indicates that the number of predicates shared by two birds is constant, regardless of how the two birds are chosen or how the predicates are set. In other words, all two birds are similar to the same degree. Thus, it is not possible to distinguish the ugly duckling from other ordinary ducklings. (End of proof.)

Watanabe continues as follows: This theorem assumes that all predicates of the same dimension have the same importance. As a result, the concept of similarity is eliminated, and it becomes impossible to group similar objects together as a class. Such a conclusion cannot be accepted at all.

To get out of this absurdity, we must admit that "some predicates are more important than others". Being similar means "sharing more important predicates", and according to this definition of similarity, we can talk about similarity because the number of shared important predicates differs depending on the pair of objects.

In the above, if we replace "predicate" with "feature", the meaning becomes clearer. In other words, in principle, it is not possible to classify objects into multiple classes simply by selecting certain features from them. In pattern recognition, adding importance to features is the essence of feature selection. For humans, this corresponds to making value judgments, and for pattern recognition, it corresponds to weighting features.

Watanabe published "Theorem of the ugly duckling" and its rigorous proof in 1961. According to him, "Some people have manifested their surprise and delight, while others grumbled that they knew something like this must be true. But when I asked the latter group of people where they had read or written it, I could get no clear answer" (Watanabe 1969).

6.2 Normalization of Features

As already mentioned in Sect. 1.2, it is important to note that the similarity between patterns must be reflected as a distance in the feature space. That is, similar patterns should be located close to each other in the feature space.

In general, feature vectors are composed of elements with different properties such as weight and length. Therefore, the positional relationship of patterns in the feature space can change drastically just by changing the way the units are taken.

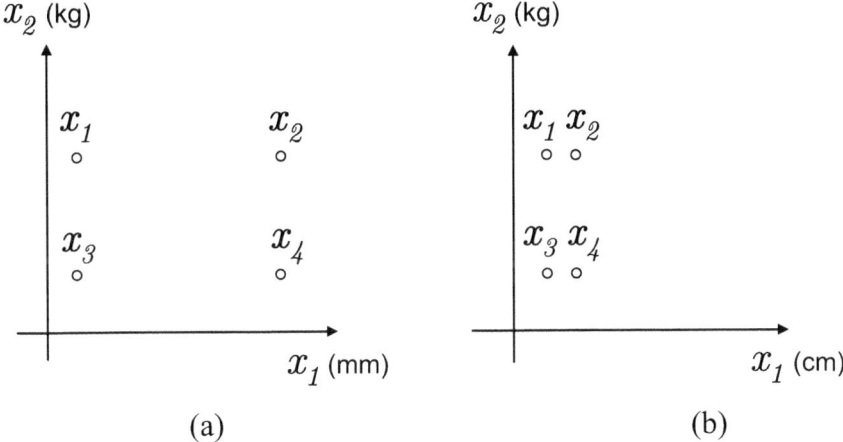

Fig. 6.2 Unit settings of coordinate axes and their effects. (**a**) Unit of x_1 axis is mm. (**b**) Unit of x_1 axis is cm

For example, suppose that two-dimensional feature vectors x_1, x_2, x_3, and x_4 are located in the feature space as shown in Fig. 6.2(a). Let us now assume that x_1 is a feature representing length, and that mm is used as the unit in Fig. 6.2(a). Also, assume that x_2 is a feature representing weight in kg. If the unit of x_1 is changed to cm, x_1 is scaled to $1/10$ as in Fig. 6.2(b). Compared to Fig. 6.2(a), it is clear that the distance relationship between the patterns changes drastically. Depending on the unit setting, a situation may arise in which a particular feature is emphasized to an extreme degree, or conversely, the feature may be completely ignored. The choice of units, or in other words, scaling has the same effect as weighting the features. In order to avoid arbitrariness in the weighting depending on the choice of units, a *normalization* process under a certain policy is necessary for each feature axis.

Let us apply the idea of minimizing the distance between patterns as a normalization method (Sebestyen 1962). Now, denote the p-th ($p = 1, 2, \ldots, n$) pattern among the n pattern set in the d-dimensional feature space by x_p and let

$$x_p = (x_{p1}, x_{p2}, \ldots, x_{pd})^t. \tag{6.2}$$

If the transformation matrix $\mathbf{A}$ for normalization is

$$\mathbf{A} = \begin{pmatrix} a_1 & & & 0 \\ & a_2 & & \\ & & \ddots & \\ 0 & & & a_d \end{pmatrix}, \tag{6.3}$$

and the pattern obtained by normalizing x_p is y_p, it can be expressed as

$$y_p = (y_{p1}, y_{p2}, \ldots, y_{pd})^t \tag{6.4}$$

$$= A^t x_p. \tag{6.5}$$

The element-wise representation is as follows:

$$y_{pj} = a_j x_{pj} \qquad (p = 1, 2, \ldots, n \quad j = 1, 2, \ldots, d). \tag{6.6}$$

Let r_p^2 be the mean square distance between the p-th pattern in the n pattern set and the other $(n - 1)$ patterns, then after normalization, the following equation holds:

$$r_p^2 = \frac{1}{n-1} \sum_{q=1}^{n} \sum_{j=1}^{d} (y_{pj} - y_{qj})^2. \tag{6.7}$$

Therefore, if the mean square distance between each pattern after normalization is denoted as R^2, then

$$R^2 = \frac{1}{n} \sum_{p=1}^{n} r_p^2 \tag{6.8}$$

$$= \frac{1}{n(n-1)} \sum_{p=1}^{n} \sum_{q=1}^{n} \sum_{j=1}^{d} (y_{pj} - y_{qj})^2. \tag{6.9}$$

Substituting Eq. (6.6), we obtain

$$R^2 = \frac{1}{n(n-1)} \sum_{p=1}^{n} \sum_{q=1}^{n} \sum_{j=1}^{d} a_j^2 (x_{pj} - x_{qj})^2 \tag{6.10}$$

$$= \frac{n}{n-1} \sum_{j=1}^{d} a_j^2 \frac{1}{n} \sum_{p=1}^{n} \sum_{q=1}^{n} \left(\frac{1}{n} x_{pj}^2 - \frac{2}{n} x_{pj} x_{qj} + \frac{1}{n} x_{qj}^2 \right) \tag{6.11}$$

$$= \frac{n}{n-1} \sum_{j=1}^{d} a_j^2 \left(\frac{1}{n} \sum_{q=1}^{n} \frac{1}{n} \sum_{p=1}^{n} x_{pj}^2 - 2 \cdot \frac{1}{n} \sum_{p=1}^{n} x_{pj} \frac{1}{n} \sum_{q=1}^{n} x_{qj} \right.$$

$$\left. + \frac{1}{n} \sum_{p=1}^{n} \frac{1}{n} \sum_{q=1}^{n} x_{qj}^2 \right) \tag{6.12}$$

$$= \frac{n}{n-1} \sum_{j=1}^{d} a_j^2 \left(\frac{1}{n} \sum_{q=1}^{n} \overline{x_j^2} - 2\overline{x_j}^2 + \frac{1}{n} \sum_{p=1}^{n} \overline{x_j^2} \right) \tag{6.13}$$

$$= \frac{2n}{n-1} \sum_{j=1}^{d} a_j^2 \left(\overline{x_j^2} - \overline{x_j}^2 \right), \tag{6.14}$$

where $\overline{x}$ denotes the ensemble mean of x. On the other hand, the variance σ_j^2 of the j-th feature x_j is expressed as follows:[5]

$$\sigma_j^2 = \frac{1}{n-1} \sum_{p=1}^{n} \left(x_{pj} - \overline{x_j} \right)^2 \tag{6.15}$$

$$= \frac{n}{n-1} \left(\overline{x_j^2} - 2\overline{x_j}^2 + \overline{x_j}^2 \right) \tag{6.16}$$

$$= \frac{n}{n-1} \left(\overline{x_j^2} - \overline{x_j}^2 \right). \tag{6.17}$$

Substituting Eq. (6.17) into Eq. (6.14), we obtain

$$R^2 = 2 \sum_{j=1}^{d} a_j^2 \sigma_j^2. \tag{6.18}$$

Now let us find a_j that minimizes Eq. (6.18) under the following constraints:

$$\prod_{j=1}^{d} a_j = 1. \tag{6.19}$$

This constraint corresponds to the condition that the volume of the unit hypercube in the feature space be kept constant before and after normalization. Let us find the extreme value of

$$L = 2 \sum_{j=1}^{d} a_j^2 \sigma_j^2 - \lambda \left(\prod_{j=1}^{d} a_j - 1 \right), \tag{6.20}$$

using Lagrange's method of undetermined multipliers. The result of partial differentiation of L by a_j is set to 0, then

---

[5] Here we use the variance as an unbiased estimator.

$$\frac{\partial L}{\partial a_j} = 0 \tag{6.21}$$

is obtained. From Eqs. (6.20) and (6.21), we obtain

$$4a_j\sigma_j^2 - \lambda \prod_{k\neq j}^{d} a_k = 0. \tag{6.22}$$

Multiplying both sides by a_j and using Eq. (6.19), we obtain

$$a_j = \frac{\sqrt{\lambda}}{2\sigma_j}. \tag{6.23}$$

Substituting again into Eq. (6.19), we obtain

$$\lambda = 4\left(\prod_{j=1}^{d} \sigma_j\right)^{2/d}. \tag{6.24}$$

Therefore a_j can be written as

$$a_j = \frac{1}{\sigma_j}\left(\prod_{k=1}^{d} \sigma_k\right)^{1/d}. \tag{6.25}$$

Since $(\,\cdot\,)$ in Eq. (6.25) is common to each feature axis, a_j is proportional to $1/\sigma_j$. That is,

$$a_j \propto \frac{1}{\sigma_j}. \tag{6.26}$$

This is an intuitively natural process in which each feature axis is normalized by its standard deviation to equalize the variance, i.e., the spread of patterns around the mean.

In this chapter, we will discuss the transformation methods of the feature space, i.e., KL expansion in Sect. 6.3 and Fisher's method in Sect. 6.4, respectively. Note that Fisher's method is invariant to the normalization process described in this section, whereas KL expansion is not (see Sect. 6.4.1.1). For example, the principal axes of the KL expansion are set 90° differently in (a) and (b) of Fig. 6.2.

6.3 KL Expansion

In this section, we describe the KL expansion, which is known as one of the dimensionality reduction methods. First, we show that the KL expansion can be derived separately based on two different evaluation criteria. Then, each of these evaluation criteria is explained in detail.

6.3.1 Criteria for Dimensionality Reduction

The *Karhunen–Loève expansion* is a method for finding the subspace that best approximates the distribution of patterns in the feature space. This technique is often used in signal processing as well as pattern recognition as one of the dimensionality reduction methods. It is usually abbreviated as the *KL expansion*, and will be referred to hereafter as such. In *multivariate analysis*, a field of statistics, *principal component analysis* is a well-known method for extracting principal components from a large number of multidimensional data. As is known, the KL expansion and principal component analysis are almost mathematically equivalent. Here, we use two evaluation criteria: the maximum variance criterion and the minimum mean square error criterion to describe the dimensionality reduction by the KL expansion.

Using the example of dimensionality reduction from a two-dimensional space (x_1, x_2) to a one-dimensional space y_1, y_2, the meaning of these two evaluation criteria is shown in Fig. 6.3. Let's compare a one-dimensional space y_1 and its orthogonal one-dimensional space y_2 using the two evaluation criteria.

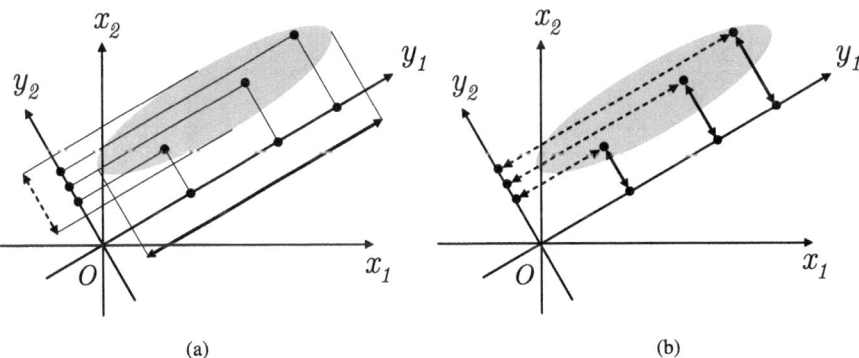

(a)   (b)

Fig. 6.3 Two types of evaluation criteria for dimensionality reduction. (**a**) Maximum variance criterion. (**b**) Minimum mean square error criterion

The maximum variance criterion (a) defines the best space as the one in which the variance of patterns (indicated by the arrows) is maximized in a one-dimensional space. The subspace y_1 that can achieve the variance indicated by the thick arrow is better than the subspace y_2 indicated by the dotted arrow.

On the other hand, the minimum mean square error criterion (b) defines the best subspace as the one that minimizes the mean square error (arrows) generated by mapping patterns in the original space to the one-dimensional space, and similarly, y_1 with the error indicated by the thick arrows is a better subspace than y_2 with the error indicated by the dotted arrows.

In Fig. 6.3, both (a) and (b) use the same y_1 and y_2. Here, there is a point to note. The axis y_1 is parallel to the principal axis of the distribution, which is the best axis for the maximum variance criterion, but not the best axis for the minimum mean square error criterion. However, they can be made to coincide by moving the origin. For details, refer to the explanation in Sect. 6.3.3.

The linear discriminant method described in Sect. 6.4 is another method of dimensionality reduction, but the effects of the KL expansion and the linear discriminant method are very different and must be used depending on the application. This is explained in detail in Sect. 6.5.1.

6.3.2 Maximum Variance Criterion

As can be seen from Fig. 6.3(a), if the pattern variance is large in the transformed $\tilde{d}$ $(< d)$-dimensional subspace, that subspace can be regarded as a space that preserves the characteristics of the pattern distribution in the original space. Therefore, using the *maximum variance criterion*, which maximizes the variance of the transformed pattern distribution, let us find the $\tilde{d}$-dimensional subspace and the transformation matrix $\mathbf{A}$ for obtaining that subspace.

Let us define an orthonormal basis consisting of $\tilde{d}$ d-dimensional vectors spanning a $\tilde{d}$-dimensional subspace, as follows:

$$\{\mathbf{u}_1, \ldots, \mathbf{u}_{\tilde{d}}\} \qquad (\tilde{d} < d). \tag{6.27}$$

From the orthonormality of the basis, we have

$$\mathbf{u}_i^t \mathbf{u}_j = \delta_{ij}, \tag{6.28}$$

where δ_{ij} is the *Kronecker delta* and defined as follows:

$$\delta_{ij} = \begin{cases} 1 & \text{if } i = j \\ 0 & \text{otherwise.} \end{cases} \tag{6.29}$$

The transformation matrix $\mathbf{A}$ from the original feature space to the subspace is given by

$$\mathbf{A} = (\mathbf{u}_1, \ldots, \mathbf{u}_{\tilde{d}}), \tag{6.30}$$

and the feature vector x is transformed to

$$y = A^t x. \tag{6.31}$$

From Eq. (6.28),

$$A^t A = I \tag{6.32}$$

is satisfied. Note that I is a $\tilde{d}$-dimensional identity matrix.

Let n be the number of patterns, m be the mean of patterns in the original feature space, and $\tilde{m}$ be the mean of patterns in the subspace; then we obtain the next equations:

$$m = \frac{1}{n} \sum_{x \in X} x \tag{6.33}$$

$$\tilde{m} = \frac{1}{n} \sum_{y \in Y} y = \frac{1}{n} \sum_{x \in X} A^t x = A^t m. \tag{6.34}$$

From the above equations, the variance $\tilde{\sigma}^2(A)$ of patterns in the subspace is

$$\tilde{\sigma}^2(A) = \frac{1}{n} \sum_{y \in Y} (y - \tilde{m})^t (y - \tilde{m}) \tag{6.35}$$

$$= \frac{1}{n} \sum_{x \in X} \left(A^t (x - m) \right)^t \left(A^t (x - m) \right) \tag{6.36}$$

$$= \frac{1}{n} \sum_{x \in X} \mathrm{tr} \left(A^t (x - m) \left(A^t (x - m) \right)^t \right) \tag{6.37}$$

$$= \mathrm{tr} \left(A^t \frac{1}{n} \sum_{x \in X} \left((x - m)(x - m)^t \right) A \right) \tag{6.38}$$

$$= \mathrm{tr}(A^t \Sigma A), \tag{6.39}$$

where Σ denotes the covariance matrix of the pattern set in the original feature space and is defined by

$$\Sigma = \frac{1}{n} \sum_{x \in X} (x - m)(x - m)^t. \tag{6.40}$$

In Eq. (6.36), X denotes the set of patterns x, and Y in Eq. (6.35) denotes the set of patterns y to which x is transformed by Eq. (6.31), and $\mathrm{tr}(X)$ denotes the trace (Appendix B) of the square matrix X. In addition, Eq. (6.37) uses the fact that the next equation holds for any vector x (Eq. (B.12) in Appendix B):

$$x^t x = \text{tr}(xx^t).$$ (6.41)

From Eq. (6.39), finding $\mathbf{A}$ that maximizes the variance $\tilde{\sigma}^2(\mathbf{A})$ is reduced to an optimization problem of finding $\mathbf{A}$ that maximizes $\text{tr}(\mathbf{A}^t \Sigma \mathbf{A})$ under the constraints of Eq. (6.32). Let Λ be a $\tilde{d}$-dimensional diagonal matrix and partial differentiation of

$$J(\mathbf{A}) \stackrel{\text{def}}{=} \text{tr}(\mathbf{A}^t \Sigma \mathbf{A}) - \text{tr}((\mathbf{A}^t \mathbf{A} - \mathbf{I})\Lambda)$$ (6.42)

by $\mathbf{A}$ and setting $\mathbf{0}$, we obtain

$$\Sigma \mathbf{A} = \mathbf{A}\Lambda.$$ (6.43)

For the partial differentiation of trace, the corresponding formula of Appendix B is used.

Here, if we let the diagonal matrix Λ be

$$\Lambda = \begin{pmatrix} \lambda_1 & & & 0 \\ & \lambda_2 & & \\ & & \ddots & \\ 0 & & & \lambda_{\tilde{d}} \end{pmatrix}$$ (6.44)

and write Eq. (6.43) as a relational expression regarding the vector $\mathbf{u}_i$ of Eq. (6.30), we get

$$\Sigma \mathbf{u}_i = \lambda_i \mathbf{u}_i \qquad (i = 1, \ldots, \tilde{d}).$$ (6.45)

The above equation is the so-called *eigenvalue problem*, where λ_i and $\mathbf{u}_i$ are called the *eigenvalue* and *eigenvector* of Σ, respectively.[6]

From Eqs. (6.32) and (6.43), we obtain

$$\mathbf{A}^t \Sigma \mathbf{A} = \Lambda,$$ (6.46)

where $\mathbf{A}$ is the matrix that diagonalizes Σ. If the d eigenvalues of the matrix Σ are λ_i ($\lambda_1 \geq \lambda_2 \geq \ldots \geq \lambda_d$), then from Eqs. (6.39) and (6.46), we have[7]

---

[6] In the derivation of the eigenvalue problem shown in Eq. (6.45), the trace was introduced and the arithmetic procedure of differentiating the trace by a matrix was applied. For a different derivation method, see Problem 6.1.

[7] The matrix Σ is of size $d \times d$ and if Σ is non-singular, then Σ has d eigenvalues ($\neq 0$) and eigenvectors. Equations (6.43) and (6.45) are expressions for $\tilde{d}$ eigenvalues and eigenvectors among them.

$$\max\{\tilde{\sigma}^2(\mathbf{A})\} = \max\{\mathrm{tr}(\mathbf{A}^t\mathbf{\Sigma}\mathbf{A})\} \tag{6.47}$$

$$= \max\{\mathrm{tr}\mathbf{\Lambda}\} \tag{6.48}$$

$$= \sum_{i=1}^{\tilde{d}}\lambda_i. \tag{6.49}$$

Thus, the transformation matrix $\mathbf{A}$ that maximizes $\tilde{\sigma}^2(\mathbf{A})$ is obtained as a matrix whose columns are the $\tilde{d}$ *orthonormal eigenvectors* corresponding to the top $\tilde{d}$ eigenvalues $\lambda_1, \ldots, \lambda_{\tilde{d}}$ of $\mathbf{\Sigma}$. In the example of Fig. 6.4, the axis of the optimal one-dimensional space obtained by the maximum variance criterion is P_a. For a concrete calculation example using real data, refer to Problem 6.2.

Fig. 6.4 Transformation of feature space by KL expansion

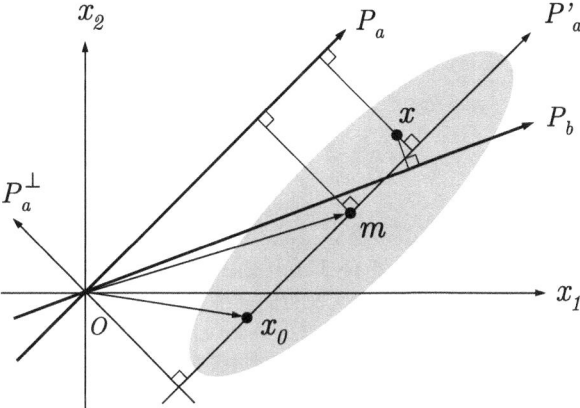

6.3.3   Minimum Mean Square Error Criterion

In this subsection, we try to find the best subspace using the *minimum mean square error criterion*. Since the basis of the transformed space is indicated by Eq. (6.27), the vector $\mathbf{y}$ ($= \mathbf{A}^t\mathbf{x} = (y_1, \ldots, y_{\tilde{d}})^t$) transformed by $\mathbf{A}$ is given by $\mathbf{A}\mathbf{y}$ ($= y_1\mathbf{u}_1 + \ldots + y_{\tilde{d}}\mathbf{u}_{\tilde{d}}$) in the original coordinate system. The distance between $\mathbf{A}\mathbf{y}$ and $\mathbf{x}$ is the error caused by the transformation $\mathbf{A}$. Therefore, $\mathbf{A}$ that minimizes this error can be regarded as the transformation that best preserves the original distribution of feature vectors (Fig. 6.3(b)). Thus, we obtain the transformation matrix $\mathbf{A}$ by minimizing the *mean square error*.

Let $\varepsilon^2(\mathbf{A})$ be the mean square error caused by the transformation $\mathbf{A}$, and using Eq. (6.32), we get

$$\varepsilon^2(\mathbf{A}) = \frac{1}{n}\sum(\mathbf{A}\mathbf{y} - \mathbf{x})^t(\mathbf{A}\mathbf{y} - \mathbf{x}) \tag{6.50}$$

$$= \frac{1}{n} \sum (\mathbf{AA}^t \mathbf{x} - \mathbf{x})^t (\mathbf{AA}^t \mathbf{x} - \mathbf{x}) \tag{6.51}$$

$$= \frac{1}{n} \sum (\mathbf{x}^t \mathbf{x} - (\mathbf{A}^t \mathbf{x})^t \mathbf{A}^t \mathbf{x}) \tag{6.52}$$

$$= \frac{1}{n} \sum (\text{tr}(\mathbf{x}\mathbf{x}^t) - \text{tr}(\mathbf{A}^t \mathbf{x}\mathbf{x}^t \mathbf{A})) \tag{6.53}$$

$$= \text{tr}\mathbf{R} - \text{tr}(\mathbf{A}^t \mathbf{R}\mathbf{A}). \tag{6.54}$$

Here, $\mathbf{R}$ is the *autocorrelation matrix* expressed by the following equation:

$$\mathbf{R} \overset{\text{def}}{=} \frac{1}{n} \sum_{x \in \mathcal{X}} \mathbf{x}\mathbf{x}^t. \tag{6.55}$$

Between the autocorrelation matrix $\mathbf{R}$ and the covariance matrix $\mathbf{\Sigma}$, the following relation holds:[8]

$$\mathbf{\Sigma} = \frac{1}{n} \sum_{x \in \mathcal{X}} (\mathbf{x} - \mathbf{m})(\mathbf{x} - \mathbf{m})^t = \mathbf{R} - \mathbf{m}\mathbf{m}^t. \tag{6.56}$$

Minimizing the mean square error is equivalent to maximizing $\text{tr}(\mathbf{A}^t \mathbf{R}\mathbf{A})$ under the constraint of Eq. (6.32). If the eigenvalues of $\mathbf{R}$ are λ_i ($\lambda_1 \geq \lambda_2 \geq \ldots \geq \lambda_d$), then from the same procedure as in the previous subsection, we have

$$\min\{\varepsilon^2(\mathbf{A})\} = \text{tr}\mathbf{R} - \sum_{i=1}^{\tilde{d}} \lambda_i, \tag{6.57}$$

and the transformation matrix $\mathbf{A}$ that minimizes $\varepsilon^2(\mathbf{A})$ is obtained as a matrix whose columns are orthonormal eigenvectors, corresponding to the top $\tilde{d}$ eigenvalues $\lambda_1, \ldots, \lambda_{\tilde{d}}$ in $\mathbf{R}$.

However, the subspace thus obtained is different from the subspace obtained in the previous subsection. For example, in the example of Fig. 6.4, the axis obtained by the maximum variance criterion is P_a, and the axis obtained by the minimum mean square error criterion is P_b. The reason for this is that the viewpoint when looking at the distribution of feature vectors is not the mean of the distribution, but

---

[8] Using Eqs. (6.33) and (6.55), Eq. (6.56) is derived from Eq. (6.40) as follows:

$$\mathbf{\Sigma} = \frac{1}{n} \sum_{x \in \mathcal{X}} (\mathbf{x} - \mathbf{m})(\mathbf{x} - \mathbf{m})^t = \frac{1}{n} \sum_{x \in \mathcal{X}} \mathbf{x}\mathbf{x}^t - 2\mathbf{m}\frac{1}{n} \sum_{x \in \mathcal{X}} \mathbf{x}^t + \mathbf{m}\mathbf{m}^t = \mathbf{R} - 2\mathbf{m}\mathbf{m}^t + \mathbf{m}\mathbf{m}^t$$

$$= \mathbf{R} - \mathbf{m}\mathbf{m}^t.$$

the origin of the space. Therefore, let us consider a transformation based on the minimum mean square error criterion after translating the origin by $\mathbf{m}$. Since the mean of patterns and the origin coincide after the translation, substituting $\mathbf{m} = \mathbf{0}$ into Eq. (6.56), we obtain $\mathbf{\Sigma} = \mathbf{R}$ and from Eq. (6.54), we have

$$\varepsilon^2(\mathbf{A}) = \text{tr}\mathbf{R} - \text{tr}(\mathbf{A}^t \mathbf{R}\mathbf{A}) \tag{6.58}$$

$$= \text{tr}\mathbf{\Sigma} - \text{tr}(\mathbf{A}^t \mathbf{\Sigma}\mathbf{A}). \tag{6.59}$$

Therefore, $\mathbf{A}$ obtained in this case is equal to $\mathbf{A}$ obtained by the maximum variance criterion.[9]

Then, is the translation $\mathbf{m}$ applied here an optimal translation? Let's find $\boldsymbol{x}_0$ and $\mathbf{A}$ that satisfy the minimum mean square error criterion, assuming the translation as $\boldsymbol{x}_0$. The pattern $\boldsymbol{x}$ is transferred to $\boldsymbol{y} = \mathbf{A}^t(\boldsymbol{x} - \boldsymbol{x}_0)$ by the transformation. Conversely, $\boldsymbol{y}$ is $\mathbf{A}\boldsymbol{y} + \boldsymbol{x}_0$ in the coordinate system of the original space. If $\mathbf{I}$ is a d-dimensional identity matrix, the following equation holds:

$$(\mathbf{A}\boldsymbol{y} + \boldsymbol{x}_0) - \boldsymbol{x} = \mathbf{A}\mathbf{A}^t(\boldsymbol{x} - \boldsymbol{x}_0) + \boldsymbol{x}_0 - \boldsymbol{x} \tag{6.60}$$

$$= (\mathbf{A}\mathbf{A}^t - \mathbf{I})(\boldsymbol{x} - \boldsymbol{x}_0). \tag{6.61}$$

Here, if we set

$$\mathbf{J} = \mathbf{I} - \mathbf{A}\mathbf{A}^t, \tag{6.62}$$

then, from Eq. (6.32)

$$\mathbf{J}^t \mathbf{J} = \mathbf{J}, \tag{6.63}$$

and the mean square error $\varepsilon^2(\mathbf{A}, \boldsymbol{x}_0)$ can be expressed as follows:

$$\varepsilon^2(\mathbf{A}, \boldsymbol{x}_0) - \frac{1}{n}\sum (\mathbf{J}(\boldsymbol{x} - \boldsymbol{x}_0))^t \, \mathbf{J}(\boldsymbol{x} - \boldsymbol{x}_0) \tag{6.64}$$

$$= \frac{1}{n}\sum (\boldsymbol{x} - \boldsymbol{x}_0)^t \mathbf{J}^t \mathbf{J}(\boldsymbol{x} - \boldsymbol{x}_0) \tag{6.65}$$

$$= \frac{1}{n}\sum (\boldsymbol{x} - \boldsymbol{x}_0)^t \mathbf{J}(\boldsymbol{x} - \boldsymbol{x}_0). \tag{6.66}$$

In general, when $\mathbf{A}$ is a $d \times m$ matrix whose columns are linearly independent $m(< d)$ d-dimensional vectors, then

---

[9] The collected patterns are often normalized to $\mathbf{m} = \mathbf{0}$ in advance, which corresponds to moving the origin by $\mathbf{m}$.

$$\mathbf{P} \overset{\text{def}}{=} \mathbf{A}(\mathbf{A}^t\mathbf{A})^{-1}\mathbf{A}^t \tag{6.67}$$

is called the *orthogonal projection matrix* to the m-dimensional subspace spanned by the column vectors of $\mathbf{A}$. Also, $\mathbf{P}x$ is called the *orthogonal projection* of x. For the derivation of the above equation, refer to Eq. (G.11) of Appendix G. Here, from Eq. (6.32),

$$\mathbf{P} = \mathbf{A}\mathbf{A}^t \tag{6.68}$$

holds.[10] Equation (6.62), i.e.,

$$\mathbf{J} = \mathbf{I} - \mathbf{A}\mathbf{A}^t = \mathbf{I} - \mathbf{P} \tag{6.69}$$

is the orthogonal projection matrix onto the *orthogonal complement*[11] ($P_a^\perp$ of Fig. 6.4) of the subspace spanned by the column vectors of $\mathbf{A}$. Transformation using orthogonal projection matrices is also used in the subspace method described in Chap. 7.

Here, partial differentiation of ε^2 by x_0 and setting it as $\mathbf{0}$ yields

$$\frac{\partial \varepsilon^2}{\partial x_0} = \frac{1}{n} \sum (2\mathbf{J}x_0 - 2\mathbf{J}x) \tag{6.70}$$

$$= 2\mathbf{J}(x_0 - \mathbf{m}) \tag{6.71}$$

$$= \mathbf{0}, \tag{6.72}$$

so we get

$$\mathbf{J}x_0 = \mathbf{J}\mathbf{m}. \tag{6.73}$$

Substituting this into Eq. (6.64) yields

$$\varepsilon^2(\mathbf{A}) = \frac{1}{n} \sum (\mathbf{J}(x - \mathbf{m}))^t \, \mathbf{J}(x - \mathbf{m}) \tag{6.74}$$

$$= \frac{1}{n} \sum \left((x - \mathbf{m})^t \mathbf{J}(x - \mathbf{m}) \right) \tag{6.75}$$

---

[10] In Eq. (6.67), the m column vectors of the matrix $\mathbf{A}$ are linearly independent. Moreover, if the column vectors are orthonormal bases of the m-dimensional subspace, i.e., $\mathbf{A}^t\mathbf{A} = \mathbf{I}_m$, then Eq. (6.67) is $\mathbf{P} = \mathbf{A}\mathbf{A}^t$. Here $m = \tilde{d}$ from Eq. (6.30), and the column vectors of $\mathbf{A}$ satisfy the orthonormal basis condition Eq. (6.32), so Eq. (6.68) holds.

[11] For a vector space V and its subspace S,

$$S^\perp = \{x \in V \mid x^t y = 0 \quad \forall y \in S\}$$

is called the orthogonal complement of S in V (see Appendix G).

$$= \frac{1}{n} \sum \left((x - m)^t (I - AA^t)(x - m) \right) \tag{6.76}$$

$$= \frac{1}{n} \sum \left(\text{tr}(x - m)(x - m)^t - \text{tr}(A^t (x - m)(x - m)^t A) \right) \tag{6.77}$$

$$= \text{tr}\Sigma - \text{tr}(A^t \Sigma A). \tag{6.78}$$

Therefore, the transformation matrix A that minimizes $\varepsilon^2(A)$, while allowing the translation of the origin, is the matrix whose columns are the orthonormal eigenvectors corresponding to the top $\tilde{d}$ eigenvalues $\lambda_1, \ldots, \lambda_{\tilde{d}}$ of Σ. This is the same process as shown in Sect. 6.3.2. The axis of the subspace obtained in this way is P_a in the example of Fig. 6.4. On the other hand, the necessary condition for x_0 is Eq. (6.73), and m is just one of the x_0 that satisfy the condition. The x_0 is an arbitrary vector whose projection onto its complementary space ($P_a^{\perp}$ in Fig. 6.4) is equal to its projection onto the complementary space of m. In Fig. 6.4, m and x_0 are such examples.[12] The transformation matrix A obtained by the minimum mean square error criterion by considering the translation x_0 of the origin as a parameter corresponds to the A obtained by the minimum mean square error criterion by assuming $x_0 = m$. Furthermore, this A also agrees with the A obtained by the maximum variance criterion described in Sect. 6.3.2.

The KL expansion, used as dimensionality reduction for pattern recognition, provides the subspace obtained by the maximum variance criterion or the minimum mean square error criterion that allows origin translation. That is, this subspace uses the eigenvectors corresponding to the top eigenvalues of the covariance matrix Σ as basis.

On the other hand, the pattern recognition method called the subspace method described in Chap. 7 uses the subspace obtained by the minimum mean square error criterion for each class distribution that does not allow origin translation, as described in the first half of this subsection. That is, this subspace uses the eigenvectors corresponding to the upper eigenvalues of the autocorrelation matrix R as basis.

In any case, the KL expansion and the linear discriminant method described later are transformations and dimensionality reduction methods based on linear transformations. If the restriction of linearity is removed, we can naturally expect a wide variety of transformations. However, it should be noted that it is the restriction of linearity that makes it possible to handle the transformations in a mathematically feasible manner. As a nonlinear method of transforming the feature space and dimensionality reduction, the use of multi-layer neural network is known. For more information, refer to the coffee break in Sect. 12.6.2.

---

[12] The concrete value of x_0 is not important, since what is needed is a basis for spanning the subspace.

Misconceptions about KL Expansion

There are often misconceptions about the KL expansion.

One of them is the misconception that "The KL expansion yields a subspace that is effective for classification". In general, unless the distribution of each class is isotropic in the feature space, the KL expansion does not provide an effective subspace for classification. It is the linear discriminant method introduced in the next section that provides an effective subspace for classification.

The other misconception is that "The KL expansion is the only expansion that best approximates the pattern set". In other words, this interpretation can be rephrased as "The KL expansion is a necessary and sufficient condition for the best approximation of a pattern set". However, the truth is that the KL expansion is a sufficient condition for the best approximation of a pattern set, not a necessary condition. The subspace obtained by the KL expansion best approximates the pattern set, but the way to express the best subspace, that is, the way to choose the basis for spanning the subspace, is not unique. Since the set of eigenvectors obtained by the KL expansion is only one of its orthogonal bases, it must be noted that only the sufficiency of the above is valid.

6.4 Linear Discriminant Method

In this section, we introduce the linear discriminant method, which is often used as a dimensionality reduction method along with the KL expansion. The linear discriminant method is widely used in pattern recognition applications due to its simplicity and high usefulness. This method is oriented toward determining the subspace suitable for classification, and this point differs from the KL expansion, which also aims at dimensionality reduction. The difference between the two is described in detail in Sect. 6.5.1.

First, a linear discriminant method for two classes, known as Fisher's method, is introduced, followed by a linear discriminant method for multiple classes. Finally, we discuss these methods from the perspective of transforming the feature space.

6.4.1 Linear Discriminant Method for Two-Class Problems (Fisher's Method)

The *linear discriminant method* is called the *discriminant analysis* in the field of statistics, and is known as a basic technique for multivariate analysis.[13] The most

───────────────────────

[13] The linear discriminant method used in pattern recognition and discriminant analysis in statistics have the same mathematical framework, although they are used for different purposes, and both originated from the paper by Fisher (1950). However, it should be noted that the definitions and derivation methods of the fundamental quantities differ among textbooks. This book has been

common use in pattern recognition is classification for two classes, which is called *Fisher's linear discriminant method*, or simply *Fisher's method*. Fisher's method is a technique, from the distribution in a d-dimensional feature space, to find the optimal one-dimensional axis (straight line) for classifying two classes. The optimal axis is the one that, when projecting the patterns, the two classes are distributed as far apart as possible.

Figure 6.5 shows an example of two classes (● and ○) distributed in a two-dimensional feature space. The direction of the axes to be obtained is represented by a vector, and two vectors w_1 and w_2 are shown in the figure. From the figure, it is clear that w_2 is superior to w_1.

Fig. 6.5 Projection to one-dimensional feature space

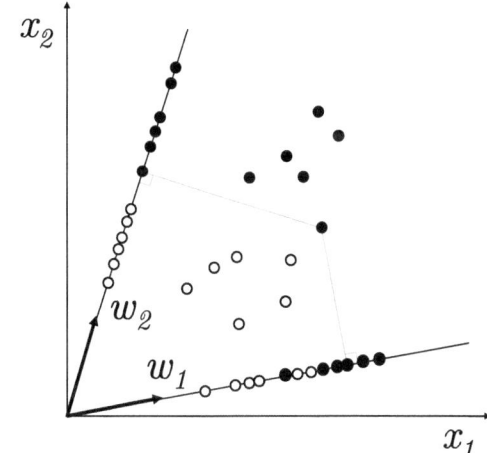

In Sect. 6.4.1.1, we describe Fisher's method based on the maximum ratio of between-class scatter to within-class scatter criterion, and in Sect. 6.4.1.2, we present a more general formulation of Fisher's method using prior probabilities and the maximum ratio of between-class variance to within-class variance criterion. In addition, in Sect. 6.4.1.3, we introduce the Mahalanobis distance, which is closely related to the ratio of between-class variance to within-class variance.

6.4.1.1 Maximum Ratio of Between-Class Scatter to Within-Class Scatter Criterion

Define the *scatter matrix* S_i as a matrix representing the variation of class ω_i as follows:

---

written with consideration for avoiding confusion even when read in conjunction with other textbooks.

$$\mathbf{S}_i \overset{\text{def}}{=} \sum_{x \in X_i} (x - \mathbf{m}_i)(x - \mathbf{m}_i)^t \qquad (i = 1, 2), \qquad (6.79)$$

where $\mathbf{m}_i$ is the mean vector of class ω_i. The scatter matrix $\mathbf{S}_i$ is defined as the sum of squares of the difference between the feature vector x belonging to class ω_i and the mean vector $\mathbf{m}_i$.

Next, using all feature vectors of the two classes, the *within-class scatter matrix* $\mathbf{S}_W$ and the *between-class scatter matrix* $\mathbf{S}_B$ are defined respectively as follows (Problem 6.6):[14]

$$\mathbf{S}_W \overset{\text{def}}{=} \mathbf{S}_1 + \mathbf{S}_2 \qquad (6.80)$$

$$= \sum_{i=1,2} \sum_{x \in X_i} (x - \mathbf{m}_i)(x - \mathbf{m}_i)^t \qquad (6.81)$$

$$\mathbf{S}_B \overset{\text{def}}{=} \sum_{i=1,2} n_i (\mathbf{m}_i - \mathbf{m})(\mathbf{m}_i - \mathbf{m})^t \qquad (6.82)$$

$$= \frac{n_1 n_2}{n} (\mathbf{m}_1 - \mathbf{m}_2)(\mathbf{m}_1 - \mathbf{m}_2)^t. \qquad (6.83)$$

Here, $\mathbf{m}$ is the mean vector of all patterns and n_i is the number of patterns in class ω_i. As shown in Eq. (6.83), $\mathbf{S}_B$ is a quantity determined by the distance between class mean vectors. Let w be a vector that represents the transformation from a d-dimensional feature space to a one-dimensional space. In this case, the pattern x transformed by w is a scalar quantity, and if this is y, it can be written as

$$y = w^t x. \qquad (6.84)$$

The class mean $\tilde{m}_i$ in the transformed space is

---

[14] Many textbooks give

$$\mathbf{S}_F \overset{\text{def}}{=} (\mathbf{m}_1 - \mathbf{m}_2)(\mathbf{m}_1 - \mathbf{m}_2)^t$$

as the definition of the between-class scatter matrix when explaining the linear discriminant method for two classes. Although this is in accordance with the definition in Fisher's original paper, it lacks extensibility to multiple classes, so we define it here in the more general form of Eq. (6.82). From Eq. (6.83), the following relation between $\mathbf{S}_F$ and $\mathbf{S}_B$ is established. The w obtained in the following discussion is the same regardless of which one is used.

$$\mathbf{S}_F = \frac{n}{n_1 n_2} \mathbf{S}_B$$

$$\tilde{m}_i = \frac{1}{n_i} \sum_{y \in \mathcal{Y}_i} y = \frac{1}{n_i} \sum_{x \in \mathcal{X}_i} \boldsymbol{w}^t \boldsymbol{x} \tag{6.85}$$

$$= \boldsymbol{w}^t \boldsymbol{m}_i \qquad (i = 1, 2), \tag{6.86}$$

where $\mathcal{Y}_i$ denotes the set of patterns belonging to ω_i in the transformed space. The within-class scatter s_W and between-class scatter s_B in the transformed space can be computed in the same way using Eqs. (6.84) and (6.86) as

$$s_W = s_1 + s_2 \tag{6.87}$$

$$= \sum_{i=1,2} \sum_{y \in \mathcal{Y}_i} (y - \tilde{m}_i)^2 \tag{6.88}$$

$$= \boldsymbol{w}^t \mathbf{S}_W \boldsymbol{w}, \tag{6.89}$$

$$s_B = \sum_{i=1,2} n_i (\tilde{m}_i - \tilde{m})^2 \tag{6.90}$$

$$= \frac{n_1 n_2}{n} (\tilde{m}_1 - \tilde{m}_2)^2 \tag{6.91}$$

$$= \boldsymbol{w}^t \mathbf{S}_B \boldsymbol{w}. \tag{6.92}$$

Here, s_i ($i = 1, 2$) is the within-class scatter of the patterns belonging to class ω_i after the transformation, defined by

$$s_i \stackrel{\text{def}}{=} \sum_{y \in \mathcal{Y}_i} (y - \tilde{m}_i)^2 \qquad (i = 1, 2) \tag{6.93}$$

in the same way as $\mathbf{S}_i$ in Eq. (6.79). Letting $\tilde{m}_i$ and $\tilde{\sigma}_i^2$ denote the class mean and the variance in a one-dimensional space after the transformation, respectively, we obtain

$$s_W = n_1 \tilde{\sigma}_1^2 + n_2 \tilde{\sigma}_2^2 \tag{6.94}$$

$$s_B = n_1 (\tilde{m}_1 - \tilde{m})^2 + n_2 (\tilde{m}_2 - \tilde{m})^2 \tag{6.95}$$

$$= \frac{n_1 n_2}{n} (\tilde{m}_1 - \tilde{m}_2)^2. \tag{6.96}$$

The basic idea of Fisher's method is to find the one-dimensional axis that maximizes the *ratio of between-class scatter to within-class scatter*. In other words, the transformation $\boldsymbol{w}$ is determined so that s_W is as small as possible and s_B is as large as possible so that the two classes are well separated in the transformed space. If we express the ratio of between-class scatter to within-class scatter as $J_S(\boldsymbol{w})$, then

$$J_S(\boldsymbol{w}) \overset{\text{def}}{=} \frac{s_B}{s_W} = \frac{n_1 n_2}{n} \cdot \frac{(\tilde{m}_1 - \tilde{m}_2)^2}{n_1 \tilde{\sigma}_1^2 + n_2 \tilde{\sigma}_2^2} \tag{6.97}$$

$$= \frac{\boldsymbol{w}^t \mathbf{S}_B \boldsymbol{w}}{\boldsymbol{w}^t \mathbf{S}_W \boldsymbol{w}}. \tag{6.98}$$

This evaluation criterion $J_S(\boldsymbol{w})$ is called *Fisher's criterion*.[15] In other words, Fisher's method is based on the *maximum ratio of between-class scatter to within-class scatter criterion*.

The problem of finding $\boldsymbol{w}$ that maximizes J_S is reduced to the optimization problem of maximizing

$$s_B = \boldsymbol{w}^t \mathbf{S}_B \boldsymbol{w}, \tag{6.99}$$

under the constraint

$$s_W = \boldsymbol{w}^t \mathbf{S}_W \boldsymbol{w} = 1. \tag{6.100}$$

If λ is a Lagrange multiplier and

$$J(\boldsymbol{w}) \overset{\text{def}}{=} \boldsymbol{w}^t \mathbf{S}_B \boldsymbol{w} - \lambda \left(\boldsymbol{w}^t \mathbf{S}_W \boldsymbol{w} - 1 \right) \tag{6.101}$$

is differentiated by $\boldsymbol{w}$ and set to $\mathbf{0}$, we obtain

$$\mathbf{S}_B \boldsymbol{w} = \lambda \mathbf{S}_W \boldsymbol{w}. \tag{6.102}$$

To derive the above equation, we apply Eq. (B.5) in Appendix B and use the fact that $\mathbf{S}_B$ and $\mathbf{S}_W$ are symmetric matrices. Therefore, if $\mathbf{S}_W$ is non-singular, then

$$\mathbf{S}_W^{-1} \mathbf{S}_B \, \boldsymbol{w} = \lambda \boldsymbol{w}. \tag{6.103}$$

So, if the maximum eigenvalue of $\mathbf{S}_W^{-1} \mathbf{S}_B$ is λ_1, then

$$\max\{J_S(\boldsymbol{w})\} = \lambda_1 \tag{6.104}$$

is obtained.[16] Also, $\boldsymbol{w}$ that maximizes J_S is found as the eigenvector corresponding to the maximum eigenvalue λ_1.

Furthermore, from Eqs. (6.83) and (6.102), we obtain

---

[15] Note that some textbooks define Fisher's criterion as $J \overset{\text{def}}{=} (\tilde{m}_1 - \tilde{m}_2)^2 / (\tilde{\sigma}_1^2 + \tilde{\sigma}_2^2)$. This criterion is equivalent to setting the prior probability to $P(\omega_1) = P(\omega_2) = 1/2$ in the more general formulation using the covariance matrix shown later in this section. See also Sect. 4.3.1 and the coffee break in Sect. 6.4.2.

[16] Note that $\mathbf{S}_B$ is positive semidefinite and its rank is at most $(d - 1)$.

$$\lambda \mathbf{S}_W \boldsymbol{w} = \mathbf{S}_B \boldsymbol{w} = \frac{n_1 n_2}{n} (\mathbf{m}_1 - \mathbf{m}_2)(\mathbf{m}_1 - \mathbf{m}_2)^t \boldsymbol{w}, \tag{6.105}$$

and noting that $(\mathbf{m}_1 - \mathbf{m}_2)^t \boldsymbol{w}$ is a scalar quantity, the following formula holds:[17]

$$\boldsymbol{w} \propto \mathbf{S}_W^{-1}(\mathbf{m}_1 - \mathbf{m}_2). \tag{6.106}$$

The feature space obtained by the transformation vector $\boldsymbol{w}$ thus found is a one-dimensional space that maximizes the ratio of between-class scatter to within-class scatter. The above is the method called Fisher's method. This criterion used in the linear discriminant method is oriented toward the classification after transformation, and this point differs from the case of the KL expansion.

Let us now consider Eq. (6.106). Suppose that the within-class scatter of each class is *isotropic*. Isotropic within-class scatter means that the distribution is not biased around a certain point (Fig. 6.6(a)), and in this case, the scatter matrix $\mathbf{S}_i$ can be written as

$$\mathbf{S}_i = \alpha_i \mathbf{I}_d \qquad (i = 1, 2), \tag{6.107}$$

where α_i is a constant and $\mathbf{I}_d$ is a d-dimensional identity matrix. At this time $\mathbf{S}_W = \mathbf{S}_1 + \mathbf{S}_2$ is also isotropic, and consequently $\mathbf{S}_W^{-1}$ can be written as

$$\mathbf{S}_W^{-1} = \alpha \mathbf{I}_d, \tag{6.108}$$

where α is a constant. Then, from Eq. (6.106),

$$\boldsymbol{w} \propto \mathbf{m}_1 - \mathbf{m}_2 \tag{6.109}$$

is obtained.

Let us verify the above by a simple example. Figure 6.6(a), (b) show examples of the distribution of two classes ω_1 and ω_2 in the two-dimensional feature space. The mean vectors $\mathbf{m}_1$ and $\mathbf{m}_2$ of the patterns in each class are identical in (a) and (b), but (a) shows the case where the within-class scatter is isotropic and (b) the case where it is not.

From the figure, it is clear that in (a), the direction of the optimal axis y coincides with $(\mathbf{m}_1 - \mathbf{m}_2)$. On the other hand, in (b), the optimal axis y deviates from the direction $(\mathbf{m}_1 - \mathbf{m}_2)$. The term to compensate for this deviation is nothing but $\mathbf{S}_W^{-1}$ in Eq. (6.106). The optimal axis direction $(\mathbf{m}_1 - \mathbf{m}_2)$ in the case of isotropic within-class scatter is the principal axis obtained from the KL expansion. Therefore, we can see that Eq. (6.106) is the product of the KL expansion term and the term that

---

[17] The vector $\boldsymbol{w}$ is meaningful only in that direction, so $\boldsymbol{w} = \mathbf{S}_W^{-1}(\mathbf{m}_1 - \mathbf{m}_2)$ is acceptable. Similarly, Eq. (6.109) is also acceptable as $\boldsymbol{w} = \mathbf{m}_1 - \mathbf{m}_2$. In fact, Fig. 6.6 was set up in such a way. However, it is easier to handle by normalizing $\boldsymbol{w}$ as $\|\boldsymbol{w}\| = 1$.

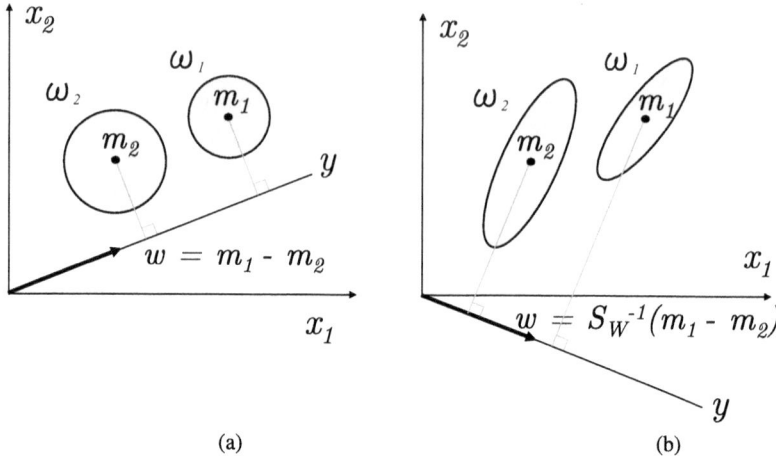

Fig. 6.6 Vector w obtained by Fisher's method. (**a**) Within-class scatter is isotropic. (**b**) Within-class scatter is not isotropic

corrects for the spread of the distribution. The details are described in Sect. 6.4.3 of this section. See also Problem 6.3.

Once the optimal axis y is obtained by Fisher's method, we need to project the patterns of classes ω_1 and ω_2 onto the y axis and separate both classes. If w obtained by Eq. (6.106) is normalized to $\|w\| = 1$, then the coordinate value y of the projection of the pattern x onto this axis is $y = w^t x$. Therefore, to separate the two classes on the y axis, an appropriate threshold $-w_0$ is set on the y axis as a decision boundary, and the two classes are classified by

$$w^t x > -w_0 \quad \text{or} \quad w^t x < -w_0. \tag{6.110}$$

This process is nothing other than classifying the two classes based on the sign of the following linear discriminant function

$$g(x) = w^t x + w_0. \tag{6.111}$$

In this case, the decision boundary defined by $g(x) = 0$ represents a hyperplane orthogonal to w that intersects the y axis at $y = -w_0$ in a d-dimensional space.

So far, we have introduced the perceptron learning rule and the Widrow–Hoff learning rule as methods for finding the optimal linear discriminant function. It can be seen that Fisher's method described here is also a method for finding the optimal linear discriminant function. Like the Widrow–Hoff learning rule, this method can be applied regardless of whether the learning patterns are linearly separable or not.

One point in Fisher's method requires special attention. That is, while the perceptron learning rule and the Widrow–Hoff learning rule can obtain both w and

w_0 of Eq. (6.111) by learning, Fisher's method can only obtain $\boldsymbol{w}$, and w_0 must be obtained separately by other methods.[18]

See Sect. 4.3.1 for a method of setting a decision boundary on the axes. To compare the perceptron learning rules with Fisher's method, Problem 2.3 and Problem 6.3 are instructive.

The importance of feature normalization was described in Sect. 6.2. However, normalization is not necessary when the optimal axes are obtained by Fisher's method. The result obtained is the same whether normalization is used or not (Problem 6.4). However, one should be aware that the KL expansion is affected by normalization. This has already been described in Sect. 6.2.

6.4.1.2 Maximum Ratio of Between-Class Variance to Within-Class Variance Criterion

In order to express Fisher's method in a more general form, let us try to formulate it using the covariance matrix and the prior probability $P(\omega_i)$ instead of the scatter matrix. The covariance matrix $\boldsymbol{\Sigma}_i$ of patterns belonging to class ω_i is defined by

$$\boldsymbol{\Sigma}_i \overset{\text{def}}{=} \frac{1}{n_i} \sum_{x \in X_i} (\boldsymbol{x} - \mathbf{m}_i)(\boldsymbol{x} - \mathbf{m}_i)^t = \frac{1}{n_i}\mathbf{S}_i. \tag{6.112}$$

Furthermore, let the *within-class covariance matrix* $\boldsymbol{\Sigma}_W$ and the *between-class covariance matrix* $\boldsymbol{\Sigma}_B$ be defined respectively as

$$\boldsymbol{\Sigma}_W \overset{\text{def}}{=} \sum_{i=1,2} P(\omega_i)\boldsymbol{\Sigma}_i \tag{6.113}$$

$$= \sum_{i=1,2} \left(P(\omega_i)\frac{1}{n_i} \sum_{x \in X_i} (\boldsymbol{x} - \mathbf{m}_i)(\boldsymbol{x} - \mathbf{m}_i)^t \right), \tag{6.114}$$

$$\boldsymbol{\Sigma}_B \overset{\text{def}}{=} \sum_{i=1,2} P(\omega_i)(\mathbf{m}_i - \mathbf{m})(\mathbf{m}_i - \mathbf{m})^t \tag{6.115}$$

$$= P(\omega_1)P(\omega_2)(\mathbf{m}_1 - \mathbf{m}_2)(\mathbf{m}_1 - \mathbf{m}_2)^t. \tag{6.116}$$

Similar quantities ϕ_W and ϕ_B can be obtained in the space transformed by the vector $\boldsymbol{w}$. From Eqs. (6.84) and (6.86), they are expressed as

---

[18] Textbooks on discriminant analysis in statistics often proceed with discussions based on the assumption that the feature vectors of each class follow a normal distribution. If we apply the linear discriminant method under this assumption, the boundary as well as the axes are uniquely determined, and as a result, we have designed a discriminant function. This corresponds to the parametric learning described in Sect. 4.1.

$$\phi_W = P(\omega_1)\tilde{\sigma}_1^2 + P(\omega_2)\tilde{\sigma}_2^2 \tag{6.117}$$

$$= \sum_{i=1,2} \left(P(\omega_i)\frac{1}{n_i}\sum_{y\in\mathcal{Y}_i}(y - \tilde{m}_i)^2 \right) \tag{6.118}$$

$$= \boldsymbol{w}^t \boldsymbol{\Sigma}_W \boldsymbol{w}, \tag{6.119}$$

$$\phi_B = P(\omega_1)P(\omega_2)(\tilde{m}_1 - \tilde{m}_2)^2 \tag{6.120}$$

$$= P(\omega_1)P(\omega_2)\boldsymbol{w}^t(\mathbf{m}_1 - \mathbf{m}_2)(\mathbf{m}_1 - \mathbf{m}_2)^t \boldsymbol{w} \tag{6.121}$$

$$= \boldsymbol{w}^t \boldsymbol{\Sigma}_B \boldsymbol{w}. \tag{6.122}$$

As in Sect. 6.4.1.1, ϕ_W and ϕ_B are scalar quantities, which are called the within-class variance and the between-class variance in the transformed one-dimensional space, respectively. As can be seen from the definitions, the within-class variance is the weighted sum of the pattern variances for each class, and the between-class variance is the weighted distance between the two class means. Therefore, for the transformed space to be effective in separating two classes, the within-class variance should be as small as possible and the between-class variance should be as large as possible.

Therefore, we define the evaluation function J_Σ, which represents the degree of separation between classes in the transformed space, as the ratio of between-class variance to within-class variance. This evaluation criterion was introduced as a feature evaluation method in Sect. 5.2. The evaluation function J_Σ can be expressed as a function of the vector $\boldsymbol{w}$ as

$$J_\Sigma(\boldsymbol{w}) \stackrel{\text{def}}{=} \frac{\phi_B}{\phi_W} = \frac{P(\omega_1)P(\omega_2)(\tilde{m}_1 - \tilde{m}_2)^2}{P(\omega_1)\tilde{\sigma}_1^2 + P(\omega_2)\tilde{\sigma}_2^2} \tag{6.123}$$

$$= \frac{\boldsymbol{w}^t \boldsymbol{\Sigma}_B \boldsymbol{w}}{\boldsymbol{w}^t \boldsymbol{\Sigma}_W \boldsymbol{w}}. \tag{6.124}$$

The above equation corresponds to Eq. (6.98). That is, Fisher's method can be generalized as a method based on the *maximum ratio of between-class variance to within-class variance criterion*.

By exactly the same procedure as in Sect. 6.4.1.1, the maximum value of $J_\Sigma(\boldsymbol{w})$ equals the maximum eigenvalue λ_1 of

$$\boldsymbol{\Sigma}_W^{-1}\boldsymbol{\Sigma}_B, \tag{6.125}$$

and the eigenvector corresponding to λ_1 is $\boldsymbol{w}$ that maximizes J_Σ, that is

$$\max\{J_\Sigma(\boldsymbol{w})\} = \lambda_1. \tag{6.126}$$

Furthermore, from Eq. (6.124), we obtain

$$w \propto \Sigma_W^{-1}(\mathbf{m}_1 - \mathbf{m}_2), \tag{6.127}$$

which is similar to Eq. (6.106). In the function J introduced in Eq. (4.40), the result of setting $k_1 = P(\omega_1)$ and $k_2 = P(\omega_2)$ corresponds to $J_\Sigma(w)$ of Eq. (6.123).

In the case of two classes ($c = 2$), using the *total covariance matrix* Σ_T ($= \Sigma_W + \Sigma_B$) (see Eq. (6.161)),

$$w \propto \Sigma_T^{-1}(\mathbf{m}_1 - \mathbf{m}_2) \tag{6.128}$$

is established.[19] In fact, placing $\mathbf{m}_d \overset{\text{def}}{=} \mathbf{m}_1 - \mathbf{m}_2$ and noting that $\mathbf{m}_d^t \Sigma_W^{-1} \mathbf{m}_d$ is a scalar quantity, the following equation holds:

$$\Sigma_T \Sigma_W^{-1} \mathbf{m}_d = (\Sigma_W + \Sigma_B) \Sigma_W^{-1} \mathbf{m}_d \tag{6.129}$$

$$= \left(\Sigma_W + P(\omega_1)P(\omega_2)\mathbf{m}_d\mathbf{m}_d^t\right) \Sigma_W^{-1} \mathbf{m}_d \tag{6.130}$$

$$= \mathbf{m}_d + P(\omega_1)P(\omega_2)\mathbf{m}_d\mathbf{m}_d^t \Sigma_W^{-1} \mathbf{m}_d \tag{6.131}$$

$$= \mathbf{m}_d + P(\omega_1)P(\omega_2)\mathbf{m}_d^t \Sigma_W^{-1} \mathbf{m}_d\mathbf{m}_d \tag{6.132}$$

$$= \left(1 + P(\omega_1)P(\omega_2)\mathbf{m}_d^t \Sigma_W^{-1} \mathbf{m}_d\right) \mathbf{m}_d. \tag{6.133}$$

From this, we obtain

$$\Sigma_W^{-1} \mathbf{m}_d = \left(1 + P(\omega_1)P(\omega_2)\mathbf{m}_d^t \Sigma_W^{-1} \mathbf{m}_d\right) \Sigma_T^{-1} \mathbf{m}_d \tag{6.134}$$

which leads to Eq. (6.128) from Eq. (6.127).

6.4.1.3 Maximum Value of J_S, J_Σ and Mahalanobis Distance

The Mahalanobis distance is essentially a quantity that represents the distance between two points x_1 and x_2 in a distribution characterized by a covariance matrix Σ. The Mahalanobis distance $D_M(x_1, x_2)$ is defined by

$$D_M^2(x_1, x_2) \overset{\text{def}}{=} (x_1 - x_2)^t \Sigma^{-1}(x_1 - x_2). \tag{6.135}$$

It can be regarded as the distance normalized by the covariance matrix Σ. The Mahalanobis distance between the means of two distributions with equal covariance matrices can be expressed as

$$D_M^2(\mathbf{m}_1, \mathbf{m}_2) \overset{\text{def}}{=} (\mathbf{m}_1 - \mathbf{m}_2)^t \Sigma^{-1}(\mathbf{m}_1 - \mathbf{m}_2). \tag{6.136}$$

---

[19] When $P(\omega_i) = n_i/n$, the total covariance matrix Σ_T coincides with the covariance matrix Σ.

By replacing Σ^{-1} with Σ_W^{-1} in this Mahalanobis distance, we can extend it to the distance between means of distributions with different covariance matrices.[20] In this book, $D_M(\mathbf{m}_i, \mathbf{m}_j)$ denotes the Mahalanobis distance between means of distributions of class ω_i and class ω_j. Figure 6.7(a)–(c) indicate three pairs of normal distributions with equal *Euclidean distance* between the means of the two distributions. Letting D_a, D_b and D_c be the Mahalanobis distances for (a), (b), and (c), respectively, they are represented as follows:

$$D_a^2 = \frac{(m_1 - m_2)^2}{\sigma_1^2}, \tag{6.137}$$

$$D_b^2 = \frac{(m_1 - m_2)^2}{\sigma_2^2}, \tag{6.138}$$

$$D_c^2 = \frac{(m_1 - m_2)^2}{P(\omega_1)\sigma_1^2 + P(\omega_2)\sigma_2^2}. \tag{6.139}$$

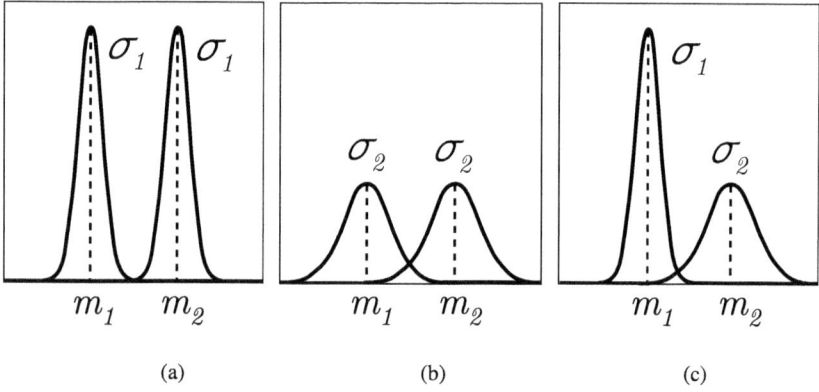

(a)   (b)   (c)

Fig. 6.7 Euclidean distance and Mahalanobis distance

It is clear from these equations that the Euclidean distances between the means are $|m_1 - m_2|$ in all cases, but for the Mahalanobis distances, the relation $D_a \geq D_c \geq D_b$ holds.

Between the Mahalanobis distance $D_M(\mathbf{m}_1, \mathbf{m}_2)$ and the maximum value of the evaluation function $J_\Sigma(\mathbf{w})$, the following equation holds:

$$\lambda_1 = \max\{J_\Sigma(\mathbf{w})\} \tag{6.140}$$

---

[20] The covariance matrix of the entire two different distributions is represented by Σ_W, which is obtained by weighting and adding the respective covariance matrices by $P(\omega_i)$, as shown in Eq. (6.113).

$$= P(\omega_1)P(\omega_2)D_M^2(\mathbf{m}_1, \mathbf{m}_2) \tag{6.141}$$

$$= P(\omega_1)D_M^2(\mathbf{m}_1, \mathbf{m}) + P(\omega_2)D_M^2(\mathbf{m}_2, \mathbf{m}). \tag{6.142}$$

Furthermore, under the constraint[21]

$$\boldsymbol{w}^t \boldsymbol{\Sigma}_W \boldsymbol{w} = 1, \tag{6.143}$$

the next equation holds (Problem 6.6):

$$D_M^2(\mathbf{m}_1, \mathbf{m}_2) = (\tilde{m}_1 - \tilde{m}_2)^2. \tag{6.144}$$

If the number of patterns in each class can be assumed to reflect the prior probability[22] of that class, that is,

$$P(\omega_i) = \frac{n_i}{n} \quad (i = 1, \ldots, c) \tag{6.145}$$

holds ($c = 2$ in this case), then since $\boldsymbol{\Sigma}_W$ and $\boldsymbol{\Sigma}_B$ are expressed as

$$\boldsymbol{\Sigma}_W = \frac{1}{n}\mathbf{S}_W, \tag{6.146}$$

$$\boldsymbol{\Sigma}_B = \frac{1}{n}\mathbf{S}_B, \tag{6.147}$$

we obtain

$$J_\Sigma(\boldsymbol{w}) = J_S(\boldsymbol{w}). \tag{6.148}$$

6.4.2 Linear Discriminant Method for Multi-Class Problems

The previous subsection described a method for converting a d-dimensional feature space consisting of two-class feature vectors into a one-dimensional space. This subsection describes a linear discriminant method that is extended to the case of multiple classes ($c > 2$). In the linear discriminant method for the multi-class case, the d-dimensional feature space is reduced to a $\tilde{d}$ ($\leq c - 1$)-dimensional space by a transformation.[23]

---

[21] This constraint corresponds to considering the matrix $\mathbf{A}$ of Eq. (6.175) as a vector $\boldsymbol{w}$ of $d \times 1$. If $\boldsymbol{\Sigma}_W = \mathbf{I}$, this constraint matches the usual normalization condition $\|\boldsymbol{w}\| = 1$.

[22] Regarding the prior probability, refer to coffee breaks in this section and in Sect. 14.2.3.

[23] Strictly speaking, $\tilde{d} < d$ must also be satisfied, but this condition is omitted because it is self-evident. The same applies hereafter.

Let us first consider the case $\tilde{d} = 1$. That is, we extend the two-class problem of Fisher's method to a multi-class problem.

Even in the multi-class case, the evaluation criterion used is the same as Eq. (6.98), but the $\sum_{i=1,2}$ of Eqs. (6.81) and (6.82) must be $\sum_{i=1}^{c}$. That is,

$$\mathbf{S}_W = \sum_{i=1}^{c} \mathbf{S}_i \tag{6.149}$$

$$= \sum_{i=1}^{c} \sum_{x \in X_i} (x - \mathbf{m}_i)(x - \mathbf{m}_i)^t, \tag{6.150}$$

$$\mathbf{S}_B = \sum_{i=1}^{c} n_i (\mathbf{m}_i - \mathbf{m})(\mathbf{m}_i - \mathbf{m})^t. \tag{6.151}$$

In the multi-class case, it is important to note that $\mathbf{S}_B$ cannot be transformed like Eq. (6.83).

In the multi-class case, Eq. (6.102) is also derived as a condition to be satisfied by $\mathbf{S}_B$ and $\mathbf{S}_W$, and it becomes an eigenvalue problem. As already mentioned, Fisher's method for two classes does not require solving the eigenvalue problem, and Eq. (6.106) can be used to directly obtain w. However, in the multi-class case, the eigenvalue problem of Eq. (6.102) must be solved.

Here, as a concrete example, let us take the case where learning patterns of five classes ($c = 5$): $\omega_1, \omega_2, \ldots, \omega_5$ are distributed in a two-dimensional ($d = 2$) feature space. The distribution of the patterns is shown in Fig. 6.8.

Fig. 6.8 Five classes distributed in a two-dimensional feature space

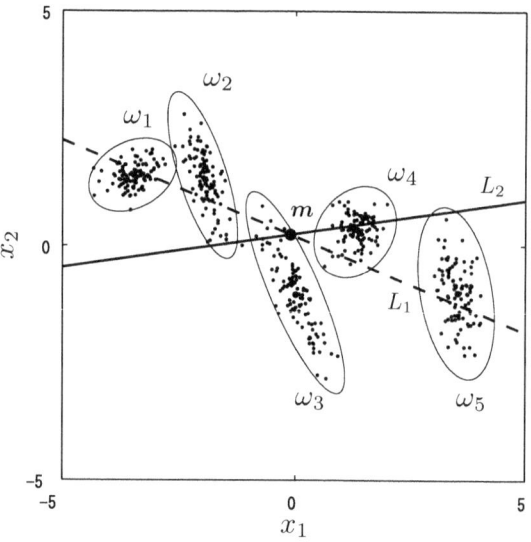

The number of patterns is 100 per class, and they are distributed according to two-dimensional normal distributions. The two-dimensional normal distributions have different means and covariance matrices for each class, and the contours of the distribution are indicated by thin lines. As is clear from the figure, these distributions are not isotropic.

The principal axis (L_1) obtained by applying the KL expansion to this data is shown by the chain line. Also, among the eigenvectors obtained by solving the eigenvalue problem of Eq. (6.102), the axis (L_2) of the eigenvector corresponding to the maximum eigenvalue is shown by the solid line. Both axes are set to pass through the mean vector **m** (a black circle in the figure) of all patterns.

The distributions of the patterns projected onto the two axes, L_1 and L_2, are shown in Fig. 6.9. On the L_1 axis, there is some overlap between classes, while on the L_2 axis, the classes are relatively well separated. Thus, it is confirmed that the linear discriminant method is also effective in classifying multiple classes.

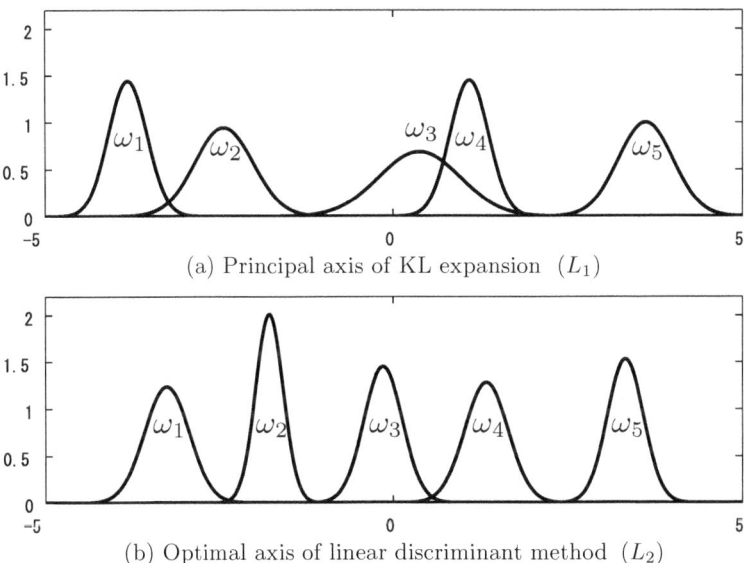

(a) Principal axis of KL expansion (L_1)

(b) Optimal axis of linear discriminant method (L_2)

Fig. 6.9 Projection onto two axes L_1 and L_2

In the linear discriminant method for multiple classes, a d-dimensional feature space is reduced to a $\tilde{d}$-dimensional space ($\tilde{d} \leq c - 1$) by a transformation. Let $d \times \tilde{d}$ matrix representing this transformation be denoted by **A**.[24] As in the two-class case, we define the within-class covariance matrix Σ_W[25] and the between-class covariance matrix Σ_B as (Problem 6.6)

---

[24] If the matrix size is $d \times 1$, it coincides with **w** in Eq. (6.84).

[25] Instead of Σ_W, its unbiased estimator is sometimes used, but since n is usually sufficiently large, there is no big difference in practical use.

$$\Sigma_W \overset{\text{def}}{=} \sum_{i=1}^{c} P(\omega_i)\Sigma_i \tag{6.152}$$

$$= \sum_{i=1}^{c} \left(P(\omega_i)\frac{1}{n_i} \sum_{x \in \mathcal{X}_i} (x - m_i)(x - m_i)^t \right), \tag{6.153}$$

$$\Sigma_B \overset{\text{def}}{=} \sum_{i=1}^{c} P(\omega_i)(m_i - m)(m_i - m)^t \tag{6.154}$$

$$= \frac{1}{2}\sum_{i=1}^{c}\sum_{j=1}^{c} P(\omega_i)P(\omega_j)(m_i - m_j)(m_i - m_j)^t \tag{6.155}$$

$$= \sum_{i=1}^{c}\sum_{j<i} P(\omega_i)P(\omega_j)(m_i - m_j)(m_i - m_j)^t. \tag{6.156}$$

The within-class scatter matrix S_W, and the between-class scatter matrix S_B can be defined in the same way, as follows:

$$S_W \overset{\text{def}}{=} \sum_{i=1}^{c}\sum_{x \in \mathcal{X}_i} (x - m_i)(x - m_i)^t \tag{6.157}$$

$$= \sum_{i=1}^{c} S_i = \sum_{i=1}^{c} n_i \Sigma_i, \tag{6.158}$$

$$S_B \overset{\text{def}}{=} \sum_{i=1}^{c} n_i(m_i - m)(m_i - m)^t. \tag{6.159}$$

Between Σ_W, Σ_B, S_W, and S_B defined in this way and the total covariance matrix Σ_T and the *total scatter matrix* S_T, the following relation holds:

$$\Sigma_T \overset{\text{def}}{=} \sum_{i=1}^{c} \left(P(\omega_i)\frac{1}{n_i} \sum_{x \in \mathcal{X}_i} (x - m)(x - m)^t \right) \tag{6.160}$$

$$= \Sigma_W + \Sigma_B \tag{6.161}$$

$$S_T \overset{\text{def}}{=} \sum_{x \in \mathcal{X}} (x - m)(x - m)^t \tag{6.162}$$

$$= S_W + S_B. \tag{6.163}$$

When $P(\omega_i) = n_i/n$ holds for the prior probability as described in the previous subsection, the following relation holds:

$$\Sigma_W = \frac{1}{n}S_W, \tag{6.164}$$

$$\Sigma_B = \frac{1}{n}S_B. \tag{6.165}$$

Therefore, since the formulation using the scatter matrix corresponds to the special case where $P(\omega_i) = n/n_i$ in the formulation using the covariance matrix, we will use the expression using the covariance matrix from now on.

─────────────── Coffee Break ───────────────

How to Determine Prior Probabilities

In the definition of the within-class covariance matrix (Eq. (6.114)), there exists a prior probability $P(\omega_i)$ as an unknown parameter. Then, when Eq. (6.145) holds, as described in Sect. 6.4.1, the subspace obtained from the maximum ratio of between-class scatter to within-class scatter criterion and the subspace obtained from the maximum ratio of between-class variance to within-class variance criterion are the same. Therefore, the former can be regarded as a special case of the latter.

In Fisher's method, the prior probability $P(\omega_i)$ does not appear in the equation, but how is the prior probability reflected in the equation? As can be seen from Eqs. (6.81) and (6.83), the sums S_W and S_B are in the form of adding each term for the number of patterns. Therefore, the class with more n_i patterns has a larger contribution to S_W and S_B, which has the same effect as if $P(\omega_i) = n_i/n$ in the covariance matrix representation. Then, as a practical matter, how valid is it to set $P(\omega_i) = n_i/n$? Unfortunately, this question depends on the problem.

In practice, the prior probability $P(\omega_i)$ is often determined by one of the following methods:

1. Setting $P(\omega_i) = n_i/n$

   If the patterns are randomly sampled from the population, n_i/n is proportional to $P(\omega_i)$, which is a very natural method. However, it is rather rare to achieve perfect *random sampling*, and in many cases there is considerable bias in the collected patterns.

2. Setting $P(\omega_i) = 1/c$

   This method is based on the premise that it is impossible to estimate prior probabilities and treats each class equally. This method is often adopted in character recognition.

3. Estimating $P(\omega_i)$ by a completely different method

   When one wants to classify a particular class more correctly, such as when the loss associated with a particular misclassification is significant, the solution is often to have more learning patterns for that class. This is equivalent to estimating a larger $P(\omega_i)$ for a particular class. The problem of how to determine the prior probability is also related to the design of the classifier, see also coffee breaks in Sect. 5.4.1, in Sect. 14.2.3, and in Sect. 15.1.3.

───

Let us find the within-class covariance matrix $\tilde{\Sigma}_W$ and the between-class covariance matrix $\tilde{\Sigma}_B$ in the transformed space. Since $y = A^t x$, the vector w of $d \times 1$ shown in Eqs. (6.119) and (6.122) is replaced by the matrix A of $d \times \tilde{d}$, resulting in

$$\tilde{\Sigma}_W = \mathbf{A}'\Sigma_W\mathbf{A}, \tag{6.166}$$

$$\tilde{\Sigma}_B = \mathbf{A}'\Sigma_B\mathbf{A}, \tag{6.167}$$

$$\tilde{\Sigma}_T = \mathbf{A}'\Sigma_T\mathbf{A}. \tag{6.168}$$

Here, we need an evaluation function $J(\mathbf{A})$ to evaluate the degree of separation between classes in the transformed space. Unlike ϕ_W and ϕ_B of Eqs. (6.119) and (6.122) defined for the two-class case, $\tilde{\Sigma}_W$ and $\tilde{\Sigma}_B$ are not scalar quantities but square matrices of order $\tilde{d}$. Therefore, the following evaluation criteria $J_i(\mathbf{A})$ ($i = 1, \ldots, 4$) are considered as candidates:

$$J_1(\mathbf{A}) \stackrel{\text{def}}{=} \frac{\text{tr}\left(\tilde{\Sigma}_B\right)}{\text{tr}\left(\tilde{\Sigma}_W\right)}, \tag{6.169}$$

$$J_2(\mathbf{A}) \stackrel{\text{def}}{=} \text{tr}\left(\tilde{\Sigma}_W^{-1}\tilde{\Sigma}_B\right), \tag{6.170}$$

$$J_3(\mathbf{A}) \stackrel{\text{def}}{=} \frac{\det\left(\tilde{\Sigma}_B\right)}{\det\left(\tilde{\Sigma}_W\right)} = \det\left(\tilde{\Sigma}_W^{-1}\tilde{\Sigma}_B\right), \tag{6.171}$$

$$J_4(\mathbf{A}) \stackrel{\text{def}}{=} \log\left(\frac{\det(\tilde{\Sigma}_T)}{\det(\tilde{\Sigma}_W)}\right). \tag{6.172}$$

Here, $\det(\mathbf{X})$ denotes the determinant of the matrix $\mathbf{X}$ (See Appendix B). The evaluation function J_4 uses $\tilde{\Sigma}_T$ instead of $\tilde{\Sigma}_B$ because $\det(\tilde{\Sigma}_B)$ can be 0.

In the case of Fisher's method, the transformation was from a d-dimensional space to a one-dimensional space, while the transformation discussed here is from a d-dimensional space to a $\tilde{d}$-dimensional space. Even in this case, the idea is the same, i.e., to aim for a transformation with as small within-class variance as possible and as small between-class variance as possible. All the evaluation functions shown in Eqs. (6.169)–(6.172) reflect the above idea in $\tilde{d}$-dimensional spaces However, since they are applied to a multidimensional space, it is not possible to simply use the ratio of between-class variance to within-class variance on a one-dimensional axis as in Eq. (6.124).

The spread of a distribution on an axis is represented by the variance of the projection of the distribution on that axis. The variance is nothing but the eigenvalue obtained from the covariance matrix Σ of the distribution. If $\tilde{d}$ eigenvalues obtained from each axis are $\lambda_1, \ldots, \lambda_{\tilde{d}}$, then the spread of distribution in a $\tilde{d}$-dimensional space can be evaluated by $\sum_{i=1}^{\tilde{d}} \lambda_i$ or $\prod_{i=1}^{\tilde{d}} \lambda_i$. Furthermore, since equations

$$\sum_{i=1}^{\tilde{d}} \lambda_i = \text{tr}(\tilde{\Sigma}), \tag{6.173}$$

$$\prod_{i=1}^{\tilde{d}} \lambda_i = \det\left(\tilde{\Sigma}\right) \tag{6.174}$$

hold, $J_1(\mathbf{A})$–$J_4(\mathbf{A})$ are valid as evaluation functions, where $\tilde{\Sigma}$ is the covariance matrix of the pattern set in the transformed space.

It is known that these maximization problems are equivalent to maximizing the numerator of $J_1(\mathbf{A})$–$J_4(\mathbf{A})$ under the condition

$$\tilde{\Sigma}_W - \mathbf{A}^t \Sigma_W \mathbf{A} = \mathbf{I}, \tag{6.175}$$

and can be attributed to exactly the same eigenvalue problem[26]

$$\Sigma_B \mathbf{A} = \Sigma_W \mathbf{A} \Lambda, \tag{6.176}$$

where Λ is a $\tilde{d}$-dimensional diagonal matrix. Therefore, the eigenvectors corresponding to the larger $\tilde{d}$ eigenvalues $\lambda_1, \ldots, \lambda_{\tilde{d}}$ of $\Sigma_W^{-1} \Sigma_B$ are the basis for the space after the transformation. In general, eigenvalues are uniquely determined in eigenvalue problems, but eigenvectors are not.[27] Therefore, it is desirable to impose some constraints on the eigenvectors so that the solution can be uniquely determined.[28] Here, the normalization condition of Eq. (6.175) is imposed as a constraint.

The maximum values of $J_i(\mathbf{A})$ shown in Eqs. (6.169)–(6.172) are

$$\max\{J_1(\mathbf{A})\} = \frac{1}{\tilde{d}} \sum_{i=1}^{\tilde{d}} \lambda_i, \tag{6.177}$$

$$\max\{J_2(\mathbf{A})\} = \sum_{i=1}^{\tilde{d}} \lambda_i, \tag{6.178}$$

$$\max\{J_3(\mathbf{A})\} = \prod_{i=1}^{\tilde{d}} \lambda_i, \tag{6.179}$$

---

[26] Since Eq. (6.161) is valid, even if the combination of $(\tilde{\Sigma}_B, \tilde{\Sigma}_W)$ in $J(A)$ (Eqs. (6.169)–(6.172)) is replaced by $(\tilde{\Sigma}_T, \tilde{\Sigma}_W)$ or $(\tilde{\Sigma}_B, \tilde{\Sigma}_T)$, the essence of the problem does not change. Because if we transform Eq. (6.176) using the relation $\Sigma_T = \Sigma_W + \Sigma_B$, we obtain $\Sigma_B \mathbf{A}(\mathbf{I} + \Lambda) = \Sigma_T \mathbf{A} \Lambda$, and Eq. (6.176) can be transformed as $\Sigma_B \mathbf{A} = \Sigma_T \mathbf{A} \Lambda'$. Therefore, Eq. (6.176) and the above equation are equivalent eigenvalue problems. Here, Λ' is a $\tilde{d}$-dimensional diagonal matrix and if (i, i) components of Λ and Λ' are λ_i and λ_i', respectively, then $\lambda_i' = \lambda_i/(1 + \lambda_i)$.

[27] For example, in the eigenvalue problem $\Sigma x = \lambda x$ for the matrix Σ, if x is an eigenvector, then ax multiplied by a constant a is also an eigenvector.

[28] Often applied is a normalization process that sets the norm of the eigenvectors to 1.

$$\max\{J_4(\mathbf{A})\} = \sum_{i=1}^{\tilde{d}} \log(\lambda_i + 1), \tag{6.180}$$

using eigenvalues of $\boldsymbol{\Sigma}_W^{-1}\boldsymbol{\Sigma}_B$, respectively. Refer to Problem 6.5 for the proof.

In terms of the dimensionality reduction of the feature space, the dimensionality of the reduced space does not necessarily have to be $\tilde{d}$ ($= c - 1$). We can choose a subspace spanned by the eigenvectors corresponding to an arbitrary number of eigenvalues (up to $\tilde{d}$) starting from the largest one. In Sect. 6.5.2, we describe how to evaluate the subspace obtained by the KL expansion using the cumulative contribution ratio. Similarly, the evaluation value J described above can be regarded as a quantity to evaluate the discriminant ability of the space.

From the viewpoint that each axis can be treated independently and its evaluation value has the *additive property*,[29] J_2 and J_4 are preferable as the evaluation values of the discriminant ability of the space. The value J_1 represents the average evaluation of each axis divided by the dimensionality of the subspace. Therefore, when a new subspace is created by adding a new eigenvector corresponding to a small eigenvalue, the value of J conversely decreases, even though the actual discriminant ability increases by that amount. In addition, for J_3, when eigenvectors corresponding to eigenvalues close to 0 are adopted, its evaluation value approaches 0. Furthermore, J_2, J_3, and J_4 are invariant to the regular linear transformation, which is desirable as an evaluation of the discriminant ability of the space. Therefore, these evaluations are not affected by the feature normalization described in Sect. 6.2.

Since Eq. (6.174) holds, J_3 is the most natural generalization for the two-class case. The following equation holds between the maximum value of J_2 and the Mahalanobis distance (Problem 6.6):

$$\max\{J_2(\mathbf{A})\}$$
$$= \sum_{i=1}^{c} P(\omega_i)D_M^2(\mathbf{m}_i, \mathbf{m}) = \sum_{i=1}^{c}\sum_{j<i} P(\omega_i)P(\omega_j)D_M^2(\mathbf{m}_i.\mathbf{m}_j). \tag{6.181}$$

No matter which J introduced here is chosen, the eigenvector that maximizes J is the same, so the obtained subspace does not depend on how J is chosen. However, when J is used to evaluate the discriminant ability of a space or to evaluate the features shown as an example in Sect. 5.2, it is necessary to choose an appropriate J. In addition, although omitted from this book, in the clustering method (see also footnote 10), which divides a set of patterns into several clusters, the evaluation values described here are used to determine the optimal clusters.

---

[29] When $J(a_1)$ and $J(a_2)$ are the evaluation values for the axes (one-dimensional space) a_1 and a_2, respectively, and $J(a_1, a_2)$ is the evaluation value for the two-dimensional subspace spanned by a_1 and a_2, the additive property means that $J(a_1, a_2) = J(a_1) + J(a_2)$ holds.

In the case of multi-class, the subspace obtained as above may not always have sufficient discriminant ability in terms of separation between classes. Therefore, care must be taken when using the linear discriminant method for multi-class. This has already been discussed in Sect. 5.2. It is more reliable to perform multi-class classification by combining Fisher's method for two-class described in Sect. 6.4.1, just as the two-class discriminant functions are extended to multi-class in Sect. 4.3.2.

6.4.3 Linear Discriminant Method and Spatial Transformations

As described so far, the linear discriminant method is a method for finding a $\tilde{d}$ ($\leq c - 1$)-dimensional subspace that maximizes the ratio of between-class variance to within-class variance by a linear transformation of the space. This is a transformation considering classification. Here, we discuss the meaning of this transformation, represented by the $d \times \tilde{d}$ matrix $\mathbf{A}$, in terms of spatial transformation. In fact, this transformation $\mathbf{A}$ can be written as $\mathbf{A} = \mathbf{A}_1 \mathbf{A}_2$ using certain special transformations $\mathbf{A}_1$ and $\mathbf{A}_2$, that is, the transformation can be divided into two stages.

Since $\mathbf{\Sigma}_W$ is a symmetric matrix, there exists a d-order square matrix $\mathbf{A}_1$ satisfying

$$\mathbf{A}_1^t \mathbf{\Sigma}_W \mathbf{A}_1 = \mathbf{I}_d. \tag{6.182}$$

Here, $\mathbf{I}_d$ is a d-dimensional identity matrix (Problem 6.7). On the other hand, since $\mathbf{\Sigma}_B$ is a positive semidefinite symmetric matrix[30] of rank $\tilde{d}$, $\mathbf{A}_1^t \mathbf{\Sigma}_B \mathbf{A}_1$ is also a positive semidefinite symmetric matrix. As a result, if the $d \times \tilde{d}$ matrix whose columns are the $\tilde{d}$ eigenvectors of $\mathbf{A}_1^t \mathbf{\Sigma}_B \mathbf{A}_1$ is $\mathbf{A}_2$, and $\tilde{d}$-dimensional diagonal matrix with corresponding eigenvalues as components is $\mathbf{\Lambda}$, then

$$\left(\mathbf{A}_1^t \mathbf{\Sigma}_B \mathbf{A}_1 \right) \mathbf{A}_2 = \mathbf{A}_2 \mathbf{\Lambda} \tag{6.183}$$

$$\mathbf{A}_2^t \mathbf{A}_2 = \mathbf{I}_{\tilde{d}} \tag{6.184}$$

holds. Using $\mathbf{A}_1$ and $\mathbf{A}_2$ thus defined and defining $\mathbf{A}$ as

$$\mathbf{A} \stackrel{\text{def}}{=} \mathbf{A}_1 \mathbf{A}_2, \tag{6.185}$$

this $\mathbf{A}$ satisfies the conditions of the eigenvalue problem presented in Eqs. (6.175) and (6.176), as shown below.

Using Eq. (6.182),

---

[30] A symmetric matrix $\mathbf{Y}$ is positive semidefinite, if there exists a matrix $\mathbf{X}$ such that $\mathbf{Y} = \mathbf{X}^t \mathbf{X}$.

$$\tilde{\boldsymbol{\Sigma}}_W = \mathbf{A}^t \boldsymbol{\Sigma}_W \mathbf{A} = \mathbf{A}_2^t (\mathbf{A}_1^t \boldsymbol{\Sigma}_W \mathbf{A}_1) \mathbf{A}_2 \tag{6.186}$$

$$= \mathbf{A}_2^t \mathbf{I}_d \mathbf{A}_2 = \mathbf{A}_2^t \mathbf{A}_2 = \mathbf{I}_{\tilde{d}} \tag{6.187}$$

holds, thus Eq. (6.175) is shown. Furthermore, since

$$(\mathbf{A}_1^t)^{-1} = \boldsymbol{\Sigma}_W \mathbf{A}_1 \tag{6.188}$$

holds from Eq. (6.182), multiplying this by Eq. (6.183) from the left yields

$$\text{the left side} = (\mathbf{A}_1^t)^{-1} \left(\mathbf{A}_1^t \boldsymbol{\Sigma}_B \mathbf{A}_1 \right) \mathbf{A}_2 = \boldsymbol{\Sigma}_B \mathbf{A}, \tag{6.189}$$

$$\text{the right side} = \boldsymbol{\Sigma}_W \mathbf{A}_1 \mathbf{A}_2 \boldsymbol{\Lambda} = \boldsymbol{\Sigma}_W \mathbf{A} \boldsymbol{\Lambda}. \tag{6.190}$$

Then, Eq. (6.176), that is,

$$\boldsymbol{\Sigma}_B \mathbf{A} = \boldsymbol{\Sigma}_W \mathbf{A} \boldsymbol{\Lambda} \tag{6.191}$$

is also shown.

The above results show that the transformation $\mathbf{A}$ can be divided into two stages, $\mathbf{A}_1$ and $\mathbf{A}_2$, as mentioned earlier. The first transformation $\mathbf{A}_1$ is a transformation that performs the normalization of the within-class variance $\boldsymbol{\Sigma}_W$ as shown by Eq. (6.182), and this operation is called *whitening*. The second transformation $\mathbf{A}_2$ can be regarded as a transformation that performs the KL expansion on the class mean $\mathbf{A}_1^t \boldsymbol{\Sigma}_B \mathbf{A}_1$ in the space normalized by $\mathbf{A}_1$, as shown by Eq. (6.183). This is equivalent to the following process. That is, ignoring the spread, one representative pattern is considered to exist at the position of each class mean, and KL expansion is performed for a total of c patterns. For details, refer to Problem 6.8.

Furthermore, from Eq. (6.187) and the equation

$$\tilde{\boldsymbol{\Sigma}}_B = \mathbf{A}^t \boldsymbol{\Sigma}_B \mathbf{A} = \mathbf{A}_2^t (\mathbf{A}_1^t \boldsymbol{\Sigma}_B \mathbf{A}_1) \mathbf{A}_2 = \mathbf{A}_2^t \mathbf{A}_2 \boldsymbol{\Lambda} = \boldsymbol{\Lambda}, \tag{6.192}$$

it can be seen that $\mathbf{A}$ is a transformation that simultaneously diagonalizes $\boldsymbol{\Sigma}_W$ and $\boldsymbol{\Sigma}_B$. This operation is called *simultaneous diagonalization* and can generally be performed on any two symmetric matrices of equal dimension.[31]

The above transformations by $\mathbf{A}_1$ and $\mathbf{A}_2$ are illustrated by concrete examples in Fig. 6.10. In this figure, (a) represents the distribution of feature vectors of two classes in a two-dimensional feature space. The process of determining the subspace from the distribution given by (a) using the linear discriminant method is described below.

First, the distributions of ω_1 and ω_2 in the feature space in (a) are superimposed so that their centers of gravity coincide as shown in (a-1), and we call this

---

[31] The simultaneous diagonalization is one of the useful tools in pattern recognition. A good explanation can be found in Fukunaga (1990).

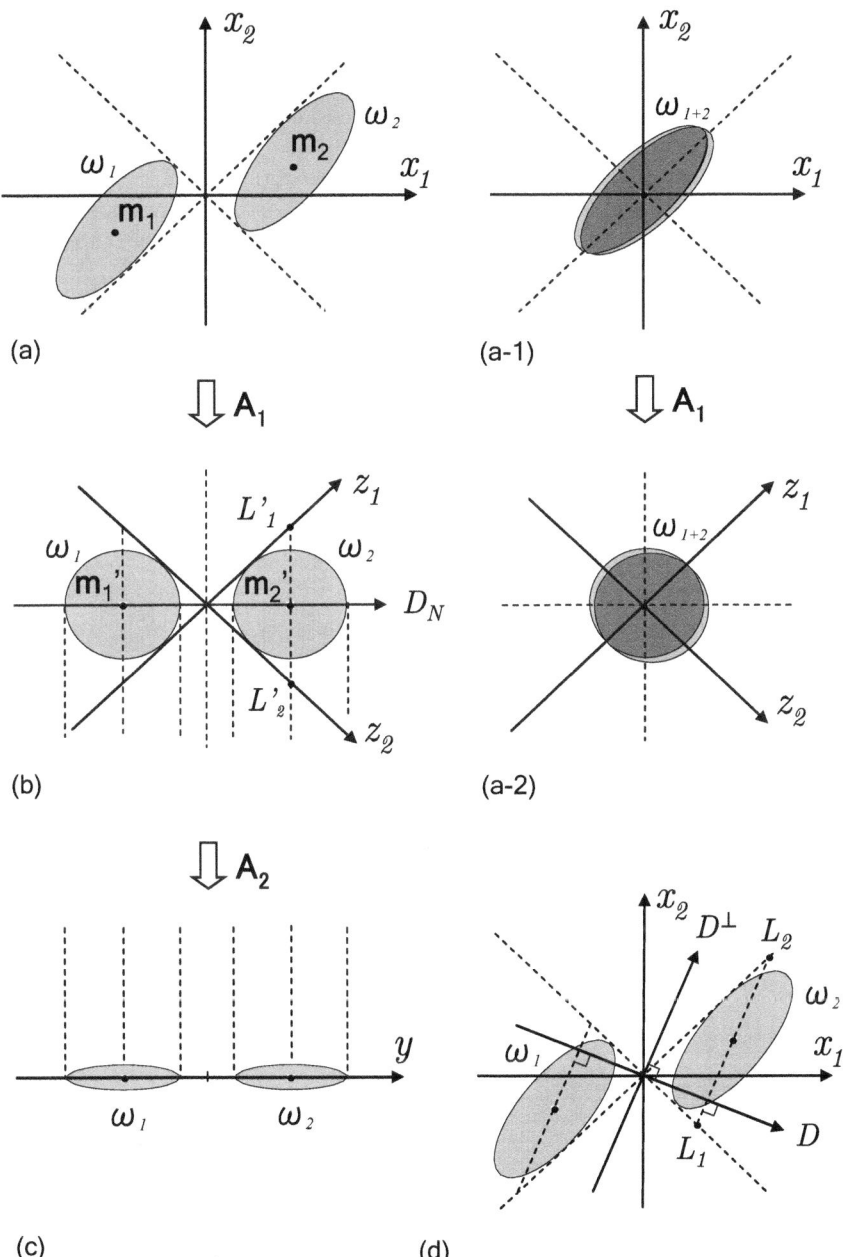

Fig. 6.10 Two-step spatial transformation in linear discriminant method. (**a**) Feature space. (**b**) Normalized space. (**c**) Discriminant space. (**d**) Feature space and discriminant axis

distribution ω_{1+2}. Find the transformation $\mathbf{A}_1$ such that the distribution ω_{1+2} is isotropic for each axis as shown in (a-2), i.e., the covariance matrix of the distribution is a constant times the identity matrix. In this example, it corresponds to a transformation that multiplies by a constant in the direction of $x_1 = x_2$ and $x_1 = -x_2$.

Next, the space (b) is obtained by applying the transformation $\mathbf{A}_1$ to the original space (a). This space is called a *normalized space*. The axes of the normalized space obtained by this transformation are z_1 and z_2. Then, the KL expansion is applied to the mean vector of each class in this normalized space, i.e., two points $\mathbf{m}_1'$ and $\mathbf{m}_2'$. Since there are only two points in this case, the required subspace is the axis D_N connecting the two points. This yields the discriminant space (c), where the axis of the space is y. This transformation corresponds to the second transformation $\mathbf{A}_2$.

Finally, consider a line segment $L_1' L_2'$ in the normalized space consisting of a sequence of points that project to the same point on y, and map $L_1' L_2'$ to the original space to obtain $L_1 L_2$ in (d). Since the line segment $L_1 L_2$ is mapped to the same point in the discriminant space, the desired discriminant axis is the axis perpendicular to it and D is obtained. If the covariance matrices and prior probabilities of the two distributions are equal, the plane perpendicular to D passing through the midpoint of the mean vectors of the two classes is determined as the optimal decision boundary $D^\perp$ separating the two classes.

6.5 How to Apply KL Expansion

As mentioned in the coffee break in Sect. 6.3.3, the KL expansion is often misunderstood in terms of its purpose and utility. In this section, we summarize points to be noted when applying the KL expansion. One is the difference from the linear discriminant method, and the other is the relationship between the KL expansion and the number of learning patterns. These points are described in detail below.

6.5.1 KL Expansion and Linear Discriminant Method

Both the KL expansion described in Sect. 6.3 and the linear discriminant method described in Sect. 6.4 are methods of dimensionality reduction for pattern recognition, but their effects are actually quite different. As already mentioned, the KL expansion is a method to find a subspace that "best approximates the overall pattern distribution", while the linear discriminant method is a method to find a subspace that "maximizes the degree of separation of the pattern distribution for each class". Therefore, these two methods must be used correctly according to the purpose.

The KL expansion is a method to reduce the dimensionality of the feature space so that it can reflect the information of the entire distribution of feature vectors as

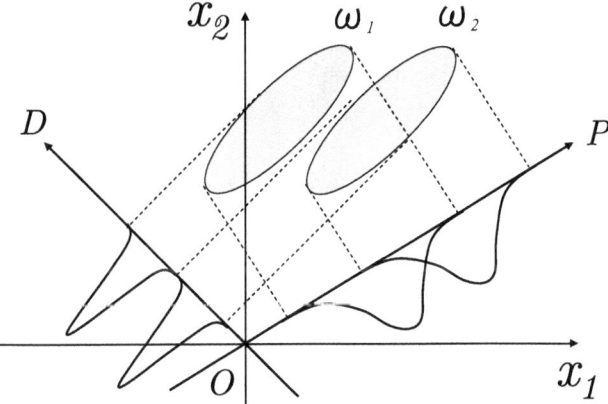

Fig. 6.11 Dimensionality reduction for representation and for discrimination

much as possible. Therefore, when multiple classes exist in the feature space, the space obtained by the KL expansion is not necessarily an effective space to classify these classes. An example is shown in Fig. 6.11. The two distributions in the figure represent the distribution of two-dimensional feature vectors of patterns belonging to the classes ω_1 and ω_2, respectively. The principal axis determined by the KL expansion is P, and the two distributions are not well separated on this axis. The axis that best separates the two distributions is D, which is quite different from P. This is because the KL expansion does not take classification into account at all.

On the other hand, the linear discriminant method is one of the methods of spatial transformation considering the *separability* of the distributions of different classes. The difference between the dimensionality reduction by the KL expansion and that by linear discriminant method is that the former is a dimensionality reduction for *representation* or *compression*, while the latter is a dimensionality reduction for *discrimination* (see also Problems 6.2 and 6.3).

Although the KL expansion is a dimension reduction method that does not take classification into consideration, it is widely used in pattern recognition processing. This is due to the following reasons.

First, advanced recognition, such as character recognition or speech recognition, usually requires high-dimensional feature vectors. Therefore, dimensionality reduction is essential as a means of escaping the curse of dimensionality.

Second, the initially selected features may contain pairs of features with correlation. In particular, there is an extremely high risk that a high-dimensional feature vector may contain a pair of highly correlated features without our knowledge. When two highly correlated features exist, the covariance matrix has eigenvalues close to 0, so reducing the dimensionality of the feature space by the KL expansion means reducing redundant information. In addition, when there are strongly correlated feature pairs, the computation error of the inverse matrix is often large, and dimensionality reduction by the KL expansion can prevent this.

However, as shown in Fig. 6.11, it should be noted that reducing the dimensionality of the feature space by the KL expansion always carries the risk of losing important information for classification.

6.5.2   KL Expansion and Number of Learning Patterns

Let us briefly summarize the points to note regarding the relation between the KL expansion and the number of learning patterns.

It has already been mentioned in Sect. 4.4 that a sufficient number of learning patterns must be prepared compared to the dimensionality of the feature space. This also applies to the KL expansion. In order to perform the KL expansion, we first need to compute the covariance matrix from the learning patterns, and then compute its eigenvalues and eigenvectors. If the number of learning patterns n is less than or equal to the dimensionality d ($n \leq d$), then ($d - n + 1$) eigenvalues are zeros.[32] In other words, even though the feature space appears to be d-dimensional, patterns are actually distributed in a smaller ($n - 1$)-dimensional subspace.

Let us now examine how the number of learning patterns affects the computation of the KL expansion in the following two experiments.

Experiment 1   First, patterns are artificially generated that are distributed according to a multivariate normal distribution in a 16-dimensional feature space. We investigate how the deviation between the principal axes obtained by the KL expansion and the correct principal axes changes as the number of patterns increases. Let the angle between the two axes be θ, and the deviation is evaluated by $\cos \theta$, i.e. cosine similarity (see Eq. (13.39)). The maximum value 1 is taken when the direction of the obtained principal axis coincides with the correct direction. The result is shown in (a) of Fig. 6.12. The horizontal axis indicates the number of patterns n, and the vertical axis indicates $\cos \theta$. In this example, when the number of patterns equals the number of dimensions ($n = 16$), $\cos \theta = 0.285$, and the obtained principal axes deviate from the correct direction by $\theta = 73.4$ degrees. Even when the number of patterns reaches about 6 times the dimensionality ($n = 100$), the error is still large : $\cos \theta = 0.581$ and $\theta = 54.5$ degrees. In any case, as mentioned above, this example shows that a sufficient number of patterns must be prepared relative to the dimensionality.

In reality, dimensionality of features ranges from hundreds to thousands, but we often see cases where the number of patterns is at most as large as the dimensionality, or sometimes even smaller. Nevertheless, why is this not a serious problem? To investigate this, the following experiment was conducted.

---

[32] If the number of patterns is less than the dimensionality, the covariance matrix will no longer be non-singular and its eigenvalues will include those that are zeros.

Experiment 2 Here, we performed the same experiment as above using feature vectors obtained from actual character patterns instead of artificial feature vectors. In the experiment, GLK16 was used (Appendix D). The number of patterns was 1000 per class, and the character "5" was selected from them. The principal axis obtained using all 1000 patterns was considered to be the correct principal axis. The result is shown in (b) of Fig. 6.12. In this case, unlike the previous experiment, the obtained principal axis almost coincides with the correct direction even though the number of patterns is relatively small. The reason for this difference is as follows.

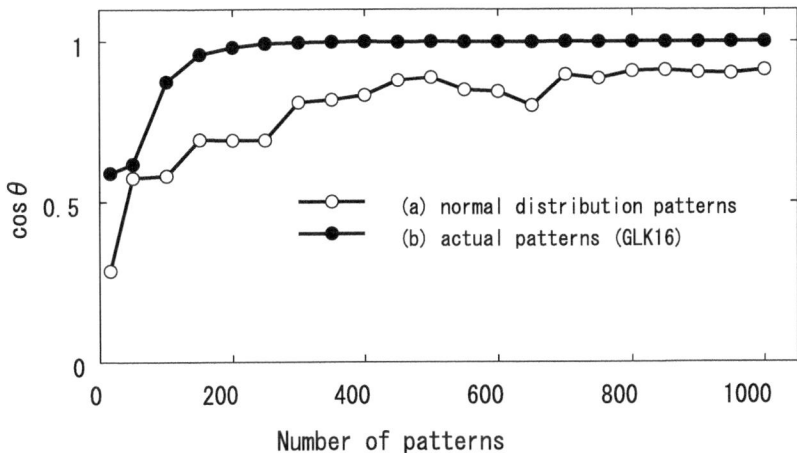

Fig. 6.12 Number of patterns and accuracy in principal axis direction

In reality, it is difficult to prepare features that are independent of each other, and they are always correlated (see Sect. 6.5.1). Even if we intend to add a new feature, it often happens that it can almost be represented by a linear combination of already prepared features. In this example as well, a significant portion of the features are considered to be correlated with each other due to the nature of Glucksman's feature.

Here, we define the contribution ratio and the cumulative contribution ratio as follows. Let d eigenvalues obtained by solving the eigenvalue problem of Eq. (6.45) be $\lambda_1, \lambda_2, \ldots, \lambda_d$ in descending order, and the corresponding eigenvectors be $\mathbf{u}_1, \ldots, \mathbf{u}_d$. In this case,

$$u_i = \lambda_i \left/ \sum_{j=1}^{d} \lambda_j \right. \tag{6.193}$$

is called the *contribution ratio* of $\mathbf{u}_i$, and

$$c_i = \sum_{j=1}^{i} \lambda_j \bigg/ \sum_{j=1}^{d} \lambda_j \qquad\qquad (6.194)$$

is called the *cumulative contribution ratio*[33] of $\mathbf{u}_1, \ldots, \mathbf{u}_i$.

Figure 6.13 shows the relationship between the number of features and the cumulative contribution ratio. The number of patterns used is 1000. The graph (b) in this figure represents GLK16, which shows that the cumulative contribution ratio reaches almost 0.99 for the first 10 features. This indicates that the patterns are actually distributed in a 10-dimensional subspace even though the feature space is 16-dimensional. Therefore, as shown in Fig. 6.12, relatively accurate principal axes were obtained even with a small number of patterns. Thus, when the apparent dimensionality is large but patterns are actually distributed in a space of smaller dimensionality, this actual dimensionality is called the *intrinsic dimensionality*.

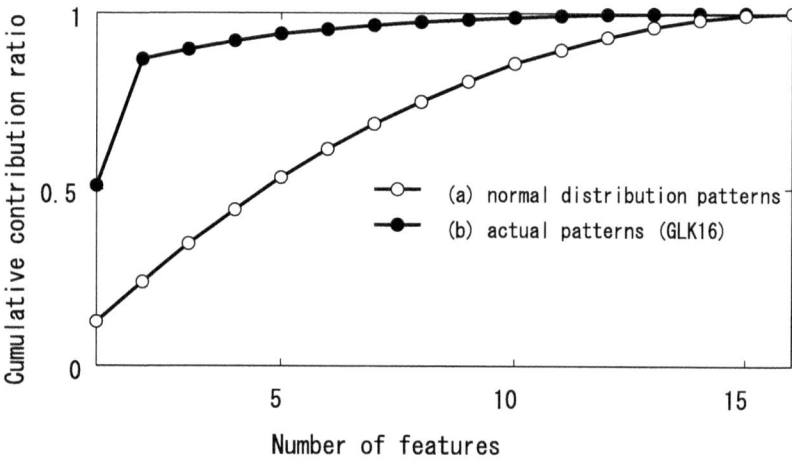

Fig. 6.13 Number of features and cumulative contribution ratio

On the other hand, a similar curve of the cumulative contribution ratio for patterns following a multivariate normal distribution is shown by (a) in the figure. In this case, unlike (b), the cumulative contribution rate does not grow rapidly, nor does it saturate in the middle of the curve. This is because the patterns used in (a) are artificial and the independence among the 16 features is high, that is, the intrinsic dimensionality is considered to be close to 16. Therefore, if the principal axes are to

---

[33] The cumulative contribution ratio indicates the ratio of the sum of the eigenvalues added up to a certain number of eigenvalues in descending order to the sum of all eigenvalues. This is a measure of how faithfully the original distribution can be represented using only a limited number of principal components, and the portion less than 100% corresponds to the error from the original distribution.

be obtained with the same accuracy, the number of patterns required in (a) is much larger than in (b).

It should be noted that even if the patterns are distributed only in a subspace with a small dimensionality, a large number of patterns relative to the dimensionality are required to confirm this fact.

Problems

6.1 * Without using the trace operation, show that $\mathbf{A} = (\mathbf{u}_1, \ldots, \mathbf{u}_{\tilde{d}})$ maximizing $\tilde{\sigma}^2(\mathbf{A})$ of Eq. (6.35) is a solution to the eigenvalue problem of Eq. (6.45).

6.2 Suppose that six learning patterns $x_1, x_2, \ldots, x_6$ are given in the two-dimensional feature space as follows (the same data as in Problem 2.3):

$$x_1 = (11, 8)^t, \ x_2 = (10, 10)^t, \ x_3 = (6, 3)^t,$$
$$x_4 = (6, 5)^t, \quad x_5 = (2, 8)^t, \quad x_6 = (1, 2)^t.$$

Now, using the vector $\mathbf{u}$, we apply the transformation

$$y = \mathbf{u}^t x \quad (x = x_1, \ldots, x_6)$$

to map the learning patterns from the two-dimensional feature space to the one-dimensional feature space.

(1) Find $\mathbf{u}$ by the KL expansion such that the distribution situation in the original space is preserved as much as possible. Here, $\mathbf{u}$ should be normalized to $\|\mathbf{u}\| = 1$.

(2) Plot the projection axis (principal axis) determined by $\mathbf{u}$ obtained above on the graph. The projection axis should pass through the mean $\mathbf{m}$ of all patterns.

6.3 Suppose that six learning patterns shown in Problem 6.2 are given in a two-dimensional feature space. Among them, x_1, x_2, x_3 belong to class ω_1 and x_4, x_5, x_6 belong to class ω_2, respectively. Now, using the vector w, let us apply the transformation

$$y = w^t x \quad (x = x_1, \ldots, x_6)$$

to separate the learning patterns of the above two classes as efficiently as possible according to the value of y.

(1) Find the vector w by Fisher's method. Assume that w is normalized to $\|w\| = 1$. Plot the obtained projection axis y on the graph and also plot the patterns $x_1, x_2, \ldots, x_6$ on the y-axis. The projection axis y should pass through the origin.

(2) Compare the above projection axis y with the principal axis of the KL expansion obtained by Problem 6.2.

(3) Set the threshold $-w_0$ indicated in Eq. (6.110) on the projection axis y, then find the decision boundary to classify the two classes, and plot it on the graph.

(4) Set the projection axis y in the direction $\mathbf{m}_1 - \mathbf{m}_2$, then project the learning patterns of the two classes onto the y axis and compare them with the results of (1).

6.4 Patterns of class ω_1 and class ω_2 are distributed in a d-dimensional feature space. We apply Fisher's method to these patterns and denote the obtained axis by a vector $\boldsymbol{w}$. Here, all patterns are mapped onto the new space using the transformation matrix $\mathbf{A}$ for normalization shown in Eq. (6.3).

(1) Apply Fisher's method in this transformed space and show that if the obtained axis is represented by a vector $\boldsymbol{v}$, the relationship $\boldsymbol{v} \propto \mathbf{A}^{-1} \boldsymbol{w}$ holds.

(2) Show that Fisher's evaluation criterion values shown in Eq. (6.98) are equivalent for $\boldsymbol{w}$ and $\boldsymbol{v}$, and that the normalization process by $\mathbf{A}$ of Eq. (6.3) is unnecessary when applying Fisher's method.

6.5 Prove Eq. (6.177) through Eq. (6.180).

6.6 Answer the following questions (1) through (4):[34]

(1) Prove that Eqs. (6.82) and (6.83) are equal.

(2) Prove that Eqs. (6.140) to (6.142) and Eq. (6.144) are valid.

(3) Prove that Eqs. (6.154) to (6.156) hold.

(4) Prove that Eq. (6.181) holds.

6.7 Show that a real symmetric matrix $\mathbf{S}$ of size $d \times d$ can be expressed using a d-dimensional square matrix $\mathbf{A}$ as

$$\mathbf{A}^t \mathbf{S} \mathbf{A} = \mathbf{I},$$

where $\mathbf{I}$ is a d-dimensional identity matrix. Similarly, show that $\mathbf{S}$ can be expressed using a d-dimensional square matrix $\mathbf{B}$ as the following equation:

$$\mathbf{S} = \mathbf{B}^t \mathbf{B}.$$

6.8 Ten learning patterns $x_1, x_2, \ldots, x_{10}$ are given in the two-dimensional feature space as follows, where x_1 to x_5 belong to class ω_1 and x_6 to x_{10} belong to class ω_2, respectively. Thus, $n_1 = n_2 = 5$, $n = n_1 + n_2 = 10$. Below, we assume $P(\omega_i) = n_i/n$ $(i = 1, 2)$.

$$x_1 = (1, 0)^t, \quad x_2 = (3, 2)^t, \ x_3 = (2, -1)^t, \ x_4 = (-1, -2)^t, \ x_5 = (0, 1)^t$$
$$x_6 = (-1, 0)^t, \ x_7 = (1, 2)^t, \ x_8 = (0, -1)^t, \ x_9 = (-3, -2)^t, \ x_{10} = (-2, 1)^t.$$

---

[34] Since these questions are interrelated, they are summarized here as questions.

(1) Plot the above patterns on a two-dimensional feature space. Next, according to the definition, find the scatter matrices $\mathbf{S}_1, \mathbf{S}_2, \mathbf{S}_B, \mathbf{S}_W, \mathbf{S}_T$ and the covariance matrices $\boldsymbol{\Sigma}_1, \boldsymbol{\Sigma}_2, \boldsymbol{\Sigma}_B, \boldsymbol{\Sigma}_W, \boldsymbol{\Sigma}_T$, respectively.

(2) Apply the KL expansion to the above patterns to obtain the principal axis and plot it on the two-dimensional feature space.

(3) Apply the linear discriminant method to find the optimal discriminant axis for classifying the above two classes and plot it on the two-dimensional feature space.

(4) Find the mean vector $\mathbf{m}_1, \mathbf{m}_2$ of each class and the total mean vector $\mathbf{m}$. From the definition of the Mahalanobis distance, verify that Eq. (6.142) holds. Calculate the Mahalanobis distance between the mean vectors of ω_1 and ω_2.

(5) Decompose the transformation matrix $\mathbf{A}$, determined by the linear discriminant method, into $\mathbf{A} = \mathbf{A}_1 \mathbf{A}_2$ as in Eq. (6.185) and find $\mathbf{A}_1$ and $\mathbf{A}_2$. Plot the original patterns on the normalized space determined by the transformation matrix $\mathbf{A}_1$. Furthermore, confirm that Eq. (6.192) holds.

(6) Show the decision boundary obtained by the linear discriminant method.

Chapter 7
Subspace Methods

Abstract In the previous chapter, we described a feature selection method based on a linear transformation of the feature space. A recognition system is constructed by adding a classifier at the latter stage. In contrast, the *subspace method* introduced in this chapter is an interesting method that uses a linear transformation of the feature space itself for classification without separating feature selection and classification. The history of this method can be traced back to the 1960s, when Watanabe et al. noted that when many feature vectors are plotted in a multidimensional feature space, they are often unevenly distributed in a subspace of very small dimension in the feature space. By utilizing this property of uneven distribution of feature vectors, it is possible to focus only on the subspace where data is distributed when performing classification.

7.1 Basics of the Subspace Method

Watanabe first constructed a subspace using feature vectors of all classes, and then proposed a method for classification using only this subspace, which he called the *SELFIC method* (Watanabe 1969). By using this subspace, the amount of data can be reduced. Although this method was proposed independently from the KL expansion, it is closely related to the KL expansion described in the previous chapter.

What happens if we focus on the distribution of one class only, instead of the distribution of all classes? In this case, it is generally possible to represent the distribution in an even lower-dimensional space. The subspace method classifies unknown patterns by preparing a low-dimensional subspace for each class and comparing which subspace best approximates the unknown pattern. The subspace is obtained independently for each class by KL expansion from the learning patterns.

Supplementary Information The online version contains supplementary material available at https://doi.org/10.1007/978-981-95-1478-6_7.

Typical subspace methods create a subspace for a class using only the learning patterns of the class of interest. Therefore, even if the subspace is optimal for representing the class, it is not necessarily optimal for classifying between classes because it does not consider the distribution of other classes. As an extension of this method, a method to create a subspace of a class while considering the distribution of other classes has also been proposed. This method is also explained in the latter half of this chapter.

7.2   The CLAFIC Method

Various improved subspace methods have been proposed (Oja 1983). However, the basic method is the *CLAFIC method* (CLAss-Featuring Information Compression) proposed by Watanabe in 1969 (Watanabe 1969). In the CLAFIC method, a subspace is created by eigenvectors for large eigenvalues of the autocorrelation matrix for each class, and this subspace is used to classify unknown patterns. If the feature vectors can be approximated by a low-dimensional subspace that differs from class to class, then the input patterns can be classified by this method.

Let the subspaces of c classes $\omega_1, \omega_2, \ldots, \omega_c$ be $\mathbf{L}_1, \mathbf{L}_2, \ldots, \mathbf{L}_c$, and its dimensionalities be $d_1, d_2, \ldots, d_c$. For each class ω_i, let d_i d-dimensional orthonormal vectors spanning subspace $\mathbf{L}_i$ be $\mathbf{u}_{i1}, \ldots, \mathbf{u}_{id_i}$. By the orthogonality of $\mathbf{u}_{ik}$ which constitutes a subspace such that

$$\mathbf{u}_{ik}^t \mathbf{u}_{il} = \delta_{kl}, \tag{7.1}$$

where δ_{kl} is the Kronecker delta shown in Eq. (6.29).

Let us focus on class ω_i and let $\mathbf{A}_i$ be the matrix representing the transformation from the d-dimensional feature space to the d_i-dimensional subspace as follows:

$$\mathbf{A}_i = (\mathbf{u}_{i1}, \ldots, \mathbf{u}_{id_i}). \tag{7.2}$$

From Eq. (7.1),

$$\mathbf{A}_i^t \mathbf{A}_i = \mathbf{I} \tag{7.3}$$

holds, where $\mathbf{I}$ is the identity matrix. The d_i-dimensional feature vector $\mathbf{A}_i^t \mathbf{x}$ projected to the subspace is $\mathbf{A}_i \mathbf{A}_i^t \mathbf{x}$ in the original d-dimensional space. That is, the transformation from the original space to the subspace $\mathbf{L}_i$ is performed by the orthogonal projection matrix as follows:

$$\mathbf{P}_i = \mathbf{A}_i \mathbf{A}_i^t = \sum_{j=1}^{d_i} \mathbf{u}_{ij} \mathbf{u}_{ij}^t \tag{7.4}$$

Fig. 7.1 Projection of vector
x onto subspace $\mathbf{L}_i$

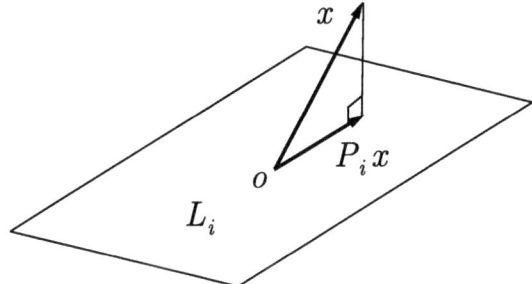

(see Fig. 7.1 and Eq. (6.68)).

Furthermore, from Eqs. (7.3) and (7.4), this can be written as follows:

$$\mathbf{P}_i \mathbf{P}_i = \mathbf{P}_i \tag{7.5}$$

$$\mathbf{P}_i^t = \mathbf{P}_i. \tag{7.6}$$

Then the orthographic projection of x onto the subspace $\mathbf{L}_i$ is $\mathbf{P}_i x$ and the square of its length $\|\mathbf{P}_i x\|^2$ is as follows:

$$\|\mathbf{P}_i x\|^2 = x^t \mathbf{P}_i \mathbf{P}_i x \tag{7.7}$$

$$= x^t \mathbf{P}_i x. \tag{7.8}$$

This length can be regarded as the similarity between the unknown vector x and the class ω_i. That is, $S_i(x)$, representing the similarity between x and the class ω_i, can be expressed as follows:

$$S_i(x) = x^t \mathbf{P}_i x. \tag{7.9}$$

Using this similarity, the classification rule can be expressed as follows:

$$\max_{i=1,\dots,c} \{S_i(x)\} = S_k(x) \quad \Longrightarrow \quad x \in \omega_k. \tag{7.10}$$

This means that the feature vector is classified as the class of the subspace with the largest projective component. By comparing Eq. (7.10) with Eq. (2.3) we see that $S_i(x)$ can be used as a discriminant function.

Although the projection matrix is useful in explaining the method, the actual computation of the subspace method can be done more efficiently using the orthonormal vectors $\mathbf{u}_{ij}$ than using the projection matrix. Equation (7.7) can be rewritten as follows:

$$\|\mathbf{P}_i x\|^2 = x^t \mathbf{P}_i x \tag{7.11}$$

$$= x^t \left(\sum_{j=1}^{d_i} \mathbf{u}_{ij}\mathbf{u}_{ij}^t \right) x \qquad (7.12)$$

$$= \sum_{j=1}^{d_i} (x^t \mathbf{u}_{ij})^2. \qquad (7.13)$$

That is, $\sum_{j=1}^{d_i}(x^t\mathbf{u}_{ij})^2$ can be computed as the similarity, and the vector x is classified as the class with the largest value.

The orthonormal vector $\mathbf{u}_{ij}$ $(j = 1, \ldots, d_i)$ is obtained as follows. First, the class autocorrelation matrix of patterns belonging to class ω_i is computed as

$$\mathbf{R}_i = \frac{1}{n_i} \sum_{x \in \mathcal{X}_i} xx^t, \qquad (7.14)$$

where n_i is the number of patterns of ω_i. Then, let $\mathbf{u}_{ij}$ be the eigenvector of this class autocorrelation matrix corresponding to λ_{ij}, where the j-th eigenvalue, sorted in descending order of eigenvalues, is λ_{ij} $(j = 1, \ldots, d)$ (Problem 7.1). This formulation is deeply related to the KL expansion described in Sect. 6.3. Specifically, it is the subspace obtained by the minimum mean square error criterion without moving the origin for patterns belonging to the class of interest. That is, when a pattern x belonging to class ω_i is projected onto a subspace of dimensionality d_i, the subspace that minimizes the mean square error relative to the original distribution is the subspace composed of the orthonormal vectors from $\mathbf{u}_{i1}$ to $\mathbf{u}_{id_i}$ obtained above. If the dimensionality of the pattern x is d_i, the size of the autocorrelation matrix in Eq. (7.14) is $d_i \times d_i$, and if d_i is large, the amount of computation required for the operation also increases. If the number of patterns n_i is smaller than d_i, this problem can be avoided and the computation becomes more efficient. See Problem 7.2 for details.

The dimensionality of each class is an important issue in the subspace method. When we calculate the eigenvalue of $\mathbf{R}_i$ using actual data, we find that the eigenvalue λ_{ij} $(\lambda_{i1} \geq \cdots \lambda_{ij} \geq \cdots \geq \lambda_{id} \geq 0)$ gradually approaches zero as j increases. The dimensionality of the subspace can be truncated to an appropriate value. However, if the dimensionality is set too small, the subspace will not be able to represent each class. On the other hand, if the dimensionality is too large, it will increase the overlap of subspaces among classes and reduce the classification power. The optimal number of dimensionalities can only be determined experimentally. In such cases, the optimal number of dimensionalities may be fixed at d_0, or a different number of dimensionalities d_i for each class ω_i. One way to determine d_i is to use the cumulative contribution ratio introduced in Eq. (6.194), which is written as follows:

$$a(d_i) = \frac{\sum_{j=1}^{d_i} \lambda_{ij}}{\sum_{j=1}^{d} \lambda_{ij}}. \tag{7.15}$$

That is, a fixed parameter κ is set for all classes, and the dimensionality d_i is chosen satisfying the following equation for each class (Oja 1983):

$$a(d_i) \le \kappa \le a(d_i + 1). \tag{7.16}$$

On the other hand, in some applications, it may be necessary to determine that an unknown vector does not belong to any class. This decision is called reject,[1] but there is no decision called reject in the classification rule given in Eq. (7.10). The following is a possible way to introduce rejects. The normalized vector x is projected to all subspaces, and when the maximum length of the vector

$$\max_i \left\{ \frac{x^t P_i x}{x^t x} \right\}, \tag{7.17}$$

is less than a certain threshold value, it is judged as rejected.

As another reject method, Watanabe used the following method. For example, in the case of a two-class problem,

$$\text{if} \quad \frac{x^t P_1 x}{x^t P_2 x} > \tau, \quad x \text{ is classified as } \omega_1; \tag{7.18}$$

$$\text{if} \quad \frac{x^t P_1 x}{x^t P_2 x} < \frac{1}{\tau}, \quad x \text{ is classified as } \omega_2; \tag{7.19}$$

$$\text{otherwise,} \quad x \text{ is rejected.} \tag{7.20}$$

Watanabe calls this τ the *fidelity value*.

7.3 Subspace Methods and Similarity Methods

As a method of character recognition, Iijima et al. proposed a *multiple similarity method* and a *compound similarity method* (Iijima et al. 1973; Iijima 1974). Although these methods are currently positioned as a variant of the subspace

---

[1] The former of the two types of rejects mentioned in footnote 2 in Chap. 1.

method, they were developed independently of the subspace method by Iijima et al. This section explains the relationship between these similarity methods and subspace methods.

7.3.1 The Multiple Similarity Method

The definition of multiple similarity is given by

$$S_i(\boldsymbol{x}) = \sum_{j=1}^{d} \frac{\lambda_{ij}(\boldsymbol{x}^t \mathbf{u}_{ij})^2}{\lambda_{i1}\boldsymbol{x}^t \boldsymbol{x}}, \tag{7.21}$$

where the meanings of the symbols are the same as those used in the subspace method described in the previous section. Also, the denominator $\boldsymbol{x}^t \boldsymbol{x}$ of this equation can be omitted since it is class-independent, but $\boldsymbol{x}^t \boldsymbol{x}$ is introduced to normalize the value. The difference between the subspace method and multiple similarity method is that each eigenvector is multiplied by a factor $\lambda_{ij}/\lambda_{i1}$ for the multiple similarity method, as can be seen by comparing the two expressions. This means that the eigenvalues are weighted accordingly (Problems 7.3 and 7.4). In the multiple similarity method, the class ω_i that maximizes this similarity is used as the classification result.

The multiple similarity method basically varies j up to d, but in general, the eigenvalue λ_{ij} of the autocorrelation matrix becomes rapidly smaller as j increases, so the value of S_i is almost the same when the similarity is calculated using the formula

$$S_i(\boldsymbol{x}) \simeq \sum_{j=1}^{d_i} \frac{\lambda_{ij}(\boldsymbol{x}^t \mathbf{u}_{ij})^2}{\lambda_{i1}\boldsymbol{x}^t \boldsymbol{x}}, \tag{7.22}$$

with an appropriate value $d_i(\ < d)$ instead of d. Therefore, for practical purposes, it is sufficient to discontinue the calculation at dimension d_i where the cumulative contribution ratio

$$a = \sum_{j=1}^{d_i} \lambda_{ij} \Big/ \sum_{j=1}^{d} \lambda_{ij}, \tag{7.23}$$

is sufficiently large. In fact, if we use the dimension d_i where the cumulative contribution ratio is large enough, the result is almost the same as if we had calculated up to dimension d. This means that the choice of the value of d_i does not affect the classification results as much as the CLAFIC method. The computational efficiency is also improved by terminating at an appropriate dimension d_i. The multiple similarity method has also been applied to speech recognition.

Let us look at the actual eigenvector patterns. Figure 7.2 shows an example of Kanji character patterns (see coffee break below) of class ω_b (meaning "book") represented by 32×32 pixels, which is deformed in various ways due to differences in font and noise. This example is the result of scanning and binarizing patterns. These character patterns are represented by 1024-dimensional feature vectors with each pixel as an element. Figure 7.3 shows four eigenvalues of the autocorrelation matrix for the 40 Kanji characters of class ω_b, in descending order of their eigenvalues.

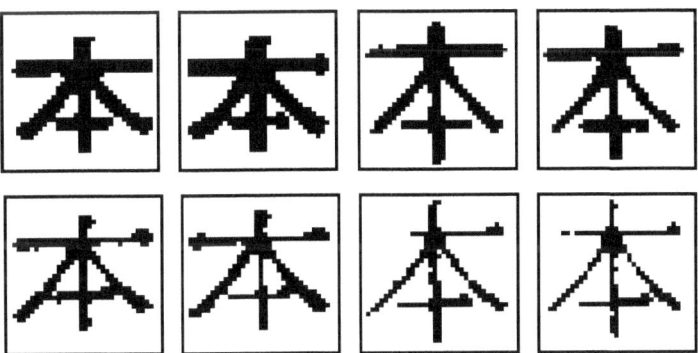

Fig. 7.2 Examples of scanned printed Kanji characters of class ω_b (meaning "book") in various fonts and noises

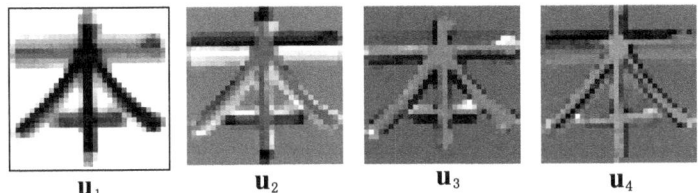

Fig. 7.3 Eigenvectors created from 40 Kanji characters of class ω_b (meaning "book")

Qualitatively, the first eigenvector shows a structure similar to the average pattern of the various characters, while the second and later eigenvectors show the variation components such as misalignment and blurring in the contour of the characters.

─────────────── **Coffee Break** ───────────────

Necessity is the Mother of Invention: Challenge to Kanji Character Recognition

In the 1970s, mainframe computers began to be used in society. This was before personal computers became popular. Alphanumeric characters could be easily input into computers using typewriters, but it was difficult to input Kanji characters into computers at that time. Kanji characters are used in Japan, along with characters native to Japan, but were originally taken from traditional Chinese characters. As one of the input methods into computers, pattern recognition of Kanji characters was focused on.

However, Kanji characters were more difficult to recognize than alphanumeric characters because there are more than 2000 classes and many different classes of Kanji characters with similar shapes. Template matching, the basic method for printed character recognition at that time, was difficult for recognition of Kanji characters. Therefore, the idea of the multiple similarity method, one of the subspace methods, was developed to extend template matching to cope with variations in character shape. In addition, there are many classes of Kanji characters that are similar in shape but differ in the presence or absence of dots or bars (Fig. 7.4).

Fig. 7.4 Examples of Kanji characters with similar shapes but different classes

The compound similarity method was developed to deal with these problems. Although this idea can be applied to other fields, it is thought that the efforts to solve the problems of Kanji character recognition at that time contributed to the development of the subspace method. Necessity is the mother of invention.

7.3.2 The Compound Similarity Method

The multiple similarity method and CLAFIC method create a subspace of a class from only the patterns of that class. Therefore, the subspace is optimal for accurately representing the class, but not optimal for classifying between classes. In other words, as described in Sect. 6.5, the subspace is not necessarily optimal for classifying between classes. As mentioned in Sect. 6.5, a good space for representing the original data is different from a space suitable for separating the data of multiple classes. The compound similarity method (Iijima 1974) proposed by Iijima et al. introduces a separation function between classes into the multiple similarity method.

For example, consider the two classes of Kanji characters, ω_t (meaning "tree") and ω_b (meaning "book") (see Fig. 7.4). The overall shapes of the two are very similar. Therefore, the similarity of $x \in \omega_t$ to the class ω_t is close to the similarity of x to the class ω_b in the multiple similarity method, which observes the entire image equally. This means that the multiple similarity method is likely to produce misclassification between the two. The difference between $x \in \omega_t$ and $x \in \omega_b$ is the presence or absence of horizontal bars. Therefore, it is expected that the separation between class ω_t and class ω_b can be improved by defining a similarity that focuses on the location of the horizontal bar in the image. Therefore, Iijima et al. extended the multiple similarity method and proposed the following compound similarity method. The compound similarity is represented by

$$S_i(x) = \sum_{j=1}^{d} \frac{\frac{\lambda_{ij}}{\lambda_{i1}}(x^t u_{ij})^2 - \mu(x^t v_i)^2}{x^t x}, \tag{7.24}$$

where μ is a parameter and v_i is the difference between the average pattern m_k of similar class ω_k and the learning pattern set of class ω_i, defined by

$$v_i = \frac{m_k - \sum_{j=1}^{d} m_k^t u_{ij} u_{ij}}{\sqrt{m_k^t m_k - \sum_{j=1}^{d} (m_k^t u_{ij})^2}}. \tag{7.25}$$

Similar to the multiple similarity, the formula for the compound similarity can be computationally efficient by terminating the addition at a suitable dimension d_i. In the above equation, v_i is the difference vector between the vector obtained by projecting the average pattern m_k of similar class onto the subspace of class ω_i, and the original vector whose magnitude is normalized to 1. Figure 7.5 shows v_i. Here, L_i represents the subspace created by the learning patterns of class ω_i, and $L_i^\perp$ represents its orthogonal complement.

Fig. 7.5 Explanation of v_i in compound similarity

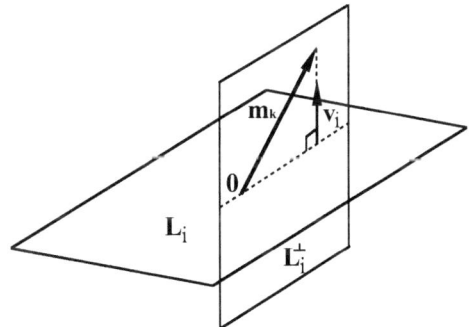

The compound similarity is supposed to emphasize the differences between similar classes. Figure 7.6 shows an example of v_i for class ω_b to class ω_t. In this example, we can see that the differences are emphasized. In other words, the compound similarity method is considered to be effective for recognition targets such as Kanji characters that have many similar classes.

Fig. 7.6 An example of $\mathbf{v}_i$ in compound similarity for class ω_b to class ω_t

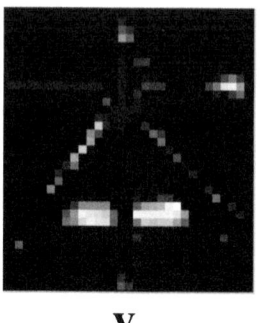

$$\mathbf{V}_1$$

7.4   The Orthogonal Subspace Method

In addition to the methods described above, several other methods have been proposed as modifications of the subspace method. In this section, we outline the *orthogonal subspace method*. These methods, like the compound similarity method, consider the relationship between classes (Oja 1983).

In the subspace method, when the subspaces of all classes are orthogonal to each other, it is called the orthogonal subspace method. This means that feature vectors belonging to one class of subspaces are orthogonal to subspaces of other classes, i.e., they give the lowest similarity. Let the d-dimensional orthonormal vectors of the two subspaces $\mathbf{L}_i$ and $\mathbf{L}_j$ be $\mathbf{u}_{i1}, \ldots, \mathbf{u}_{id_i}$ and $\mathbf{u}_{j1}, \ldots, \mathbf{u}_{jd_j}$, respectively, it would constitute a subspace such that

$$\mathbf{u}_{ik}^t \mathbf{u}_{jl} = \delta_{kl}\delta_{ij}, \tag{7.26}$$

for all k, l and i, j. where δ_{kl} is the Kronecker delta as in Eq. (7.1).

In other words, in the orthogonal subspace method, in addition to the condition that each basis is a set of orthonormal vectors, the condition (Eq. (7.26)) that each basis between classes is orthogonal is added when creating the subspace of each class. This condition is rather strict and it is generally impossible to create this basis when there are multiple classes. However, it is possible in the case of two classes.

First, let us consider the case of the c classes. Let the autocorrelation matrix for each class be $\mathbf{R}_1, \ldots, \mathbf{R}_c$, and the prior probabilities of each class are $P(\omega_1), \ldots, P(\omega_c)$. Then the matrix

$$\mathbf{R}_0 = P(\omega_1)\mathbf{R}_1 + \cdots + P(\omega_c)\mathbf{R}_c, \tag{7.27}$$

is the autocorrelation matrix of the whole distribution. Since the matrix $\mathbf{R}_0$ is a real symmetric matrix, it can be expressed as

$$\mathbf{A}\mathbf{R}_0\mathbf{A}^t = \mathbf{I}, \tag{7.28}$$

using an n-dimensional square matrix $\mathbf{A}$ (see also Problem 6.7). That is,

$$P(\omega_1)\mathbf{A}\mathbf{R}_1\mathbf{A}^t + \cdots + P(\omega_c)\mathbf{A}\mathbf{R}_c\mathbf{A}^t = \mathbf{I}. \tag{7.29}$$

Here, for example, if only for the two-class case, $P(\omega_1)\mathbf{A}\mathbf{R}_1\mathbf{A}^t$ and $P(\omega_2)\mathbf{A}\mathbf{R}_2\mathbf{A}^t$ have the same eigenvectors. In addition, the eigenvalues λ_1 and λ_2 of $\mathbf{A}\mathbf{R}_1\mathbf{A}^t$ and $\mathbf{A}\mathbf{R}_2\mathbf{A}^t$ have the relationship as follows:

$$\lambda_1 + \lambda_2 = 1. \tag{7.30}$$

This means that the eigenvector for the largest eigenvalue of $\mathbf{A}\mathbf{R}_1\mathbf{A}^t$ can be considered the most important basis vector for one class, but the least important basis vector for the other class.

By using the basis created by the above method, the subspace of each class is constructed. In the classification stage, similar to the CLAFIC method, the similarity between the input pattern and the subspace of each class is calculated, and the class with the largest value is selected as the classification result.

7.5   The Learning Subspace Method

The subspace method described so far calculates eigenvectors from the autocorrelation matrix. That is, instead of looking at individual patterns, the subspace that minimizes the mean square error of the entire pattern is determined. However, this subspace is not always optimal for separating classes. Since misclassifications usually occur near the boundaries between classes, minimizing the mean square error of the entire pattern does not mean that the boundaries of this classification are optimal. Therefore, Teuvo Kohonen et al. proposed a method for sequentially finding subspaces to minimize the misclassification rate for learning patterns, which they called the *learning subspace method* (Oja 1983).

Suppose that the learning pattern x of class ω_i is misclassified as class ω_j different from ω_i. To avoid this misclassification, let $\mathbf{Z}$ be the subspace of class ω_j, and let $\mathbf{Z}'$ be the subspace obtained by slightly rotating $\mathbf{Z}$ as follows:

$$\mathbf{Z}' = (\mathbf{I} + \gamma x x^t)\mathbf{Z}, \tag{7.31}$$

where γ is a parameter that indicates how much the effect of misclassification of x is reflected in the rotation of the subspace. For example, if

$$\gamma = -(x^t x)^{-1}, \tag{7.32}$$

then

$$\mathbf{I} + \gamma x x^t = \mathbf{I} - \frac{x x^t}{x^t x}. \tag{7.33}$$

In this case,

$$\left(\mathbf{I} - \frac{xx^t}{x^t x}\right) x = 0, \tag{7.34}$$

so that x and the subspace after rotation are orthogonal. However, this modification is too large for just one learning pattern, so in practice the value of γ is determined so that the modification is smaller than this. Next, the subspace $\mathbf{Z}'$ is orthogonalized after the rotation. In practice, the *Gram–Schmidt orthogonalization* method, for example, can be used here. This process is repeated iteratively until there are no more learning patterns, thus completing the creation of the learning subspace. Although this iterative computation is expensive due to the nonlinear nature of the process, efficient computation methods have been proposed. In the classification phase, similar to the CLAFIC method, the similarity between the input pattern and the subspace is calculated, and the class with the largest similarity value is output as the classification result. This method has been applied to phoneme classification and its effectiveness has been demonstrated.

This chapter describes the subspace method. The subspace method basically creates a subspace for each class, and classifies unknown patterns based on which subspace the patterns can be approximated accurately. As shown in Eq. (7.13), the subspace method can be considered as the *template matching* where templates are extended to subspaces, and is applicable to many applications that are suitable for template matching. Various improved subspace methods have been proposed, including those described above. In general, however, it is difficult to determine the superiority or inferiority among them. It is not uncommon for a simple CLAFIC method to achieve higher recognition accuracy than other improved and more complex methods. In other words, it is difficult to say which method has higher recognition accuracy in actual patterns, because the distributions within and among classes differ in complex ways depending on the recognition target.

As a further development of the subspace method, the parametric eigenspace method (Murase and Nayar 1995) has been proposed, in which the deformation of a pattern is represented by a manifold in the subspace. This method not only classifies unknown patterns by simply projecting them into the subspace, but also can represent pattern deformations more accurately and evaluate the degree of deformation at the same time. Parametric eigenspace methods have been applied to object recognition and video recognition. Furthermore, the kernel subspace method, which combines the subspace method with the kernel method, has been proposed (see Sect. 11.6).

Problems

7.1 Let n d-dimensional patterns belonging to class ω be y_k ($k = 1, \ldots, n$). Using the multiple similarity method or CLAFIC method, the vectors $\mathbf{u}_j$ ($j = 1, \ldots, d'$)

spanning the subspace of ω are obtained by solving the eigenvalue problem of the autocorrelation matrix of y_k (see Sect. 7.2). On the other hand, these $\mathbf{u}_j$ are known to be orthonormal vectors whose vector lengths have a maximum mean square when the n patterns are projected onto their subspaces. Verify this using Lagrange's method of undetermined multipliers, for the simple case where the dimensionality of the subspace is 1 ($d' = 1$).

7.2 Here is a $d \times n$ matrix $\mathbf{X}$. Assume that $d \geq n$. If the eigenvalues and eigenvectors of matrix $\mathbf{X}^t \mathbf{X}$ are λ_i and $\mathbf{e}_i$, respectively, show that the nonzero eigenvalues and corresponding eigenvectors of matrix $\mathbf{X}\mathbf{X}^t$ are λ_i and $\mathbf{X}\mathbf{e}_i / \sqrt{\lambda_i}$, respectively ($i = 1, 2, \ldots, n$).

7.3 Let y_k ($k = 1, \ldots, n$) denote the n d-dimensional patterns used to compute the subspace of class ω in the multiple similarity method and CLAFIC method and let d' d-dimensional orthonormal vectors ($d' < d$) spanning this subspace be $\mathbf{u}_j$ ($j = 1, \ldots, d'$), show the relation between y_k and $\mathbf{u}_j$ (see Sect. 7.2).

7.4 The multiple similarity is the similarity between the input vector x and the weighted orthonormal vector $\mathbf{u}_j$ with eigenvalue λ_i, as explained in Sect. 7.3, and can be simplified and written as follows:

$$S(x) = \sum_{j=1}^{d} \lambda_j (x^t \mathbf{u}_j)^2.$$

Explain what kind of similarity is equivalent to this similarity $S(x)$ using x and y_k ($k = 1, \ldots, n$).

Part II
Nonlinear Classification

Chapter 8
From Linear to Nonlinear

Abstract The linear discriminant functions that we have discussed so far, have the advantage of being simple, easy to implement and computationally inexpensive. So, they are effective for linearly separable data. However, linear separability is not always expected for real data. In this and subsequent chapters, we will attempt to upgrade discriminant functions from linear to nonlinear in order to deal with data that are linearly nonseparable. The basic process is to transform patterns in a given feature space to another space by means of a nonlinear transformation. The classification process applied on the transformed space is nothing but the linear process introduced so far, while complex nonlinear processes can be realized on the original feature space.

8.1 Linearly Nonseparable Distributions

An example of linearly nonseparable data is shown in Fig. 8.1. The figure shows the distribution of learning patterns of two classes ω_1 and ω_2 in the two-dimensional feature space. In the figure, patterns of class ω_1 and ω_2 are indicated by $\bigcirc$ and $\triangle$, respectively. The number of patterns is 200 for each class, for a total of 400 patterns. Since this data will be used frequently from now on, its structure is described below.

In Fig. 8.1, the classes ω_1 and ω_2 are both mixture distributions of two kinds of multivariate normal distributions. Now, if we denote the multivariate normal distribution as $N(x; \mu, \Sigma)$ which is defined by d-dimensional vector x as a random variable, μ as a mean vector, and Σ as a covariance matrix, then,

$$N(x; \mu, \Sigma) = \frac{1}{(2\pi)^{d/2}|\Sigma|^{1/2}} \exp\left[-\frac{1}{2}(x - \mu)^t \Sigma^{-1}(x - \mu)\right], \quad (8.1)$$

_____

Supplementary Information The online version contains supplementary material available at https://doi.org/10.1007/978-981-95-1478-6_8.

Fig. 8.1 Linearly
nonseparable learning
patterns and the decision
boundary by the Bayes
discriminant function

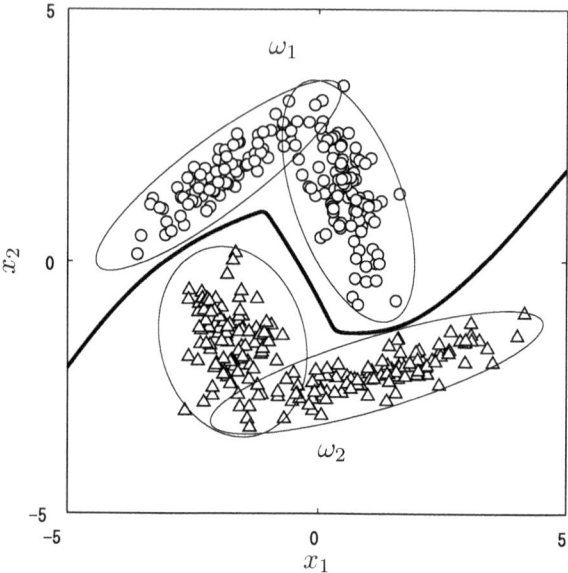

where $|\mathbf{\Sigma}|$ denotes the determinant of $\mathbf{\Sigma}$. Since the data shown in the figure are
two-dimensional, $d = 2$ in the above equation.[1] Denoting prior probabilities of the
classes ω_i by $P(\omega_i)$, the probability density functions by $p(\mathbf{x}|\omega_i)$ $(i = 1, 2)$ and the
probability density function for all patterns by $p(\mathbf{x})$, we have

$$p(\mathbf{x}) = P(\omega_1)p(\mathbf{x}|\omega_1) + P(\omega_2)p(\mathbf{x}|\omega_2), \tag{8.2}$$

$$p(\mathbf{x}|\omega_1) = P_{11} \, \mathcal{N}(\mathbf{x}; \boldsymbol{\mu}_{11}, \mathbf{\Sigma}_{11}) + P_{21} \, \mathcal{N}(\mathbf{x}; \boldsymbol{\mu}_{21}, \mathbf{\Sigma}_{21}), \tag{8.3}$$

$$p(\mathbf{x}|\omega_2) = P_{12} \, \mathcal{N}(\mathbf{x}; \boldsymbol{\mu}_{12}, \mathbf{\Sigma}_{12}) + P_{22} \, \mathcal{N}(\mathbf{x}; \boldsymbol{\mu}_{22}, \mathbf{\Sigma}_{22}). \tag{8.4}$$

In the above equation, P_{11}, P_{21} represent the mixing ratio of two normal distributions, $P_{11} + P_{21} = 1$, and P_{12}, P_{22} are defined similarly, $P_{12} + P_{22} = 1$. In this data,
we set

$$P(\omega_1) = P(\omega_2) = 1/2, \tag{8.5}$$

$$P_{11} = P_{21} = 1/2, \tag{8.6}$$

$$P_{12} = P_{22} = 1/2. \tag{8.7}$$

The mean vector and the covariance matrix for each normal distribution were set as
follows:

---

[1] The case $d = 1$ is simply called the *normal distribution*.

$$\mu_{11} = (-2.0, \quad 1.7)^t, \qquad \Sigma_{11} = \begin{pmatrix} 0.659 & 0.453 \\ 0.453 & 0.399 \end{pmatrix}, \tag{8.8}$$

$$\mu_{21} = (0.6, \quad 1.2)^t, \qquad \Sigma_{21} = \begin{pmatrix} 0.197 & -0.185 \\ -0.185 & 0.638 \end{pmatrix}, \tag{8.9}$$

$$\mu_{12} = (1.2, \quad -2.2)^t, \qquad \Sigma_{12} = \begin{pmatrix} 1.236 & 0.360 \\ 0.360 & 0.168 \end{pmatrix}, \tag{8.10}$$

$$\mu_{22} = (-1.7, \quad -1.6)^t, \qquad \Sigma_{22} = \begin{pmatrix} 0.242 & -0.045 \\ -0.045 & 0.398 \end{pmatrix}. \tag{8.11}$$

The four ellipses indicated by the thin line in Fig. 8.1 represent the shape of the normal distributions that constitute the distributions of the classes ω_1 and ω_2.

If the information about the probability distribution of the data shown in Fig. 8.1 is known, then the classification method based on the Bayes decision rule is optimal, as described in Sects. 4.1 and 5.3. This minimizes the error probability. The discriminant function based on the Bayes decision rule is the Bayes discriminant function and is denoted by Eq. (4.6). Therefore, by setting $g_i(x)$ in Eq. (2.3) to

$$g_i(x) = P(\omega_i)p(x|\omega_i), \tag{8.12}$$

classification by the discriminant function method is possible.

For the two-class problem to determine whether the pattern x belongs to ω_1 or ω_2, the Bayes discriminant function is

$$g_0(x) = g_1(x) - g_2(x) \tag{8.13}$$
$$= P(\omega_1)p(x|\omega_1) - P(\omega_2)p(x|\omega_2), \tag{8.14}$$

and classification is performed by the following formula:

$$\begin{cases} g_0(x) > 0 & \Longrightarrow \quad x \in \omega_1 \\ g_0(x) < 0 & \Longrightarrow \quad x \in \omega_2. \end{cases} \tag{8.15}$$

Therefore, the decision boundary separating the two classes is expressed by

$$g_0(x) = 0. \tag{8.16}$$

For the data shown in Fig. 8.1, let us find the decision boundary by the Bayes discriminant function. From Eqs. (8.5), (8.14) and (8.16), the decision boundary by the Bayes discriminant function is

$$p(x|\omega_1) - p(x|\omega_2) = 0. \tag{8.17}$$

The decision boundary obtained is indicated by the thick line in the figure. As the figure shows, the Bayes discriminant function in this example is a nonlinear discriminant function. In general, Bayes discriminant functions do not always correctly classify all learning patterns. In this example, however, the learning patterns of the two classes can be separated without misclassifications, as can be seen in the figure.

In this section, the optimal discriminant function is obtained assuming that the probability distribution of the learning patterns is known. Hereafter, when using the data of Fig. 8.1 in experiments, we assume that the information shown in Eqs. (8.2) to (8.11) is unknown, unless otherwise noted. That is, we assume that only the individual patterns x generated according to an unknown probability distribution, and their belonging classes, are known.

8.2   Applying Linear Discriminant Functions

In this section, we would like to clarify the extent to which the linear discriminant functions can be applied to linearly nonseparable learning patterns. When applying linear discriminant functions to linearly nonseparable data, we must give up the idea of completely separating the classes. As specific examples of the linear discriminant functions, we take Fisher's method introduced in Sect. 6.4, and the minimum square error learning introduced in Sect. 3.1.

8.2.1   Applying Fisher's Method

In classifying the two classes ω_1 and ω_2, the linear discriminant function

$$g(x) = w_0 + w^t x \tag{8.18}$$

is meaningful only for its sign. Therefore, there is no loss of generality by assuming that w is normalized such that

$$\|w\| = 1. \tag{8.19}$$

The classification rule of Eq. (2.21) is expressed by the following equation:

$$\begin{cases} w^t x > -w_0 & \Longrightarrow & x \in \omega_1 \\ w^t x < -w_0 & \Longrightarrow & x \in \omega_2. \end{cases} \tag{8.20}$$

Now consider a one-dimensional axis y set in the direction of w, passing through the origin in the d-dimensional feature space. Then $w^t x$ is the coordinate value of the projection of the pattern x onto the axis y, and Eq. (8.20) means that the classification is performed by setting the threshold $-w_0$ for this coordinate value. This has already been explained using Fig. 2.7 in Sect. 2.3.3. To facilitate the classification process, the projection axis y should be such that patterns of different classes are as far apart as possible and patterns of the same class are as concentrated as possible. Fisher's method is to determine w based on such an evaluation criterion.

Let $\mathbf{m}_1$ and $\mathbf{m}_2$ be the mean vectors of classes ω_1 and ω_2, respectively, and $\mathbf{S}_W$ be the within-class scatter matrix defined by Eq. (6.81) in the original d-dimensional space. Then, the weight vector w, as shown in Eq. (6.106), is expressed as

$$w = a \, S_W^{-1}(\mathbf{m}_1 - \mathbf{m}_2). \tag{8.21}$$

The term a in the above equation is a constant to satisfy the normalization condition of Eq. (8.19). However, w_0 in Eq. (8.20) cannot be obtained by Fisher's method and must be obtained by other methods. Let us apply Fisher's method to the linearly nonseparable data shown in Fig. 8.1. The upper figure of Fig. 8.2 shows the result of projecting all patterns onto the projection axis y obtained by Fisher's method, and the lower figure shows the enlarged area near $y = -1$ to 0.

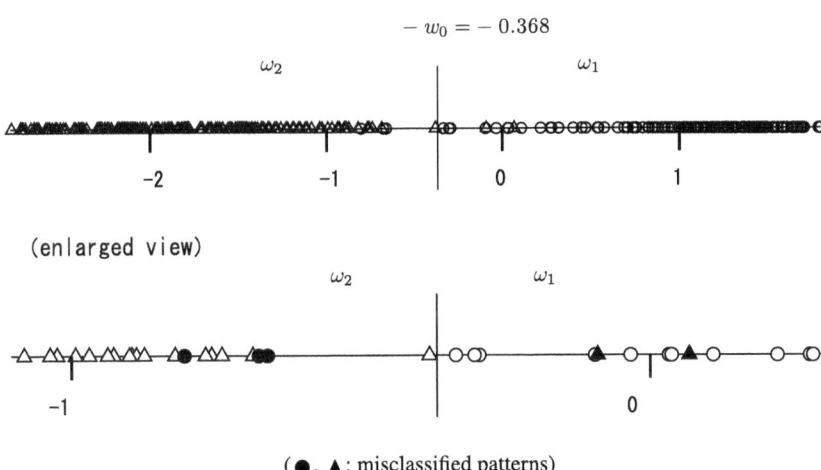

Fig. 8.2 Learning patterns projected onto Fisher's axis

The figure shows that the two classes overlap on this axis, making separation impossible. To obtain the decision boundary, it is necessary to determine $- w_0$. Therefore, we set $- w_0 = -0.368$ to minimize the number of misclassifications on this projection axis. The figure shows the position of $- w_0$. As a result, the number of misclassifications are 3 for class ω_1 and 2 for class ω_2, respectively. Total of five misclassified patterns are filled in black in the lower figure. The calculated weight vector w indicates the normal direction of the decision boundary. Thus, the decision boundary $g(x) = w_0 + w^t x = 0$ is a hyperplane (a straight line in this example) passing through $y = -w_0$ and orthogonal to the projection axis y. In Fig. 8.3, the projection axis y is indicated by a thin line and the decision boundary by a thick

Fig. 8.3 The decision
boundary by Fisher's method

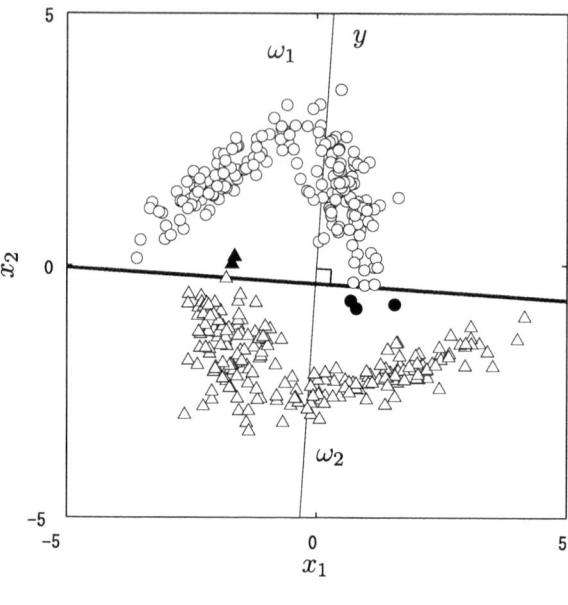

($\bullet$, $\blacktriangle$: misclassified patterns)

line, respectively. It can be seen that they are orthogonal. In the figure, the five misclassified patterns are blacked out as well.

8.2.2   *Applying the Minimum Square Error Learning*

In the minimum square error learning, when designing the linear discriminant function $g(x) = w_0 + w^t x$ of Eq. (8.18), we aim to keep the value of $g(x_k)$ for each learning pattern as close as possible to the teaching signal b_k. Since we are dealing with a two-class problem, Eq. (3.36) is used as the teaching signal. The square error between $g(x_k)$ and the teaching signal b_k is calculated by Eq. (3.38) and the $\mathbf{w}$ that minimizes it is found. The result is shown in Eq. (3.41). Using the pattern matrix $\mathbf{X}$ of $n \times (d + 1)$ defined by Eq. (3.16) and the n-dimensional column vector $\mathbf{b}$ in Eq. (3.39), the optimal $\mathbf{w}$ is obtained as

$$\mathbf{w} = (w_0, \, \boldsymbol{w}^t)^t = (\mathbf{X}^t \mathbf{X})^{-1} \mathbf{X}^t \mathbf{b}, \tag{8.22}$$

where the matrix $\mathbf{X}^t \mathbf{X}$ is assumed to be non-singular.

Figure 8.4 shows the result of applying this method to the linearly nonseparable learning patterns shown in Fig. 8.1. As before, decision boundary is indicated by a thick line, and misclassified patterns are blacked out. There are eight misclassified patterns: six patterns for class ω_1 and two patterns for ω_2.

The minimum square error learning is closely related to Fisher's method. The augmented weight vector $\mathbf{w}$ is $(w_0, \mathbf{w}^t)^t$, which is the weight vector $\mathbf{w}$ plus w_0, as shown in Eq. (8.22). The $\mathbf{w}$ of Eq. (8.22) obtained by the minimum square error learning coincides with the $\mathbf{w}$ of Eq. (8.21) obtained by Fisher's method. This is confirmed by the fact that the decision boundaries of Figs. 8.3 and 8.4 show the same slope. For the proof, see Problem 8.1. The difference between the two methods is that only $\mathbf{w}$ can be determined by Fisher's method, while $\mathbf{w}$ and w_0 can be determined simultaneously by the minimum square error learning, as shown in Eq. (8.22).

Fig. 8.4 The decision boundary by the minimum square error learning

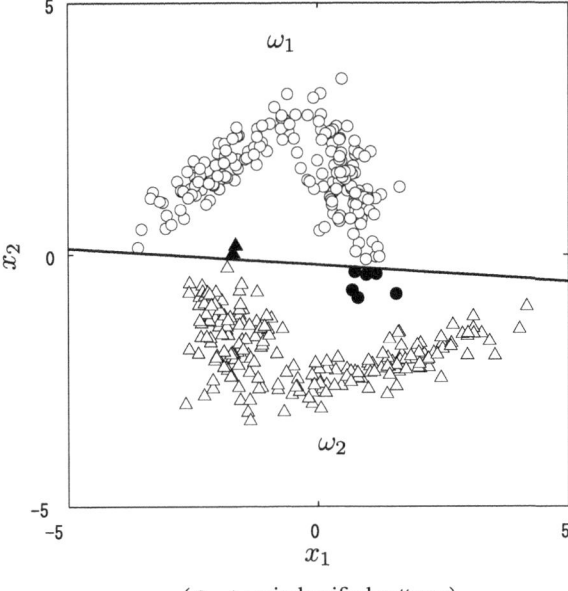

($\bullet$, $\blacktriangle$: misclassified patterns)

The weight coefficient w_0 obtained by the minimum square error learning does not necessarily minimize the number of misclassifications. In fact, by Fisher's method, the number of misclassifications could be reduced to 5 by adjusting the value of w_0 on the projection axis. On the other hand, in the minimum square error learning, the number of misclassifications was 8, even though the same $\mathbf{w}$ as Fisher's method was obtained.

In the above, the methods introduced in Sects. 8.2.1 and 8.2.2 can be applied regardless of whether the learning patterns are linearly separable or not, since the evaluation measure is the degree of separation between the classes. It should be noted, however, that these methods do not use the number of misclassifications as their evaluation measure, so even if the learning patterns are linearly separable, it is not always possible to obtain a decision boundary that can correctly separate them. It should be also noted that there is no guarantee that they can minimize the number of misclassifications when the patterns are linearly nonseparable.

8.3   Applying Piecewise Linear Discriminant Functions

In order to achieve high classification accuracy for linearly nonseparable distributions, a nonlinear discriminant function must be introduced. We have shown in Sect. 2.8 that even for linearly nonseparable distributions, if there is no overlap in the distributions, the piecewise linear discriminant function can be used to separate the classes. The piecewise linear discriminant function can be regarded as a nonlinear discriminant function. However, since the piecewise linear discriminant function is composed of several linear discriminant functions, it should be regarded as an extension of the linear discriminant function rather than a purely nonlinear discriminant function. The linear discriminant function is implemented by the nearest neighbor rule with one prototype per class. In contrast, the piecewise linear discriminant function can be constructed using multiple prototypes per class, that is, multiple linear discriminant functions per class, as shown in Fig. 2.18. The most extreme method is the complete storage scheme in which all learning patterns are prototypes, and the results are shown in Fig. 8.5.

Fig. 8.5 The decision boundary by the complete storage scheme

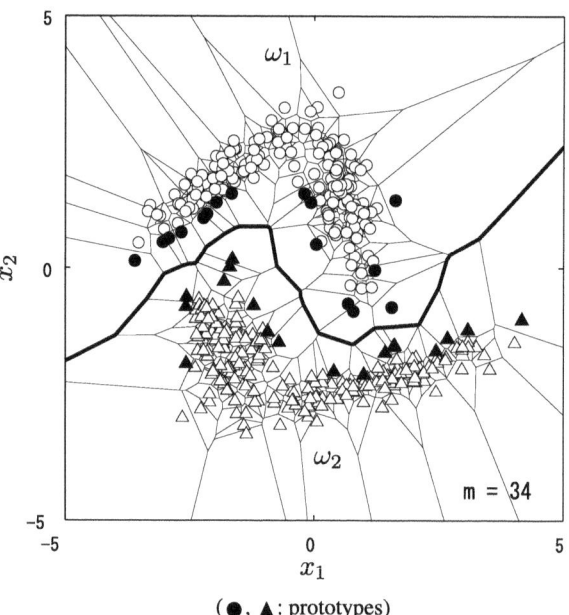

($\bullet$, $\blacktriangle$: prototypes)

The thin lines represent the Voronoi diagram and the thick lines show the decision boundary. It is clear from the figure that the two classes are correctly separated by the complex decision boundary. The prototypes contributing to the setting of the decision boundary are indicated by $\bullet$ and $\blacktriangle$ in the figure, with 16 patterns for the class ω_1 and 18 patterns for the class ω_2, for a total of 34 patterns. Therefore, the number of prototypes used in the nearest neighbor rule can be reduced from 400 to

34. Since these prototypes contribute to the decision boundary setting, they play the same role as the component vectors described in Sect. 2.5 (see also Fig. 1.10).

Here, we further reduced the number of prototypes to show that it is possible to separate classes with only two prototypes for each class, for a total of four prototypes. The result is shown in Fig. 8.6. As before, the prototypes are indicated by ● and ▲, and the decision boundary is indicated by thick lines.

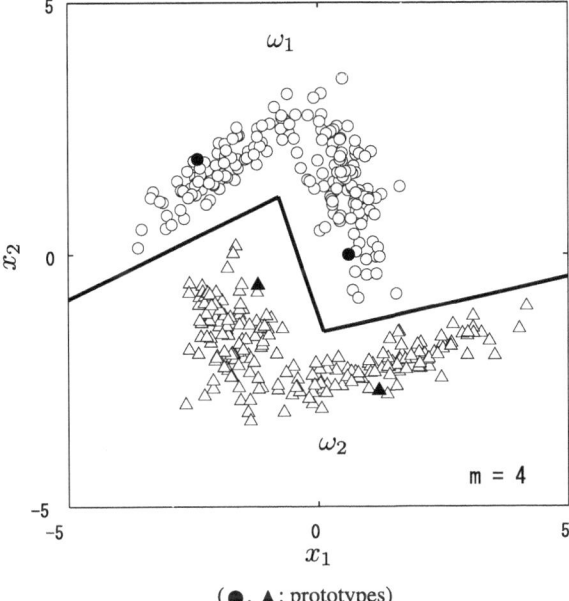

Fig. 8.6 The decision boundary by the piecewise linear discriminant function

(●, ▲: prototypes)

To reduce memory capacity and computational complexity, it is desirable to keep the number of prototypes as small as possible. If the feature space is only two-dimensional, as in this example, it is not so difficult to visually select the appropriate and minimum necessary prototypes and determine the decision boundary. However, such an intuitive method cannot be adopted for high-dimensional feature spaces. Instead, an efficient method without visual inspection for prototype reduction is desired. In the next section, we introduce the condensed nearest neighbor rule, known as the classical method, for this purpose.

8.4 Condensed Nearest Neighbor Rule

In the nearest neighbor rule, only patterns near the decision boundary are valid as prototypes and play an important role. Therefore, if only patterns near the decision boundary can be extracted as prototypes, the classifier design is more efficient. Therefore, we define a *border ratio* as a measure of how close each pattern is to the decision boundary as follows.

Fig. 8.7 Border ratio

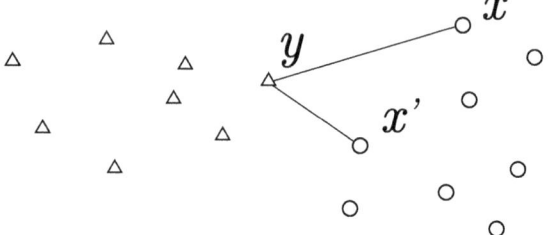

As shown in Fig. 8.7, consider a pattern x (marked with ○) belonging to a certain class. First, let y be the pattern (marked with △) that is the nearest neighbor of x among the patterns belonging to a class different from x. Next, let x' be the pattern that is the nearest neighbor of y among the patterns belonging to the same class as x.

The border ratio $r(x)$ of the pattern x is defined as

$$r(x) = \frac{\|x' - y\|}{\|x - y\|}. \tag{8.23}$$

As is clear from the definition, $\|x' - y\|$ never exceeds $\|x - y\|$, so $0 < r(x) \leq 1$. Here $r(x) = 1$ when $x' = x$.

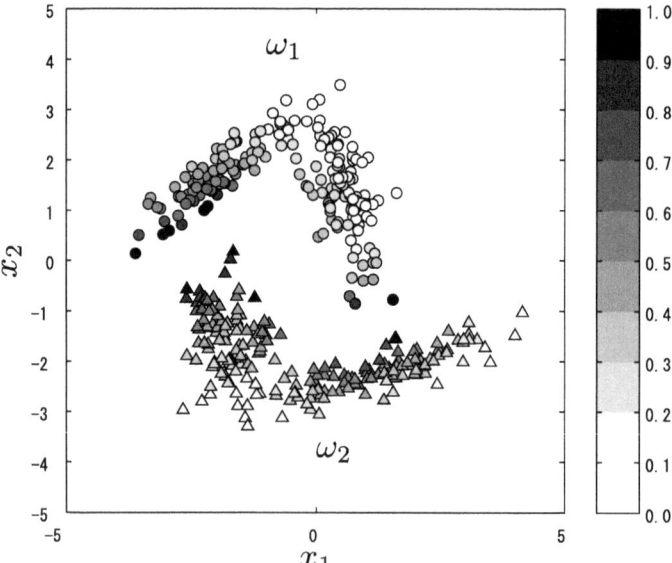

Fig. 8.8 Border ratio calculation results (2 classes)

Figure 8.8 shows the results of quantizing the border ratio of Eq. (8.23) to 10 levels for the patterns of the two classes used in Fig. 8.1. In the figure, the patterns with larger border ratios are filled in darker, and it can be seen that the closer the pattern is to the boundary, the larger the border ratio is. The explanation here is based on the case of two classes as an example, but even if the number of classes is increased, the border ratio can be calculated in the same way (Problem 8.2).

As already mentioned, only prototypes near the decision boundary contribute to determining the decision boundary. Therefore, if we can eliminate unnecessary prototypes far from the decision boundary, we can reduce the amount of storage and computational cost.

One of the promising methods is known as the *condensed nearest neighbor rule* (Hart 1968). This process starts from an initial state with one prototype, selects learning patterns one by one, and the selected pattern is classified by the nearest neighbor rule. If the pattern is misclassified, it is registered as a prototype, and this operation is repeated. This method produces a piecewise linear discriminant function. The algorithm is shown below.

Condensed Nearest Neighbor Rule

Step 1 A pattern list and a prototype list are prepared.[2] As an initial setting, all learning patterns are registered in the pattern list in a specific order, and patterns are selected according to this order in the following process. The prototype list is left empty.

Step 2 The first pattern in the pattern list is moved to the prototype list.

Step 3 The next pattern in the pattern list is classified by the nearest neighbor rule. Classification is performed by using the prototypes registered in the prototype list. If the pattern is correctly classified, it remains in the pattern list. If the pattern is misclassified, it is moved from the pattern list to the prototype list. In the same way, the next pattern in the pattern list is selected and the above process is repeated.

Step 4 After reaching the last pattern in the pattern list, return to the first pattern in the list and repeat the classification process of Step 3.

Step 5 The process terminates when one of the following conditions is satisfied. Otherwise, the classification process of Step 3 is repeated:

(1) when the pattern list is empty;
(2) when there are no more patterns to move from the pattern list to the prototype list.

The prototypes obtained in the above process are the final set of prototypes by the condensed nearest neighbor rule. They play the same role as the component vectors, as described in Sect. 8.3. Unfortunately, there is no guarantee that the prototype set

---

[2] In the original paper Hart (1968), the pattern list is called GRABBAG and the prototype list is called STORE.

obtained by the algorithm is minimal or optimal. Also, the results depend on the initial conditions and are not uniquely determined. If the algorithm converges to the state of (1) in Step 5, it is nothing but the complete storage scheme, and thus cannot be effective as the condensed nearest neighbor rule.

The order in which patterns are initially registered in the pattern list can be determined arbitrarily. However, it is considered more efficient to give priority to patterns near the decision boundary. For this purpose, the border ratio $r(x)$ described in the previous section should be calculated for all learning patterns. Then, they are arranged alternately from the two classes in descending order of the border ratio to form the initial pattern list.[3]

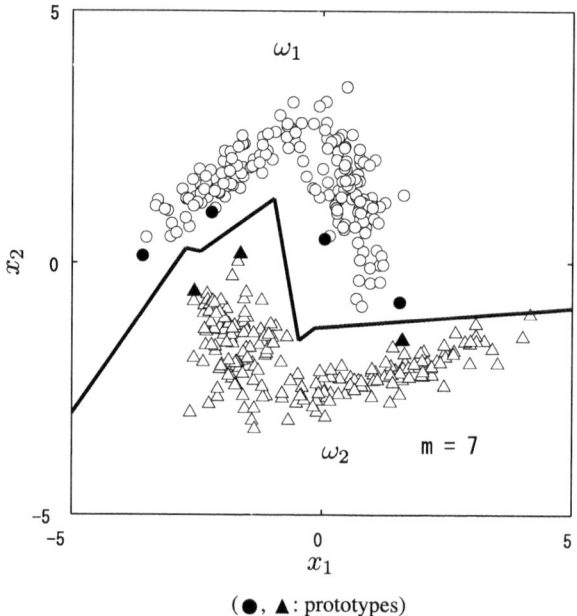

Fig. 8.9 Results of applying the condensed nearest neighbor rule (1)

($\bullet$, $\blacktriangle$: prototypes)

The experimental results of applying the condensed nearest neighbor rule to the data of Fig. 8.1 are shown in Fig. 8.9. The figure shows that four patterns from class ω_1 and three patterns from class ω_2, a total of seven patterns ($m = 7$) were selected as prototypes. In other words, 400 patterns were represented by seven prototypes. This means that the memory capacity and computational complexity for classification can be reduced to $7/400 = 0.018$, which is efficient. In the figure, the patterns selected as prototypes are indicated by $\bullet$ and $\blacktriangle$ as before. The decision boundary determined using these prototypes is indicated by thick lines in the same figure.

---

[3] In the original paper, no such sorting by the border ratio was performed.

Fig. 8.10 Results of
applying the condensed
nearest neighbor rule (2)

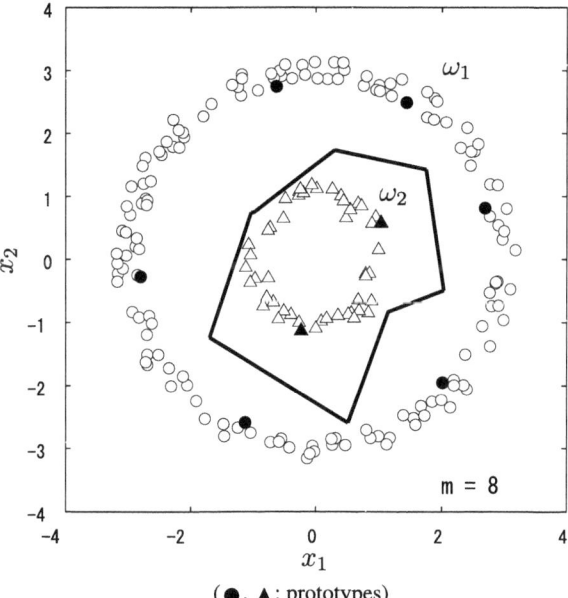

($\bullet$, $\blacktriangle$: prototypes)

Similarly, the result of applying the condensed nearest neighbor rule to other data is shown in Fig. 8.10. The figure shows patterns of two classes distributed concentrically, with class ω_1 on the outside and class ω_2 on the inside. The number of patterns of class ω_1 and ω_2 are 150 and 50, respectively, for a total of 200 patterns. The number of selected prototypes m is $6 + 2 = 8$, and the reduction effect is $8/200 = 0.04$. The representation of prototypes and decision boundary is the same as before. A simpler example is shown in Problem 8.3 for reference.

In the above, we discussed the case where the patterns are linearly nonseparable, and showed the limitations of the linear discriminant function together with experimental examples. When the patterns are linearly nonseparable, it is naturally impossible for the linear discriminant function to achieve zero misclassifications. However, if a certain degree of misclassification can be tolerated, the linear discriminant function is superior in terms of processing volume and ease of implementation, and may still be an effective method.

The piecewise linear discriminant function, an advanced form of the linear discriminant function, can be used to achieve zero misclassifications even when patterns are linearly nonseparable. Unfortunately, the piecewise linear discriminant function does not have an effective learning method comparable to the perceptron learning rule. However, although not as efficient as the perceptron learning rule, the condensed nearest neighbor rule can be used to obtain a decision boundary with zero misclassifications in a finite number of iterations. The method is based on the iterative method, and the results are subject to chance. Therefore, the disadvantage is that the decision boundary is not uniquely determined and there is no guarantee that it is optimal.

8.5 Generalized Linear Discriminant Functions

We have discussed linear discriminant functions and piecewise linear discriminant functions. Now, let us consider nonlinear discriminant functions by removing the "linear" restriction that has been assumed so far.

Consider the case where two classes distributed on a d-dimensional feature space are classified by a linear discriminant function. In this case, for a feature vector

$$x = (x_1, \ldots, x_d)^t, \tag{8.24}$$

a linear discriminant function

$$g(x) = \mathbf{w}^t \mathbf{x} = \sum_{j=0}^{d} w_j x_j \qquad (x_0 \equiv 1) \tag{8.25}$$

is applied as shown in Eq. (2.19), and the two classes are classified by Eq. (2.21).

Here, taking the linear discriminant function for $d = 2$ as an example, Eq. (8.25) is represented as

$$g(x) = w_0 + w_1 x_1 + w_2 x_2. \tag{8.26}$$

A discriminant function

$$g(x) = w_0 + w_1 x_1 + w_2 x_2 + w_3 x_1 x_2 + w_4 x_1^2 + w_5 x_2^2, \tag{8.27}$$

with a second-order term added to Eq. (8.26), or a discriminant function

$$g(x) = w_0 + w_1 x_1 + w_2 x_2 + w_3 x_1 x_2 + w_4 x_1^2 + w_5 x_2^2$$
$$+ w_6 x_1^2 x_2 + w_7 x_1 x_2^2 + w_8 x_1^3 + w_9 x_2^3 \tag{8.28}$$

with a third-order term added, etc. are no longer linear discriminant functions, but are nonlinear discriminant functions. Equation (8.27) is called a *quadratic discriminant function*; Eq. (8.28) is called a *cubic discriminant function*.

In the following, more general discriminant functions including linear discriminant functions are considered so that complex discriminant functions such as those shown above can also be handled. We define a function $\Phi(x)$ such that[4]

$$\Phi(x) = \sum_{j=1}^{D} w_j \phi_j(x), \tag{8.29}$$

---

[4] So far, $w_0, w_1, \ldots$ were used for the weight coefficients with indexes starting from 0, but $w_1, w_2, \ldots$ will be used thereafter to be consistent with the notation used in the subsequent chapters.

where $w_1, \ldots, w_D$ are D weight coefficients and $\phi_1(x), \ldots, \phi_D(x)$ are D functions of x. Such a function is called the Φ *function* (Cover 1964; Nilsson 1965).

In the above equation, if $d = 2$ and we set $\phi_i(x)$ $(i = 1, 2, 3)$ as

$$\phi_1(x) = x_0 = 1, \quad \phi_2(x) = x_1, \quad \phi_3(x) = x_2, \tag{8.30}$$

then $D = 3$ and we obtain a linear discriminant function of Eq. (8.26).

If we add

$$\phi_4(x) = x_1 x_2, \quad \phi_5(x) = x_1^2, \quad \phi_6(x) = x_2^2 \tag{8.31}$$

to the above equation, then $D = 6$ and we obtain a quadratic discriminant function of Eq. (8.27).

If we further add

$$\phi_7(x) = x_1^2 x_2, \quad \phi_8(x) = x_1 x_2^2, \quad \phi_9(x) = x_1^3, \quad \phi_{10}(x) = x_2^3, \tag{8.32}$$

then $D = 10$ and we obtain a cubic discriminant function of Eq. (8.28).

In the same way, *polynomial discriminant functions* with higher-order terms can be constructed. The above shows the case of $d = 2$, but polynomial discriminant functions can be defined for $d > 2$. In general, if we set up a polynomial discriminant function of order p for a d-dimensional feature vector x, D is expressed as a function of d and p as

$$D = f(d, p) = {}_{d+p}C_p. \tag{8.33}$$

Since $f(2, 2) = {}_4C_2 = 6$ and $f(2, 3) = {}_5C_3 = 10$ in the above equation, the results are consistent with those already checked. For proofs, refer to Problems 8.4 and 8.5.

The Φ function, expressed as a polynomial discriminant function, can be regarded as a kind of series expansion, and in principle, any function can be defined. This indicates that the Φ function can approximate complex decision boundaries in the feature space with arbitrary accuracy. The Φ function described above uses the polynomials of $x_1, \ldots, x_d$ as its component $\phi_j(x)$, but it is not limited to this and can be any single-valued real function.

As is clear from Eq. (8.29), the Φ function is nonlinear with respect to the components $x_1, \ldots, x_d$ of x, but linear with respect to $\phi_1(x), \ldots, \phi(x)$. For this reason, the Φ function is called the *generalized linear discriminant function*. The classification using this function is called the *generalized linear discriminant function method*.

Here, the D-dimensional vector $\phi(x)$ is defined as follows:

$$\phi(x) = (\phi_1(x), \ldots, \phi_D(x))^t. \tag{8.34}$$

The above equation can be regarded as a transformation equation for mapping a vector x in a d-dimensional space to a vector $\phi(x)$ in a D-dimensional space by the *nonlinear transformation*. In contrast to the d-dimensional *original space*, this D-dimensional space is called Φ *space*. In the case of $D = 10$ shown in Eqs. (8.30), (8.31), and (8.32), the following equation is obtained:

$$\boldsymbol{\phi}(\boldsymbol{x}) = (\phi_1(\boldsymbol{x}), \ldots, \phi_{10}(\boldsymbol{x}))^t$$
$$= (1, \ x_1, \ x_2, \ x_1 x_2, \ x_1^2, \ x_2^2, \ x_1^2 x_2, \ x_1 x_2^2, \ x_1^3, \ x_2^3)^t. \tag{8.35}$$

The effect of such nonlinear transformations is obvious. For example, let us consider two classes distributed concentrically around the origin in a two-dimensional ($d = 2$) original space, as shown in Fig. 8.10. These distributions can be written as

$$\begin{cases} x_1^2 + x_2^2 \approx 9 & (\boldsymbol{x} \in \omega_1) \\ x_1^2 + x_2^2 \approx 1 & (\boldsymbol{x} \in \omega_2), \end{cases} \tag{8.36}$$

and are linearly nonseparable. If we define $\phi_1(\boldsymbol{x}) = x_1^2$, $\phi_2(\boldsymbol{x}) = x_2^2$ with $D = 2$ and apply the nonlinear transformation

$$\boldsymbol{x} = (x_1, x_2)^t \quad \rightarrow \quad \boldsymbol{\phi}(\boldsymbol{x}) = (\phi_1(\boldsymbol{x}), \phi_2(\boldsymbol{x}))^t, \tag{8.37}$$

then the distribution in Φ space is

$$\begin{cases} \phi_1 + \phi_2 \approx 9 & (\boldsymbol{x} \in \omega_1) \\ \phi_1 + \phi_2 \approx 1 & (\boldsymbol{x} \in \omega_2). \end{cases} \tag{8.38}$$

In other words, it is clear that in the new feature space the two classes are transformed into almost two linear distributions, which are linearly separable. The situation is illustrated in Fig. 8.11, where (a) is the distribution in the original space (x_1, x_2) before the transformation and (b) is the distribution in Φ space (ϕ_1, ϕ_2) after the transformation.

In Eq. (8.34), if

$$D = d + 1, \tag{8.39}$$

$$\phi_1(\boldsymbol{x}) = 1, \quad \phi_2(\boldsymbol{x}) = x_1, \quad \ldots, \quad \phi_D(\boldsymbol{x}) = x_d, \tag{8.40}$$

then

$$\boldsymbol{\phi}(\boldsymbol{x}) = (1, x_1, \ldots, x_d)^t \tag{8.41}$$

$$= \mathbf{x}, \tag{8.42}$$

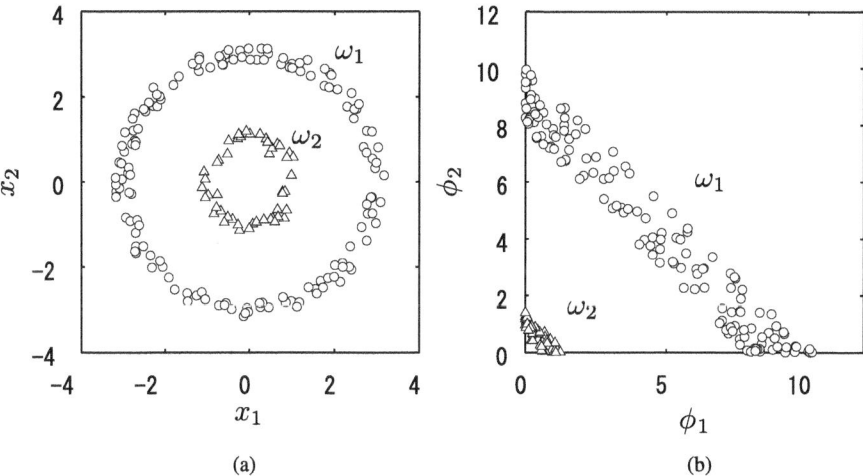

Fig. 8.11 Effects of nonlinear transformations. (**a**) Original space. (**b**) Φ space

so the transformation from the feature vector x to the augmented feature vector $\mathbf{x}$ is also represented by Eq. (8.34), and the augmented feature vector $\mathbf{x}$ is also included as a special case of $\boldsymbol{\phi}(x)$.

After the nonlinear transformation of Eq. (8.34) from x in a d-dimensional space to $\boldsymbol{\phi}(x)$ in a D-dimensional space, the method of Chap. 2, which assumes linearity, is directly applicable. As a discriminant function, we can use the Φ function of Eq. (8.29). That is, instead of the d-dimensional weight vector w, a D-dimensional weight vector

$$\mathbf{w} = (w_1, \ldots, w_D)^t \tag{8.43}$$

is used. As a result, from Eqs. (8.29) and (8.34), the discriminant function $g(x)$ can be written as

$$g(x) = \Phi(x) \tag{8.44}$$

$$= \sum_{j=1}^{D} w_j \phi_j(x) \tag{8.45}$$

$$= \mathbf{w}^t \boldsymbol{\phi}(x). \tag{8.46}$$

Instead of Eq. (2.21), the following formula is used for classification:

$$\begin{cases} g(x) = \mathbf{w}^t \boldsymbol{\phi}(x) > 0 \implies x \in \omega_1 \\ g(x) = \mathbf{w}^t \boldsymbol{\phi}(x) < 0 \implies x \in \omega_2. \end{cases} \tag{8.47}$$

As before, the decision boundary separating the classes ω_1 and ω_2 is expressed by the following equation:

$$g(\boldsymbol{x}) = 0. \tag{8.48}$$

We have introduced the perceptron learning rule in Chap. 2, the minimum square error learning in Chap. 3, and Fisher's method in Chap. 6 as methods for obtaining linear discriminant functions in d-dimensional spaces. These methods can be applied directly to the D-dimensional space after the nonlinear transformation described in this section. Even if the decision boundary defined in Φ space is a linear decision boundary (hyperplane), it is a nonlinear decision boundary in the original d-dimensional space. Therefore, it is expected that more advanced classification systems can be realized by applying the previous methods in Φ space. However, the necessary conditions of $\phi_j(\boldsymbol{x})$ that should be prepared to realize the discriminant function $g(\boldsymbol{x})$ are generally unknown.

In order to confirm what has been described so far, we introduce, in the following, the results of applying the perceptron learning rule, Fisher's method, and the minimum square error learning in Φ space. Support vector machines can also be extended from d-dimensional spaces to D-dimensional Φ spaces. The details will be described in Chap. 10.

8.6   Primal and Dual Perceptrons

Let us apply the perceptron learning rule to the Φ space. If the two classes are linearly separable in the transformed D-dimensional space, then the perceptron learning rule will arrive at the correct weights $w_1, \ldots, w_D$ in a finite number of iterations.

The perceptron learning rule in the transformed D-dimensional space is accomplished by replacing $\mathbf{x}$ with $\boldsymbol{\phi}(\boldsymbol{x})$. In other words, the perceptron in Φ space is a generalization of the perceptron in the original space. Henceforth, the perceptron in Φ space will simply be referred to as the perceptron. When it is necessary to distinguish between the two, we will refer to the perceptron in the original space as the *simple perceptron*, and the perceptron in the Φ space as the *generalized perceptron*.

For the two-class problem, the algorithm introduced in Sect. 2.3.2 is as follows.

Primal Perceptron Learning Rule

Step 1   For the feature vector $\boldsymbol{x}$, set up a function $\boldsymbol{\phi}(\boldsymbol{x})$ that performs a nonlinear transformation from d-dimensional to D-dimensional space.

Step 2   Prepare n learning patterns $\boldsymbol{x}_1, \ldots, \boldsymbol{x}_n$ whose belonging classes are known, together with $b_1, \ldots, b_n$ of teaching signals defined by Eq. (2.29).

Step 3 Set the initial value of the weight vector $\mathbf{w}$.

Step 4 Select one pattern $\mathbf{x}_k$ $(k = 1, \ldots, n)$ from the learning patterns, calculate the following $g(\mathbf{x}_k)$, and perform classification using Eq. (8.47):

$$g(\mathbf{x}_k) = \mathbf{w}^t \boldsymbol{\phi}(\mathbf{x}_k). \tag{8.49}$$

Step 5 Depending on the classification result, $\mathbf{w}$ is modified and replaced with a new weight vector $\mathbf{w}'$ as follows. Note that ρ is a positive constant.

$$\begin{cases} \mathbf{w}' - \mathbf{w} + \rho \, b_k \boldsymbol{\phi}(\mathbf{x}_k) & (b_k \, g(\mathbf{x}_k) \leq 0) \\ \mathbf{w}' = \mathbf{w} & \text{(otherwise)}. \end{cases} \tag{8.50}$$

Step 6 If all the learning patterns are correctly classified, the process terminates. Otherwise, return to Step 4 and repeat the process with another $\mathbf{x}_k$.

The above process of finding a solution by repeatedly updating $\mathbf{w}$ is the basic learning method of the perceptron, which we will call the *primal perceptron*. In order to investigate what kind of discriminant function is finally obtained, the same procedure as in Sect. 2.5 can be used. Let the constant in the perceptron learning rule be $\rho = 1$, and the initial value of the weight vector be $\mathbf{w} = \mathbf{0}$. As a result, the final weight vector obtained after convergence can be expressed as follows by replacing $\mathbf{x}_k$ in Eq. (2.39) with $\boldsymbol{\phi}(\mathbf{x}_k)$:

$$\mathbf{w} = \sum_{k=1}^{n} \alpha_k b_k \boldsymbol{\phi}(\mathbf{x}_k). \tag{8.51}$$

If the above equation holds, then from Eq. (8.46), the discriminant function is

$$g(\mathbf{x}) = \sum_{k=1}^{n} \alpha_k b_k \boldsymbol{\phi}(\mathbf{x}_k)^t \boldsymbol{\phi}(\mathbf{x}). \tag{8.52}$$

The α_k in Eq. (8.52) is an element of the error counter vector and represents the number of times that the pattern $\mathbf{x}_k$ is misclassified during the learning iterations. In other words, only the patterns $\mathbf{x}_k$ that are $\alpha_k \neq 0$ constitute the discriminant function as component vectors. Equation (8.52) shows that the generalized linear discriminant function can be computed as a linear sum of $\boldsymbol{\phi}(\mathbf{x}_k)^t \boldsymbol{\phi}(\mathbf{x})$, i.e., the inner product of the learning pattern and the input pattern.[5] If we replace $\mathbf{x}$ in the original space with $\boldsymbol{\phi}(\mathbf{x})$ in the Φ space, the three points mentioned in Sect. 2.5 are still valid here.

---

[5] Here, $\boldsymbol{\phi}(\mathbf{x})$ instead of $\mathbf{x}$ is also called an input pattern, and similarly $\boldsymbol{\phi}(\mathbf{x}_k)$ instead of $\mathbf{x}_k$ is called a learning pattern.

In representing the discriminant function $g(x)$, Eq. (8.46) uses the weight vector $\mathbf{w}$ as a parameter, while Eq. (8.52) uses the error counter vector $\boldsymbol{\alpha}$. As already mentioned in Sect. 2.5, the ability to express the same content in two different forms is called *duality*. If one of the two forms of duality is denoted by a *primal representation*, the other is denoted by a *dual representation*.

The algorithm introduced as the learning rule for the primal perceptron is a learning method that iteratively updates the weight vector $\mathbf{w}$ based on Eq. (8.46).

On the other hand, this algorithm can be described as a learning method that iteratively updates $\boldsymbol{\alpha}$ based on the dual representation of Eq. (8.52). Such a perceptron is called the *dual perceptron* as opposed to the primal perceptron (Aizerman et al. 1964; Guyon et al. 1992; Boser et al. 1992). Below is the dual perceptron learning rule for the two classes.

Dual Perceptron Learning Rule

Step 1 For the feature vector x, set up a function $\boldsymbol{\phi}(x)$ that performs a nonlinear transformation from d-dimensional to D-dimensional space.

Step 2 Prepare n learning patterns $x_1, \ldots, x_n$ whose belonging classes are known, together with $b_1, \ldots, b_n$ of teaching signals defined by Eq. (2.29).

Step 3 Initialize the error counter vector $\boldsymbol{\alpha}$ as follows:

$$\boldsymbol{\alpha} = (\alpha_1, \alpha_2, \ldots, \alpha_n)^t = \mathbf{0}. \tag{8.53}$$

Step 4 Select one pattern x_k $(k = 1, \ldots, n)$ from the learning patterns, calculate the following $g(x_k)$ according to Eq. (8.52), and perform classification using Eq. (8.47).

$$g(x_k) = \sum_{i=1}^{n} \alpha_i b_i \boldsymbol{\phi}(x_i)^t \boldsymbol{\phi}(x_k). \tag{8.54}$$

Step 5 Depending on the classification result, α_k is modified and replaced with new α_k' as follows:

$$\begin{cases} \alpha_k' = \alpha_k + 1 & (b_k\, g(x_k) \le 0) \\ \alpha_k' = \alpha_k & (\text{otherwise}). \end{cases} \tag{8.55}$$

Step 6 If all the learning patterns are correctly classified, the process terminates. Otherwise, return to Step 4 and repeat the above process with another pattern.

Equation (8.53) corresponds to setting the initial weight vector to $\mathbf{w} = \mathbf{0}$ in Step 3 of the primal perceptron. Also, Eq. (8.55) corresponds to setting $\rho = 1$ in Step

5 of the primal perceptron. Therefore, the result obtained by applying the dual perceptron is the same as the primal perceptron when $\mathbf{w}$ is initially set to $\mathbf{w} = \mathbf{0}$ and $\rho = 1$. However, the amount of computation of $g(\mathbf{x}_k)$ required in the learning process is larger for the dual perceptron (Eq. (8.54)), than for the primal perceptron (Eq. (8.49)). Despite this, the reason why we introduce the dual perceptron here is that this algorithm is closely related to the potential function method of Chap. 9 and the kernel method of Chap. 11. Details are described in the following chapters.

Examples of running the primal perceptron and the dual perceptron are shown in Problem 8.6. Refer to it as a concrete example to understand the difference between the two.

8.7 Experiments on Generalized Linear Discriminant Function Methods

We have introduced so far the perceptron learning rule, Fisher's method, and the minimum square error learning as methods for finding linear discriminant functions. When the learning patterns are linearly nonseparable, the perceptron learning rule cannot be applied, but Fisher's method and the minimum square error learning can be applied, and we have experimentally shown the results of their application in Sect. 8.2.

As already mentioned, in Φ space, the above linear discriminant functions become generalized linear discriminant functions. The generalized linear discriminant functions obtained by Fisher's method and the minimum square error learning can exhibit higher classification ability than the linear discriminant functions in the original space. Moreover, the simple perceptron becomes a generalized perceptron in Φ space and is guaranteed to converge if the learning patterns are linearly separable in Φ space, which gives it higher performance than the simple perceptron. In the following, we will experimentally verify this.

In the experiments, we target linearly nonseparable learning patterns shown in Fig. 8.1. The cubic discriminant function is used as the discriminant function, and the Φ function of Eq. (8.29) is set as follows:

$$\boldsymbol{\phi}(\mathbf{x})$$

$$= (\phi_1(\mathbf{x}), \ldots, \phi_{10}(\mathbf{x}))^t \tag{8.56}$$

$$= (1, \sqrt{3}x_1, \sqrt{3}x_2, \sqrt{6}x_1 x_2, \sqrt{3}x_1^2, \sqrt{3}x_2^2, \sqrt{3}x_1^2 x_2, \sqrt{3}x_1 x_2^2, x_1^3, x_2^3)^t. \tag{8.57}$$

The above equation is a transformation from two dimensions ($d = 2$) to ten dimensions ($D = 10$). However, it differs from Eq. (8.35) in that each element is multiplied by a factor of $\sqrt{3}$ or $\sqrt{6}$. The reason for using these factors is described in Sect. 9.5.2. The ten weight coefficients $w_1, \ldots, w_{10}$ in Eq. (8.43) are to be obtained by learning. These experimental conditions are the same for the subsequent experiments.

8.7.1   Primal Perceptrons in Φ Space

In the following, we perform experiments using the primal perceptron. Similar results can be obtained using the dual perceptron.

The initial value of the weight vector was set to $\mathbf{w} = \mathbf{0}$. Next, the learning patterns were transformed into ten-dimensional vectors ($D = 10$) of Eq. (8.57) and the perceptron learning rule was applied in Φ space. The learning rate was set to $\rho = 1$. As a result, the learning converged after 2406 iterations, and all learning patterns were correctly classified. In the iterative process, patterns were alternately selected and classified from the two classes in order of descending border ratio, similar to the method used in the condensed nearest neighbor rule experiments. The resulting decision boundary is a hyperplane in the ten-dimensional Φ space. The decision boundary in the original two-dimensional feature space is shown as the thick line in Fig. 8.12. The decision boundary is no longer a straight line in the original feature space, but a complex curve, indicating that the two classes are correctly separated. Thus, the learning patterns in Fig. 8.1 are linearly nonseparable in the original space, but is linearly separable in Φ space by Eq. (8.57).

Fig. 8.12 The decision boundary obtained by the perceptron

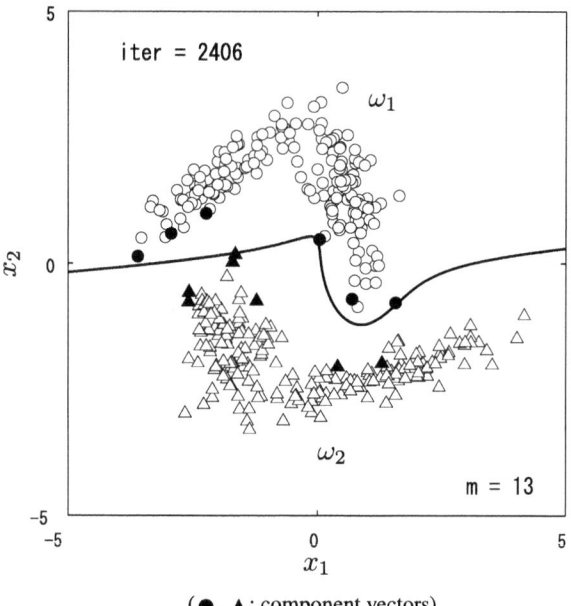

($\bullet$, $\blacktriangle$: component vectors)

Here, let us denote by m ($\leq n$) the number of learning patterns that resulted in $\alpha_k \neq 0$ in Eq. (8.52), i.e., the number of component vectors. In this experiment, $m = 13$ consisting of six and seven patterns for classes ω_1 and ω_2, respectively, which are indicated by $\bullet$ and $\blacktriangle$ in the figure. The number of learning patterns contributing to the construction of the final discriminant function is small ($m = 13$) compared to the total number of learning patterns ($n = 400$).

The experiment with the condensed nearest neighbor rule introduced in Fig. 8.9 showed that only the seven patterns used as prototypes contribute to the determination of the decision boundary. Thus, both methods have in common that only a part of the learning patterns are involved in the construction of the discriminant function. The process of obtaining the final decision boundary is shown in Fig. 8.13.

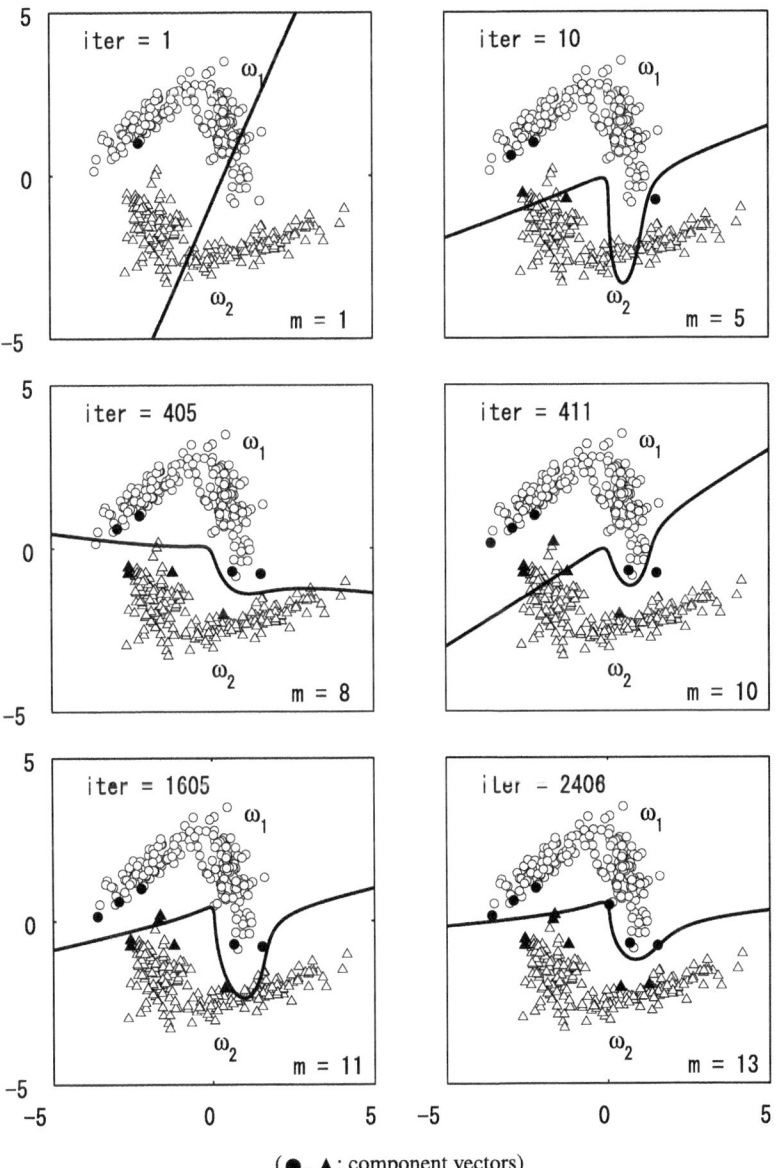

($\bullet$, $\blacktriangle$: component vectors)

Fig. 8.13 Convergence process of the decision boundary by the Φ function and the perceptron

The number of iterations is indicated in the upper left corner of each graph, and the number of component vectors m is indicated in the lower right corner.

In a subsequent experiment, let us try to obtain the decision boundary for the two concentrically distributed classes shown in Fig. 8.10 in the same way as before. In the experiment, $\rho = 1$ and the initial value of **w** was set to **w** = **0**. The results are shown in Fig. 8.14. For this data, convergence was achieved after 785 iterations, indicating that the two classes were still correctly separated. The number of component vectors was 47 in total, 8 for class ω_1 and 39 for class ω_2, which are blacked out in the figure.

Fig. 8.14 The decision boundary obtained by the perceptron (data of Fig. 8.10)

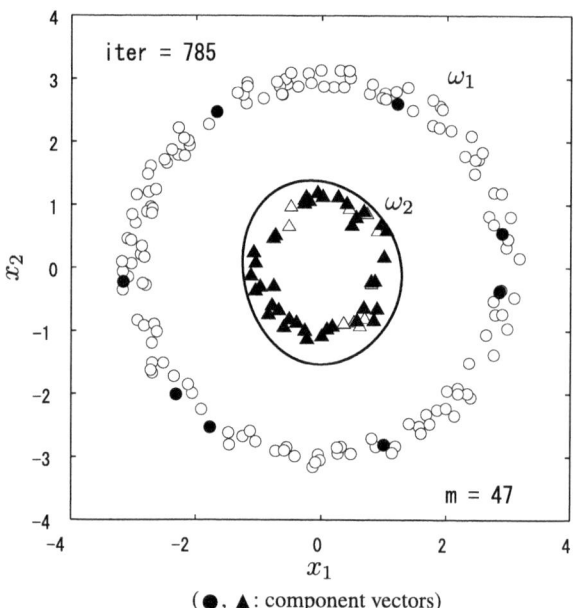

($\bullet$, $\blacktriangle$: component vectors)

8.7.2   Fisher's Method in Φ Space

Let us apply Fisher's method introduced in Sect. 8.2.1 to patterns in Φ space.

The result of projecting the learning patterns in the ten-dimensional feature space ($D = 10$) (after nonlinear transformation by Eq. (8.57)) onto the projection axis y obtained by Fisher's method is shown in Fig. 8.15.

The threshold value $-w_0 = -0.728$ shown in the figure allows the two classes to be separated without misclassifications, indicating that the learning patterns are linearly separable in Φ space. The difference is obvious when compared to Fig. 8.2 which shows the projection of patterns in the original two-dimensional feature space. As already mentioned in Sect. 8.2, Fisher's method cannot automatically determine the threshold $-w_0$ which minimizes misclassifications, so $-w_0$ must be

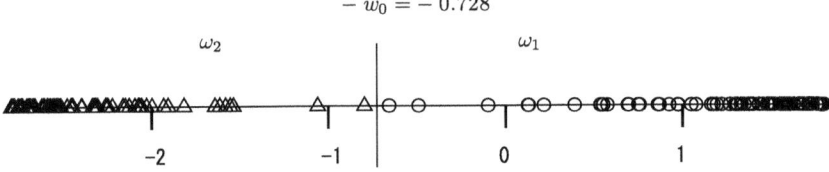

$$-w_0 = -0.728$$

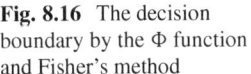

Fig. 8.15 Learning patterns in Φ space (10 dimensions) projected onto Fisher's axes

Fig. 8.16 The decision
boundary by the Φ function
and Fisher's method

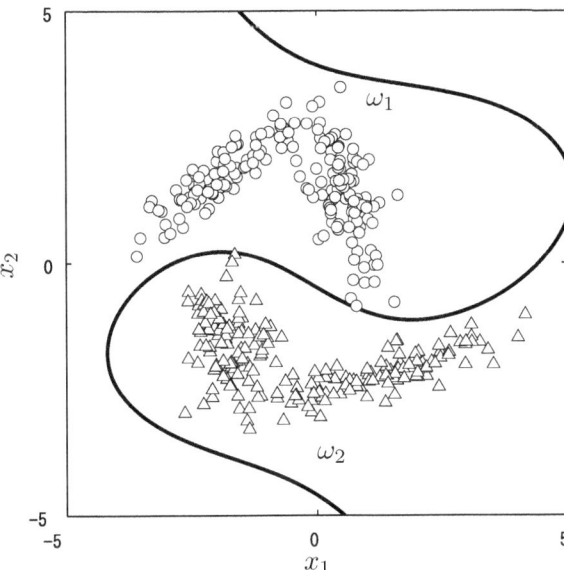

obtained separately by other means. The above $-w_0 = -0.728$ is the value obtained by visual observation, and its position on the projection axis is shown in Fig. 8.15. The hyperplane orthogonal to the projection axis of the figure is the decision boundary in the Φ space. The thick line of Fig. 8.16 is the decision boundary drawn on the original two-dimensional feature space which is the hyperplane that is orthogonal to the projection axis and passes through at $-w_0 = -0.728$ in the Φ space. As is clear from the figure, it can be confirmed that the two classes are correctly separated by the nonlinear decision boundary. However, the margin for separation is not sufficient.

8.7.3 Minimum Square Error Learning in Φ Space

Next, let's apply the minimum square error learning introduced in Sect. 8.2.2 in the Φ space. For that, we replace $\mathbf{x}_k$ in Eq. (3.38) with $\boldsymbol{\phi}(\boldsymbol{x}_k)$ and find $\mathbf{w}$ that minimizes

$$J(\mathbf{w}) = \frac{1}{2} \sum_{k=1}^{n} \left(\mathbf{w}^t \boldsymbol{\phi}(\boldsymbol{x}_k) - b_k \right)^2 . \tag{8.58}$$

By using

$$\mathbf{X} = (\boldsymbol{\phi}(\boldsymbol{x}_1), \boldsymbol{\phi}(\boldsymbol{x}_2), \dots, \boldsymbol{\phi}(\boldsymbol{x}_n))^t \tag{8.59}$$

instead of Eq. (3.16) as the pattern matrix $\mathbf{X}$, and using Eq. (3.39) as the column vector $\mathbf{b}$, we obtain $\mathbf{w}$ by the following equation, similar to Eq. (3.41):

$$\mathbf{w} = (\mathbf{X}^t \mathbf{X})^{-1} \mathbf{X}^t \mathbf{b}. \tag{8.60}$$

The decision boundary in the Φ space obtained by applying the minimum square error learning is drawn in the original two-dimensional space as shown by the thick line in Fig. 8.17.

Fig. 8.17 The decision boundary by the Φ function and the minimum square error learning

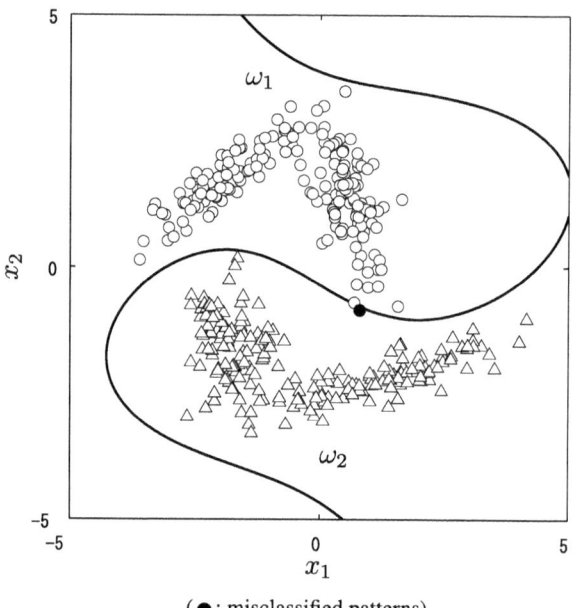

($\bullet$: misclassified patterns)

As already mentioned in Sect. 8.2.2, the minimum square error learning and Fisher's method are closely related, and the weight vectors $\boldsymbol{w}$ obtained by both methods are the same. In the minimum square error learning, the value of w_0 as well as $\boldsymbol{w}$ is automatically determined. However, the decision boundary obtained by the minimum square error learning does not necessarily minimize the number of misclassifications. In fact, Fig. 8.17 shows that one pattern (black color) is misclassified even though the learning patterns are linearly separable in the Φ space.

————————————————— **Coffee Break** —————————————————

Subsequent History of the Φ Function as a Universal Discriminant Function

Generalized linear discriminant functions, or Φ functions, can represent arbitrary functions. Therefore, the Φ function is a universal discriminant function that can set any complex decision boundary. Moreover, since the Φ function can be realized by simply replacing the d-dimensional vector x with a D-dimensional vector $\phi(x)$, existing methods such as the perceptron learning rule can be applied without any modification. Its effectiveness has been demonstrated in the experiments in this chapter. However, after the generalized linear discriminant function was proposed, the Φ function was dismissed as a classical method and never used. Why is this?

The title of this chapter, "From Linear to Nonlinear", must have been a great dream of many researchers for a long time. It can be said that a number of beautiful formulations were possible because of the constraint of "linearity". In contrast, the "nonlinear" world is so vast. Therefore, the reason why it was not used is that it was not possible to give a definite guideline on how to choose the nonlinear function $\phi(x)$.

However, the situation changed when the potential function method proposed later was shown to be equivalent to the Φ function under certain conditions. This is because the choice of the nonlinear function $\phi(x)$ has been replaced by the choice of a much more prospective potential function. As a result, even though the nonlinear transformation by $\phi(x)$ is actually applied, the equivalent process can be realized without explicitly specifying $\phi(x)$.

The potential function method was followed over the years by support vector machines and the kernel method.

8.8 Problems

8.1 Show that w of Eq. (8.21) obtained by Fisher's method and w of Eq. (8.22) obtained by the minimum square error learning are the same.

8.2 † The following 12 patterns $x_1, x_1, \ldots, x_{12}$ are distributed in a two-dimensional feature space, where x_1, x_2, x_3, x_4 belong to class ω_1, x_5, x_6, x_7, x_8 to class ω_2, and $x_9, x_{10}, x_{11}, x_{12}$ to class ω_3, respectively:

$$\left.\begin{array}{l} x_1 = (5,9)^t, \ x_2 = (9,9)^t, \ x_3 = (8,8)^t, \ x_4 = (6,7)^t \quad \in \omega_1 \\ x_5 = (2,6)^t, \ x_6 = (4,5)^t, \ x_7 = (3,4)^t, \ x_8 = (1,3)^t \quad \in \omega_2 \\ x_9 = (8,4)^t, \ x_{10} = (9,1)^t, \ x_{11} = (7,2)^t, \ x_{12} = (8,2)^t \quad \in \omega_3 \end{array}\right\}.$$

Plot these patterns on a two-dimensional feature space and find the border ratio of each pattern.

8.3 † Apply the condensed nearest neighbor rule to the data in the previous problem, and illustrate the prototypes obtained and the decision boundaries set by them. The patterns are given repeatedly from x_1 to x_{12} in this order.

8.4 Let k be a nonzero integer, and $_d H_k$ be the number of integer solutions $(r_1, r_2, \ldots, r_d)$, satisfying

$$\begin{cases} r_1 + r_2 +, \ldots, + r_d = k \\ r_j \geq 0 \quad (j = 1, \ldots, d). \end{cases} \tag{8.61}$$

Show that $_d H_k$ is expressed as

$$_d H_k = {}_{d+k-1}C_k.$$ (8.62)

8.5 Prove that Eq. (8.33) holds in the following two ways.

(1) Show that the following equation holds and derive Eq. (8.33) from it (Proof 1). (Hint: Use Eq. (8.62) in the derivation.)

$$f(d, p) = \sum_{k=0}^{p} {}_{d+k-1}C_k.$$ (8.63)

(2) After showing that the following equation holds, show by mathematical induction that Eq. (8.33) holds (Proof 2):

$$f(d, p) = f(d - 1, p) + f(d, p - 1).$$ (8.64)

8.6 [†]   The following six learning patterns $x_1, \ldots, x_6$ are distributed in a two-dimensional feature space, of which x_1, x_2, x_3 belong to class ω_1 and x_4, x_5, x_6 to class ω_2, respectively:

$$\left. \begin{array}{l} x_1 = (1, 5)^t, \ x_2 = (3, 2)^t, \ x_3 = (4, 3)^t \\ x_4 = (5, 6)^t, \ x_5 = (2, 4)^t, \ x_6 = (6, 1)^t \end{array} \right\}.$$

Let us find the decision boundary for classifying both classes by means of a cubic discriminant function.

(1) After mapping patterns to the ten-dimensional Φ space by Eq. (8.35), the primal perceptron learning rule (Sect. 8.6) is applied. The learning patterns are repeatedly given from x_1 to x_6 in this order. Set the initial value of the weight vector to

$$w = (w_1, \ldots, w_{10})^t = 0$$

and set the learning rate to $\rho = 1$. Then, show the final weight vector w after learning. Also, show the value of w for the first and the last five epochs leading to convergence. In addition, illustrate the decision boundary determined by the weight vector.

(2) Apply the dual perceptron learning rule (Sect. 8.6) to obtain a cubic discriminant function for classifying the two classes. Assuming that the learning patterns are given repeatedly from x_1 to x_6 in this order, show the final error counter vector α and the error counter vector αs for the first and the last five epochs leading to convergence.

Chapter 9
Potential Function Methods

Abstract In the previous chapter, we discussed nonlinear discriminant functions. The process introduced can be summarized as follows.

(1) Map the learning patterns from the d-dimensional to the D-dimensional feature space by nonlinear transformation. (2) Set a linear discriminant function in the transformed D-dimensional space to obtain the decision boundary for class separation. (3) Transform the obtained decision boundary to the decision boundary in the original d-dimensional space.

Although the above process ultimately sets up a nonlinear discriminant function in the d-dimensional space, the main operation is to find a linear discriminant function in the D-dimensional space. In other words, it can be said that the nonlinear discriminant function is obtained indirectly through the D-dimensional space. Therefore, this chapter introduces the potential function method as a typical method for obtaining nonlinear discriminant functions directly in the original d-dimensional space.

9.1 Principles of Potential Function Method

The potential function method considers the *electrostatic potential* introduced by a charged particle as a discriminant function. An overview of the method is given below.

Now, when a charged particle with unit charge is at position x', the electrostatic potential at position x is expressed as $K(x, x')$. This is called the *potential function*. As before, the following two-class problem of ω_1 and ω_2 is considered. There are n patterns belonging to one of these classes, and the k-th pattern is x_k ($k = 1, \ldots, n$). The patterns are assumed to be generated according to the probability distribution of each class. There is a charged particle with charge q_k at x_k in the feature space, with $q_k > 0$ for patterns of class ω_1 and $q_k < 0$ for patterns of class ω_2. Then, the

Supplementary Information The online version contains supplementary material available at https://doi.org/10.1007/978-981-95-1478-6_9.

electrostatic potential $g(x)$ at position x brought about by these n charged particles can be expressed as

$$g(x) = \sum_{k=1}^{n} q_k K(x, x_k), \tag{9.1}$$

using the potential function, i.e. the electrostatic potential $K(x, x_k)$ with unit charge.

The potential function $K(x, x_k)$ is used in physics as

$$\begin{cases} \underset{x}{\mathrm{argmax}} \ K(x, x_k) = x_k \\ \lim_{\|x - x_k\| \to \infty} K(x, x_k) = 0, \end{cases} \tag{9.2}$$

such that it is maximum at $x = x_k$ and decreases monotonically as x moves away from x_k and approaches 0. However, the potential function applied to pattern recognition need not necessarily satisfy these conditions. Details are described later.

For a pattern x of unknown class, the following formula is used for classification:

$$\begin{cases} g(x) > 0 & \Longrightarrow & x \in \omega_1 \\ g(x) < 0 & \Longrightarrow & x \in \omega_2 \end{cases} \tag{9.3}$$

The decision boundary separating both classes is $g(x) = 0$. Such a discrimination method is called the *potential function method*, and was actively studied by M. A. Aizerman and others in the 1960s (Aizerman et al., 1964). In the multi-class case ($c > 2$), by setting the charge $q_k > 0$ for all x_k, the electrostatic potentials for each class $g_1(x), g_2(x), \ldots, g_c(x)$ are obtained. The classification should be

$$\max_{i=1,\ldots,c} \{g_i(x)\} = g_k(x) \quad \Longrightarrow \quad x \in \omega_k. \tag{9.4}$$

If $K(x, x_k)$ has the property as a probability density function, it is possible to estimate the original probability distribution from n patterns. That is, using the patterns of classes ω_1 and ω_2, the probability distribution can be estimated as

$$\begin{cases} p(x, \omega_1) = a \sum_{x_k \in \omega_1} K(x, x_k) \\ p(x, \omega_2) = a \sum_{x_k \in \omega_2} K(x, x_k), \end{cases} \tag{9.5}$$

where a is a positive constant. This method is known as the *Parzen window*. For details of the method, see Appendix E.

Here, the charge q_k of Eq. (9.1) is set to $q_k = 1$ for patterns of class ω_1 and $q_k = -1$ for patterns of class ω_2. Then, from Eqs. (9.1) and (9.5), we can write $g(x)$

as

$$g(x) = \sum_{k=1}^{n} q_k K(x, x_k)$$

$$= \sum_{x_k \in \omega_1} K(x, x_k) - \sum_{x_k \in \omega_2} K(x, x_k)$$

$$= \big(p(x, \omega_1) - p(x, \omega_2)\big)/a$$

$$= \big(P(\omega_1)p(x|\omega_1) \quad P(\omega_2)p(x|\omega_2)\big)/a. \tag{9.6}$$

As shown in Eq. (8.14), it can be seen that $g(x)$ in the above equation serves as a discriminant function based on the Bayes decision rule, i.e., the Bayes discriminant function.

Often used as a potential function is the following *Gaussian function*:

$$K(x, x_k) = \exp\left[-\frac{\|x - x_k\|^2}{2\sigma^2}\right], \tag{9.7}$$

where σ is a positive constant and is a parameter that controls the spread of the Gaussian function.[1] In the following, let us draw $g(x)$ of Eq. (9.1) for the two classes of linearly nonseparable distributions of Fig. 8.1, using the above potential function. The result is shown in Fig. 9.1. Here, $q_k = 1$ for patterns of class ω_1 and $q_k = -1$ for patterns of class ω_2. In the figure, the contour lines of $g(x)$ and the decision boundary $g(x) = 0$ separating the two classes are indicated by thin and thick lines, respectively. The parameter σ was varied in four ways: $\sigma = 0.2, 1.0, 2.0, 4.0$.

The potential function method has the following two problems:

Problem 1 The discriminant function $g(x)$ obtained as $q_k = 1$ or -1 ($k = 1, \ldots, n$) in Eq. (9.1), does not always correctly classify all learning patterns. (for example, Fig. 9.1 with $\sigma = 1.0$, $\sigma = 2.0$ and $\sigma = 4.0$).

Problem 2 To classify an unknown pattern x by the potential function method, it is necessary to calculate the sum of $K(x, x_k)$ for n learning patterns $x_1, \ldots, x_n$, as shown in Eq. (9.1), which requires a huge amount of computation.

In the following, we will discuss these issues and show how to solve them.

---

[1] This equation corresponds to the multivariate normal distribution of Eq. (8.1), with $\Sigma = \sigma^2 I_d$ for the covariance matrix, where I_d is a d-dimensional *identity matrix*. Such Σ is called isotropic.

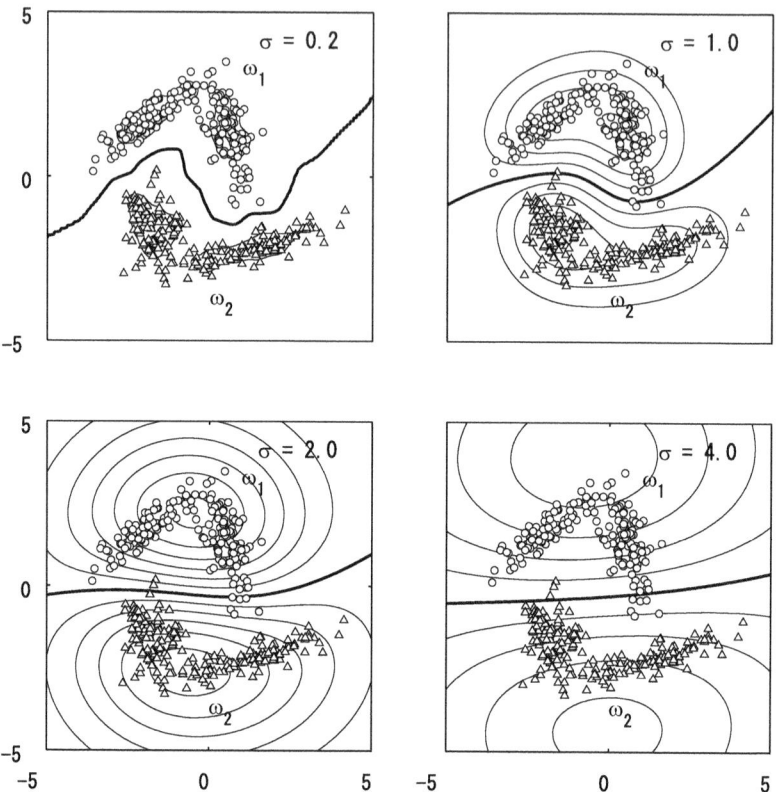

Fig. 9.1 Decision boundaries by potential function method (effect of parameter σ)

9.2 Potential Function Method as a Discriminative Model

Let us compare the potential function method with the nearest neighbor rule (Sect. 1.3) based on the complete storage scheme. Both methods have in common that the discriminant function $g(x)$ is constructed from all learning patterns. Naturally, the nearest neighbor rule based on the complete storage scheme can correctly classify all learning patterns. The potential function method can also achieve the same performance if the parameter σ is set to a small value (for example, $\sigma = 0.2$ in Fig. 9.1). This is because if σ is small, the function $K(x, x_k)$ will have a sharp peak at x_k, and the value of $K(x, x_k)$ is approximately determined by x_k, which is the nearest neighbor of x. In this case, the potential function method has the same effect as the nearest neighbor rule with the prototype set at x_k. In fact, the decision boundary for $\sigma = 0.2$ in Fig. 9.1 is almost equal to the decision boundary by the complete storage scheme shown in Fig. 8.5. However, if σ is extremely small, the estimation accuracy of the obtained probability distribution decreases.

The method described in the previous section is based on the idea of the *generative model*. That is, the original probability distribution is estimated with high accuracy, and the discriminant function is obtained based on it. Therefore, if the estimation accuracy of the probability distribution is low due to inappropriate setting of the parameter σ, the resulting discriminant function cannot achieve sufficient performance. If σ is set appropriately and the probability distribution of each class can be estimated accurately, the decision boundary can be found as the difference between the probability density functions of the two classes. However, to estimate the original probability density functions with high accuracy, a large number of learning patterns that reflect the actual probability distributions must be prepared. To classify an unknown pattern x by the potential function method, the product-sum operations between q_k and $K(x, x_k)$ $(k = 1, \ldots, n)$ are required, as shown in Eq. (9.1). Therefore, as the number of learning patterns increases, this process requires a huge amount of computation.

On the other hand, if the final goal is to separate classes for the given learning patterns, the process of strictly reproducing the probability distribution is not necessary. Therefore, a method to directly obtain the discriminant function without going through the estimation of the probability distribution can be considered. Such a method is based on the idea of the *discriminative model* and contrasts with the aforementioned generative model.

Suppose that learning patterns $x_1, \ldots, x_n$ were classified by the discriminant function of Eq. (9.1), and x_k was misclassified. If x_k belongs to ω_1, we can expect to improve the discriminant function by increasing q_k by a constant amount of charge to make $g(x) + \rho K(x, x_k)$, and if x_k belongs to ω_2, by decreasing q_k by a constant amount of charge to make $g(x) - \rho K(x, x_k)$. Note that ρ is the learning rate and is a positive constant. This method is nothing but the error-correction method introduced in Sect. 2.3, because it modifies the discriminant function when the pattern is not correctly classified. Denoting the discriminant function after such modification as $g(x)'$, the process described above can be summarized as follows:

$$
\begin{cases}
g(x)' = g(x) + \rho b_k K(x, x_k) & (b_k\, g(x_k) \leq 0) \\
g(x)' = g(x) & (\text{otherwise}),
\end{cases}
\tag{9.8}
$$

where b_k is the teaching signal shown in Eq. (2.29). If a discriminant function that can correctly classify all learning patterns is obtained by repeating the above process, Problem 1 can be solved. The details are described in the next section.

The form of the discriminant function obtained by repeating the process of Eq. (9.8) can be calculated by the procedure introduced in Sect. 8.6. That is, we initialize $g(x) = 0$ by setting $q_k = 0$ $(k = 1, \ldots, n)$ in Eq. (9.1) and further set $\rho = 1$. As a result, the final discriminant function obtained by the potential function method is expressed as follows in the same way as Eq. (8.52):

$$
g(x) = \sum_{k=1}^{n} \alpha_k b_k K(x, x_k).
\tag{9.9}
$$

By comparing with the above equation, we can see that q_k in Eq. (9.1) is $q_k = \alpha_k b_k$. The α_k in the above equation is an element of the error counter vector shown in Eq. (2.42) and represents the number of times the pattern x_k is not correctly classified in the iterative learning process. Therefore, only the pattern x_k that is $\alpha_k \neq 0$ constitutes the discriminant function as a component vector and contributes to the decision boundary determination.

It is clear from Eq. (9.9) that if the number of component vector m is small compared to n, the potential function values only need to be computed for a small number of learning patterns. In that case, the computational complexity of $g(x)$ is reduced compared to Eq. (9.1), and Problem 2 is solved.

9.3 Learning by Potential Function Method

To incorporate Eq. (9.8) as a potential function learning algorithm, we can write it in terms of updating the error counter vector α, similar to the dual perceptron learning rule in Sect. 8.6. As a result, the potential function learning algorithm can be written as follows.

Potential Function Learning Algorithm

Step 1 Define the potential function $K(x, x_k)$.

Step 2 Prepare n learning patterns $x_1, \ldots, x_n$ whose belonging classes are known, together with $b_1, \ldots, b_n$ of teaching signals.

Step 3 Initialize the error counter vector α as follows:

$$\alpha = (\alpha_1, \alpha_2, \ldots, \alpha_n)^t = \mathbf{0}. \tag{9.10}$$

Step 4 Select one pattern x_k ($k = 1, \ldots, n$) from the learning patterns, calculate the following $g(x_k)$ according to Eq. (9.9), and perform classification using Eq. (9.3):

$$g(x_k) = \sum_{i=1}^{n} \alpha_i b_i K(x_k, x_i). \tag{9.11}$$

Step 5 Depending on the classification result, α_k is modified and replaced with new α_k' as follows:

$$\begin{cases} \alpha_k' = \alpha_k + 1 & (b_k\, g(x_k) \leq 0) \\ \alpha_k' = \alpha_k & (\text{otherwise}). \end{cases} \tag{9.12}$$

Step 6 If all the learning patterns are correctly classified, the process terminates. Otherwise, return to Step 4 and repeat the process with another pattern.

Equation (9.12) is equivalent to Eq. (9.8) for the modification process with $\rho = 1$. Each time a modification is made according to Eq. (9.12), the discriminant function is updated in the direction of improvement. However, repeating this process does not guarantee that the algorithm will converge to a discriminant function $g(x)$ that correctly classifies all learning patterns. Then, what conditions must be satisfied to obtain a correct discriminant function?

Suppose that the following equation holds (Duda and Hart 1973):

$$K(x, x_k) = \phi(x_k)^t \phi(x). \tag{9.13}$$

As already explained in Chap. 8, $\phi(x)$ is a D-dimensional vector in Φ space defined by

$$\phi(x) = (\phi_1(x), \ldots, \phi_D(x))^t. \tag{9.14}$$

Then Eq. (9.9) can be written as

$$g(x) = \sum_{k=1}^{n} \alpha_k b_k \phi(x_k)^t \phi(x). \tag{9.15}$$

which coincides with Eq. (8.52). If the discriminant function of the potential function method can be written in the form of the above equation, then the potential function method is equivalent to the generalized linear discriminant function method. That is, if Eq. (9.13) holds, then the potential function learning algorithm is consistent with the dual perceptron learning rule described in Sect. 8.6. Therefore, if the learning patterns are linearly separable in Φ space, the potential function learning algorithm converges after a finite number of iterations, and a discriminant function $g(x)$ that can correctly classify all learning patterns is obtained. In this case, Problem 1 listed in Sect. 9.1 has been solved. That is, if the learning patterns are linearly separable in Φ space and Eq. (9.13) holds, then the potential function method can yield $g(x)$ that correctly classifies all the learning patterns.

However, $K(x, x_k)$ satisfying Eq. (9.13) does not necessarily satisfy the physical condition of Eq. (9.2). It was mentioned in Sect. 9.1 that the potential functions used in pattern recognition, unlike physical potential functions, do not necessarily satisfy Eq. (9.2).

Then, can the Gaussian function, which has been used as the potential function so far, be expressed as the inner product of two vectors as in Eq. (9.13)? Fortunately, the Gaussian function $K(x, x_k)$ of Eq. (9.7), called the Gaussian kernel, can be expressed as the inner product of infinite-dimensional ($D = \infty$) vectors. See Problem 9.1 for the proof. The Gaussian function also satisfies Eq. (9.2).

From now on, the potential function $K(x, x_k)$ is assumed to satisfy Eq. (9.13). Such potential functions are not limited to Gaussian functions. Other examples of potential functions are introduced in Chap. 11 as kernel functions. Kernel functions are equivalent to potential functions.

9.4 Experiments on the Potential Function Method

In this section, we present the experiments using the potential function method shown in Sect. 9.3. In our experiments, we used as learning patterns the linearly nonseparable patterns in a two-dimensional space shown in Fig. 8.1. The Gaussian function of Eq. (9.7) was used as the potential function and set to $\sigma = 2.0$. The learning patterns to be classified were selected in descending order of the border ratio $r(x)$, and were selected alternately from the two classes.

The potential function learning converged at 955 iterations, yielding results of Fig. 9.2. In the figure, the $g(x)$ contours and decision boundary $g(x) = 0$ are indicated by thin and thick lines, respectively. The decision boundary confirms that the two classes of learning patterns are correctly separated. In the final discriminant function of Eq. (9.9), the number of patterns with $\alpha_k \neq 0$, i.e., the number of component vectors, was 7 for class ω_1 and 8 for class ω_2, totaling 15. They are indicated by ● and ▲ in the figure, respectively, and the number of component vectors m is shown in the lower right corner. The progress toward convergence is also shown in Fig. 9.3. The number of iterations "iter" and the number of component vectors m are shown in the upper left and lower right of each figure, respectively. It can be seen that the number of component vectors increases as the number of iterations increases. The bottom right figure shows the converged state (the same as Fig. 9.2).

Fig. 9.2 Contours and decision boundary by potential function method $(\sigma = 2.0)$

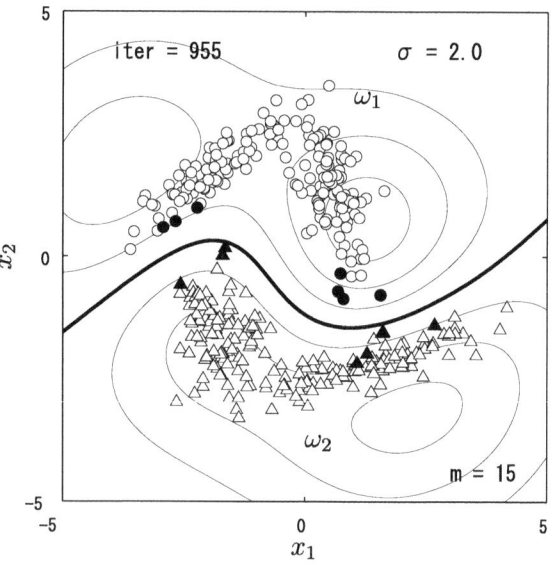

(●, ▲: component vectors)

As above, we used the Gaussian function as the potential function, and the experimental results for other learning patterns are shown in Fig. 9.4. This data is

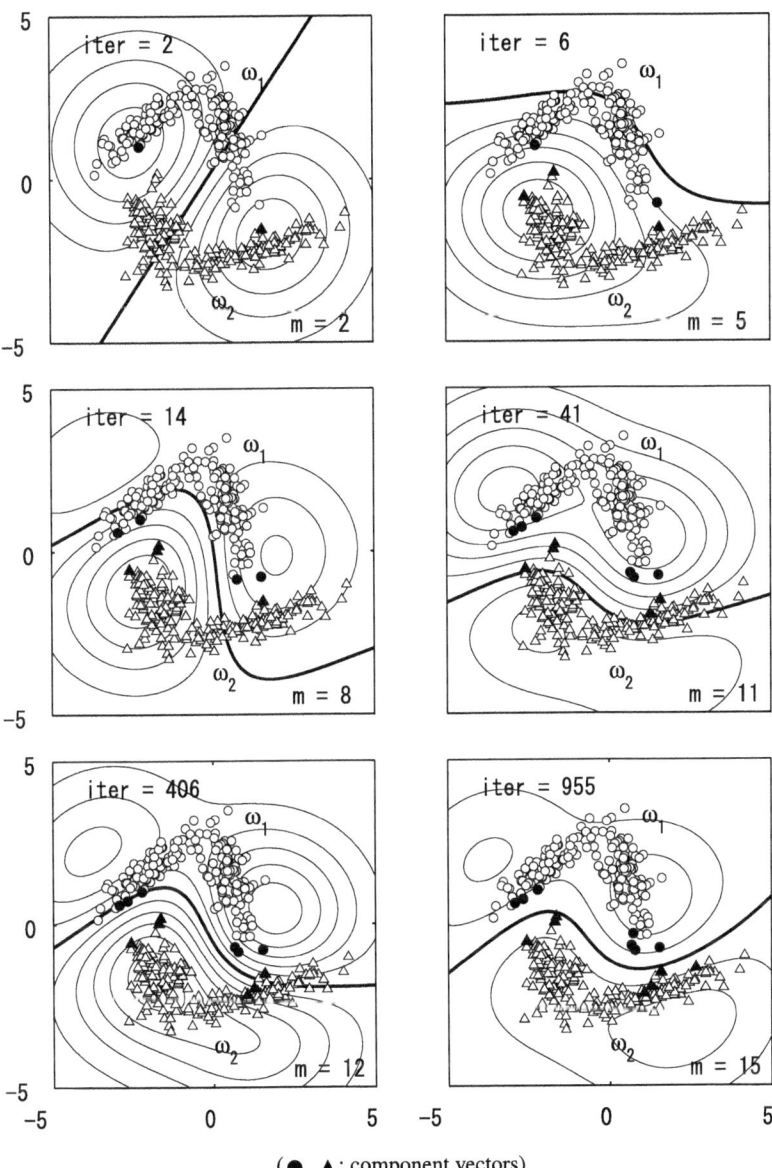

($\bullet$, $\blacktriangle$: component vectors)

Fig. 9.3 Learning process of decision boundaries by potential function method ($\sigma = 2.0$)

the learning patterns shown in Fig. 8.10, and the number of component vectors was 5 and 3 for the classes ω_1 and ω_2, respectively, for a total of 8 patterns ($m = 8$). As before, they are indicated by $\bullet$ and $\blacktriangle$ in the figure, respectively.

A simpler example is given as Problem 9.2 (1), refer to it.

Fig. 9.4 Decision boundary by potential function method (data from Fig. 8.10)

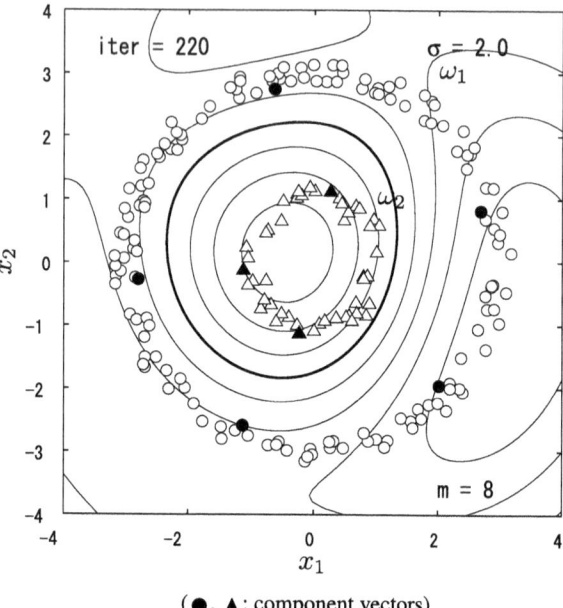

(●, ▲ : component vectors)

9.5   Potential Function Method and Perceptron

The fact that the perceptron has a duality was discussed in Sect. 8.6 (Boser et al. 1992; Guyon et al. 1992). In this section, we first show that a similar duality holds for the potential function method. Then, we show that this duality leads to the concept of the kernel trick, which plays an important role in support vector machines and kernel methods which will be described later.

9.5.1   Duality of Potential Function Method

The discriminant function of the potential function method shown in Eq. (9.9) is represented by the error counter vector α as a parameter. As a result, the modification of the potential function method is a process of repeatedly updating α as shown in Eq. (9.12).

On the other hand, if the potential function $K(x, x_k)$ is represented by Eq. (9.13), then the next equation holds:

$$g(x) = \sum_{k=1}^{n} \alpha_k b_k K(x, x_k) = \sum_{k=1}^{n} \alpha_k b_k \phi(x_k)^t \phi(x). \tag{9.16}$$

By placing

$$\mathbf{w} = \sum_{k=1}^{n} \alpha_k b_k \boldsymbol{\phi}(\boldsymbol{x}_k) \tag{9.17}$$

in Eq. (9.16), the weight $\mathbf{w}$ of the generalized linear discriminant function, shown by Eq. (8.51), is obtained. As a result, $g(\boldsymbol{x})$ can be denoted by

$$g(\boldsymbol{x}) = \mathbf{w}^t \boldsymbol{\phi}(\boldsymbol{x}), \tag{9.18}$$

with the weight $\mathbf{w}$ as a parameter. In other words, the above equation is a dual representation for Eq. (9.9), corresponding to Eq. (8.46) in the generalized linear discriminant function method. Therefore, the modification of the potential function method can be realized by an iterative process of updating $\mathbf{w}$ as in the primal perceptron of Eq. (8.50).

As is clear from the above, the potential function method and the perceptron are linked by a duality relation, and the perceptron can be regarded as a form of the potential function method (Aizerman et al. 1964). The relationship between the two is shown in Fig. 9.5.

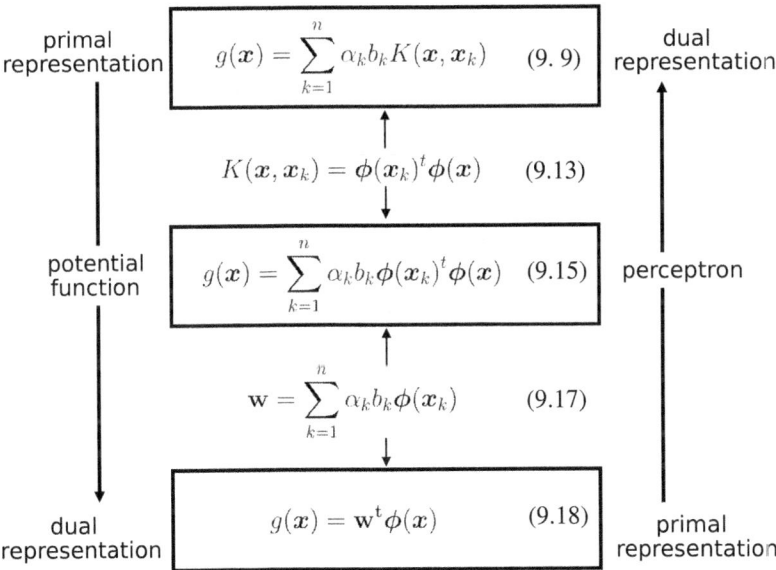

Fig. 9.5 Duality between potential function method and perceptron

As can be seen in the figure, Eqs. (9.13) and (9.17) play an important role in the realization of duality. The former shows that the potential function is represented by the inner product of two vectors, while the latter shows that the weight vector $\mathbf{w}$ is represented as a product-sum operation of the component vectors. The primal representation Eq. (9.9) of the potential function method is then transformed via Eq. (9.13) to Eq. (9.15), and then via Eq. (9.17) to the dual representation Eq. (9.18) of the potential function method. Equation (9.18) is also the primal representation of the perceptron. The inverse can be traced from the primal representation of the perceptron to its dual representation.

9.5.2 Duality and Kernel Trick

In the following, we will discuss the duality in more detail with specific examples, referring to Fig. 9.5.

When Eq. (9.18) of the primal representation of the perceptron is used as the discriminant function, it is **w** that is updated in the learning process as shown in Eq. (8.50), and it is sufficient to always keep the latest **w**. As is clear from the equation form, the classification of a pattern x is completed with a single inner product calculation in a D-dimensional space. To do so, a nonlinear function $\phi(x)$ must first be set, as shown in Step 1 of the primal perceptron learning rule (Sect. 8.6). However, there is a wide range of nonlinear functions to choose from, and it is difficult to find an appropriate $\phi(x)$. This tendency is especially remarkable when the dimensionality D becomes large.

In the following, as a concrete example to understand the duality, we will take up again the cubic discriminant function in a two-dimensional feature space, which was introduced in Chap. 8. We use Eq. (8.57) as a nonlinear transformation of $x \rightarrow \phi(x)$ to realize such a cubic discriminant function. The two-dimensional learning patterns of Fig. 8.1 was transformed into ten-dimensional vectors ($D = 10$) by the nonlinear transformation of Eq. (8.57), and the weight vector **w** was obtained by the primal perceptron learning rule. The result was already shown in Fig. 8.12.

Let us find the inner product $\phi(x)^t \phi(y)$ for two-dimensional vectors $x = (x_1, x_2)$ and $y = (y_1, y_2)$ using $\phi(x)$ shown in Eq. (8.57). The result is as follows:

$$\phi(x)^t \phi(y) = 1 + 3x_1 y_1 + 3x_2 y_2 + 6x_1 x_2 y_1 y_2 + 3x_1^2 y_1^2 + 3x_2^2 y_2^2$$

$$\qquad + 3x_1^2 x_2 y_1^2 y_2 + 3x_1 x_2^2 y_1 y_2^2 + x_1^3 y_1^3 + x_2^3 y_2^3 \qquad (9.19)$$

$$= (1 + x_1 y_1 + x_2 y_2)^3 \qquad (9.20)$$

$$= (1 + x^t y)^3. \qquad (9.21)$$

The reason for setting the nonlinear transformation $\phi(x)$ as Eq. (8.57) is to be able to derive Eq. (9.21). Since Eq. (9.21) can be expressed in the form of inner product shown in Eq. (9.13), the third-degree polynomial of Eq. (9.21) can be used as the potential function as follows:[2]

---

[2] Note that $K(x, x_k)$ in Eq. (9.22) does not satisfy Eq. (9.2).

$$K(x, x_k) = (1 + x_k^t x)^3.$$
(9.22)

As a result, the discriminant function can be expressed as Eq. (9.9) of Fig. 9.5, i.e., the dual representation of the perceptron and the primal representation of the potential function method. In this case, the learning is performed by repeatedly modifying the error counter vector α. The result is the same as when modifying w. Equation (9.22) is a kind of the polynomial kernel, which will be introduced in detail in Chap. 11. See also Problem 9.2 (2).

From the above, we can see that the following points should be noted. That is, the right-hand side of Eq. (9.13) is a function of a D dimensional vector $\phi(x)$, while the left-hand side is a function of a d-dimensional vector x. In other words, via Eq. (9.13), we can replace the perceptron's operation in a D-dimensional space (Eq. (9.18)) with that of the potential function method in a d-dimensional space (Eq. (9.9)). In Eq. (9.9), there is no need to define the specific form of the nonlinear transformation $\phi(x)$, only the form of $K(x, x_k)$ should be defined. This is the Step 1 operation in the potential function learning algorithm (Sect. 9.3).

On the other hand, in Eq. (9.9), it is the error counter vector α that is updated during learning. Let m be the number of the component vectors, i.e., the number of patterns that $\alpha_k \neq 0$, then to classify a pattern x, $K(x, x_k)$ must be calculated for m patterns x_k and added. Therefore, as m increases, the amount of computation increases and the learning efficiency decreases.

Even though Eq. (9.9) has this drawback, it is advantageous to avoid the process of nonlinear transformation $\phi(x)$. This trick, later called the *kernel trick*, will play an important role in support vector machines and kernel methods.

9.6 Potential Function Method and Condensed Nearest Neighbor Rule

The potential function method has many points in common with the condensed nearest neighbor rule introduced in Sect. 8.4. The simplest implementation of the nearest neighbor rule is the complete storage scheme, which registers all learning patterns as prototypes. However, this method is not realistic because it requires a large amount of memory and computation when the number of learning patterns increases. One of the methods devised to solve this problem is the condensed nearest neighbor rule.

Based on the idea that only patterns near the decision boundary contribute to classification, the condensed nearest neighbor rule selects as prototypes, only a few learning patterns distributed near the decision boundary. The condensed nearest neighbor rule, as already shown in Sect. 8.4, selects a learning pattern at each iteration and classifies it using the prototypes registered at this iteration. If a pattern is not correctly classified, the operation of additionally registering the pattern as a prototype is repeated. The final result obtained by this method is multiple prototypes that contribute to classification and a decision boundary determined by them. This decision boundary constitutes a piecewise linear discriminant function. However, it is not uniquely determined which patterns are selected as prototypes and how many prototypes there will be.

The potential function method performs a similar process. That is, a learning pattern is selected and classified based on the electrostatic potential (discriminant function) that has been formed so far, and if it is not classified correctly, the discriminant function is modified by increasing or decreasing a certain amount of charge at the location of the pattern. The process starts with an initial state in which there are no charged particles at all. The final result is learning patterns with nonzero charge and the electrostatic potential that they constitute. In the two-class case, the decision boundary is the boundary where the electrostatic potential is zero, i.e., $g(x) = 0$.

Patterns with nonzero charge correspond to the prototypes of the condensed nearest neighbor rule and contribute to the determination of the decision boundary. However, it is not uniquely determined which patterns have nonzero charge and how many such patterns there are. This is also the case with the condensed nearest neighbor rule. If the parameter σ of the potential function is set small, the potential function method becomes almost equivalent to the condensed nearest neighbor rule, as explained in Sect. 9.2.

To confirm this, the results of the experiment with the parameter set to a small value of $\sigma = 0.2$ are shown in Fig. 9.6. The drawing method is the same as before. The number of component vectors was 4 for class ω_1 and 3 for class ω_2, totaling

Fig. 9.6 Contours and decision boundary by potential function method ($\sigma = 0.2$)

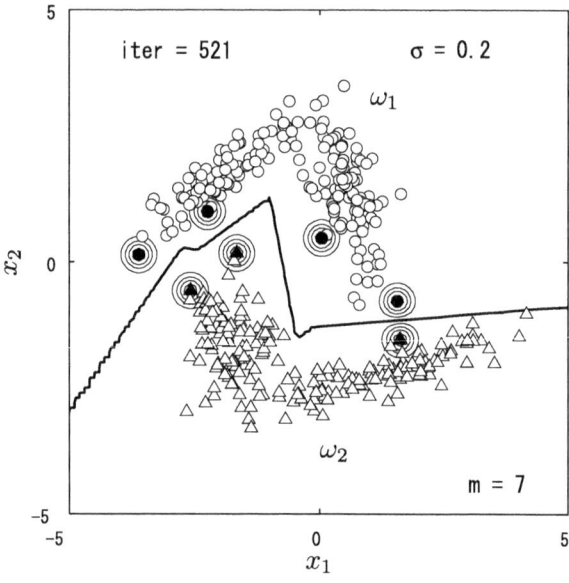

($\bullet$, $\blacktriangle$: component vectors)

7 ($m = 7$). They are indicated by ● and ▲ in the figure, respectively. Comparing the results with Fig. 8.9 obtained by applying the condensed nearest neighbor rule to the same data, we can see that very similar results are obtained. Comparing the component vectors of Fig. 9.6 with the prototypes of Fig. 8.9, we can also see that their positions and numbers are the same, and the decision boundaries are almost the same.

Problems

9.1 Using the series expansion formula

$$e^x = \sum_{k=0}^{\infty} x^k/k!,$$

show that $K(x, x_k)$ in Eq. (9.7) is represented by the inner product of infinite-dimensional vectors.

9.2 † Using the potential function learning algorithm, we wish to find a discriminant function that can correctly classify the six learning patterns listed in Problem 8.6. In learning, the learning patterns are repeatedly given from x_1 to x_6 in this order.

(1) Using the Gaussian function of Eq. (9.7) as the potential function, illustrate the decision boundary obtained by learning. The parameter of the Gaussian function should be set to $\sigma = 2.0$.
(2) Using the third-degree polynomial (polynomial kernel) of Eq. (9.22) as the potential function, illustrate the decision boundary obtained by learning. Also, compare the obtained results with those of Problem 8.6.

Chapter 10
Support Vector Machines

Abstract The support vector machine is a supervised learning algorithm that performs two-class classification, and its emergence is one of the greatest milestones in the history of pattern recognition and machine learning. The basis of the support vector machine is the concept of the margin introduced in Chap. 2. By formulating margin maximization as a convex quadratic programming problem, it is now possible to obtain discriminant functions with high generalization ability. Furthermore, by introducing the kernel trick introduced in Chap. 9, support vector machines can be used in high-dimensional spaces, thus extending their range of application. Today, support vector machines have established themselves as a general-purpose pattern classification method that can be used with confidence for many problems, and have become one of the indispensable tools in all areas of information processing. In this chapter, we explain the functions and characteristics of support vector machines, taking into account their historical background.

10.1 Birth of Support Vector Machines

The basic idea of the support vector machine (SVM)[1] was proposed by Bernhard Boser, Isabelle Guyon, and Vladimir Vapnik (Guyon et al. 1992; Boser et al. 1992) in 1992. Then, in 1995, the soft margin technology (Cortes and Vapnik 1995) was added to the machine in order to apply it to linearly nonseparable distributions, resulting in the SVM that is widely used today. After extensive experiments by Bernhard Schölkopf et al. and the release of the SVM-Light program package by Thorsten Joachims et al. and its application to *text classification* (Joachims 2002), SVMs have spread rapidly since the mid-1990s. Behind the spread of SVMs are the favorable characteristics described below.

Supplementary Information The online version contains supplementary material available at https://doi.org/10.1007/978-981-95-1478-6_10.

---

[1] Hereafter, the support vector machine is referred to as SVM.

(1) High *generalization ability* far superior to existing classification methods:
    While most of the existing methods show a decrease in classification perfor-
    mance when the dimensionality of the feature space d is increased under a
    constant number of learning patterns n, the SVM can maintain classification
    performance even in high dimensional feature spaces. This makes the SVM
    a powerful tool when a large number of features are used, such as in text
    classification and natural language processing, or when learning patterns are
    not sufficient.
(2) Ability to determine complex decision boundaries with nonlinear discriminant
    functions:
    The process is equivalent to a nonlinear transformation of the feature space to a
    higher dimension, and can generate a nonlinear discriminant function with high
    classification ability.
(3) Convergence to a globally optimal solution:
    Since a globally optimal solution is always obtained without falling into a
    locally optimal solution, the performance evaluation and parameter tuning of
    the discriminant function are easy. This characteristic contributes greatly to the
    efficiency of research and development and system implementation.
(4) Unique solution determination:
    Since the computation is the direct method, and it is expressed as a convex
    quadratic programming problem, the solution is uniquely determined. In this
    respect, it has an advantage over iterative methods such as the potential function
    method and the neural network. Methods with many local solutions, such as
    neural networks, require repeated trials because the convergence result depends
    on the initial values.

In this chapter, the basic idea of the SVM is first introduced. Then, the linear
SVM, which is the basic type of the SVM, and the nonlinear SVM using the kernel
method, are introduced with some experiments. To understand the SVM presented
in this chapter, knowledge of optimization under inequality constraints is necessary.
In particular, knowledge of the convex quadratic programming problem, the KKT
conditions, primal and dual problems are essential. These are explained in detail
using examples in Appendix F. Readers who do not necessarily have sufficient
knowledge in this field are recommended to read through the Appendix first.

10.2 Margin Maximization

In Chap. 2, we dealt with class separation for linearly separable learning patterns
and introduced the perceptron learning rule. In that learning rule, the variation
of the decision boundary obtained is large, and the decision boundary cannot be
determined uniquely even when the same learning patterns are used. As a result,
it is not uncommon for the learning process to terminate with the acquisition of a
decision boundary with no margin, i.e., a decision boundary that does not provide
sufficient distance from the learning patterns. As a solution to this problem, we
introduced the concept of a margin in Sect. 2.7 and explained a method to obtain

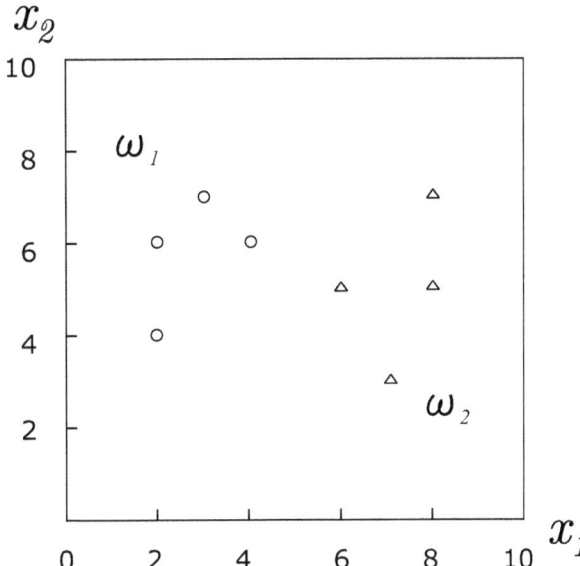

Fig. 10.1 Learning patterns in two-dimensional feature space

a decision boundary with a margin. However, it was pointed out that the decision boundary still cannot be determined uniquely, and that there remains a issue of the margin setting as well.

The SVM dealt with in this chapter can uniquely determine the optimal decision boundary among the infinite number of decision boundaries that can be realized by a linear discriminant function. In the following, we assume that a total of n learning patterns belonging to either of class ω_1 or ω_2 are to be separated. Suppose that the linear discriminant function with a margin, introduced in Sect. 2.7, is obtained by learning. Then, as shown in Fig. 2.14, hyperplanes H_1 and H_2, which are separated from each other by R, can be obtained so that the learning patterns are not contained within the space between them. As shown in Fig. 2.14, there is at least one pattern of class ω_1 on the hyperplane H_1, and similarly, at least one pattern of class ω_2 on H_2. We have already mentioned that this R is called the margin. An optimal classification system can be achieved by maximizing the margin R, which indicates the range where no learning pattern exists. However, the method introduced in Sect. 2.7 is a trial-and-error procedure to obtain as large a margin as possible, and the final R obtained is not necessarily the largest. In the following, we describe how to maximize the margin R using a simple example.

Figure 10.1 shows the distribution of learning patterns of two classes ω_1 and ω_2 in the two-dimensional feature space. Clearly these patterns are linearly separable. Figure 10.2 shows an example of hyperplanes H_1 and H_2 with margin R in this space, which are indicated by thin lines in the figure.[2] From the definition of the

---

[2] In this example, we are dealing with a two-dimensional feature space, so the hyperplanes are straight lines.

margin, there is no pattern between H_1 and H_2, which are parallel to each other and separated by R. Also, the hyperplanes H_1 and H_2 contain at least one pattern of class ω_1 and ω_2, respectively. They are indicated by ● and ▲ in the figure, respectively.[3]

Fig. 10.2 An example of a decision boundary with margin

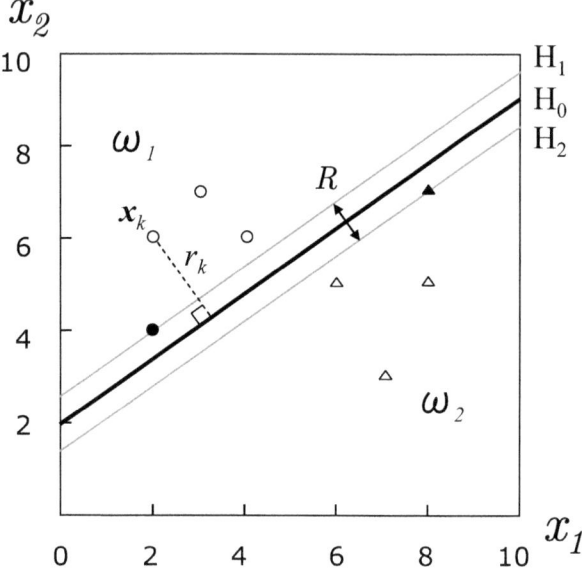

As shown in Fig. 10.2, if we set a hyperplane H_0 equidistant from H_1 and H_2 as a decision boundary, H_0 can correctly separate both classes and keep all patterns more than $R/2$ away from the decision boundary. That is, let r_k be the distance between x_k and the decision boundary, then from Eq. (2.47) the following equation holds:[4]

$$r_k = \frac{b_k(w_0 + w^t x_k)}{\|w\|} \geq \frac{R}{2} \qquad (k = 1, \dots, n). \tag{10.1}$$

As before, b_k is the teaching signal denoted by Eq. (2.29). The patterns on H_1 and H_2 are the ones for which the distance r_k from the decision boundary is the minimum value $R/2$, and these patterns are indicated by ● and ▲ in the figure.

As is clear from the above, the goal is to find w_0 and w that maximize R of Eq. (10.1). Note that w_0 and w, which define the decision boundary, are not uniquely determined. This is because the same decision boundary can be defined using $\widehat{w}_0 = \beta w_0$ and $\widehat{w} = \beta w$ multiplied by a constant β instead of w_0 and w. Also, the value of r_k is not changed by using such $\widehat{w}_0$ and $\widehat{w}$. So, after dividing Eq. (10.1) by $R/2$

---

[3] Here we show an example where there is one pattern on each hyperplane, but in general there are multiple patterns.

[4] In the example of Fig. 2.14, the decision boundary H_0 is not located equidistant from H_1 and H_2, as shown in Eq. (2.48). After obtaining the hyperplanes H_1 and H_2, if H_0 is moved so that $R_1 = R_2 = R/2 = 0.675$, then Eq. (10.1) holds as well.

to obtain

$$b_k \left(\frac{2}{R\|w\|} w_0 + \frac{2}{R\|w\|} w^t x_k \right) \geq 1, \tag{10.2}$$

we put $\beta = 2/(R\|w\|)$ to obtain

$$b_k(\widehat{w}_0 + \widehat{w}^t x_k) \geq 1 \qquad (k = 1, \ldots, n). \tag{10.3}$$

If $\widehat{w}_0$ and $\widehat{w}$ are replaced by w_0 and w, respectively, the above equation is rewritten as

$$b_k g(x_k) = b_k(w_0 + w^t x_k) \geq 1 \qquad (k = 1, \ldots, n). \tag{10.4}$$

The following formula is obtained for each class:.

$$g(x_k) = w_0 + w^t x_k \begin{cases} \geq 1 & (x_k \in \omega_1) \\ \leq -1 & (x_k \in \omega_2) \end{cases} \qquad (k = 1, \ldots, n). \tag{10.5}$$

That is, all learning patterns satisfy Eq. (10.4). In particular, the equality holds for patterns ● and ▲ on the hyperplanes H_1 and H_2 as in the following equation:

$$b_k g(x_k) = b_k(w_0 + w^t x_k) = 1. \tag{10.6}$$

For these patterns, from Eq. (10.1), the equation

$$r_k = \frac{b_k(w_0 + w^t x_k)}{\|w\|} = \frac{1}{\|w\|} \tag{10.7}$$

holds and the margin R is obtained as

$$R = \frac{2}{\|w\|}. \tag{10.8}$$

From the above, it can be seen that the problem of finding the optimal decision boundary under the criterion of margin maximization can be formulated as the problem of finding w_0 and w that maximize $2/\|w\|$ under the constraint of Eq. (10.4). This process can be expressed as follows. That is:

"At the position that separates both classes, allocate the largest region between the parallel hyperplanes H_1 and H_2 that contains no pattern."

Figure 10.3 is an example of setting up these hyperplanes according to the above criteria, and the difference is obvious when compared to the margin of Fig. 10.2.

The patterns ● and ▲ that are on H_1 and H_2 and closest to H_0 play an important role here. These are the patterns that determine H_1, H_2, and finally H_0.

Fig. 10.3 Decision boundary
with maximum margin

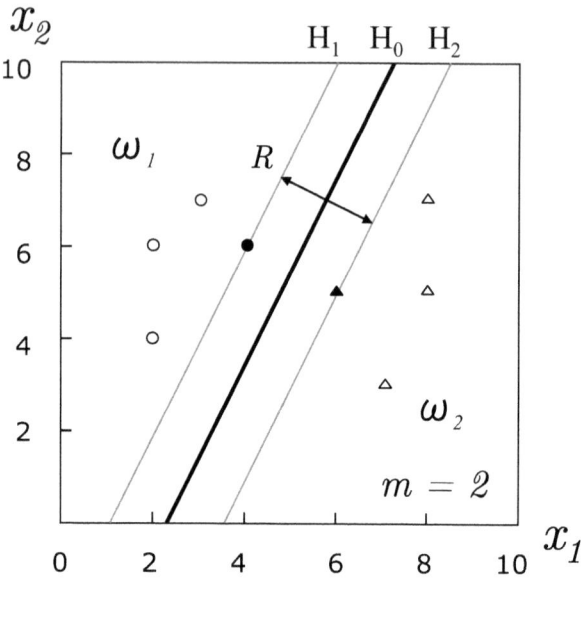

($\bullet$, $\blacktriangle$: support vectors)

Thus, the criterion of margin maximization yields learning patterns that contribute to the decision boundary determination, and these patterns are called the
support vector. In Fig. 10.3, the number of support vectors is two, one for each class,
and is shown as $m = 2$ in the lower right corner of the figure. Patterns other than
support vectors do not contribute to the determination of the decision boundary H_0.
Thus, the decision boundary determined by the support vectors is used to separate
the classes. The classifier that achieves class separation by the decision boundary
determined by the support vectors is called the *support vector machine (SVM)*.

Therefore, the following formulation is used to achieve margin maximization. In
fact, rather than maximizing $2/\|w\|$ in Eq. (10.8), it is mathematically easier to treat
it as a problem of minimizing $\|w\|^2/2$, i.e., finding w as follows:[5]

$$\min_{w} \left\{ \frac{1}{2} \|w\|^2 \right\}. \tag{10.9}$$

In the following, we seek the optimal decision boundary according to Eq. (10.9).
The coefficient $1/2$ is introduced in the above equation to simplify the subsequent
notation of the calculation formulas.

---

[5] It appears that w_0 is not involved in the optimization in Eq. (10.9), but this point will be clarified
in Sect. 10.3.1 Eq. (10.43).

——————————————————— **Coffee Break** ———————————————————

Generalization Ability of the SVM

"In order to obtain a classifier with excellent generalization ability through learning, the relationship $d \ll n$ must be established between the number of dimensions d of the feature space and the number of learning patterns n."

The above is a guideline that has been regarded as a golden rule for designers of classifiers and was also emphasized in Sect. 4.4. However, this condition does not necessarily need to be followed as long as the SVM is used. This is because the SVM has inherently high generalization ability (Guyon et al. 1992). What is the reason for this high generalization ability of the SVM?

When the number of learning patterns is small compared to the dimensionality of the feature space, the distribution of patterns becomes sparse in the feature space, and the constraints for determining the decision boundary become loose. As a result, in a learning method such as the perceptron, where convergence is achieved as long as the number of misclassifications is zero, the decision boundary obtained by chance from among countless candidates is likely to be a hyperplane with no margin.

On the other hand, in case of the SVM, even if the distribution of patterns is sparse, the optimal decision boundary that satisfies the strict condition of margin maximization is selected. This is the reason for the high generalization ability of the SVM. Nevertheless, the fact still remains in the SVM: the more learning patterns, the higher the generalization performance. We should be careful not to become so accustomed to the benefits of the SVM that we become indifferent to the relationship between the number of learning patterns and the dimensionality of the feature space.

10.3 Linear SVMs

We have introduced so far the basic idea of classifying learning patterns in a d-dimensional space by a linear discriminant function obtained by the SVM. In this section, we discuss specific methods for applying the SVM to the cases in which learning patterns in a d-dimensional space are linearly separable and those in which they are not. Since in both cases, a linear discriminant function is used for classification, we call this method the *linear SVM*.

10.3.1 Linear SVMs for Linearly Separable Learning Patterns

In the following, we first apply the linear SVM to linearly separable learning patterns in d-dimensional spaces. The optimization problem described at the end of the previous section can be rewritten as follows. Hereafter, the optimization problem is abbreviated as Optprob.

Optprob. 10.1 Minimization problem on w and w_0 (1)

When a scalar w_0 and a d-dimensional vector w satisfy the following n conditions:

$$b_k (w_0 + w^t x_k) - 1 \geq 0 \qquad (k = 1, \dots, n), \tag{10.10}$$

find w ($= w^*$) and w_0 ($= w_0^*$) that minimize

$$f(w) = \frac{1}{2} \| w \|^2. \tag{10.11}$$

The above is an optimization problem under inequality constraints described in Appendix F. Since Eq. (10.10), which represents the constraint condition, is a linear function of w, and $f(w)$ of Eq. (10.11) is a quadratic function of w and a convex function, this problem is a convex quadratic programming problem.

To solve this problem, Lagrange's method of undetermined multipliers is applied. We introduce an n-dimensional column vector λ whose elements are Lagrange's undetermined multipliers λ_k (≥ 0), as shown in the equation below:

$$\lambda = (\lambda_1, \dots, \lambda_n)^t. \tag{10.12}$$

The Lagrangian function $L(w, w_0, \lambda)$ can be written as[6]

$$L(w, w_0, \lambda) = \frac{1}{2} \| w \|^2 - \sum_{k=1}^{n} \lambda_k \big(b_k (w_0 + w^t x_k) - 1 \big). \tag{10.13}$$

As mentioned in Appendix F, a necessary and sufficient condition for the existence of the solutions w^* and w_0^* of the problem is that the KKT conditions hold. That is, the KKT conditions are the existence of $\lambda = \lambda^* = (\lambda_1^*, \dots, \lambda_n^*)^t$ for which the following equations hold for the solution $(w, w_0) = (w^*, w_0^*)$:

$$\frac{\partial L}{\partial w} = 0, \tag{10.14}$$

$$\frac{\partial L}{\partial w_0} = 0, \tag{10.15}$$

$$b_k (w_0 + w^t x_k) - 1 \geq 0, \tag{10.16}$$

$$\lambda_k \geq 0, \tag{10.17}$$

---

[6] Comparing Eq. (10.13) with Eq. (F.2), the signs of the second term on the right side of both equations are reversed. This is because the direction of the inequalities that indicate the constraint conditions are reversed for Eqs. (10.10) and (F.1).

$$\lambda_k \left(b_k(w_0 + \boldsymbol{w}^t \boldsymbol{x}_k) - 1 \right) = 0 \tag{10.18}$$

$$(k = 1, \ldots, n).$$

Computing Eq. (10.14) yields

$$\frac{\partial L}{\partial \boldsymbol{w}} = \boldsymbol{w} - \sum_{k=1}^{n} \lambda_k b_k \boldsymbol{x}_k = \boldsymbol{0}, \tag{10.19}$$

and thus

$$\boldsymbol{w} = \sum_{k=1}^{n} \lambda_k b_k \boldsymbol{x}_k. \tag{10.20}$$

Next, computing Eq. (10.15) yields

$$\frac{\partial L}{\partial w_0} = -\sum_{k=1}^{n} \lambda_k b_k = 0. \tag{10.21}$$

From these results, the following shows that the Lagrangian function can be expressed as a function $L(\boldsymbol{\lambda})$ of only $\boldsymbol{\lambda}$.

Rewriting Eq. (10.13), we obtain

$$L(\boldsymbol{\lambda}) = L(\boldsymbol{w}, \ w_0, \ \boldsymbol{\lambda})$$

$$= \frac{1}{2} \|\boldsymbol{w}\|^2 - \sum_{k=1}^{n} \lambda_k b_k \boldsymbol{w}^t \boldsymbol{x}_k - \sum_{k=1}^{n} \lambda_k (b_k w_0 - 1). \tag{10.22}$$

Using Eq. (10.20), the second term on the right side of Eq. (10.22) is expressed as

$$\sum_{k=1}^{n} \lambda_k b_k \boldsymbol{w}^t \boldsymbol{x}_k = \boldsymbol{w}^t \sum_{k=1}^{n} \lambda_k b_k \boldsymbol{x}_k$$

$$= \|\boldsymbol{w}\|^2. \tag{10.23}$$

Using Eq. (10.21), the third term on the right side of Eq. (10.22) is expressed as

$$\sum_{k=1}^{n} \lambda_k (b_k w_0 - 1) = w_0 \sum_{k=1}^{n} \lambda_k b_k - \sum_{k=1}^{n} \lambda_k \tag{10.24}$$

$$= -\sum_{k=1}^{n} \lambda_k. \tag{10.25}$$

Using these results and Eq. (10.20), $L(\lambda)$ of Eq. (10.22) can be written as

$$L(\lambda) = \sum_{k=1}^{n} \lambda_k - \frac{1}{2} \|\boldsymbol{w}\|^2 \tag{10.26}$$

$$= \sum_{k=1}^{n} \lambda_k - \frac{1}{2} \left(\sum_{i=1}^{n} \lambda_i b_i \boldsymbol{x}_i \right)^t \left(\sum_{j=1}^{n} \lambda_j b_j \boldsymbol{x}_j \right) \tag{10.27}$$

$$= \sum_{k=1}^{n} \lambda_k - \frac{1}{2} \sum_{i=1}^{n} \sum_{j=1}^{n} \lambda_i \lambda_j b_i b_j \boldsymbol{x}_i^t \boldsymbol{x}_j. \tag{10.28}$$

Here, we introduce vector and matrix notation for a more concise formulation. First, define an n-dimensional column vector $\mathbf{1}_n$ whose elements are all 1, as follows:

$$\mathbf{1}_n = (\overbrace{1, \ 1, \ \ldots, \ 1}^{n})^t. \tag{10.29}$$

Also, if we define an n-dimensional column vector $\mathbf{b}$ whose elements are n learning patterns' teaching signals as

$$\mathbf{b} = (b_1, \ \ldots, \ b_n)^t, \tag{10.30}$$

then Eq. (10.21) is expressed as

$$\lambda^t \mathbf{b} = 0. \tag{10.31}$$

We further set

$$h_{ij} = b_i b_j \boldsymbol{x}_i^t \boldsymbol{x}_j \qquad (i, j = 1, \ldots, n), \tag{10.32}$$

and define an $n \times n$ matrix $\mathbf{H} = (h_{ij})$ whose (i, j) component is h_{ij}. Using these, $L(\lambda)$ of Eq. (10.28) is expressed as

$$L(\lambda) = \lambda^t \mathbf{1}_n - \frac{1}{2} \lambda^t \mathbf{H} \lambda. \tag{10.33}$$

Therefore, Optprob. 10.1 is attributed to its dual problem, Optprob. 10.2, as shown below.

Optprob. 10.2 Maximization problem on λ (1)
Suppose that λ of Eq. (10.12) and $\mathbf{b}$ of Eq. (10.30) satisfy conditions

$$\lambda^t \mathbf{b} = 0, \tag{10.34}$$

$$\boldsymbol{\lambda} \geq \mathbf{0}, \tag{10.35}$$

and the matrix $\mathbf{H} = (h_{ij})$ is represented by

$$h_{ij} = b_i b_j \boldsymbol{x}_i^t \boldsymbol{x}_j \qquad (i, j = 1, \ldots, n). \tag{10.36}$$

Under these conditions, find $\boldsymbol{\lambda}\ (= \boldsymbol{\lambda}^*)$ that maximizes

$$L(\boldsymbol{\lambda}) = \boldsymbol{\lambda}^t \mathbf{1}_n - \frac{1}{2} \boldsymbol{\lambda}^t \mathbf{H} \boldsymbol{\lambda}. \tag{10.37}$$

The above problem shows that the minimization problem for $\boldsymbol{w}$ and w_0 shown in Optprob. 10.1 is replaced by the equivalent maximization problem for $\boldsymbol{\lambda}$. For the reason, see Appendix F.

The optimization problem shown above is a typical convex quadratic programming problem, and there are many libraries available to find its solution. By solving this dual problem, the solution to the original primal problem can be easily obtained as follows.

If the solution $\boldsymbol{\lambda}^*$ of Optprob. 10.2 is

$$\boldsymbol{\lambda}^* = (\lambda_1^*, \ldots, \lambda_n^*), \tag{10.38}$$

then from Eq. (10.20), $\boldsymbol{w}^*$ is obtained as

$$\boldsymbol{w}^* = \sum_{k=1}^{n} \lambda_k^* b_k \boldsymbol{x}_k. \tag{10.39}$$

It should be noted here that there exists a complementarity condition of Eq. (10.18), which is one of the KKT conditions, mentioned in Appendix F. As shown in Eqs. (F.9) and (F.10), the complementarity condition of Eq. (10.18) asserts that for the optimal solutions $\boldsymbol{w}^*$, w_0^* and λ_k^*, one of the following holds:

$$\lambda_k^* > 0 \text{ and } b_k(w_0^* + \boldsymbol{w}^{*t}\boldsymbol{x}_k) - 1 = 0, \tag{10.40}$$

$$\lambda_k^* = 0 \text{ and } b_k(w_0^* + \boldsymbol{w}^{*t}\boldsymbol{x}_k) - 1 > 0. \tag{10.41}$$

In other words, the above formula indicates that for $\lambda_k^* > 0$, Eq. (10.16) is satisfied by equality, and for $\lambda_k^* = 0$, Eq. (10.16) is satisfied by inequality. It is clear from Eq. (10.39) that only $\boldsymbol{x}_k$s corresponding to $\lambda_k^* > 0$ contribute to the determination of $\boldsymbol{w}^*$. We have already mentioned in Sect. 10.2 that such $\boldsymbol{x}_k$ is called the support vector. That is, for a support vector $\boldsymbol{x}_k$,

$$b_k(w_0^* + \boldsymbol{w}^{*t}\boldsymbol{x}_k) - 1 = 0 \tag{10.42}$$

holds as shown in Eq. (10.40). Multiplying both sides of the above equation by b_k and using the fact that $b_k^2 = 1$, w_0^* can be obtained as

$$w_0^* = b_s - \boldsymbol{w}^{*t} \boldsymbol{x}_s \tag{10.43}$$

$$= b_s - \sum_{k=1}^{n} \lambda_k^* b_k \boldsymbol{x}_k^t \boldsymbol{x}_s, \tag{10.44}$$

where $\boldsymbol{x}_s$ is an arbitrary support vector and b_s is its teaching signal.[7]
    From the above, the discriminant function $g(\boldsymbol{x})$ to be obtained is

$$g(\boldsymbol{x}) = w_0^* + \boldsymbol{w}^{*t} \boldsymbol{x} \tag{10.45}$$

$$= w_0^* + \sum_{k=1}^{n} \lambda_k^* b_k \boldsymbol{x}_k^t \boldsymbol{x}, \tag{10.46}$$

and the classification rule for unknown patterns is

$$\begin{cases} g(\boldsymbol{x}) > 0 & \implies \quad \boldsymbol{x} \in \omega_1 \\ g(\boldsymbol{x}) < 0 & \implies \quad \boldsymbol{x} \in \omega_2. \end{cases} \tag{10.47}$$

Substituting Eq. (10.43) into Eq. (10.45) yields

$$g(\boldsymbol{x}) = b_s - \boldsymbol{w}^{*t} \boldsymbol{x}_s + \boldsymbol{w}^{*t} \boldsymbol{x}$$

$$= b_s + \boldsymbol{w}^{*t} (\boldsymbol{x} - \boldsymbol{x}_s). \tag{10.48}$$

Thus, when $\boldsymbol{x}$ is a support vector, putting $\boldsymbol{x} = \boldsymbol{x}_s$ gives

$$g(\boldsymbol{x}_s) = b_s = \begin{cases} 1 & (\boldsymbol{x} \in \omega_1) \\ -1 & (\boldsymbol{x} \in \omega_2). \end{cases} \tag{10.49}$$

Therefore, $g(\boldsymbol{x}_s)$ is 1 or -1 depending on the class to which $\boldsymbol{x}_s$ belongs. That is, the support vector $\boldsymbol{x}_s$ lies on the hyperplane H_1 or H_2 that determines the margin.
    Figure 10.4 shows the results of applying the linear SVM to the linearly separable learning patterns shown in Fig. 2.10. In the figure, the thick line indicates the hyperplane H_0 that shows the decision boundary, and the thin lines indicate the hyperplanes H_1 and H_2 that define the margin, respectively. In the figure, the support vectors are indicated by ● and ▲ marks as before, and three vectors ($m = 3$),

---

[7] Multiple support vectors can be obtained, and the value of w_0^* calculated by Eq. (10.43) should be the same no matter which support vector is used. In reality, however, due to computational errors, there will be minute variations in these values. Therefore, to ensure stability of the solution, it is desirable to calculate w_0^* for all support vectors and use their average value.

Fig. 10.4 Result of applying SVM

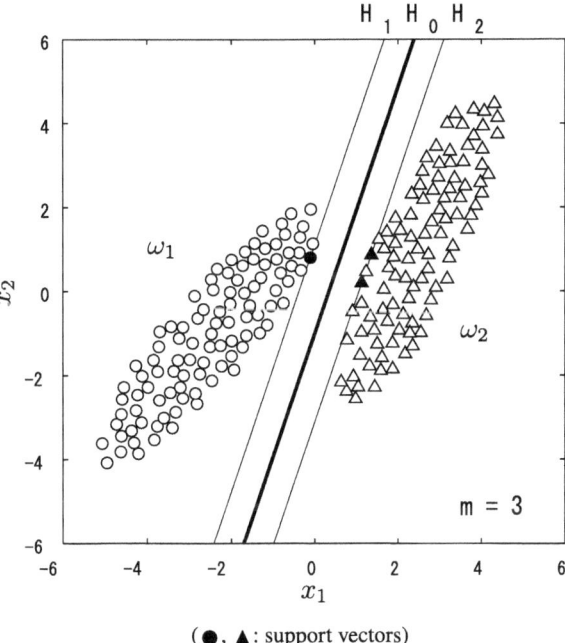

($\bullet$, $\blacktriangle$: support vectors)

one from class ω_1 and two from class ω_2, are selected. It is confirmed that the support vectors of classes ω_1 and ω_2 lie on the hyperplanes H_1 and H_2, respectively, and determine the maximum margin. The margin obtained in this experiment was $R = 2/\|w\| = 1.359$. The pattern closest to the decision boundary is the support vector. In this case, the distance between the support vector and the decision boundary is $1/\|w\| = 1.359/2 \approx 0.680$, which is the maximum.[8]

Here, we would like to take up again the perceptron introduced in Chap. 2 and point out its similarity to the linear SVM. Let's compare Eqs. (10.39) and (2.39) of the weight vector. The following is a reiteration of Eq. (2.39):

$$\mathbf{w} = \sum_{k=1}^{n} \alpha_k b_k \mathbf{x}_k. \tag{10.50}$$

The form of the two is the same, with the difference that Eq. (10.39) uses the weight vector w and the feature vector x and Eq. (10.50) uses the augmented weight vector $\mathbf{w}$ and the augmented feature vector $\mathbf{x}$. The α_k and λ_k that appear in both equations correspond, and only $\mathbf{x}_k$ which is $\alpha_k \neq 0$ in Eq. (10.50), and x_k which is $\lambda_k \neq 0$ in Eq. (10.39), contribute to the construction of the weight vector. The

---

[8] Therefore, the setting of $\delta = 0.670$ in the series of experiments shown in Fig. 2.13 was consequently appropriate.

vector $\boldsymbol{x}_k$ with such a function has been referred to as the component vector but in this chapter, the support vector corresponds to it. In each case, a small number of component vectors or support vectors are selected from the learning patterns, and the weight is constructed by linear combination of these vectors. The coefficient α_k in Eq. (10.50) is an element of the error counter vector which represents the number of times $\boldsymbol{x}_k$ is misclassified during the learning process, and takes a non-negative integer value. On the other hand, λ_k^* in Eq. (10.39) is generally a non-negative real value. See also Problem 10.2.

As mentioned above, Points 1, 2 listed in Sect. 2.5 for the perceptron learning rule are also applicable to the SVM. On the other hand, Point 3 does not apply to the SVM, and the decision boundary is uniquely determined. This is because the SVM uses the *direct method* of solving the convex quadratic programming problem, whereas the perceptron learning rule uses the iterative method. Therefore, the SVM provides a means to solve Point 3, the problem with perceptrons.

10.3.2   Linear SVMs for Linearly Nonseparable Learning Patterns

We applied the linear SVM to linearly separable learning patterns in Sect. 10.3.1. In the following, we describe how to apply the same method to linearly nonseparable learning patterns in a d-dimensional space. For linearly nonseparable learning patterns, w_0 and $\boldsymbol{w}$ satisfying Eq. (10.5) do not exist. Therefore, we relax the restriction on the discriminant function and introduce the following equation instead of Eq. (10.5):

$$g(\boldsymbol{x}_k) = w_0 + \boldsymbol{w}^t \boldsymbol{x}_k \begin{cases} \geq \quad 1 - \xi_k & (\boldsymbol{x}_k \in \omega_1) \\ \leq -1 + \xi_k & (\boldsymbol{x}_k \in \omega_2) \end{cases} \qquad (k = 1, \ldots, n). \qquad (10.51)$$

The above equation can be summarized as

$$b_k \, g(\boldsymbol{x}_k) = b_k (w_0 + \boldsymbol{w}^t \boldsymbol{x}_k) \geq 1 - \xi_k \qquad (k = 1, \ldots, n), \qquad (10.52)$$

where $\xi_k (\geq 0)$ is a value set for each pattern. The above is a technique called the *soft margin*. Naturally, if $\xi_k = 0$, the above equation agrees with Eq. (10.4).

When applying the SVM to linearly nonseparable learning patterns, as in the linearly separable case, three hyperplanes are set up. They are a hyperplane H_0 corresponding to $g(\boldsymbol{x}) = 0$, and hyperplanes H_1 and H_2 corresponding to $g(\boldsymbol{x}) = 1$ and $g(\boldsymbol{x}) = -1$ respectively. Unlike the linearly separable case, the two classes cannot be separated by the hyperplane H_0, and patterns exist between the hyperplanes H_1 and H_2. Even in such a situation, it is common to the linearly separable case that under certain conditions we aim to maximize the margin, i.e., the distance $2/\|\boldsymbol{w}\|$ between the hyperplanes H_1 and H_2. Specifically, the following optimization problem is to be solved.

Optprob. 10.3 Minimization problem on w and w_0 (2)

When scalars w_0, ξ_k and a d-dimensional vector w satisfy the following $2n$ inequalities:

$$b_k (w_0 + w^t x_k) - (1 - \xi_k) \geq 0 \qquad (k = 1, \ldots, n), \qquad (10.53)$$

$$\xi_k \geq 0 \qquad (k = 1, \ldots, n), \qquad (10.54)$$

find $w \ (= w^*)$, $w_0 \ (= w_0^*)$ and $\xi_k \ (= \xi_k^*)$ that minimize

$$f(w) = \frac{1}{2}\|w\|^2 + c_{pe} \cdot \sum_{k=1}^{n} \xi_k. \qquad (10.55)$$

This problem can be thought of as an extension of Optprob. 10.1 to the linearly nonseparable case. The first term of Eq. (10.55) is the same as Eq. (10.11) and is the term for controlling the margin. The second term is the *penalty* for patterns lying beyond a hyperplane H_1 or H_2 defining the margin, where c_{pe} (> 0) is a constant that determines the weight for the penalty. The penalty is explained using Fig. 10.5.

The figure shows four patterns x_1, x_2, x_3 and x_4 belonging to the class ω_1. Also shown are the hyperplanes H_0, H_1, H_2 corresponding to $g(x) = 0, 1, -1$, respectively. In the figure, patterns in the area above H_0 are classified as ω_1, while those below are as ω_2.

Fig. 10.5 Penalty for linearly nonseparable patterns

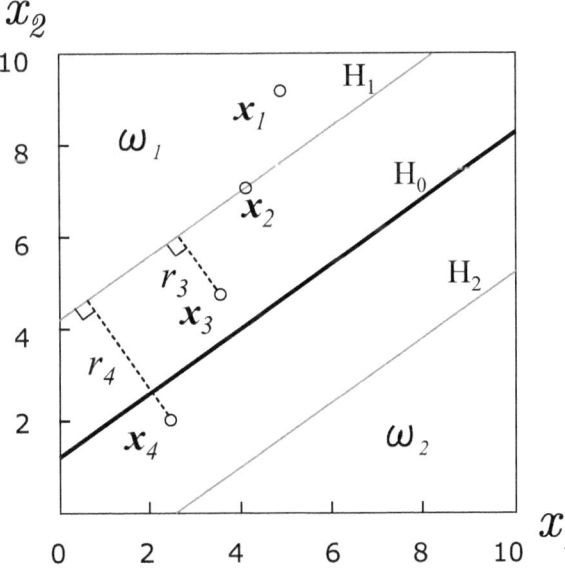

Since the pattern x_1 is above the hyperplane H_1, $g(x_1) > 1$ holds and $\xi_1 = 0$ from Eq. (10.51). The pattern x_2 is on the hyperplane H_1, so $g(x_2) = 1$ and $\xi_2 = 0$ as well. Both of these patterns are correctly classified. The pattern x_3 lies between the hyperplanes H_0 and H_1. Therefore, although x_3 is correctly classified, it cannot satisfy $g(x_3) \geq 1$, so $0 < \xi_3 < 1$ from Eq. (10.51). Since the pattern x_4 is below the hyperplane H_0, it is misclassified with $g(x_4) < 0$, so $\xi_4 > 1$ from Eq. (10.51). Similarly, ξ_k can be set for patterns of class ω_2, and ξ_k can be written as follows for both classes:

$$\xi_k = 0 \qquad \text{if} \qquad b_k\, g(x_k) \geq 1, \tag{10.56}$$

$$0 < \xi_k < 1 \qquad \text{if} \qquad 0 < b_k\, g(x_k) < 1, \tag{10.57}$$

$$\xi_k \geq 1 \qquad \text{if} \qquad b_k\, g(x_k) \leq 0. \tag{10.58}$$

Noting that Eqs. (10.53) and (10.54) are written as

$$\xi_k \geq 1 - b_k\, (w_0 + w^t x_k) \qquad (k = 1, \ldots, n), \tag{10.59}$$

$$\xi_k \geq 0 \qquad (k = 1, \ldots, n), \tag{10.60}$$

then, ξ_k can be expressed by the following equation:

$$\xi_k = \max\{1 - b_k\, g(x_k),\; 0\} \qquad (k = 1, \ldots, n). \tag{10.61}$$

Consider x_k satisfying Eq. (10.57) or (10.58). Let r_k be the distance between x_k and the hyperplane H_1 (when $x_k \in \omega_1$) or H_2 (when $x_k \in \omega_2$), we can confirm the following equation:

$$r_k = \xi_k / \|w\|. \tag{10.62}$$

For example, the values of r_3 and r_4 shown in Fig. 10.5 are $\xi_3 / \|w\|$ and $\xi_4 / \|w\|$ respectively. See Problem 10.1 for the derivation of the above equation.

From the above, we can see that ξ_k is an indicator of how close the pattern x_k is to other classes beyond the hyperplane H_1 or H_2 from its own class. In other words, ξ_k in Eq. (10.61) can be considered as the loss caused by the pattern x_k. Since $\xi_k > 1$ when the pattern x_k is misclassified, $\sum_{k=1}^{n} \xi_k$ represents the upper bound of the number of misclassifications. Therefore, it is reasonable to use the sum of ξ_k as the penalty. The weight c_{pe} for the penalty used in Eq. (10.55) has to be determined by trial and error.

The solution of Optprob. 10.3 is basically the same as the linearly separable case described in the previous section. First, in addition to λ introduced in the linearly separable case as the undetermined Lagrangian multiplier, γ is introduced as in the following equations:

$$\lambda = (\lambda_1, \ldots, \lambda_n)^t \tag{10.63}$$

$$\gamma = (\gamma_1, \ldots, \gamma_n)^t. \tag{10.64}$$

We also define an n-dimensional column vector ξ whose elements are $\xi_k \ (\geq 0)$ as follows:

$$\xi = (\xi_1, \ldots, \xi_n)^t. \tag{10.65}$$

The Lagrangian function $L(w, w_0, \xi, \lambda, \gamma)$ can be written as

$$L(w, w_0, \xi, \lambda, \gamma)$$

$$= \frac{1}{2} \|w\|^2 + c_{pe} \sum_{k=1}^{n} \xi_k$$

$$- \sum_{k=1}^{n} \lambda_k \left(b_k (w_0 + w^t x_k) - 1 + \xi_k \right) - \sum_{k=1}^{n} \gamma_k \xi_k. \tag{10.66}$$

As in the previous section, the KKT conditions can be written as follows. That is, a necessary and sufficient condition for the existence of a solution $(w, w_0) = (w^*, w_0^*)$ is the existence of $\lambda = \lambda^* = (\lambda_1^*, \ldots, \lambda_m^*)^t$ satisfying the following equations:

$$\frac{\partial L}{\partial w} = 0, \tag{10.67}$$

$$\frac{\partial L}{\partial w_0} = 0, \tag{10.68}$$

$$\frac{\partial L}{\partial \xi_k} = 0, \tag{10.69}$$

$$b_k (w_0 + w^t x_k) - 1 + \xi_k \geq 0, \tag{10.70}$$

$$\xi_k \geq 0, \tag{10.71}$$

$$\lambda_k \geq 0, \tag{10.72}$$

$$\gamma_k \geq 0, \tag{10.73}$$

$$\lambda_k \left(b_k (w_0 + w^t x_k) - 1 + \xi_k \right) = 0, \tag{10.74}$$

$$\gamma_k \xi_k = 0 \tag{10.75}$$

$$(k = 1, \ldots, n).$$

The calculation result of Eq. (10.67) is the same as that of Eq. (10.20), and the following equation is obtained:

$$w = \sum_{k=1}^{n} \lambda_k b_k x_k. \tag{10.76}$$

Also, the calculation result of Eq. (10.68) is the same as that of Eq. (10.21), and the following equation is obtained:

$$\sum_{k=1}^{n} \lambda_k b_k = 0. \tag{10.77}$$

The calculation of Eq. (10.69) yields the following equation:

$$\frac{\partial L}{\partial \xi_k} = c_{pe} - \lambda_k - \gamma_k = 0. \tag{10.78}$$

Using Eqs. (10.76) and (10.77), the third term on the right side of Eq. (10.66) is

$$\sum_{k=1}^{n} \lambda_k \left(b_k (w_0 + w^t x_k) - 1 + \xi_k \right)$$

$$= w^t \sum_{k=1}^{n} \lambda_k b_k x_k + w_0 \sum_{k=1}^{n} \lambda_k b_k + \sum_{k=1}^{n} \lambda_k (-1 + \xi_k) \tag{10.79}$$

$$= \|w\|^2 + \sum_{k=1}^{n} \lambda_k (-1 + \xi_k), \tag{10.80}$$

so, substituting the above equation and Eq. (10.78) into Eq. (10.66), the Lagrangian function $L(w^t, w_0, \xi, \lambda, \gamma)$ is expressed as

$$L(w^t, w_0, \xi, \lambda, \gamma)$$

$$= \sum_{k=1}^{n} \left(\lambda_k + (c_{pe} - \lambda_k - \gamma_k) \xi_k \right) - \frac{1}{2} \|w\|^2 \tag{10.81}$$

$$= \sum_{k=1}^{n} \lambda_k - \frac{1}{2} \|w\|^2, \tag{10.82}$$

which coincides with $L(\lambda)$ in Eq. (10.26). As already shown, $L(\lambda)$ can be expressed in a vector form as in Eq. (10.33). Therefore, the following equation holds:

$$L(w^t, w_0, \xi, \lambda, \gamma) = L(\lambda)$$

$$= \lambda^t 1_n - \frac{1}{2} \lambda^t H \lambda. \tag{10.83}$$

On the other hand, from Eq. (10.77), we obtain the following constraint for λ:

$$\lambda' \mathbf{b} = 0. \tag{10.84}$$

The above equation is the same as Eq. (10.31). Furthermore, from Eq. (10.78) and from Eq. (10.71) to (10.73), we obtain the following equation:

$$0 \leq \lambda_k \leq c_{pe} \qquad (k = 1, \ldots, n). \tag{10.85}$$

By expressing the above equation in vector notation, the following equation is obtained in addition to Eq. (10.84) as a constraint on λ_k,

$$\mathbf{0} \leq \lambda \leq c_{pe} \cdot \mathbf{1}_n, \tag{10.86}$$

where $\mathbf{1}_n$, $\mathbf{b}$ and $\mathbf{H} = (h_{ij})$ are shown in Eqs. (10.29), (10.30) and (10.32) respectively. As mentioned above, the results obtained in the previous section can be used even in linearly nonseparable case.

To summarize the above, Optprob. 10.3 is attributed to Optprob. 10.4 below, which is its dual problem.

Optprob. 10.4   Maximization problem on λ (2)
Suppose that λ of Eq. (10.12) and $\mathbf{b}$ of Eq. (10.30) satisfy conditions

$$\lambda' \mathbf{b} = 0, \tag{10.87}$$

$$\mathbf{0} \leq \lambda \leq c_{pe} \cdot \mathbf{1}_n, \tag{10.88}$$

and the matrix $\mathbf{H} = (h_{ij})$ is represented by

$$h_{ij} = b_i b_j x_i' x_j \qquad (i, j = 1, \ldots, n). \tag{10.89}$$

Under these conditions, find $\lambda \ (= \lambda^*)$ that maximizes

$$L(\lambda) = \lambda' \mathbf{1}_n - \frac{1}{2} \lambda' \mathbf{H} \lambda. \tag{10.90}$$

The above is a convex quadratic programming problem in which Eq. (10.35) of the constraint in Optprob. 10.2 is changed to Eq. (10.88). Let the solution λ^* of Optprob. 10.4 be

$$\lambda^* = (\lambda_1^*, \ldots, \lambda_n^*), \tag{10.91}$$

then $\boldsymbol{w}^*$ is obtained from Eq. (10.76) as

$$\boldsymbol{w}^* = \sum_{k=1}^{n} \lambda_k^* b_k \boldsymbol{x}_k. \tag{10.92}$$

Here, as in Eqs. (10.40) and (10.41), either of the following equations holds from the complementarity condition Eq. (10.74):

$$\lambda_k^* > 0 \quad \text{and} \quad b_k(w_0^* + \boldsymbol{w}^{*t}\boldsymbol{x}_k) - 1 + \xi_k = 0, \tag{10.93}$$

$$\lambda_k^* = 0 \quad \text{and} \quad b_k(w_0^* + \boldsymbol{w}^{*t}\boldsymbol{x}_k) - 1 + \xi_k > 0. \tag{10.94}$$

Similarly, from the complementarity condition Eq. (10.75), either of the following equations holds:

$$\gamma_k > 0 \quad \text{and} \quad \xi_k = 0, \tag{10.95}$$

$$\gamma_k = 0 \quad \text{and} \quad \xi_k > 0. \tag{10.96}$$

As already mentioned in Optprob. 10.2, only support vectors corresponding to $\lambda_k^* > 0$, contribute to the determination of $\boldsymbol{w}^*$. Let $\boldsymbol{x}_s$ be the support vector and b_s and ξ_s be the corresponding b_k and ξ_k, respectively, then from Eq. (10.93), the following equation holds:

$$b_s(w_0^* + \boldsymbol{w}^{*t}\boldsymbol{x}_s) - 1 + \xi_s = 0. \tag{10.97}$$

From this, w_0^* can be expressed, using the support vector $\boldsymbol{x}_s$, as

$$w_0^* = b_s(1 - \xi_s) - \boldsymbol{w}^{*t}\boldsymbol{x}_s. \tag{10.98}$$

In the same way as Eq. (10.46), the discriminant function $g(\boldsymbol{x})$ is obtained from Eqs. (10.92) and (10.98) as

$$g(\boldsymbol{x}) = w_0^* + \boldsymbol{w}^{*t}\boldsymbol{x} \tag{10.99}$$

$$= w_0^* + \sum_{k=1}^{n} \lambda_k^* b_k \boldsymbol{x}_k^t \boldsymbol{x}. \tag{10.100}$$

Let us consider λ_k^* in the following three ways.

(1) When $\lambda_k^* = 0$:

The pattern $\boldsymbol{x}_k$ is not a support vector. Since $\gamma_k = c_{pe} > 0$ from Eq. (10.78), $\xi_k = 0$ holds from Eq. (10.95). Also, since $\lambda_k^* = 0$, Eq. (10.94) gives $b_k(w_0^* + \boldsymbol{w}^{*t}\boldsymbol{x}_k) > 1$. In Fig. 10.5, among the patterns $\boldsymbol{x}_1, \dots, \boldsymbol{x}_4$ belonging to ω_1 ($b_k = 1$), the pattern $\boldsymbol{x}_1$ satisfies this condition.

(2) When $0 < \lambda_k^* < c_{pe}$:

The pattern x_k is a support vector. Since $0 < \gamma_k < c_{pe}$ from Eq. (10.78), $\xi_k = 0$ holds from Eq. (10.95). Since $\lambda_k^* > 0$, Eq. (10.93) gives $b_k(w_0^* + w^{*t}x_k) = 1$. In Fig. 10.5, the pattern x_2 on the hyperplane H_1 satisfies this condition.

(3) When $\lambda_k^* = c_{pe}$:

The pattern x_k is a support vector. Since $\gamma_k = 0$ from Eq. (10.78), $\xi_k > 0$ holds from Eq. (10.96). Also, since $\lambda_k^* > 0$, Eq. (10.93) gives $b_k(w_0^* + w^{*t}x_k) = 1 - \xi_k < 1$. If $0 < \xi_k < 1$, x_k is correctly classified, but if $\xi_k > 1$, it is misclassified. In Fig. 10.5, the pattern x_3 located between the hyperplanes H_1 and H_0 corresponds to the former condition, and the pattern x_4 in the region of ω_2 beyond the hyperplane H_0 corresponds to the latter.

The above can be summarized as the following four relationships among λ_k^*, γ_k, ξ_k and x_k, where $g(x)$ is represented by Eq. (10.99):

$$\lambda_k^* = 0, \qquad \gamma_k = c_{pe}, \qquad \xi_k = 0, \qquad b_k\, g(x) > 1, \qquad (10.101)$$

$$0 < \lambda_k^* < c_{pe}, \quad 0 < \gamma_k < c_{pe}, \qquad \xi_k = 0, \qquad b_k\, g(x) = 1, \qquad (10.102)$$

$$\lambda_k^* = c_{pe}, \qquad \gamma_k = 0, \qquad 0 < \xi_k < 1, \quad 0 < b_k\, g(x) < 1, \qquad (10.103)$$

$$\lambda_k^* = c_{pe}, \qquad \gamma_k = 0, \qquad \xi_k > 1, \qquad b_k\, g(x) < 0. \qquad (10.104)$$

In Fig. 10.5, the patterns $x_1, \ldots, x_4$ are examples that satisfy from Eq. (10.101) to (10.104), respectively. As is clear from the above, to find w_0^*, we should find λ_k^* satisfying Eq. (10.102) and substitute the corresponding x_k into Eq. (10.98) as the support vector x_s. Since $\xi_k = 0$ as shown in Eq. (10.102), from Eq. (10.98), w_0^* can be obtained as

$$w_0^* = b_s - w^{*t}x_s \qquad (10.105)$$

$$= b_s - \sum_{k=1}^{n} \lambda_k^* b_k x_k^t x_s, \qquad (10.106)$$

where x_s is the support vector that satisfies Eq. (10.102) and b_s is its teaching signal. Equation (10.106) is the same as Eq. (10.43). For the actual calculation, see footnote 7 in this Chapter.

In the following, we will experiment with the linear SVM using linearly nonseparable learning patterns. The data used are the two-class two-dimensional learning patterns shown in Fig. 8.1. The results are shown in Figs. 10.6 and 10.7. The weights for the penalty are set to $c_{pe} = 0.1$ for Fig. 10.6 and $c_{pe} = 10.0$ for Fig. 10.7, respectively. By solving the convex quadratic programming problem, the linear discriminant function $g(x)$ of Eq. (10.99) is obtained. In each figure, the hyperplane H_0 indicating the decision boundary $g(x) = 0$ is drawn with a thick line, and the hyperplanes H_1 indicating $g(x) = 1$ and H_2 indicating $g(x) = -1$ are drawn with thin lines.

In each figure, support vectors are filled in black and non-support vectors are drawn in white. Taking Fig. 10.6 as an example, there are a total of 37 support vectors and 363 non-support vectors. The non-support vectors satisfy Eq. (10.101). Each support vector satisfies one of Eqs. (10.102) through (10.104). The breakdown is as follows.

There are 3 support vectors satisfying Eq. (10.102), which lie on the hyperplanes H_1 or H_2. There are 29 support vectors satisfying Eq. (10.103), which lie in the own-class region between the hyperplanes H_0 and H_1, or between H_0 and H_2. There are 5 support vectors satisfying Eq. (10.104), which are misclassified because they lie beyond the hyperplane H_0 in the region of the other class.

The distance between H_1 and H_2 becomes smaller if the weight c_{pe} for the penalty is larger, because patterns will be suppressed from crossing the hyperplane H_1 or H_2 and approaching the region of the other class. The distance between both hyperplanes was 1.454 in Fig. 10.6, whereas in Fig. 10.7 it is 0.693, less than $1/2$. As a result, the number of patterns between the hyperplanes H_1 and H_2 decreases in Fig. 10.7 compared to Fig. 10.6, and the number of the support vectors, i.e., the patterns contributing to the determination of the decision boundary, also decreases. In fact, for Fig. 10.7, the number of the support vectors is 15, which is less than for Fig. 10.6.

Fig. 10.6 Decision boundary obtained by SVM($c_{pe} = 0.1$)

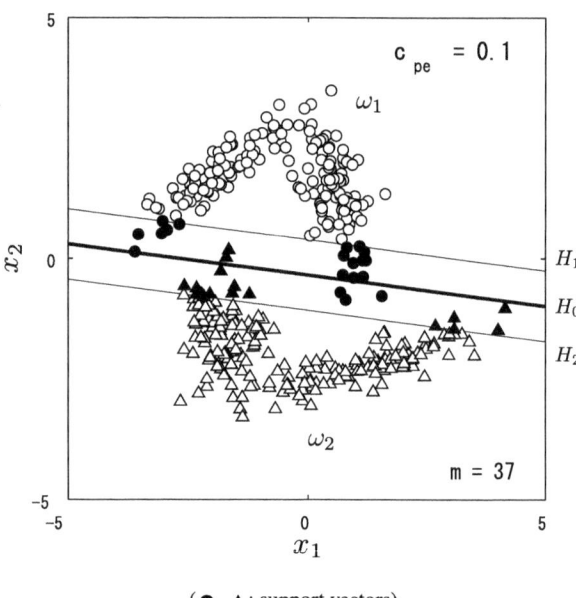

($\bullet$, $\blacktriangle$: support vectors)

Let's look again at Eqs. (10.101) through (10.104). When the learning patterns are linearly separable, all the support vectors for defining the decision boundary H_0 lie on the hyperplane H_1 or H_2. This can be verified with Fig. 10.4.

On the other hand, when the distribution is linearly nonseparable, the support vectors obtained by introducing the soft margin do not necessarily lie on H_1 or H_2. For example, the support vectors of Eq. (10.102) are on H_1 or on H_2, but those of Eq. (10.103) are neither on H_1 nor on H_2. In addition, we should note that patterns that are misclassified, as in Eq. (10.104), are also involved in the determination of the decision boundary H_0 as support vectors. This is evident from Figs. 10.6 and 10.7.

Fig. 10.7 Decision boundary obtained by SVM($c_{pe} = 10.0$)

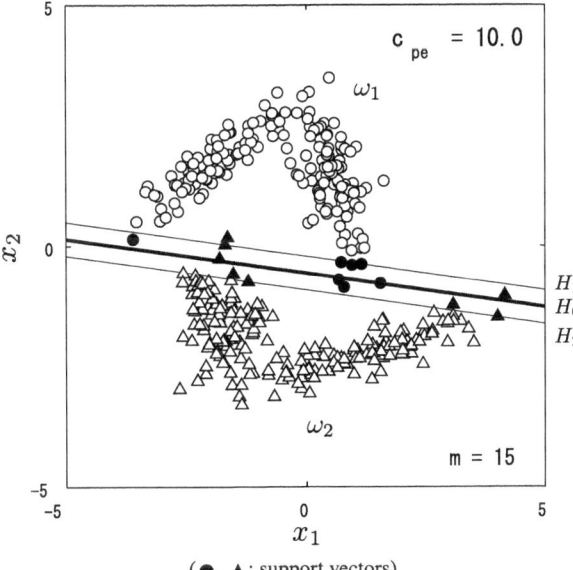

($\bullet$, $\blacktriangle$: support vectors)

Equation (10.101) indicates that a pattern with a large distance from the decision boundary H_0 cannot be a support vector. In other words, the support vectors are selected from among patterns that are close to the decision boundary H_0.

A simpler example of applying the linear SVM to linearly nonseparable learning patterns is given in Problem 10.3.

10.4 Nonlinearization of SVM by Φ Function

In the previous section, we described the linear SVM. Since the decision boundary obtained by the linear SVM is a hyperplane, misclassifications are inevitable for linearly nonseparable patterns. Therefore, to achieve high classification performance, it is necessary to realize complex decision boundaries with nonlinear discriminant functions.

In Chap. 8, we introduced the nonlinear transformation from x in the original d-dimensional space to $\phi(x)$ in the D-dimensional space. We have shown with experiments (see Sect. 8.7) that in the transformed Φ space, linear operations used in the original space are also applicable. Similarly, the SVM introduced in this chapter can be processed in the Φ space. The result of applying the linear SVM in the transformed Φ space is equivalent to applying the nonlinear discriminant function in the original d-dimensional space. The classifier constructed in this way is the *nonlinear SVM* and is explained below.

10.4.1   Applying the Nonlinear Function $\phi(x)$

Nonlinearization of the SVM is generally performed by the kernel method described in the next chapter. The kernel method and its application to SVMs will be described in detail in the next chapter, so here we attempt to nonlinearize the SVM by the Φ function as a preparation.

By applying a nonlinear transformation to

$$x = (x_1, \ldots, x_d)^t \tag{10.107}$$

in the original d-dimensional space,

$$\phi(x) = (\phi_1(x), \ldots, \phi_D(x))^t \tag{10.108}$$

in a D-dimensional space is obtained. Therefore, to apply the linear SVM in the D-dimensional space, we should replace x with $\phi(x)$. Then $\mathbf{H} = (h_{ij})$ in Eqs. (10.37) and (10.90) becomes

$$h_{ij} = b_i b_j \phi(x_i)^t \phi(x_j) \qquad (i, j = 1, \ldots, n), \tag{10.109}$$

instead of Eq. (10.32) and the optimization problem to be solved has exactly the same form as that of Optprobs. 10.2 and 10.4. Also, the desired discriminant function is the following, instead of Eqs. (10.46) and (10.100):

$$g(x) = w_0^* + \sum_{k=1}^{n} \lambda_k^* b_k \phi(x_k)^t \phi(x). \tag{10.110}$$

Here, w_0^* in the above equation can be obtained as follows. Corresponding to Eqs. (10.39) and (10.92), the following holds:

$$w^* = \sum_{k=1}^{n} \lambda_k^* b_k \phi(x_k). \tag{10.111}$$

Using the above formula, corresponding to Eqs. (10.43) and (10.105),

$$w_0^* = b_s - \boldsymbol{w}^{*t}\boldsymbol{\phi}(\boldsymbol{x}_s) \tag{10.112}$$

$$= b_s - \sum_{k=1}^{n} \lambda_k^* b_k \boldsymbol{\phi}(\boldsymbol{x}_k)^t \boldsymbol{\phi}(\boldsymbol{x}_s) \tag{10.113}$$

is obtained, where $\boldsymbol{x}_s$ is the support vector satisfying $0 < \lambda_k^* < c_{pe}$ from Eq. (10.102) and b_s is its teaching signal (see footnote 7 in this Chapter.)

As before, we show the experimental results using the learning patterns shown in Fig. 8.1. The experiment is conducted by setting up a cubic discriminant function, on a two-dimensional feature space. To realize such a cubic discriminant function, Eq. (8.57) is used as a nonlinear transformation of $\boldsymbol{x} \rightarrow \boldsymbol{\phi}(\boldsymbol{x})$.

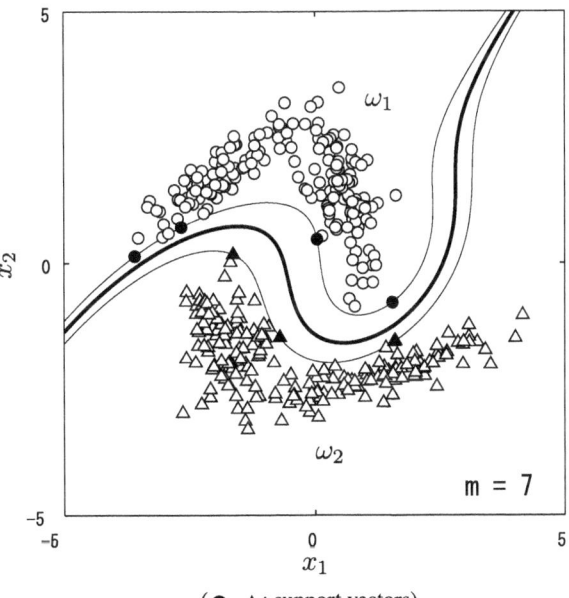

Fig. 10.8 Decision boundary by nonlinear SVM using function $\boldsymbol{\phi}(\boldsymbol{x})$

($\bullet$, $\blacktriangle$: support vectors)

To obtain the discriminant function, compute the matrix $\mathbf{H}$ using Eq. (10.109), solve the dual problem of Optprob. 10.2 and substitute $\boldsymbol{\lambda}^*$ into Eq. (10.110). The result is shown in Fig. 10.8. The decision boundary is indicated by a thick line, and the boundaries corresponding to $g(\boldsymbol{x}) = \pm 1$ are indicated by thin lines. It can be seen that the two classes are correctly separated by this decision boundary. There are a total of seven support vectors, four for the class ω_1 and three for the class ω_2, which are indicated by $\bullet$ and $\blacktriangle$ respectively in the figure. These support vectors are on the thin line $g(\boldsymbol{x}) = \pm 1$.

In Chap. 8, Fig. 8.12 shows the decision boundary obtained by applying the non-linear transformation using the same Eq. (8.57) and applying the primal perceptron learning rule. There is almost no margin between the decision boundary and the patterns in Fig. 8.12. On the other hand, the margin between them is secured due to the effect of the margin maximization in Fig. 10.8. Also, the number of support vectors is 7, less than 13 which is the number of the component vectors of Fig. 8.12.

10.4.2 Applying Kernel Tricks

In the above experiments, the function $\boldsymbol{\phi}(\boldsymbol{x})$ of the nonlinear transformation was set explicitly in the form shown in Eq. (8.57) and the operations were performed in D-dimensional space ($D = 10$). Here the kernel trick introduced in Sect. 9.5.2 can be applied.

That is, using Eqs. (8.57) and (9.21), we can set up the potential function $K(\boldsymbol{x}_i, \boldsymbol{x}_j)$ as in the following equation:

$$K(\boldsymbol{x}_i, \boldsymbol{x}_j) = \boldsymbol{\phi}(\boldsymbol{x}_i)^t \boldsymbol{\phi}(\boldsymbol{x}_j) \tag{10.114}$$

$$= (1 + \boldsymbol{x}_i^t \boldsymbol{x}_j)^3. \tag{10.115}$$

As a result, $\mathbf{H} = (h_{ij})$ can be obtained by the following equation instead of Eq. (10.109):

$$h_{ij} = b_i b_j K(\boldsymbol{x}_i, \boldsymbol{x}_j) \tag{10.116}$$

$$= b_i b_j (1 + \boldsymbol{x}_i^t \boldsymbol{x}_j)^3 \qquad (i, j = 1, \ldots, n). \tag{10.117}$$

From the above, Eqs. (10.110) and (10.113) can be rewritten as follows:

$$g(\boldsymbol{x}) = w_0^* + \sum_{k=1}^n \lambda_k^* b_k K(\boldsymbol{x}, \boldsymbol{x}_k), \tag{10.118}$$

$$w_0^* = b_s - \sum_{k=1}^n \lambda_k^* b_k K(\boldsymbol{x}_s, \boldsymbol{x}_k). \tag{10.119}$$

In Eq. (10.119), $\boldsymbol{x}_s$ is the support vector satisfying $0 < \lambda_k^* < c_{pe}$ as in Eq. (10.113), and b_s is its teaching signal.

When using the kernel trick, the matrix $\mathbf{H}$ is computed by Eq. (10.116) and the dual problem of Optprob. 10.2 is solved to obtain the discriminant function of Eq. (10.118). While Eq. (10.109) requires defining the function $\boldsymbol{\phi}(\boldsymbol{x})$ and computing the inner product of D-dimensional vectors, Eq. (10.116) requires only computing between d-dimensional vectors ($d = 2$). Naturally, the result agrees with that of Fig. 10.8.

The function $K(x, x_k)$ in Eq. (10.118) corresponds to the kernel function which will be fully introduced in Chap. 11. We have already mentioned that the kernel function expressed in the form of Eq. (10.115) in particular is called the polynomial kernel. The kernel function $K(x, x_k)$ can express the inner product $\phi(x_k)^t \phi(x)$ between D-dimensional vectors as a function of d-dimensional vectors x and x_k. Such $K(x, x_k)$ is not limited to polynomial kernels but, as mentioned in Sect. 9.3, the Gaussian function of Eq. (9.7)

$$K(x, x_k) = \exp\left[-\frac{\|x - x_k\|^2}{2\sigma^2}\right] \qquad (10.120)$$

is called the Gaussian kernel and can be expressed as an inner product such as Eq. (9.13). The conditions that must be satisfied by the kernel function will be described in detail in Chap. 11.

The experimental result of applying the nonlinear SVM to the learning patterns of Fig. 8.1 using the Gaussian kernel is shown in Fig. 10.9. The view of the figure is the same as before. The parameter of the function $K(x, x_k)$ is set to $\sigma = 2.0$. The decision boundary indicated by the thick line shows that both classes are correctly separated. The number of the support vectors is 6 in total, 4 and 2 for the classes ω_1 and ω_2, respectively, and they are on the thin lines corresponding to $g(x) = \pm 1$.

Fig. 10.9 Decision boundary by nonlinear SVM using Gaussian kernel (1)

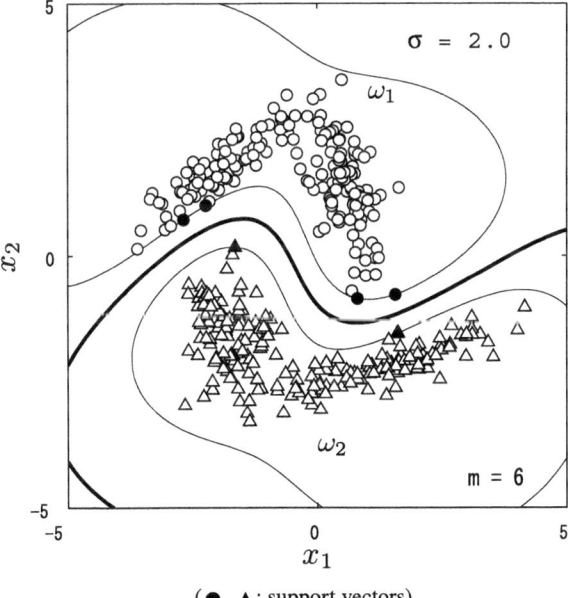

($\bullet$, $\blacktriangle$: support vectors)

The same Gaussian kernel was used to apply the nonlinear SVM to the learning patterns of Fig. 8.10 under the same conditions as above, and the experimental results are shown in Fig. 10.10.

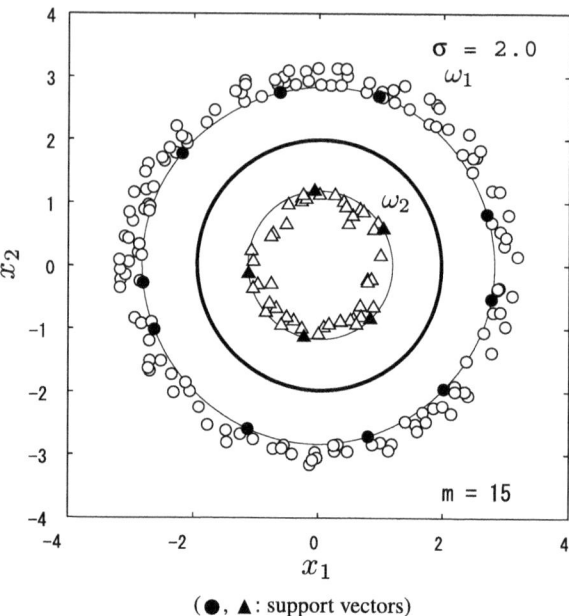

Fig. 10.10 Decision
boundary by nonlinear SVM
using Gaussian kernel (2)

($\bullet$, $\blacktriangle$: support vectors)

The number of the support vectors is 15 in total, 10 and 5 for the classes ω_1 and ω_2, respectively, and they lie on thin lines of approximately concentric circles corresponding to $g(x) = \pm 1$. As a result, the decision boundary is also a concentric circle, and the margin of separation between classes is larger than in the case of Figs. 8.14 and 9.4. This result confirms the effect of the margin maximization by the nonlinear SVM.

Comparing the above results with the experimental results of the potential function method in Chap. 9, we can see the relation between the component vectors and the support vectors. Equation (10.118) obtained by the nonlinear SVM corresponds to Eq. (9.1), and the form of the discriminant function is the same as that of the potential function method.[9] However, the method of determining the discriminant function differs significantly between the two in the following points.

That is, the algorithm for the potential function method is an iterative method as described in Sect. 9.3, and in some cases it may take a long time to converge. However, the computation for the nonlinear SVM is a direct method, which is easier to handle than the potential function method because it is reduced to solving a convex quadratic programming problem. Also, unlike the component vectors of the potential function method, the support vectors are uniquely determined for the nonlinear SVM. A simpler example is given as Problem 10.4.

---

[9] The presence or absence of the constant term w_0^* is not an essential difference.

As mentioned above, the kernel trick is extremely important in that it does not require defining the concrete form of $\boldsymbol{\phi}(\boldsymbol{x})$ and eliminates the need for vector operations in high-dimensional spaces. Furthermore, the kernel trick attracted attention due to its connection with the SVM.[10] The greatest achievement of Boser et al., the originators of the SVM, is that they combined the two ideas of the margin maximization and the kernel method to create a classifier with high performance.

10.4.3 SVM Optimization as a Regularization Problem

Let us now look at the SVM optimization problem from a different perspective. Using Eq. (10.61), the minimization problem of $f(\boldsymbol{w})$ shown in Eq. (10.55) can be rewritten as the minimization problem of $F(\boldsymbol{w})$ as follows:

$$\boldsymbol{w}^* = \operatorname*{argmin}_{\boldsymbol{w}} F(\boldsymbol{w}), \tag{10.121}$$

$$F(\boldsymbol{w}) \stackrel{\text{def}}{=} \frac{c'}{2}\|\boldsymbol{w}\|^2 + \frac{1}{n}\sum_{k=1}^{n}\max\{1 - b_k(w_0 + \boldsymbol{w}^t\boldsymbol{x}_k),\ 0\}. \tag{10.122}$$

In Eq. (10.122), c' is a positive constant and, like c_{pe} in Eq. (10.55), is a parameter that determines the balance between the first and the second terms.[11] The second term $\max\{\cdot\}$ of the function $F(\boldsymbol{w})$ corresponds to Eq. (10.61) and exceeds 1 when $\boldsymbol{x}_k$ is misclassified, representing the loss of misclassification. Thus, the second term represents the average loss caused by all patterns. If the dimensionality of the space, that is, the dimensionality of $\boldsymbol{x}$, is large enough, there exists $\boldsymbol{w}$ that does not cause misclassification for the learning patterns and $b_k(w_0 + \boldsymbol{w}^t\boldsymbol{x}_k)$ is always positive. In this case, the second term of Eq. (10.122) can be set to 0 as long as the norm $\boldsymbol{w}$ is sufficiently large. Therefore, the first term of Eq. (10.122) serves to suppress $\boldsymbol{w}$ so that $\|\boldsymbol{w}\|$ does not grow unnecessarily large. The degree of suppression is determined by the size of the parameter c'.

Although Eq. (10.55) is difficult to understand because its second term is not explicitly represented by $\boldsymbol{w}$, if we rewrite it as Eq. (10.122), we see that this expression is the so-called regularization (see Sect. 15.2.2). The first term of Eq. (10.122) is called the regularization term. As shown in Eq. (10.8), the size of the margin is inversely proportional to $\|\boldsymbol{w}\|$, so adding a regularization term that makes $\|\boldsymbol{w}\|$ smaller is nothing but the margin maximization. The optimal $\boldsymbol{w}$ can be obtained quickly by partially differentiating $F(\boldsymbol{w})$ of Eq. (10.122) and placing $\mathbf{0}$ as

---

[10] Although the term "kernel trick" was not used, it should be remembered that the basic idea and its importance were already clear in the 1960s, when the potential function method was being actively studied (Aizerman et al. 1970).

[11] $c_{pe} \to 0$ corresponds to $c' \to \infty$ and $c_{pe} \to \infty$ corresponds to $c' \to 0$, respectively.

follows:

$$\frac{\partial F(\boldsymbol{w})}{\partial \boldsymbol{w}} = c'\boldsymbol{w} - \frac{1}{n}\sum_{k=1}^{n} b_k \boldsymbol{x}_k = \boldsymbol{0}. \tag{10.123}$$

The above argument is of course valid for nonlinear SVMs, in which case the problem is to minimize $F(\mathbf{w})$ in the following equation by replacing x with $\boldsymbol{\phi}(x)$:

$$F(\mathbf{w}) = \frac{c'}{2}\|\mathbf{w}\|^2 + \frac{1}{n}\sum_{k=1}^{n} \max\{1 - b_k \mathbf{w}^t \boldsymbol{\phi}(\boldsymbol{x}_k),\ 0\} \tag{10.124}$$

This method cannot be combined with the kernel method introduced in Chap. 11, but it can be combined with the generalized linear discriminant function method if Φ is explicitly given.

10.5 Considerations for Using SVMs

As already mentioned in Sect. 10.1, while the SVM has various advantages, there are also some points to be considered when using it, as shown below.

(1) Since the basic formulation of the SVM is based on solving a two-class problem, it cannot be applied to a multi-class problem as it is. The standard method for solving the multi-class problem is to prepare a large number of two-class classifiers. For example, if the number of classes is c, one possible method is to set c two-class problems, with one class and the other $(c - 1)$ classes, and apply the SVM. Another possible method is to consider all class pairs, apply the SVM to $c(c - 1)/2$ two-class problems, and then use the majority voting scheme. See Sect. 4.3.2 for more details on multi-class problems.

(2) One should be aware of the computational complexity of solving the convex quadratic programming problem. Fast algorithms for solving the convex quadratic programming problems have been proposed, but they have to deal with the matrix $\mathbf{H}$ whose elements are h_{ij} of Eq. (10.32) or (10.109). It is important to note that when the number of learning patterns is n, the size of the square matrix $\mathbf{H}$ is $n \times n$. In standard data analysis such as the principal component analysis or the discriminant analysis, the computational complexity usually depends on the dimensionality of the feature space. On the other hand, in SVM optimization, the computational complexity depends on the number of patterns not on the dimensionality, as shown in Optprob. 10.2. Therefore, in the case of the SVM, a large number of learning patterns is advantageous in terms of improving classification performance, but disadvantageous in terms of increased computational and memory requirements.

(3) The optimal value of the penalty term c_{pe} must be determined in advance by humans. Such a parameter is called the hyperparameter,[12] and care must be taken in the optimization procedure. However, the behavior when c_{pe} is changed can be grasped intuitively, and the change in classification performance with respect to c_{pe} is expected to be unimodal, so the optimal value is relatively easy to find.

(4) If there is a large bias in the number of learning patterns depending on the class, it may be better to change c_{pe} for each class. This may be related to how to estimate the prior probability of each class. Similarly, if the learning patterns can be assigned importance, it is possible to change c_{pe} for each learning pattern.

(5) The kernel method described in the next chapter is used to design the nonlinear SVM, but there is no clear guideline for selecting the kernel function. This is described later.

────────────────────── **Coffee Break** ──────────────────────

The Long Journey of the SVM

The core technologies of the SVM proposed by Boser et al. in their 1992 paper are the margin maximization and the kernel trick. The idea of the margin had already been proposed and discussed in the 1960s and 1970s in the process of advancing the perceptron (Duda and Hart 1973). The kernel trick was also introduced by Aizerman in his 1964 paper (Aizerman et al. 1964) on the potential function method. Moreover, the theory itself which the kernel trick is based on has a long history of more than a century (Schölkopf 2000; Mercer 1909).

Although the margin and the kernel trick had been introduced and there seemed to be enough clues to encourage the idea of the SVM, it took 30 years for this research by Boser et al. to be published. Why did it take so long?

The potential function method was long unused and forgotten until Boser et al. revived. The main reason for this was that the computational power at that time was not sufficient to handle potential functions by the iterative method, which required a large amount of computation. The difficulty of using iterative methods also applies to the margin determination of the time. It is possible, though speculative, that the researchers of the time underestimated the utility of the margin.

Everyone understands that the larger the margin, the better the classifier. However, researchers at that time may not have realized that the effect of a large margin can be demonstrated only in a high-dimensional feature space. As already mentioned in the coffee break in Sect. 10.2, the SVM has a high generalization ability. With a large margin, the SVM can be expected to be effective even when the dimensionality of the feature space is extremely large relative to the number of patterns, or in other words, when patterns are sparsely distributed in the feature space. On the other hand, learning with a small number of patterns in a high-dimensional feature space causes over-fitting, which has long been regarded as a taboo because of the curse of dimensionality (Sect. 4.4). At the time, there must have been some hesitation in challenging this taboo. Even if such a challenge had

---
[12] The hyperparameters are parameters that need to be adjusted by humans prior to learning, apart from the intrinsic parameters that can be determined by learning from data. For example, the weights of a neural network are the intrinsic parameters, while the number of internal layers and the number of units are the hyperparameters. For the hyperparameters, see also Sect. 4.5.1.

been attempted, processing in extremely high dimensions would have been impossible with the poor computing power of the time.

Boser et al. are credited with reaffirming the latent power of the margin, formulating the margin maximization as a convex quadratic programming problem and showing how to solve it as a direct method. Not only that, they also made it possible to process in high-dimensional spaces, where SVMs are at their most powerful. In other words, the kernel trick replaces the enormous computation in high-dimensional spaces with an equivalent simple computation using kernels. Thus, by skillfully combining the well-known techniques of the margin and the kernel trick, they have paved the way to a solution to this problem. Their achievement is extremely great.

The story of the SVM continues. The SVM was completed in 1995 with the soft margin proposal by Cortes and Vapnick (Cortes and Vapnik 1995). However, it was not until the late 1990s and the 2000s that the SVM was taken up by many researchers and published in numerous papers and books. It took 30 years for the idea of the SVM to be perfected, but why did it take so many years for this method to then achieve a boom? It is assumed that few researchers were aware of the importance of this method at the time of the publication of the historic paper by Boser et al. What, then, triggered the boom?

First, we must mention the contribution of a series of empirical studies by Schölkopf and other researchers at Bell Labs, where Vapnick was affiliated.

The second is an application of the SVM to text classification by Joachims (2002). In document processing, the dimensionality of the feature space is extremely large compared to the number of samples because whether a particular word appears in a document or not is represented as a 0/1 feature. Joachims showed that the SVM works effectively even in such cases. In other words, he challenged the aforementioned taboo and broke through it. The fact that the SVM, which was developed in the field of pattern recognition, has transcended its field and demonstrated its power in the field of natural language processing provoked an extremely strong response.

The third is the release of a program called SVM-Light by Joachims (1998). Although it is now commonplace to release source code through GitHub, this was rare at the time. The release and popularization of SVM-Light led to the recognition of the value of the SVM across research fields.

As a matter of course, no matter how good an idea is, its value will never be known unless it is verified, and it will never see the light of day unless its value is recognized. The long journey of the SVM is an eloquent testament to this fact.

Problems

10.1 Derive Eq. (10.62).

10.2 [†] Using the linear SVM, find a discriminant function $g(x)$ that can correctly classify the linearly separable learning patterns listed in Problem 2.5. Also, illustrate the decision boundary H_0 determined by the obtained discriminant function, and the hyperplanes H_1 and H_2 that determine the margin, then show the value of the margin R. Furthermore, compare these results with those of Problems 2.5 and 2.6.

10.3 [†] Using the linear SVM, find a discriminant function $g(x)$ for the linearly nonseparable learning patterns listed in Problem 8.6. Here, the penalty of Eq. (10.55) is set to $c_{pe} = 0.1$ Also, illustrate the decision boundary H_0 with $g(x) = 0$, the hyperplane H_1 with $g(x) = 1$, and the hyperplane H_2 with $g(x) = -1$. Furthermore, show the results when the penalty is set to $c_{pe} = 10$ in the same way as above.

10.4 [†] Assuming that the nonlinear SVM is used to find a discriminant function that can correctly classify the linearly nonseparable learning patterns listed in Problem 8.6, perform the following:

(1) Illustrate the decision boundary obtained when the Gaussian kernel of Eq. (9.7) is used as the kernel function. Here, the parameter of the Gaussian kernel should be set to $\sigma = 2.0$. Also, compare the obtained results with those of Problem 9.2(1).

(2) Illustrate the decision boundary obtained when using the third-degree polynomial kernel of Eq. (9.22) as the kernel function. Also compare the obtained results with those of Problem 9.2(2).

Chapter 11
Kernel Method

Abstract The *kernel method* is a versatile technique for extending linear model processing algorithms to nonlinear models. It became well-known in conjunction with the popularization of the support vector machine, a method of pattern recognition. Its most significant feature is that by introducing a function known as the kernel function, it can compute the inner product of vectors nonlinearly mapped to high-dimensional spaces using operations in the low-dimensional original space. As a result, it became possible to naturally extend linear processing algorithms for multivariate data to nonlinear ones. Currently, the kernel method is becoming one of the fundamental techniques not only in pattern recognition and machine learning but also in computer vision, natural language processing, data mining, and bioinformatics, among other broad application fields.

11.1 Kernel Function

The kernel method, in terms of its capability to transform x into $\phi(x)$ using the nonlinear transformation ϕ, is in fact analogous to the generalized linear discriminant function method discussed in Chap. 8. Furthermore, it also relates to the potential function method in Chap. 9. However, the distinct feature of the kernel method is that it does not describe the function ϕ in an explicit form; instead, it defines ϕ through a special function known as the *kernel function*. The kernel function is defined as follows.

If there exists a function $\phi(x)$ ($\mathbb{R}^d \to \mathbb{R}^D$), and the relation

$$k(x, y) = \phi(x)^t \phi(y) \tag{11.1}$$

holds for all x and y in $\mathbb{R}^d$, a function $k(x, y)$ is called the kernel function. Therefore, whether a function $k(x, y)$ of x and y is a kernel function or not is

Supplementary Information The online version contains supplementary material available at https://doi.org/10.1007/978-981-95-1478-6_11.

determined by whether a function $\boldsymbol{\phi}$ satisfying Eq. (11.1) exists or not. This means that all computations that can be performed using the inner product operation can be computed by replacing the inner product with the kernel function $k(\cdot)$. Moreover, once the form of $k(\cdot)$ is determined, computations can be made without providing the explicit form of $\boldsymbol{\phi}$.

An important point here is that while the right side of Eq. (11.1) is the inner product of D-dimensional vectors, the kernel function $k(\boldsymbol{x}, \boldsymbol{y})$ is a function of d-dimensional vectors. When calculating the inner product of vectors $\boldsymbol{\phi}(\boldsymbol{x})$ and $\boldsymbol{\phi}(\boldsymbol{y})$, there's no need to directly compute the inner product in the space of $\boldsymbol{\phi}$; it's sufficient to just compute $k(\boldsymbol{x}, \boldsymbol{y})$. Hence, when D is much larger than d (i.e., $D \gg d$), there's a significant difference in computational cost. In other words, despite the mapping $\boldsymbol{\phi}$ determined by the kernel function $k(\cdot)$ being a nonlinear mapping from a d-dimensional space to a D-dimensional space, high-dimensional nonlinear operations like $\boldsymbol{\phi}(\boldsymbol{x})^t \boldsymbol{\phi}(\boldsymbol{y})$ can be realized by the low-dimensional linear operation $k(\boldsymbol{x}, \boldsymbol{y})$. This mechanism, as mentioned in Sect. 9.5.2, is called the "kernel trick."

Let's first look at simple examples of kernel functions.

Example 1 Consider any two-dimensional vectors $\boldsymbol{x} = (x_1, x_2)^t$ and $\boldsymbol{y} = (y_1, y_2)^t$. Let's define the mapping $\boldsymbol{\phi}$ from a two-dimensional space $\mathbb{R}^2$ to a three-dimensional space $\mathbb{R}^3$ ($d = 2, ; D = 3$) as follows:

$$\boldsymbol{\phi}(\boldsymbol{x}) = (x_1^2, x_2^2, \sqrt{2}x_1x_2)^t. \tag{11.2}$$

Now, computing the inner product of $\boldsymbol{\phi}(\boldsymbol{x})$ and $\boldsymbol{\phi}(\boldsymbol{y})$, we get

$$\boldsymbol{\phi}(\boldsymbol{x})^t \boldsymbol{\phi}(\boldsymbol{y}) = (x_1^2, x_2^2, \sqrt{2}x_1x_2)(y_1^2, y_2^2, \sqrt{2}y_1y_2)^t \tag{11.3}$$

$$= x_1^2 y_1^2 + x_2^2 y_2^2 + 2x_1x_2y_1y_2 \tag{11.4}$$

$$= (\boldsymbol{x}^t \boldsymbol{y})^2. \tag{11.5}$$

Hence, given that $\boldsymbol{\phi}$ satisfying Eq. (11.1) is provided in Eq. (11.2), the function $k_1(\boldsymbol{x}, \boldsymbol{y})$ defined as

$$k_1(\boldsymbol{x}, \boldsymbol{y}) = (\boldsymbol{x}^t \boldsymbol{y})^2 \tag{11.6}$$

is indeed a kernel function.

Example 2 Consider the mapping $\boldsymbol{\phi}$; ($d = 2, ; D = 10$), that is, a mapping from a two-dimensional space to a ten-dimensional space. As already shown in Eqs. (8.57) and (9.21), defining $\boldsymbol{\phi}(\boldsymbol{x})$ as

$\boldsymbol{\phi}(\boldsymbol{x})$

$$= (1, \sqrt{3}x_1, \sqrt{3}x_2, \sqrt{6}x_1x_2, \sqrt{3}x_1^2, \sqrt{3}x_2^2, \sqrt{3}x_1^2x_2, \sqrt{3}x_1x_2^2, x_1^3, x_2^3)^t, \tag{11.7}$$

the inner product of $\phi(x)$ and $\phi(y)$ is

$$\phi(x)^t \phi(y) = (1 + x^t y)^3. \tag{11.8}$$

Thus, similar to Example 1, since ϕ that satisfies Eq. (11.1) is given by Eq. (11.7), the function $k_2(x, y)$ defined as

$$k_2(x, y) = (1 + x^t y)^3 \tag{11.9}$$

is indeed a kernel function. This corresponds to the polynomial kernel discussed in Sect. 11.3.

Example 3 Consider the following mapping ϕ from a d-dimensional space to a $D(= d^2)$-dimensional space,

$$\phi(x) = (x_i x_j)_{i,j=1}^d. \tag{11.10}$$

Here, $(x_i x_j)_{i,j=1}^d$ refers to the d^2-dimensional vector formed by calculating $x_i x_j$ for all possible combinations of i, j $(i, j = 1, \ldots, d)$ given the d-dimensional vector $x = (x_1, x_2, \ldots, x_d)^t$. The order of the vector components can be arbitrary. For instance, when $d = 2$, it becomes

$$\phi(x) = (x_1^2, x_1 x_2, x_2 x_1, x_2^2)^t. \tag{11.11}$$

For this case, the inner product between $\phi(x)$ and $\phi(y)$ is

$$\phi(x)^t \phi(y) = ((x_i x_j)_{i,j=1}^d)^t ((y_i y_j)_{i,j=1}^d) \tag{11.12}$$

$$= \sum_{i,j=1}^d x_i x_j y_i y_j \tag{11.13}$$

$$= \sum_{i=1}^d x_i y_i \sum_{j=1}^d x_j y_j \tag{11.14}$$

$$= (x^t y)^2. \tag{11.15}$$

Thus, the function $k_3(x, y)$ defined as

$$k_3(x, y) = (x^t y)^2 \tag{11.16}$$

is indeed a kernel function. Equation (11.16) has the same form as Eq. (11.6) in Example 1, but the form of $\phi(x)$ is different between Eqs. (11.11) and (11.2). That is, for a given kernel function $k(\cdot)$, ϕ satisfying Eq. (11.1) is not necessarily uniquely determined.

11.2   Conditions for Kernel Functions

One of the essential points in applying the kernel method is choosing an appropriate kernel function. In this section, we discuss the conditions for a function $k(x, y)$ to be a kernel function, i.e., the conditions for the existence of a ϕ that satisfies Eq. (11.1). The mathematically defined Mercer's condition is well-known. Additionally, it is possible to determine the conditions for a kernel function based on the positive definiteness of matrices.

11.2.1   Kernel Functions Satisfying Mercer's Condition

Now, for any d-dimensional vectors x, y belonging to a subset S of $\mathbb{R}^d$, if

$$\iint_{S \times S} k(x, y)g(x)g(y)dx dy \geq 0, \tag{11.17}$$

holds for all functions $g(x)$ that satisfy

$$\int (g(x))^2 \, dx < \infty, \tag{11.18}$$

then the function $k(x, y)$ is said to satisfy *Mercer's condition*, and can be written in the form,

$$k(x, y) = \sum_i \phi_i(x)\phi_i(y) = \phi(x)^t \phi(y). \tag{11.19}$$

From Eq. (11.1), it follows that $k(x, y)$ is a kernel function.

For example, let's define the function $k(x, y)$ using a natural number p as

$$k(x, y) \overset{\text{def}}{=} (x^t y)^p. \tag{11.20}$$

Here, let x and y be any d-dimensional vectors contained in S, and let p_i ($1 \leq i \leq d$) be the non-negative integers satisfying

$$p_1 + p_2 + \cdots + p_d = p, \tag{11.21}$$

$$0 \leq p_i \leq p. \tag{11.22}$$

Now, expanding $(x^t y)^p = (x_1 y_1 + \cdots + x_d y_d)^p$ using the binomial theorem, we get

$$\iint_{S\times S} (x^t y)^p g(x)g(y)dxdy \tag{11.23}$$

$$= \sum_{p_1+\cdots+p_d=p} \frac{p!}{p_1!\cdots p_d!} \iint_{S\times S} x_1^{p_1}\cdots x_d^{p_d} y_1^{p_1}\cdots y_d^{p_d} g(x)g(y)dxdy \tag{11.24}$$

$$= \sum_{p_1+\cdots+p_d=p} \frac{p!}{p_1!\cdots p_d!} \left(\int_S x_1^{p_1}\cdots x_d^{p_d} g(x)dx\right)^2 \geq 0. \tag{11.25}$$

Thus, $k(x, y) = (x^t y)^p$ satisfies Mercer's condition. Therefore, we can confirm that the $k(x, y)$ defined in Eq. (11.20) is a kernel function, implying a function ϕ satisfying $(x^t y)^p = \phi(x)^t \phi(y)$ exists. Equations (11.6) and (11.16) correspond to the case $p = 2$.

11.2.2 Kernel Functions Satisfying Positive Semidefiniteness

Instead of the aforementioned integral calculations, it is also possible to determine the conditions for a kernel function based on the positive definiteness of matrices. For a real matrix $\mathbf{A}$ of size $n \times n$ which is symmetric, if it satisfies

$$y^t \mathbf{A} y \geq 0 \tag{11.26}$$

for any n-dimensional column vector y where $y \neq 0$, then $\mathbf{A}$ is said to be *positive semidefinite*. If the above equation holds without the equality, then it is said to be *positive definite*.[1]

Now, given n d-dimensional patterns $x_1, \cdots, x_n$ and a function $k(x_i, x_j)$, the (i, j) component k_{ij} is defined as

$$k_{ij} = k(x_i, x_j), \tag{11.27}$$

and we denote the $n \times n$ square matrix given by this as $\mathbf{K} = (k_{ij})$. The necessary and sufficient condition for the function $k(\cdot)$ to be a kernel function is that $\mathbf{K}$ is positive semidefinite. This condition is equivalent to $\mathbf{K}$ being symmetric and all its eigenvalues being non-negative.

---

[1] Note that the positive semidefiniteness and definiteness of a matrix are defined for symmetric matrices. Also, when $\mathbf{A}$ and y are complex matrices and complex vectors respectively, the Hermitian property of $\mathbf{A}$ can be derived from Eq. (11.26), but for real matrices and real vectors, the symmetric property of $\mathbf{A}$ cannot be inferred from the same equation.

For the proof, refer to Problems 11.1 and 11.2. Such defined kernel functions are called *positive semidefinite kernel*, and it is also said that the function $k(\cdot)$ is positive semidefinite.

When the above mentioned $k(\cdot)$ is a kernel function, due to the existence of a function $\boldsymbol{\phi}(\cdot)$ such that

$$k(\boldsymbol{x}_i, \boldsymbol{x}_j) = \boldsymbol{\phi}(\boldsymbol{x}_i)^t \boldsymbol{\phi}(\boldsymbol{x}_j), \tag{11.28}$$

from Eq. (11.27), we can write

$$k_{ij} = \boldsymbol{\phi}(\boldsymbol{x}_i)^t \boldsymbol{\phi}(\boldsymbol{x}_j). \tag{11.29}$$

Generally, using any vectors $\boldsymbol{x}_i, \boldsymbol{x}_j$ $(i, j = 1, \ldots, n)$,[2] the (i, j) component of the matrix defined by

$$f_{ij} = \boldsymbol{x}_i^t \boldsymbol{x}_j \tag{11.30}$$

is called the *Gram matrix*.[3]

Therefore, if the $\mathbf{K}$ defined by Eq. (11.27) is a Gram matrix, then from Eq. (11.29), the function $k(\cdot)$ is a kernel function.

Furthermore, $k(\cdot)$ might be given not as a functional form but as specific values. For instance, while we might not know the values of $\boldsymbol{x}_i$ and $\boldsymbol{x}_j$, there are cases where the similarity between $\boldsymbol{x}_i$ and $\boldsymbol{x}_j$ can be measured by some method. As an example, the multidimensional scaling method introduced in the coffee break in Sect. 11.5 is one such analytical method used in these situations. A matrix of $n \times n$ that uses the similarity between the two vectors $\boldsymbol{x}_i$ and $\boldsymbol{x}_j$ as its (i, j) component is referred to as a *similarity matrix*. If we consider the inner product as one of the measures of similarity between vectors, there might be a structure similar to Eq. (11.28) underlying the similarity matrix. The positive semidefiniteness of the similarity matrix allows us to investigate this. However, it's important to note that even if positive semidefiniteness is confirmed, it only holds for the chosen set of n learning patterns and might not generally hold for any arbitrary $\boldsymbol{x}$.

---

[2] In Eq. (11.30), since $\boldsymbol{x}_i$ and $\boldsymbol{x}_j$ are arbitrary vectors, they can be replaced with $\boldsymbol{\phi}(\boldsymbol{x}_i)$ and $\boldsymbol{\phi}(\boldsymbol{x}_j)$ respectively. By setting $f_{ij} = \boldsymbol{\phi}(\boldsymbol{x}_i)^t \boldsymbol{\phi}(\boldsymbol{x}_j)$, it becomes more general. Equation (11.30) can be considered as a special case where $\boldsymbol{\phi}(\boldsymbol{x}_i) = \boldsymbol{x}_i$ and $\boldsymbol{\phi}(\boldsymbol{x}_j) = \boldsymbol{x}_j$.

[3] In general, the $n \times n$ matrix $\mathbf{A}^t \mathbf{A}$ obtained from a $m \times n$ matrix $\mathbf{A}$ is called the Gram matrix for $\mathbf{A}$. However, in the field of pattern recognition, taking $\mathbf{X} = (\boldsymbol{x}_1, \ldots, \boldsymbol{x}_n)^t$, we set $\mathbf{A} = \mathbf{X}^t$ and define $\mathbf{XX}^t$ as the Gram matrix corresponding to $\mathbf{X}^t$. On the other hand, since the (i, j) component of the matrix $\mathbf{XX}^t$ is the inner product $\boldsymbol{x}_i^t \boldsymbol{x}_j$, this matrix is also called the inner product matrix of $\mathbf{X}$. Refer to Eq. (G.27) for more information on the inner product matrix.

11.2.3 Composite Kernel Functions

If a certain function satisfies the conditions of a kernel function, various functions derived or composed from it will also meet the conditions of a kernel function. The following functions $k(x, y)$ are kernel functions, where $k_1(x, y)$ and $k_2(x, y)$ are kernel functions:

$$k(x, y) = ak_1(x, y) \quad (a > 0), \tag{11.31}$$

$$k(x, y) = k_1(x, y) + k_2(x, y), \tag{11.32}$$

$$k(x, y) = k_1(x, y) \cdot k_2(x, y), \tag{11.33}$$

$$k(x, y) = \exp(k_1(x, y)), \tag{11.34}$$

$$k(x, y) = p(k_1(x, y)), \tag{11.35}$$

$$k(x, y) = f(x)k_1(x, y)f(y), \tag{11.36}$$

$$k(x, y) = k_1(\phi(x), \phi(y)), \tag{11.37}$$

$$k(x, y) = x^t A y. \tag{11.38}$$

In the above equations, a is a constant, $p(\cdot)$ is a polynomial with non-negative coefficients, $f(\cdot)$ is an arbitrary function, A is a positive semidefinite matrix, and $\phi(x)$ is a mapping from x to $\mathbb{R}^D$. Equation (11.34) can be transformed into the form of Eq. (11.32) by utilizing a Taylor expansion.

11.3 Examples of Kernel Functions

In this section, we introduce specific examples of kernel functions, namely polynomial kernel, Gaussian kernel, all-subsets kernel, and ANOVA kernel. Among these, the polynomial kernel and Gaussian kernel are the most commonly used in practice. The polynomial kernel is expressed as a polynomial with non-negative coefficients related to the inner product $x^t y$. The examples illustrated in the previous section, Examples 1, 2, and 3, fall under this category.

11.3.1 Polynomial Kernel

A function $k_p(x, y)$ defined by the following equation is called a p-degree *polynomial kernel*:

$$k_p(x, y) = (c_0 + x^t y)^p \tag{11.39}$$

$$= \sum_{i=0}^{p} {}_pC_i (x^t y)^i \cdot c_0^{p-i}. \tag{11.40}$$

In the above equation, c_0 is a constant. The mapping ϕ determined by the polynomial kernel $k_p(x, y)$ is from dimensionality d to dimensionality D. To determine the dimensionality D of the vector $\phi(x)$, we need to expand $(x_1 + \cdots + x_d + c_0)^p$ and count the number of independent terms $x_1^{i_1} \cdots x_d^{i_d}$. Denoting this number by $f(d, p)$, $f(d, p)$ is the number of combinations of $(i_1, \ldots, i_d)$ that satisfy

$$\begin{cases} 0 \le i_j \le p \quad (1 \le j \le d) \\ \sum_{j=1}^{d} i_j \le p. \end{cases} \tag{11.41}$$

As shown in Eq. (8.33), $f(d, p)$ is represented as follows:[4]

$$f(d, p) = {}_{d+p}C_p. \tag{11.42}$$

For the proof, refer to Problem 8.5.

11.3.2   Gaussian Kernel

The kernel function $k_g(x, y)$ given by the following equation is called the *Gaussian kernel*:

$$k_g(x, y) = \exp\left(-\frac{\|x - y\|^2}{2\sigma^2}\right). \tag{11.43}$$

Equation (11.43) can be transformed as

$$\exp\left(-\frac{\|x - y\|^2}{2\sigma^2}\right) = \exp\left(-\frac{x^t x}{2\sigma^2}\right) \exp\left(\frac{x^t y}{\sigma^2}\right) \exp\left(-\frac{y^t y}{2\sigma^2}\right). \tag{11.44}$$

From Eqs. (11.33) and (11.34), it can be confirmed that $k_g(x, y)$ is a kernel function. In this case, the dimensionality of $\phi(x)$ becomes infinite. For the proof, refer to Problem 9.1. In applications utilizing kernel functions, such as the nonlinear SVM

---

[4] However, one element of $\phi(x)$ becomes a constant that doesn't depend on x, so the actual dimensionality of $\phi(x)$ is $f(d, p) - 1$. For specific examples, refer to Example 2 in Sect. 11.1 and Fig. 11.4.

introduced in the next section, the polynomial kernel and the Gaussian kernel are often used.

11.3.3 All-Subsets Kernel

The all-subsets kernel $k_s(x, y)$ is one of the polynomial kernels and can be expressed in the inner product form shown in Eq. (11.1). In this case, the elements of $\phi(x)$ consist of products of any combination from $\{x_1, \ldots, x_d\}$ and the constant 1. For instance, when $d = 2$ and $x = (x_1, x_2)^t$,

$$\phi(x) = (1, \ x_1, \ x_2, \ x_1 x_2)^t \tag{11.45}$$

serves as an example. When $d = 3$ and $x = (x_1, x_2, x_3)^t$,

$$\phi(x) = (1, \ x_1, \ x_2, \ x_3, \ x_1 x_2, \ x_1 x_3, \ x_2 x_3, \ x_1 x_2 x_3)^t \tag{11.46}$$

becomes an example. Note that the order of the vector components does not matter.

Let's try to formulate the above in a more general form. Let $I = \{1, \ldots, d\}$ be a set of d natural numbers from 1 to d, and A be a subset of I. Since A is determined by which of the d natural numbers it includes and which it doesn't, the number of possibilities for A, including the empty set ϕ, is $2^d (= D)$. For example, if $d = 2$, then $D = 2^2 = 4$ and

$$A \in \{\phi, \{1\}, \{2\}, \{1, 2\}\}. \tag{11.47}$$

If $d = 3$, then $D = 2^3 = 8$ and

$$A \in \{\phi, \{1\}, \{2\}, \{3\}, \{1, 2\}, \{1, 3\}, \{2, 3\}, \{1, 2, 3\}\}. \tag{11.48}$$

Next, let's define $\phi_A(x)$ as follows:

$$\phi_A(x) = \begin{cases} 1 & \text{(when A is the empty set } \phi) \\ \displaystyle\prod_{i \in A} x_i & \text{(otherwise).} \end{cases} \tag{11.49}$$

For instance, when $d = 3$, A can be any of the options in Eq. (11.48). If $A = \phi$, then $\phi_A(x) = 1$, and if $A = \{1, 2, 3\}$, then $\phi_A(x) = x_1 x_2 x_3$. Using the above formula, $\phi(x)$ can be represented as a D-dimensional column vector as follows:

$$\phi(x) = (\phi_1(x), \ \phi_2(x), \ \ldots, \ \phi_D(x))^t \tag{11.50}$$

$$= (\phi_A(x))_{A \subseteq I}^t \qquad (\in \mathbb{R}^D; \ D = 2^d). \tag{11.51}$$

Equation (11.51) indicates that the D-dimensional column vector $\boldsymbol{\phi}(x)$ is composed of the elements $\phi_A(x)$ for all subsets A. By using Eqs. (11.49) and (11.51), formulas like Eqs. (11.45) and (11.46) can be easily derived.

The kernel function defined by $\boldsymbol{\phi}$ in Eq. (11.51) is given by

$$k_s(x, y) = \boldsymbol{\phi}(x)^t \boldsymbol{\phi}(y) \tag{11.52}$$

and is called the *all-subsets kernel*. Here, considering the sum of the elements $\phi_i(x)$ $(i = 1, \ldots, D)$ of $\boldsymbol{\phi}(x)$ shown in Eq. (11.50), we find

$$\sum_{i=1}^{D} \phi_i(x) = \sum_{A \subseteq I} \prod_{i \in A} x_i \tag{11.53}$$

$$= (1 + x_1)(1 + x_2) \cdots (1 + x_d) \tag{11.54}$$

$$= \prod_{i=1}^{d} (1 + x_i). \tag{11.55}$$

For instance, when $d = 3$, based on Eq. (11.46), we have

$$\sum_{i=1}^{8} \phi_i(x) = 1 + x_1 + x_2 + x_3 + x_1 x_2 + x_1 x_3 + x_2 x_3 + x_1 x_2 x_3 \tag{11.56}$$

$$= (1 + x_1)(1 + x_2)(1 + x_3). \tag{11.57}$$

From this, we can confirm that Eq. (11.55) holds.

Noting that Eq. (11.49) yields

$$\phi_A(x)\phi_A(y) = \begin{cases} 1 & \text{(when A is the empty set}\phi) \\ \prod_{i \in A} x_i \prod_{i \in A} y_i = \prod_{i \in A} x_i y_i & \text{(otherwise)}, \end{cases} \tag{11.58}$$

we can derive the following equation from Eqs. (11.51) and (11.52):

$$k_s(x, y) = \boldsymbol{\phi}(x)^t \boldsymbol{\phi}(y) \tag{11.59}$$

$$= \sum_{A \subseteq I} \phi_A(x)\phi_A(y) \tag{11.60}$$

$$= \sum_{A \subseteq I} \prod_{i \in A} x_i y_i \tag{11.61}$$

$$= \prod_{i=1}^{d} (1 + x_i y_i). \tag{11.62}$$

The transformation from Eqs. (11.61) to (11.62) is similar to the transformation from Eqs. (11.53) to (11.55). If we try to directly calculate the inner product $\boldsymbol{\phi}(\boldsymbol{x})^t\boldsymbol{\phi}(\boldsymbol{y})$ from $\boldsymbol{\phi}$, we need to determine each of the $D(= 2^d)$ elements of $\boldsymbol{\phi}(\boldsymbol{x})$ and $\boldsymbol{\phi}(\boldsymbol{y})$, and then compute their inner product. However, by using the kernel function $k_s(\boldsymbol{x}, \boldsymbol{y})$ in the form of Eq. (11.62), we can see that it can be computed with at most $3d$ product-sum operations.

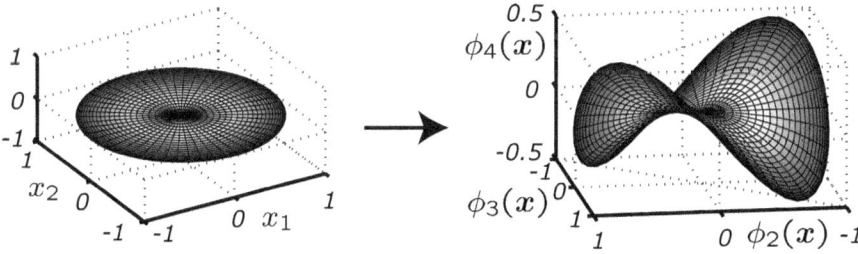

Fig. 11.1 An example of nonlinear transformation using the all-subsets kernel ($d = 2$)

In general, the $\boldsymbol{\phi}$ determined by $k_s(\boldsymbol{x}, \boldsymbol{y})$ becomes a complex nonlinear transformation. Figure 11.1 illustrates how points inside the circle $x_1^2 + x_2^2 = 1$ are mapped by $\boldsymbol{\phi}$ when $d = 2$. In the D-dimensional space of the vector $\boldsymbol{\phi}(\boldsymbol{x})$, it forms a curved surface containing the lines $\phi_2(\boldsymbol{x}) = \phi_4(\boldsymbol{x})$ and $\phi_3(\boldsymbol{x}) = \phi_4(\boldsymbol{x})$. Dividing the space into two by the plane $\phi_4(\boldsymbol{x}) = 0$, points where $\phi_4(\boldsymbol{x}) > 0$ correspond to the 1st and 3rd quadrants in the original space, and points where $\phi_4(\boldsymbol{x}) < 0$ correspond to the 2nd and 4th quadrants.

The all-subsets kernel can be more generally expressed as

$$k_s(\boldsymbol{x}, \boldsymbol{y}) = \prod_{i=1}^{d}(c_i + x_i y_i), \tag{11.63}$$

and in this case, $\boldsymbol{\phi}_A(\boldsymbol{x})$ is represented by the following formula:

$$\phi_A(\boldsymbol{x}) = \begin{cases} \displaystyle\prod_{i=1}^{d}\sqrt{c_i} & \text{(when A is the empty set } \phi) \\[2em] \displaystyle\left(\prod_{i\notin A}\sqrt{c_i}\right)\prod_{i\in A} x_i & \text{(otherwise).} \end{cases} \tag{11.64}$$

The all-subsets kernel takes all subsets of $x_1, \ldots, x_d$ that characterize the input object as a new set of features. This idea is also employed in the kernel methods for structured data, which will be discussed later (see the coffee break in Sect. 11.7). Specifically, this concept is utilized when applying kernel methods to strings or graphs.

11.3.4 ANOVA Kernel

In the all-subsets kernel, we considered all subsets A of $1, \ldots, d$ as elements of ϕ_A in $\boldsymbol{\phi}(\boldsymbol{x})$. On the other hand, for the ANOVA kernel, we focus only on subsets A of size l ($\in \mathbb{N}$), that is, those with l elements. Let $|A| = l$ denote that the number of elements in subset A is l, and let A_l denote the subset with l elements. Using these, we define $\boldsymbol{\phi}$ as follows:

$$\boldsymbol{\phi}(\boldsymbol{x}) = (\phi_A(\boldsymbol{x}))^t_{|A|=l} \qquad (\in \mathbb{R}^D; \ D = {}_d C_l), \tag{11.65}$$

$$\phi_{A_l}(\boldsymbol{x}) = \prod_{i \in A_l} x_i. \tag{11.66}$$

First, find $\phi_{A_l}(\boldsymbol{x})$ for all subsets A_l of $\{1, \ldots, d\}$ whose number of elements is l. The $D(= {}_d C_l)$-dimensional column vector with them as elements corresponds to the term $(\phi_A(\boldsymbol{x}))^t_{|A|=l}$ of Eq. (11.65). Specifically, $\boldsymbol{\phi}(\boldsymbol{x})$ is a vector that has products of $x_i (i = 1, \ldots, d)$ of order l as its elements. For instance, when $d = 3$ and $l = 2$, we have

$$A_l \in 1, 2, 2, 3, 3, 1, \tag{11.67}$$

$$\boldsymbol{\phi}(\boldsymbol{x}) = (x_1 x_2, ; x_2 x_3, ; x_3 x_1)^t \tag{11.68}$$

The kernel function defined by this $\boldsymbol{\phi}$, termed the l-th order *ANOVA kernel*, is denoted by k_a (Vapnik 1995), which is expressed as

$$k_a(\boldsymbol{x}, \boldsymbol{y}) = \boldsymbol{\phi}(\boldsymbol{x})^t \boldsymbol{\phi}(\boldsymbol{y}) \tag{11.69}$$

$$= \sum_{|A|=l} \phi_{A_l}(\boldsymbol{x}) \phi_{A_l}(\boldsymbol{y}) \tag{11.70}$$

$$= \sum_{1 \le i_1 < \cdots < i_l \le d} (x_{i_1} y_{i_1})(x_{i_2} y_{i_2}) \cdots (x_{i_l} y_{i_l}) \tag{11.71}$$

$$= \sum_{1 \le i_1 < \cdots < i_l \le d} \prod_{j=1}^{l} (x_{i_j} y_{i_j}). \tag{11.72}$$

Figure 11.2 illustrates the mapping of points on the surface of the sphere $x_1^2 + x_2^2 + x_3^2 = 1$ when $d = 3$ and $l = 2$ through $\boldsymbol{\phi}$. Points on the x_1, x_2, and x_3 axes are mapped to the origin in the space of $\boldsymbol{\phi}(\boldsymbol{x})$, and $\boldsymbol{\phi}(\boldsymbol{x})$ appears to have a shape that's indented from three directions.

The expression in Eq. (11.72) is merely a representation of the kernel function $k_a(\boldsymbol{x}, \boldsymbol{y})$ in the form of an inner product operation and requires a vast amount of computation in this form. However, an efficient computational method can be

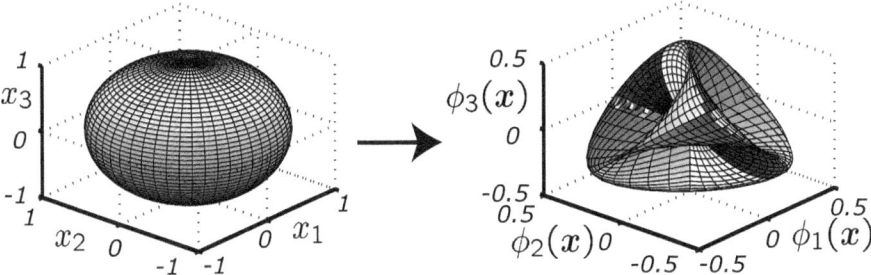

Fig. 11.2 Example of nonlinear transformation using the ANOVA kernel ($d = 3, l = 2$)

applied to evaluate $k_a(\boldsymbol{x}, \boldsymbol{y})$. The m-dimensional vectors obtained by taking the first m elements of vectors $\boldsymbol{x}$ and $\boldsymbol{y}$, where $m(\leq d)$, are denoted by $\boldsymbol{x}_{1:m}$ and $\boldsymbol{y}_{1:m}$, respectively. The l-th order ANOVA kernel taking these as arguments is denoted by K_l^m. Specifically, defining

$$\boldsymbol{x}_{1:m} = (x_1, \ldots, x_m)^t, \tag{11.73}$$

$$\boldsymbol{y}_{1:m} = (y_1, \ldots, y_m)^t, \tag{11.74}$$

$$K_l^m \overset{\text{def}}{=} k_a(\boldsymbol{x}_{1:m}, ; \boldsymbol{y}_{1:m}) = \sum_{1 \leq i_1 < \cdots < i_l \leq m} \prod_{j=1}^{l} (x_{i_j} y_{i_j}), \tag{11.75}$$

we obtain the next recursive formula:

$$K_l^m = \begin{cases} 1 & (\text{if } l = 0) \\ 0 & (\text{if } l > m) \\ K_l^{m-1} + K_{l-1}^{m-1} x_m y_m & (\text{otherwise}). \end{cases} \tag{11.76}$$

This recursive formula holds. The first term of the equation in the bottom row of Eq. (11.76) corresponds to the terms in K_l^m that do not include $x_m y_m$, and the second term corresponds to those that do include $x_m y_m$ (see Problem 11.3).

In the subsequent sections, we will introduce typical examples utilizing kernel methods, such as the nonlinear SVM, kernel principal component analysis, and kernel subspace methods.

11.4 Nonlinear Support Vector Machines

With the support vector machines described in the previous chapter being validated and recognized as a versatile machine learning technique, interest in the kernel method, one of its core technologies, has increased. Applying the kernel method to existing methods to achieve nonlinearity is referred to as *kernelization*.

11.4.1   Kernelization of Support Vector Machines

Let's recall the Optprobs. 10.2 and 10.4 of the SVM mentioned in the previous chapter. The SVM based on the margin maximization criterion is represented by Eqs. (10.37) and (10.90). Restated, it is the following equation:

$$L(\lambda) = \lambda^t \mathbf{1}_n - \frac{1}{2}\lambda^t \mathbf{H}\lambda. \tag{11.77}$$

The matrix $\mathbf{H}$ in the above equation, as indicated in Eqs. (10.36) and (10.89), has its (i, j) component h_{ij} defined as

$$h_{ij} = b_i b_j x_i^t x_j \qquad (i, j = 1, \ldots, n). \tag{11.78}$$

This is an n-dimensional square matrix.

Here, let's revisit the content described in Sect. 10.4.1. Similar to the generalized linear discriminant function method described in Chap. 8, let us consider the mapping ϕ from dimensionality d to dimensionality D, and instead of the d-dimensional vectors x_i ($i = 1, \ldots, n$), consider the D-dimensional vectors $\phi(x_i)$ ($i = 1, \ldots, n$) as the new n feature vectors.

When we apply a linear SVM to the feature vectors $\phi(x_i)$, the matrix $\mathbf{H}$ that should be maximized in Eq. (11.77) replaces Eq. (11.78) with

$$h_{ij} = b_i b_j \phi(x_i)^t \phi(x_j) \qquad (i, j = 1, \ldots, n). \tag{11.79}$$

This matrix $\mathbf{H} = (h_{ij})$ becomes the new representation.

This summarizes the content described in Sect. 10.4.1, and the above equation has already been shown in Eq. (10.109).

Therefore, if there is a kernel function $k(x, y)$ with d-dimensional vectors x and y as inputs and

$$k(x_i, x_j) \equiv \phi(x_i)^t \phi(x_j) \tag{11.80}$$

holds, then Eq. (11.79) becomes

$$h_{ij} = b_i b_j k(x_i, x_j) \qquad (i, j = 1, \ldots, n), \tag{11.81}$$

and Optprob. 10.4 can be expressed as Optprob. 11.1 shown below.

Optprob. 11.1 Maximization problem on λ (3) Suppose that λ of Eq. (10.12) and **b** of Eq. (10.30) satisfy conditions

$$\lambda^t \mathbf{b} = 0, \tag{11.82}$$

$$\mathbf{0} \leq \lambda \leq c_{pe} \cdot \mathbf{1}_n, \tag{11.83}$$

and the matrix $\mathbf{H} = (h_{ij})$ is represented by

$$h_{ij} = b_i b_j k(\mathbf{x}_i, \mathbf{x}_j) \qquad (i, j = 1, \ldots, n). \tag{11.84}$$

Under these conditions, find $\lambda \, (= \lambda^*)$ that maximizes

$$L(\lambda) = \lambda^t \mathbf{1}_n - \frac{1}{2} \lambda^t \mathbf{H}' \lambda. \tag{11.85}$$

By solving this, the optimal $\mathbf{w} \ (= \mathbf{w}^*)$, $w_0 \ (= w_0^*)$, and $\lambda \ (= \lambda_i^*)$ can be determined. The optimal discriminant function $g(\mathbf{x})$, $\mathbf{w}^*$, and w_0^* are shown in Eqs. (10.110), (10.111), and (10.113) respectively, using $\boldsymbol{\phi}(\mathbf{x})$. Here, as demonstrated below, we express the discriminant function using the kernel function $k(\cdot)$ without utilizing $\boldsymbol{\phi}(\mathbf{x})$. Namely, the discriminant function $g(\mathbf{x})$ is given by substituting Eq. (11.80) into Eq. (10.110), as follows:

$$g(\mathbf{x}) = w_0^* + \sum_{i=1}^n \lambda_i^* b_i k(\mathbf{x}_i, \mathbf{x}). \tag{11.86}$$

The w_0^* in the above equation is obtained by substituting Eq. (11.80) into Eq. (10.113), as follows:

$$w_0^* = b_s - \sum_{i=1}^n \lambda_i^* b_i k(\mathbf{x}_i, \mathbf{x}_s), \tag{11.87}$$

where $\mathbf{x}_s$ is a support vector and b_s is its teaching signal. Equation (11.86) corresponds to the potential function method of Eq. (9.9), and the $k(\mathbf{x}, \mathbf{y})$ in Eq. (11.80) is equivalent to the potential function $K(\mathbf{x}, \mathbf{x}_k)$ in Eq. (9.13). In this way, when actually using SVMs, by choosing an appropriate kernel function $k(\cdot)$, $\mathbf{H}$ is determined from Eq. (11.84), eliminating the need to calculate $\boldsymbol{\phi}(\mathbf{x})$. Moreover, an optimal solution can be obtained even if the form of $\boldsymbol{\phi}(\mathbf{x})$ is unknown. The obtained solution is a hyperplane in the space of $\boldsymbol{\phi}(\mathbf{x})$ that maximizes the margin , and in the space of $\mathbf{x}$, it becomes a nonlinear boundary dependent on the form of $\boldsymbol{\phi}$.

In the following, we present the results of classification experiments using the nonlinear SVM with the specific kernel functions described in Sect. 11.3. In the experiment, we focus on the two-class data shown in Fig. 11.3, and attempt to apply the nonlinear SVM using a kernel function.

Fig. 11.3 The overlapping of two normal distributions and the decision boundary determined by the Bayesian classification function

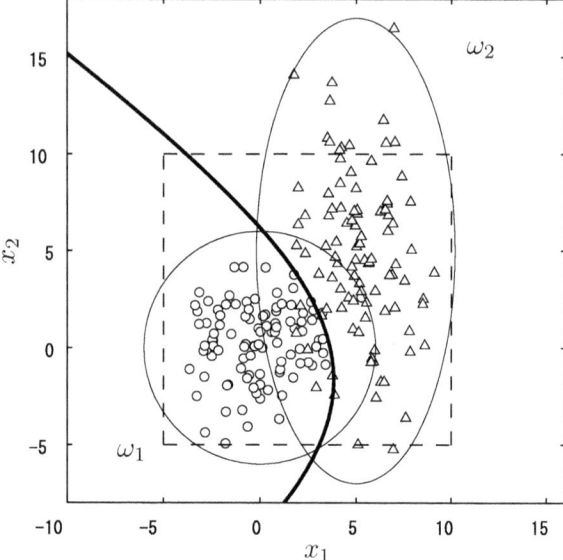

The figure shows artificial patterns generated under the assumption that class ω_i $(i = 1, 2)$ follows a two-dimensional normal distribution with the mean vector $\boldsymbol{\mu}_i$ and the covariance matrix $\boldsymbol{\Sigma}_i$ as shown in the following equations:

$$\boldsymbol{\mu}_1 = (0, \ 0)^t, \qquad\qquad \boldsymbol{\Sigma}_1 = \begin{pmatrix} 4 & 0 \\ 0 & 4 \end{pmatrix}, \qquad\qquad (11.88)$$

$$\boldsymbol{\mu}_2 = (5, \ 5)^t, \qquad\qquad \boldsymbol{\Sigma}_2 = \begin{pmatrix} 3 & 0 \\ 0 & 16 \end{pmatrix}. \qquad\qquad (11.89)$$

As before, the patterns of classes ω_1 and ω_2 are represented by $\bigcirc$ and $\triangle$ respectively, with each class containing 100 patterns, totaling 200 patterns.

In the figure, the shape of the normal distribution is indicated by a thin line. The classification rule that minimizes the error probability, as discussed in Sect. 4.1, is the Bayes decision rule. Assuming equal prior probabilities for both classes, the decision boundary determined by the Bayes discriminant function is represented by the thick line in the figure. The optimal boundary for separating two normal distributions is a quadratic curve. The number of patterns misclassified by this decision boundary was 7.

11.4.2 Example of Nonlinear SVM with Polynomial Kernel

Figure 11.4 shows the results using the polynomial kernel as described in Eq. (11.9).

The figure particularly enlarges the square area enclosed by the chain line in Fig. 11.3, that is, the region where the two distributions overlap, to examine how the SVM sets the decision boundary.

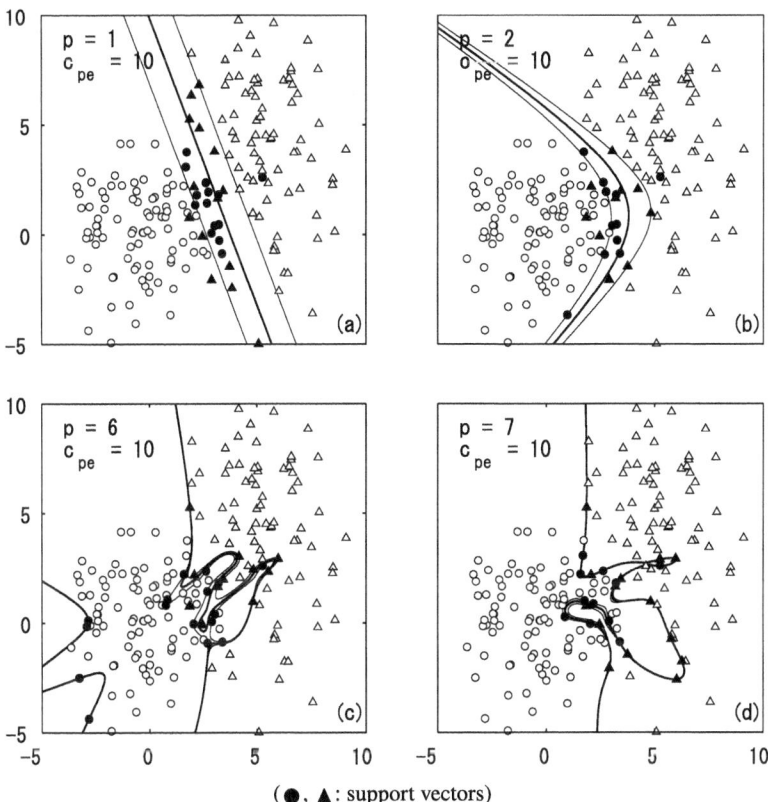

(●, ▲ : support vectors)

Fig. 11.4 Example of classification by nonlinear SVM (polynomial kernel)

In the configuration, there are the coefficient c_{pe} for the penalty used in Eq. (10.55) and the parameter p for the polynomial kernel. The effect of the penalty has already been clarified by comparing Figs. 10.6 and 10.7. Therefore, in the experiment, as shown in figures (a) to (d), the penalty was fixed at $c_{pe} = 10$, and the parameter p was varied to 1, 2, 6, 7. Each thick line indicates the decision boundary obtained by the SVM, that is, the boundary where $g(x) = 0$, and the thin line indicates the position corresponding to the margin, that is, the place where $g(x) = \pm 1$. As before, the support vectors are denoted by ● and ▲.

As can be seen from the figure, the decision boundary in the 2-dimensional original space is a straight line in (a) and a quadratic curve in (b). Therefore, (b) becomes close to the decision boundary of the Bayes decision rule, which is shown as a quadratic curve in Fig. 11.3. As the value of the parameter p increases from (a) to (d), a more complex decision boundary is formed, and the number of misclassified patterns decreases to 9, 7, 2, 0. When the value of the parameter p is made extremely large, the phenomenon of over-fitting is observed, leading to the formation of an overly complex boundary. For instance, in the lower left region of (c), a decision boundary is formed even though there are no patterns belonging to class ω_2. In (d), in order to achieve zero misclassification, a complex and unnatural decision boundary is formed, resulting in a typical case of over-fitting.

11.4.3 Example of Nonlinear SVM Using Gaussian Kernel

Figure 11.5 presents the results when using the Gaussian kernel to the same data. The interpretation of the figure is identical to Fig. 11.4. Similar to Fig. 11.4, the overlapping region of both distributions is magnified.

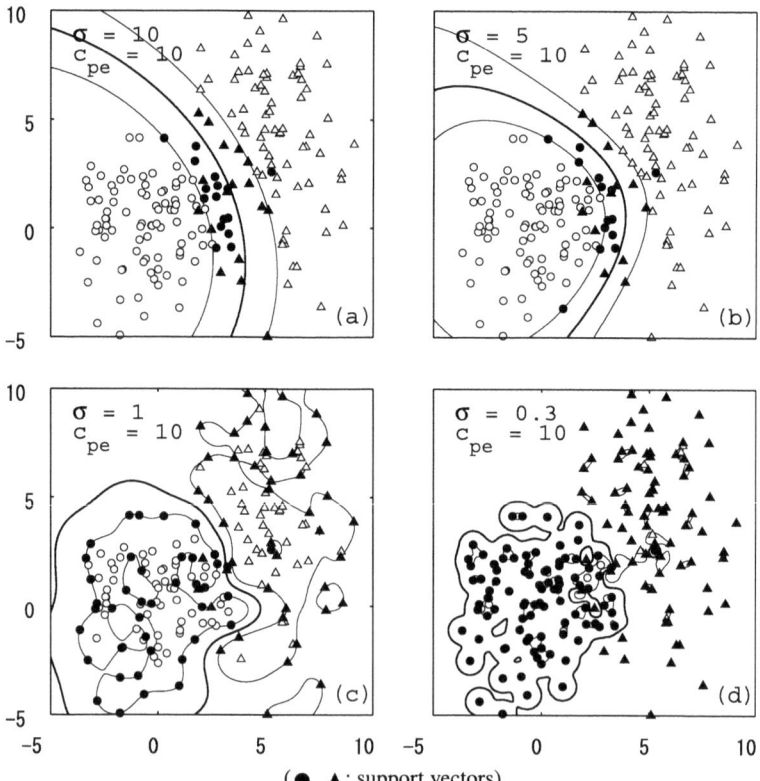

($\bullet$, $\blacktriangle$: support vectors)

Fig. 11.5 Example of classification by nonlinear SVM (Gaussian kernel)

As before, the penalty was fixed at $c_{pe} = 10$, and the parameter σ of the Gaussian kernel was changed to 10, 5, 1, 0.3 as shown in figures (a) to (d). As seen in figure (a), when the parameter σ is set to a large value, a global decision boundary is formed, and the number of misclassified patterns is also 8, which is close to the decision boundary by the Bayes decision rule shown in Fig. 11.3. Conversely, as σ is reduced in (b), (c), and (d), the decision boundary becomes influenced by local distributions, and it can be seen that, as in the case of the polynomial kernel, a complex decision boundary is formed. The number of misclassified patterns in figures (b) and (c) are 7 and 5 respectively, and when σ is extremely small as in (d), a decision boundary with zero misclassification is formed. In this case, many patterns contribute to the formation of the decision boundary as support vectors.

Since the data used here is generated from a two-dimensional normal distribution, the optimal decision boundary, when the number of patterns is infinitely increased, approaches the decision boundary based on the Bayes decision rule and can be described by a quadratic curve. Therefore, figures (c) and (d) can be regarded as examples of over-fitting where boundaries are more complex than necessary.

11.4.4 Generalization Ability and Regularization

As already mentioned in the coffee break in Sect. 10.2, the SVM often possesses high classification performance even when the learning patterns are high-dimensional vectors and there is not a sufficient number of patterns. This outstanding generalization ability was one of the biggest advantages not found in classification algorithms proposed before the SVM, and in that sense, it was revolutionary. When there are a total of n patterns in two classes, if the dimensionality D of $\phi(x)$ is greater than $n-1$, except in special cases, there always exists a hyperplane that can correctly separate the two classes as shown in Fig. 4.3. Therefore, the fact that we're mapping x to an ultra-high dimensionality using the kernel method can also be seen as transforming the learning patterns into a linearly separable situation. The seemingly simple optimization criterion of maximizing the margin is a significant factor in this high generalization ability, as mentioned in the coffee break. On the other hand, the decision boundary in the ultra-high-dimensional space defined by the kernel method corresponds to finding a tiny gap between the two classes. As seen in Fig. 11.4(c) and (d), depending on the penalty c_{pe}, it can easily cause over-fitting . As stated in Sect. 10.4.3, SVM optimization can be viewed as a regularization problem, and it can be interpreted that over-fitting is suppressed by adding an appropriate regularization term.

11.5 Kernel Principal Component Analysis

In this section, we introduce *kernel principal component analysis* abbreviated as *kernel PCA*, as an example of utilizing the kernel method for dimensionality reduction. Similar to the SVM, kernel PCA replaces the inner product computations in the algorithms of existing linear methods with kernel functions and achieves nonlinear dimensionality reduction by leveraging the kernel trick.

PCA (principal component analysis) is a technique of multivariate analysis. It identifies the axis that aligns most closely with the distribution of numerous data points represented as multivariate data (multidimensional vectors). Specifically, this method discovers the axis along which the projection of patterns has the largest variance. This technique is fundamentally equivalent to the technology known in signal processing as the Karhunen–Loève expansion, abbreviated as the KL expansion. PCA is employed not only as a tool for analyzing multivariate data, information compression, and visualization but also serves as the mathematical foundation for the subspace method, one of the pattern recognition techniques.

Figure 11.6 shows examples of applying PCA to two-dimensional pattern distributions. If the patterns are distributed approximately on a straight line, as shown in (a), they can be approximated by a one-dimensional space along that line v_1. However, since PCA is a linear analytical method, it cannot efficiently represent nonlinear pattern distributions. For instance, applying PCA to a two-dimensional distribution cannot represent nonlinear distribution shapes like in (b), and for isotropic distributions like in (c), it cannot provide a meaningful axis. However, just like with the nonlinear SVM, PCA can be made nonlinear using the kernel method, allowing it to function effectively even for the aforementioned distributions. This type of PCA is referred to as kernel PCA (Schölkopf et al. 1998).

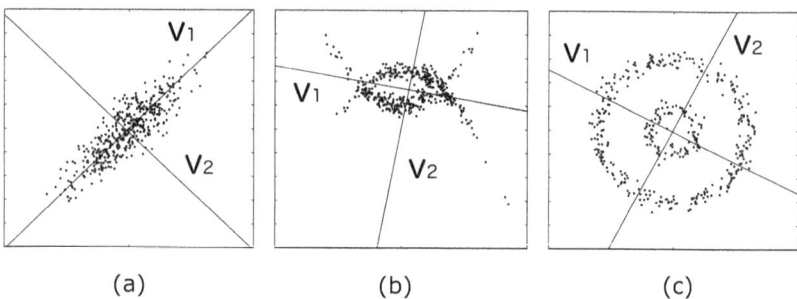

(a)   (b)   (c)

Fig. 11.6 The first principal component obtained by linear PCA (v_1) and the second principal component (v_2)

The traditional PCA used as a linear analysis method is called *linear PCA*, while the nonlinear version of PCA is referred to as *nonlinear PCA*.

In the following, we explain the mathematical procedure in linear PCA. First, from the patterns $x_1, \ldots, x_n$ in the d-dimensional space, we calculate their covariance matrix $\mathbf{C}_X$ and compute the eigenvalues and eigenvectors of $\mathbf{C}_X$. We denote the d eigenvalues of the covariance matrix $\mathbf{C}_X$ as λ_i ($\lambda_1 \geq \cdots \geq \lambda_d$), and the eigenvector $\mathbf{v}_i$ corresponding to λ_i is called the ith principal component. The d'-dimensional space spanned by the eigenvectors $\mathbf{v}_1, \ldots, \mathbf{v}_{d'}$ corresponding to the top d' eigenvalues best approximates the distribution of the patterns $x_1, \ldots, x_n$ in a $d' (< d)$-dimensional space.

In the following discussion, as we will introduce various matrix definitions and formulations using pattern matrices, for details, refer to Appendix G. Given the patterns $x_1, \ldots, x_n$, let's denote the average vector as $\bar{x}$. The covariance matrix $\mathbf{C}_X$ can be represented in relation to the pattern matrix $\mathbf{X} = (x_1, \ldots, x_n)^t \in \mathbb{R}^{n \times d}$. According to the equation in the appendix Eq. (G.19), using the mean deviation matrix, denoted as $\mathbf{X}_M = (x_1 - \bar{x}, \ldots, x_n - \bar{x})^t$, we can express the covariance matrix as

$$\mathbf{C}_X = \frac{1}{n} \mathbf{X}_M^t \mathbf{X}_M. \tag{11.90}$$

To extend the traditional principal component analysis, known as linear PCA, to nonlinear PCA using kernel methods, we can try a similar procedure by transforming x using the nonlinear transformation ϕ to $\phi(x)$. In other words, in linear PCA, if we consider the pattern matrix formed by $\phi(x_i)$ ($i = 1, \ldots, n$) as $\mathbf{X}_\phi$, we can replace x_i and $\mathbf{X}$ with

$$x_i \Rightarrow \phi(x_i), \tag{11.91}$$

$$\mathbf{X} = (x_1, \ldots, x_n)^t \Rightarrow \mathbf{X}_\phi \overset{\text{def}}{=} (\phi(x_1), \ldots, \phi(x_n))^t \in \mathbb{R}^{n \times D}. \tag{11.92}$$

Thus, by replacing them in this manner, we can solve the problem. The solution method will be explained below. There are two important points to note about the algorithm for kernel PCA.

Firstly, the subject of the PCA is the n vectors $\phi(x_i)$ ($i = 1, \ldots, n$) of dimensionality D. However, it's essential to understand that in actual computation, we never directly use $\phi(x_i)$. By leveraging the kernel trick, equivalent calculations can be performed using the kernel function $k(\cdot)$ without explicitly providing $\phi(\cdot)$.

Secondly, when solving the problem using the pattern matrix $\mathbf{X}$, instead of the usual use of $\mathbf{X}^t\mathbf{X}$, kernel PCA employs $\mathbf{X}\mathbf{X}^t$. If the matrix $\mathbf{X}$ has a size of $n \times d$, then while $\mathbf{X}^t\mathbf{X}$ is a $d \times d$ matrix, $\mathbf{X}\mathbf{X}^t$ results in an $n \times n$ matrix. If n is much smaller than d, then using the latter can lead to significant reductions in computational complexity. This algorithm is commonly used, and sometimes it's also referred to as the kernel trick. For further details, readers might also refer to Problem 11.4 (Boyd and Vandenberghe 2018).

In addition to $\mathbf{X}_M$ ($\in \mathbb{R}^{n \times d}$) and $\mathbf{C}_X$ ($\in \mathbb{R}^{d \times d}$) of Eq. (11.90), we denote the mean deviation matrix derived from $\mathbf{X}_\phi$ as $\mathbf{X}_{\phi M}$ ($\in \mathbb{R}^{n \times D}$) and the covariance matrix as $\mathbf{C}_{X_\phi}$ ($\in \mathbb{R}^{D \times D}$). Furthermore, the inner product matrices ($\in \mathbb{R}^{n \times n}$) for $\mathbf{X}, \mathbf{X}_\phi, \mathbf{X}_M$, and $\mathbf{X}_{\phi M}$ are denoted as $\mathbf{Q}_X, \mathbf{Q}_{X_\phi}, \mathbf{Q}_{X_M}$, and $\mathbf{Q}_{\mathbf{x}_{\phi M}}$, respectively (refer to Eq. (G.27) in Appendix G). The correspondence between these matrices can be summarized as follows:

$$\mathbf{X}_M = \mathbf{X} - \frac{1}{n}\mathbf{1}_{nn}\mathbf{X} \quad \Rightarrow \quad \mathbf{X}_{\phi M} = \mathbf{X}_\phi - \frac{1}{n}\mathbf{1}_{nn}\mathbf{X}_\phi, \tag{11.93}$$

$$\mathbf{C}_X = \frac{1}{n}\mathbf{X}_M^t\mathbf{X}_M \quad \Rightarrow \quad \mathbf{C}_{X_\phi} = \frac{1}{n}\mathbf{X}_{\phi M}^t\mathbf{X}_{\phi M}, \tag{11.94}$$

$$\mathbf{Q}_X = \mathbf{X}\mathbf{X}^t \quad \Rightarrow \quad \mathbf{Q}_{X_\phi} = \mathbf{X}_\phi\mathbf{X}_\phi^t = \mathbf{K}(\mathbf{X}, \mathbf{X}), \tag{11.95}$$

$$\mathbf{Q}_{X_M} = \mathbf{X}_M\mathbf{X}_M^t \quad \Rightarrow \quad \mathbf{Q}_{\mathbf{x}_{\phi M}} = \mathbf{X}_{\phi M}\mathbf{X}_{\phi M}^t. \tag{11.96}$$

The matrix $\mathbf{1}_{nn}$ of Eq. (11.93) is an $n \times n$ matrix where all the elements are 1. In Eq. (11.95), the matrix $\mathbf{K}(\mathbf{X}, \mathbf{X})$ has its (i, j) component k_{ij} defined as the inner product $\phi(\mathbf{x}_i)^t\phi(\mathbf{x}_j)$. Therefore, as shown in Eq. (11.29), it is a Gram matrix of size $n \times n$.

In kernel methods, what is given a priori is the kernel function, and the function $\phi(\cdot)$ is not explicitly provided. Therefore, it's not possible to directly compute the covariance matrix $\mathbf{C}_{X_\phi}$. On the other hand, it's important to note that $\mathbf{Q}_{X_\phi}$ can be determined as it is provided in the form of a Gram matrix.

Therefore, based on the equations from the appendix, namely Eqs. (G.28)–(G.31), if we rewrite $\mathbf{Q}_{X_M}$ using $\mathbf{Q}_X$, we get

$$\mathbf{Q}_{X_M} = \mathbf{X}_M\mathbf{X}_M^t \tag{11.97}$$

$$= (\mathbf{X} - \frac{1}{n}\mathbf{1}_{nn}\mathbf{X})(\mathbf{X} - \frac{1}{n}\mathbf{1}_{nn}\mathbf{X})^t \tag{11.98}$$

$$= (\mathbf{I}_n - \frac{1}{n}\mathbf{1}_{nn})\mathbf{X}\mathbf{X}^t(\mathbf{I} - \frac{1}{n}\mathbf{1}_{nn}) \tag{11.99}$$

$$= \mathbf{J}_n\mathbf{Q}_X\mathbf{J}_n, \tag{11.100}$$

where the matrix $\mathbf{J}_n$ is defined as follows:

$$\mathbf{J}_n \stackrel{\text{def}}{=} \mathbf{I}_n - \frac{1}{n}\mathbf{1}_{nn}. \tag{11.101}$$

Similar to Eq. (11.100), the term $\mathbf{Q}_{\mathbf{x}_{\phi M}}$ can be re-expressed as follows:

$$\mathbf{Q}_{\mathbf{x}_{\phi M}} = \mathbf{J}_n\mathbf{Q}_{X_\phi}\mathbf{J}_n \tag{11.102}$$

$$= \mathbf{J}_n\mathbf{K}(\mathbf{X}, \mathbf{X})\mathbf{J}_n. \tag{11.103}$$

From the equation above, it can be seen that $\mathbf{Q_{X_{\phi M}}}$ can be computed using the Gram matrix $\mathbf{K(X, X)}$, and its eigenvalues and eigenvectors can also be determined. Notably, by observing that

$$\mathbf{Q_{X_{\phi M}}} = \mathbf{X}_{\phi M}\mathbf{X}_{\phi M}^t \tag{11.104}$$

$$n\,\mathbf{C}_{X_\phi} = \mathbf{X}_{\phi M}^t\mathbf{X}_{\phi M}, \tag{11.105}$$

we can perform a singular value decomposition (SVD) on the matrix $\mathbf{X}_{\phi M}$ (for details on SVD, refer to Appendix H).

Let the eigenvalues of the matrix $\mathbf{X}_{\phi M}\mathbf{X}_{\phi M}^t$ be represented as $\lambda_i(\lambda_1 \geq \cdots \geq \lambda_n)$. There are $r(= \text{rank}(\mathbf{X}_{\phi M}))$ positive eigenvalues, and they coincide with the positive eigenvalues of $\mathbf{X}_{\phi M}^t\mathbf{X}_{\phi M}$. Let the orthonormal eigenvectors corresponding to the eigenvalue λ_i for the matrices $\mathbf{X}_{\phi M}\mathbf{X}_{\phi M}^t$ and $\mathbf{X}_{\phi M}^t\mathbf{X}_{\phi M}$ be $\mathbf{u}_i$ and $\mathbf{v}_i$, respectively. If we define the matrices $\mathbf{U}$, $\mathbf{V}$, and $\mathbf{\Lambda}^{1/2}$ as

$$\mathbf{U} = (\mathbf{u}_1, \ldots, \mathbf{u}_r) \qquad\qquad (\in \mathbb{R}^{n \times r}) \tag{11.106}$$

$$\mathbf{V} = (\mathbf{v}_1, \ldots, \mathbf{v}_r) \qquad\qquad (\in \mathbb{R}^{D \times r}) \tag{11.107}$$

$$\mathbf{\Lambda}^{1/2} = \begin{pmatrix} \sqrt{\lambda_1} & & 0 \\ & \ddots & \\ 0 & & \sqrt{\lambda_r} \end{pmatrix} \qquad (\in \mathbb{R}^{r \times r}), \tag{11.108}$$

then we can express them as

$$\mathbf{X}_{\phi M} = \mathbf{U}\mathbf{\Lambda}^{1/2}\mathbf{V}^t. \tag{11.109}$$

As indicated in Eq. (11.105), since $\mathbf{X}_{\phi M}^t\mathbf{X}_{\phi M}$ is n times the covariance matrix derived from the pattern vectors $\{\boldsymbol{\phi}(\mathbf{x}_1), \ldots, \boldsymbol{\phi}(\mathbf{x}_n)\}$, $\mathbf{v}_i$ signifies the ith principal component of the pattern set. However, it's important to note that what can be derived from the Gram matrix is $\mathbf{U}$, not $\mathbf{V}$. Now, let's consider a transformation $\boldsymbol{\phi}$ that first nonlinearly transforms a d-dimensional vector into a D-dimensional vector and then further translates it such that the origin becomes the centroid of $\boldsymbol{\phi}(\mathbf{x}_i)$ $(i = 1, \ldots, n)$. This transformation is denoted as $\boldsymbol{\phi}_M$. Specifically, for a d-dimensional vector $\mathbf{y}$, we have

$$\boldsymbol{\phi}_M(\mathbf{y}) = \boldsymbol{\phi}(\mathbf{y}) - \frac{1}{n}\sum_{i=1}^{n}\boldsymbol{\phi}(\mathbf{x}_i) \tag{11.110}$$

$$= \boldsymbol{\phi}(\mathbf{y}) - \frac{1}{n}\mathbf{X}_\phi^t\mathbf{1}_n. \tag{11.111}$$

Using the transformation $\boldsymbol{\phi}_M$ of the equation above, the $\mathbf{X}_{\phi_M}$ in Eq. (11.93) can be written as

$$\mathbf{X}_{\phi_M} = \left(\boldsymbol{\phi}_M(\boldsymbol{x}_1), \ \ldots, \ \boldsymbol{\phi}_M(\boldsymbol{x}_n)\right)^t. \tag{11.112}$$

As shown in the appendix Eq. (H.11), due to the properties of singular value decomposition, we have

$$\mathbf{v}_i = \frac{1}{\sqrt{\lambda_i}} \mathbf{X}_{\phi_M}^t \mathbf{u}_i. \tag{11.113}$$

Thus, the length of the projection of the vector $\boldsymbol{\phi}_M(\boldsymbol{y})$ onto $\mathbf{v}_i$ can be determined as their dot product, given by the following equation:

$$\mathbf{v}_i^t \boldsymbol{\phi}_M(\boldsymbol{y}) = \frac{1}{\sqrt{\lambda_i}} \mathbf{u}_i^t \mathbf{X}_{\phi_M} \boldsymbol{\phi}_M(\boldsymbol{y}). \tag{11.114}$$

In general, given matrices composed of l and m d-dimensional column vectors, we can represent them as

$$\mathbf{Y} = (\boldsymbol{y}_1, \ldots, \boldsymbol{y}_l)^t \quad (\in \mathbb{R}^{l \times d}), \tag{11.115}$$

$$\mathbf{Z} = (\mathbf{z}_1, \ldots, \mathbf{z}_m)^t \quad (\in \mathbb{R}^{m \times d}). \tag{11.116}$$

From Eq. (11.92), we obtain

$$\mathbf{Y}_\phi = \left(\boldsymbol{\phi}(\boldsymbol{y}_1), \ \ldots, \ \boldsymbol{\phi}(\boldsymbol{y}_l)\right)^t, \tag{11.117}$$

$$\mathbf{Z}_\phi = (\boldsymbol{\phi}(\mathbf{z}_1), \ \ldots, \ \boldsymbol{\phi}(\mathbf{z}_m))^t \tag{11.118}$$

The matrix of size $l \times m$, whose (i, j) component is the kernel function $k(\boldsymbol{y}_i, \mathbf{z}_j) = \boldsymbol{\phi}(\boldsymbol{y}_i)^t \boldsymbol{\phi}(\mathbf{z}_j)$, is called the *kernel matrix*, denoted as $\mathbf{K}(\mathbf{Y}, \mathbf{Z})$:

$$\mathbf{K}(\mathbf{Y}, \mathbf{Z}) = \mathbf{Y}_\phi \mathbf{Z}_\phi^t. \tag{11.119}$$

Similarly, from Eq. (11.112), the following is derived

$$\mathbf{Y}_{\phi_M} = \left(\boldsymbol{\phi}_M(\boldsymbol{y}_1), \ \ldots, \ \boldsymbol{\phi}_M(\boldsymbol{y}_l)\right)^t, \tag{11.120}$$

$$\mathbf{Z}_{\phi_M} = \left(\boldsymbol{\phi}_M(\mathbf{z}_1), \ \ldots, \ \boldsymbol{\phi}_M(\mathbf{z}_m)\right)^t. \tag{11.121}$$

Therefore, the matrix $\mathbf{G}_X(\mathbf{Y}, \mathbf{Z})$, which has inner products of $\boldsymbol{\phi}_M(\boldsymbol{y}_i)$ and $\boldsymbol{\phi}_M(\mathbf{z}_j)$ as its (i, j) components, and is of size $l \times m$, is equivalent to $\mathbf{K}(\mathbf{Y}_{\phi_M}, \mathbf{Z}_{\phi_M})$. Using the definition of Eq. (11.119), we get

$$\mathbf{G}_X(\mathbf{Y}, \mathbf{Z}) = \mathbf{K}(\mathbf{Y}_{\phi M}, \mathbf{Z}_{\phi M}) \tag{11.122}$$

$$= \mathbf{Y}_{\phi M} \mathbf{Z}_{\phi M}^t \tag{11.123}$$

$$= \left(\mathbf{Y}_\phi - \frac{1}{n}\mathbf{1}_{ln}\mathbf{X}_\phi\right)\left(\mathbf{Z}_\phi - \frac{1}{n}\mathbf{1}_{mn}\mathbf{X}_\phi\right)^t \tag{11.124}$$

$$= \mathbf{K}(\mathbf{Y}, \mathbf{Z}) - \frac{1}{n}\mathbf{K}(\mathbf{Y}, \mathbf{X})\mathbf{1}_{nm} - \frac{1}{n}\mathbf{1}_{ln}\mathbf{K}(\mathbf{X}, \mathbf{Z})$$

$$+ \frac{1}{n^2}\mathbf{1}_{ln}\mathbf{K}(\mathbf{X}, \mathbf{X})\mathbf{1}_{nm}. \tag{11.125}$$

In the above equation, $\mathbf{K}(\mathbf{Y}, \mathbf{Z})$, $\mathbf{K}(\mathbf{Y}, \mathbf{X})$, $\mathbf{K}(\mathbf{X}, \mathbf{Z})$, and $\mathbf{K}(\mathbf{X}, \mathbf{X})$ are all kernel matrices. Note that the kernel matrix $\mathbf{K}(\mathbf{X}, \mathbf{X})$ is also the Gram matrix.

From Eqs. (11.107) and (11.120), we have

$$\mathbf{V} = (\mathbf{v}_1, \dots, \mathbf{v}_i, \dots, \mathbf{v}_r), \tag{11.126}$$

$$\mathbf{Y}_{\phi M} = \left(\boldsymbol{\phi}_M(\mathbf{y}_1), \dots, \boldsymbol{\phi}_M(\mathbf{y}_j), \dots, \boldsymbol{\phi}_M(\mathbf{y}_l)\right)^t. \tag{11.127}$$

Here, we determine the length of the projection of the vector $\boldsymbol{\phi}_M(\mathbf{y}_j)$ onto $\mathbf{v}_i$ for all $\boldsymbol{\phi}_M(\mathbf{y}_j)$ ($j = 1, \dots, l$) and $\mathbf{v}_i$ ($i = 1, \dots, r$). To do this, we need to compute the matrix that has the inner product $\mathbf{v}_i^t \boldsymbol{\phi}_M(\mathbf{y}_j)$ as its (i, j) component, as indicated in Eq. (11.114). It is evident that this matrix is $\mathbf{V}^t \mathbf{Y}_{\phi M}^t$. We can derive it as follows: Note that, in the following, we will represent $(\boldsymbol{\Lambda}^{1/2})^{-1}$ as $\boldsymbol{\Lambda}^{-1/2}$.

From Eqs. (11.123) and (11.109), we have

$$\mathbf{G}_X(\mathbf{X}, \mathbf{Y}) = \mathbf{X}_{\phi M} \mathbf{Y}_{\phi M}^t \tag{11.128}$$

$$= \mathbf{U}\boldsymbol{\Lambda}^{1/2}\mathbf{V}^t \mathbf{Y}_{\phi M}^t. \tag{11.129}$$

Therefore, by multiplying both sides of Eq. (11.129) by $\boldsymbol{\Lambda}^{-1/2}\mathbf{U}^t$ from the left, we obtain

$$\mathbf{V}^t \mathbf{Y}_{\phi M}^t = \boldsymbol{\Lambda}^{-1/2}\mathbf{U}^t \mathbf{G}_X(\mathbf{X}, \mathbf{Y}). \tag{11.130}$$

The above matrix $\mathbf{G}_X(\mathbf{X}, \mathbf{Y})$ can be determined using the general formula Eq. (11.125) as follows:

$$\mathbf{G}_X(\mathbf{X}, \mathbf{Y}) = \mathbf{X}_{\phi M} \mathbf{Y}_{\phi M}^t \tag{11.131}$$

$$= \left(\mathbf{X}_\phi - \frac{1}{n}\mathbf{1}_{nn}\mathbf{X}_\phi\right)\left(\mathbf{Y}_\phi - \frac{1}{n}\mathbf{1}_{ln}\mathbf{X}_\phi\right)^t \tag{11.132}$$

$$= \mathbf{K}(\mathbf{X}, \mathbf{Y}) - \frac{1}{n}\mathbf{K}(\mathbf{X}, \mathbf{X})\mathbf{1}_{nl} - \frac{1}{n}\mathbf{1}_{nn}\mathbf{K}(\mathbf{X}, \mathbf{Y})$$

$$+ \frac{1}{n^2}\mathbf{1}_{nn}\mathbf{K}(\mathbf{X}, \mathbf{X})\mathbf{1}_{nl}. \tag{11.133}$$

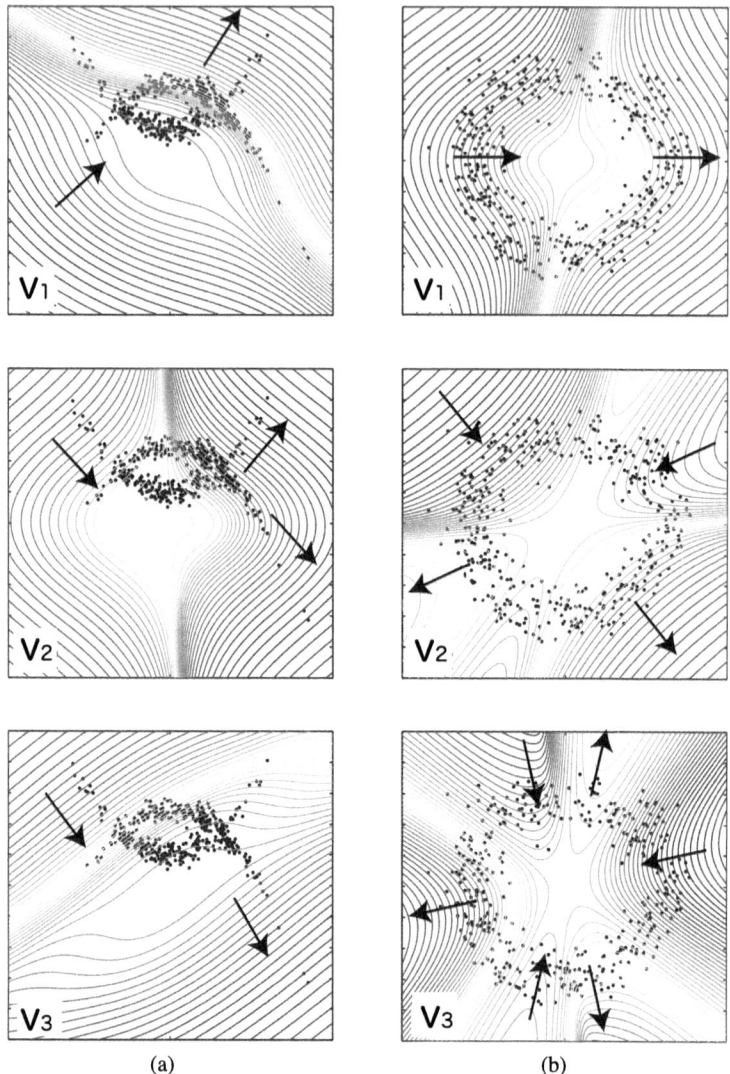

Fig. 11.7 Nonlinear PCA using the polynomial kernel ($p = 3$). (**a**) Distribution 1. (**b**) Distribution 2

Using the results above, the projection of $\boldsymbol{\phi}_M(\boldsymbol{y})$ onto $\mathbf{v}_i$ can be calculated from the kernel matrices $\mathbf{K}(\mathbf{X}, \mathbf{Y})$ and $\mathbf{K}(\mathbf{X}, \mathbf{X})$. That is, as mentioned in Sect. 11.5, even without explicitly giving $\phi(\cdot)$, the problem can be solved solely by the kernel function $k(\cdot)$. This corresponds to performing linear PCA in the space of $\boldsymbol{\phi}_M(\boldsymbol{x})$. Since $\boldsymbol{\phi}$ is a nonlinear mapping defined through the kernel function, this corresponds to conducting nonlinear PCA.

Figures 11.7 and 11.8 are computational examples of kernel PCA using polynomial and Gaussian kernels, respectively. They are applied to two types of

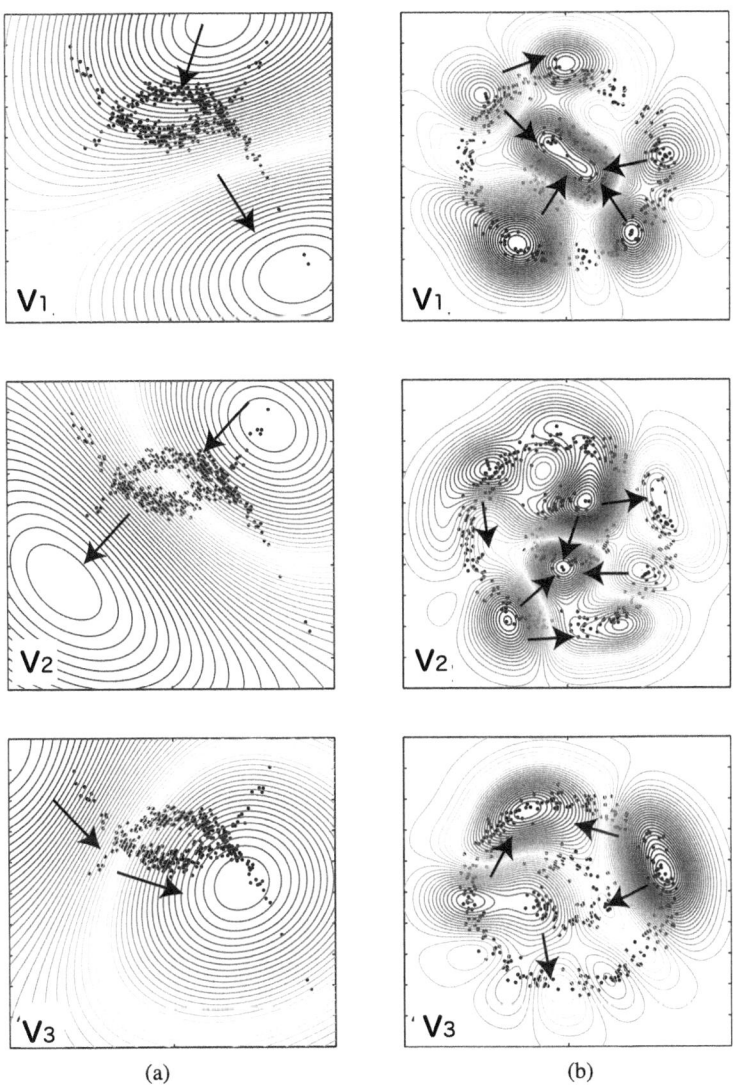

Fig. 11.8 Nonlinear PCA using the Gaussian kernel ($\sigma = 40$). (**a**) Distribution 1. (**b**) Distribution 2

distributions, (a) and (b), and contours perpendicular to the first, second, and third principal components are displayed. Each direction of the arrows indicates the direction of the higher contours. The contours represent the values of Eq. (11.114). In Fig. 11.8, the contours are equidistant in terms of height, while in Fig. 11.7, the resolution around $\mathbf{v}_i^t \boldsymbol{\phi}_M(\mathbf{y}) = 0$ is displayed to be higher. In these figures, the main axes in the space of $\phi(\mathbf{x})$ correspond to directions perpendicular to the contours. As these are axes in a space transformed by a complex nonlinear operation, even a slight change in position in the original two-dimensional space can change the direction of the axes. From these figures, it is evident that kernel PCA can provide a diverse representation of pattern distributions. However, whether this kernel PCA is useful or not needs to be evaluated based on the specific objective in question.

So far, as we've seen with examples in the SVM and PCA, methods based on inner product computations can be nonlinearized using kernel methods. They can be combined with multivariate analysis methods other than PCA. There are achievements with this method in various contexts, for instance, Fisher's method (Baudat and Anouar 2000; Mika et al. 1999), which performs dimensionality reduction suitable for pattern class discrimination, and *CCA* (*Canonical Correlation Analysis*) (Lai and Fyfe 2000; Akaho 2001; Bach and Jordan 2003), which finds correlations among multivariate sets. In the next section, we will discuss the kernel subspace method, which uses the kernel method for the subspace method. The subspace method is a pattern recognition method using PCA (Maeda and Murase 1999; Tsuda 1999).

Coffee Break

Another Dimensionality Reduction Method: Multidimensional Scaling

Principal Component Analysis (PCA) is commonly used as a standard method for projecting patterns represented as multidimensional vectors onto a lower dimensionality. Particularly, when projecting onto a two-dimensional plane, it becomes possible to visualize the distribution of patterns, making it effective. Another similar method frequently used for the same purpose is *multidimensional scaling (MDS)*. While PCA aims to find the lower-dimensional representation (or the main axes of distribution) when given n d-dimensional vectors, $x_1, \ldots, x_n$, the MDS method finds the lower-dimensional representation of d-dimensional vectors $x_1, \ldots, x_n$ using the *dissimilarity d_{ij}* provided for every pair x_i and x_j.

In reality, even if we cannot represent each of $x_1, \ldots, x_n$ as multidimensional vectors, there are cases where we can quantify the degree of difference or similarity between each instance as mere examples. For instance, when the similarity of various products is quantified by subjective evaluation, or when the genetic similarity between organisms defined by gene sequences is available. Furthermore, in data measured by multiple variables, the correlation coefficient between two variables is also an index expressing similarity between variables. MDS is frequently used in fields like psychology and marketing. Depending on the context, instead of dealing with the degree of difference between instances, one might deal with the *similarity s_{ij}* (where $s_{ij} \leq s_{ii}$). In such cases, one can convert the similarity s_{ij} into a dissimilarity d_{ij} using some method. The following equation,

$$d_{ij} = (s_{ii} - 2s_{ij} + s_{jj})^{1/2}, \tag{11.134}$$

represents one such standard method.

Now, let's represent the n patterns $y_1, \ldots, y_n$ as column vectors of dimensionality d'. We can collectively represent them as the pattern matrix $\mathbf{Y} = (y_1, \ldots, y_n)^t$ ($\in \mathbb{R}^{n \times d'}$). MDS can be described as a method to determine $\mathbf{Y}$ when given an $n \times n$ matrix $\mathbf{D} = (d_{ij})$, where d_{ij} is the (i, j) component. In other words, this method allows us to obtain the low-dimensional representations $y_1, \ldots, y_n$ of the patterns $x_1, \ldots, x_n$ that underlie the observed or measured matrix $\mathbf{D}$. This matrix $\mathbf{D}$ is called the *distance matrix*. An important point to note here is that the definition of distance behind $\mathbf{D}$ is not necessarily self-evident. However, it is assumed that $d_{ii} = 0$ and $d_{ij} = d_{ji}$ $(i \neq j)$.

First, let's consider the special case where the distance matrix $\mathbf{D}$ is defined by the Euclidean distance. Here, using the matrix $\mathbf{D}^{(2)} = (d_{ij}^2)$, where d_{ij}^2 is the (i, j) component, and the $n \times n$ matrix $\mathbf{J}_n$ defined in Eq. (11.101) (see Appendix G), we define the matrix $\mathbf{L}$ as follows:

$$\mathbf{L} \stackrel{\text{def}}{=} -\frac{1}{2} \mathbf{J}_n \mathbf{D}^{(2)} \mathbf{J}_n \qquad (\in \mathbb{R}^{n \times n}). \tag{11.135}$$

In this case, the (i, j) component l_{ij} of the matrix $\mathbf{L}$ can be verified as

$$l_{ij} = -\frac{1}{2} \left(d_{ij}^2 - \frac{1}{n} \sum_k d_{ik}^2 - \frac{1}{n} \sum_k d_{kj}^2 + \frac{1}{n^2} \sum_k \sum_l d_{kl}^2 \right). \tag{11.136}$$

For the proof, refer to Problem 11.6. The transformation from the distance matrix $\mathbf{D}$ to $\mathbf{L}$ as shown above is called the *Young–Householder transformation*. If the distance matrix $\mathbf{D}$ is set based on the Euclidean distance, then $\mathbf{L}$ is positive semidefinite, and as mentioned in Sect. 11.2.2, all n eigenvalues of $\mathbf{L}$ are non-negative. Conversely, if $\mathbf{L}$ is positive semidefinite, the matrix $\mathbf{D}$ defined by the Euclidean distance can be derived from $\mathbf{L}$. For proof, refer to Problem 11.7.

However, in general, there's no guarantee that $\mathbf{D}$ is defined by the Euclidean distance, and the $\mathbf{D}$ obtained through observation and other means may be overlaid with noise. Therefore, the matrix $\mathbf{L}$ might not be positive semidefinite, and its eigenvalues could possibly include negative values. Therefore, MDS considers such general $\mathbf{D}$ and uses the following procedure.

Select the d' eigenvalues of the matrix $\mathbf{L}$ in decreasing order as $\lambda_1, \ldots, \lambda_{d'}$ $(\lambda_1 \geq \cdots \geq \lambda_{d'} > 0)$. Here, only the positive eigenvalues are selected. The matrix formed by the eigenvectors $\mathbf{u}_i$ corresponding to the eigenvalue λ_i can be written as $\mathbf{U}_{d'}$ such that

$$\mathbf{U}_{d'} \stackrel{\text{def}}{=} (\mathbf{u}_1, \ldots, \mathbf{u}_{d'}) \qquad (\in \mathbb{R}^{n \times d'}). \tag{11.137}$$

Here, consider the $d' \times d'$ matrix $\mathbf{\Lambda}_{d'}$ with eigenvalues $\lambda_1, \ldots, \lambda_{d'}$ as its diagonal components, and the matrix $\mathbf{\Lambda}_{d'}^{1/2}$ with the square root of eigenvalues as its diagonal components as indicated in Eq. (11.108). At this time, the $n \times d'$ matrix $\mathbf{Y}$ can be determined as

$$\mathbf{Y} = \mathbf{U}_{d'} \mathbf{\Lambda}_{d'}^{1/2} \qquad (\in \mathbb{R}^{n \times d'}). \tag{11.138}$$

Matrix $\mathbf{Y}$ can be considered as the d'-dimensional representation of n patterns that underlie the distance matrix $\mathbf{D}$. Often, the focus is on the top two eigenvalues, i.e., when $d' = 2$, and $\mathbf{Y}$ is computed to visualize the patterns in a two-dimensional plane. MDS aims to find $\mathbf{Y}$ such that the distance matrix constructed from the Euclidean distances between $y_1, \ldots, y_n$ approximates $\mathbf{D}$ as closely as possible. For more details, refer to Problem 11.8 (Mardia et al. 1979) and the example in Problem 11.9.

11.6   Kernel Subspace Method

The kernel PCA mentioned in the previous section, focusing on the fact that the length of projection onto principal vectors can be expressed using inner products, can be considered a method that nonlinearizes PCA by utilizing kernel methods. Similarly, it is possible to consider a nonlinear version of the subspace method, which is one of the pattern recognition techniques. The *kernel subspace method* is a prime example of where kernel PCA can be effectively utilized. In this section, we will first describe an overview of the linear subspace method before introducing kernel PCA.

11.6.1   Linear Subspace Methods

The term "subspace method" generally refers to the *linear subspace method*, also known as class featuring information compression. In this context, subspace methods determine subspaces that characterize each class from the learning patterns. These class-specific subspaces are then used for the classification of unknown patterns. In the broader sense, we can consider subspace methods as techniques that design subspaces to distinguish different classes based on learning patterns and utilize these class-specific subspaces for the classification of unknown patterns. These broadly defined linear subspace methods include the CLAFIC method, and also the *projection distance method* (Ikeda et al. 1983) (Fig. 11.9).

The CLAFIC method determines class-specific subspaces by performing KL expansion for each class, using the origin of the feature space as the starting point. It classifies the unknown pattern x as the class whose subspace results in the minimum distance $\overline{xP_i}$ from pattern x to the class subspaces (Fig. 11.9(a)). The figure illustrates an example where the dimensionality of the original space is $d = 2$, and the dimensionality of each class's subspace is $d' = 1$. Since the origin of the feature space and the origin of the class subspaces are the same, this classification rule is equivalent to determining the class of x whose projection vector OP_i onto the class subspace has the maximum length. Furthermore, it is equivalent to making the decision based on the cosine similarity (described later in Eq. (13.39)) between x and its projection vector. The multiple similarity method, which evaluates patterns using weighted lengths instead of just the length of the projection vector, can be considered a variant of this approach (Oja 1983).

The projection distance method determines class-specific subspaces by performing KL expansion separately for each class distribution, using the centroid of each class as the origin. It classifies x as the class whose subspace results in the minimum projection distance from x to the class subspaces (Fig. 11.9(b)).

As shown in Fig. 11.9, the pattern x is classified as ω_1 in the CLAFIC method, while it is classified as ω_2 in the projection distance method. The primary difference

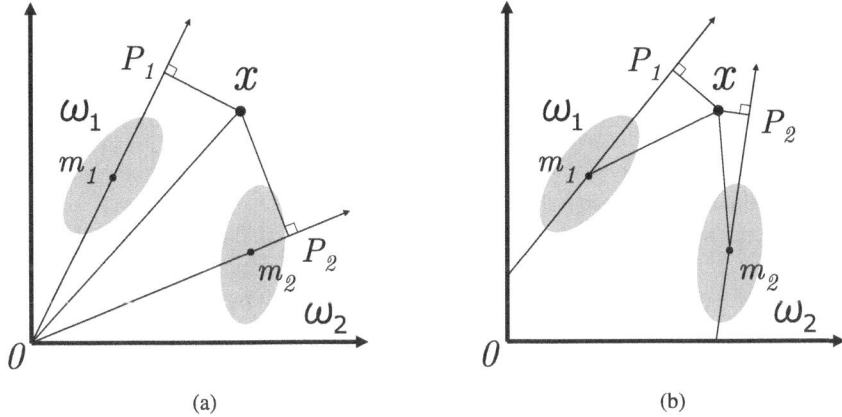

Fig. 11.9 Two kinds of linear subspace methods. (**a**) CLAFIC method. (**b**) Projection distance method

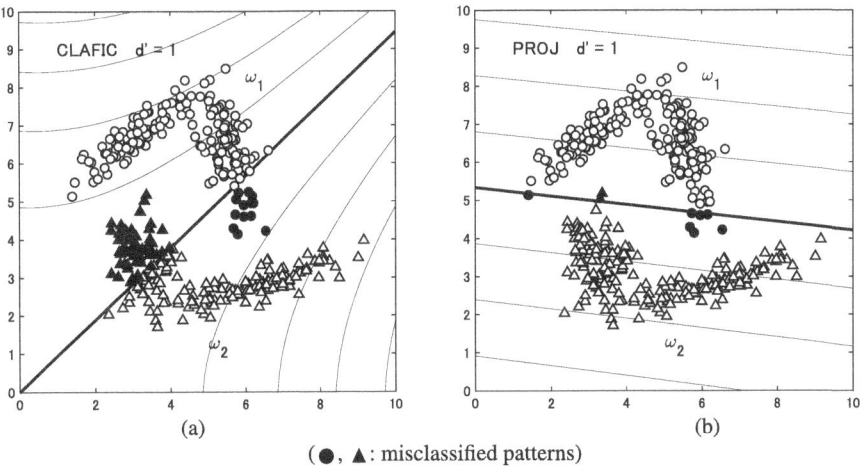

(●, ▲ : misclassified patterns)

Fig. 11.10 Linear subspace method (1). (**a**) CLAFIC method. (**b**) Projection distance method

between these two methods is that the CLAFIC method places the origin for constructing class subspaces at the origin of the feature space, while the projection distance method places it at the centroid of each class. Therefore, in the CLAFIC method, the classification result depends on the relative positions between the origin of the feature space and the distributions of each class, while it does not depend on these positions in the projection distance method. Additionally, it is clear that the CLAFIC method does not work effectively when distributions span across the origin of the feature space.

Figure 11.10 shows the results of applying two linear subspace methods to patterns of two classes ($c = 2$) ω_1 and ω_2 in a two-dimensional feature space ($d = 2$) as depicted in Fig. 8.1. In (a), the CLAFIC method is applied, while (b) shows the results of the projection distance method.

As mentioned earlier, when the distributions span across the origin of the feature space, the CLAFIC method cannot be effective. Therefore, all patterns in Fig. 8.1 were translated in advance by (5, 5).

The dimensionalities of the subspaces are all $d' = 1$. In the figures, the difference in distance to the two class subspaces is represented as thin solid contour lines, and the decision boundary where the distance difference becomes 0 is shown as a thick solid line. Patterns that are misclassified by this decision boundary are marked with ● and ▲. The number of errors is 70 patterns (error rate of 17.50%) for the CLAFIC method and 9 patterns (error rate of 2.25%) for the projection distance method.

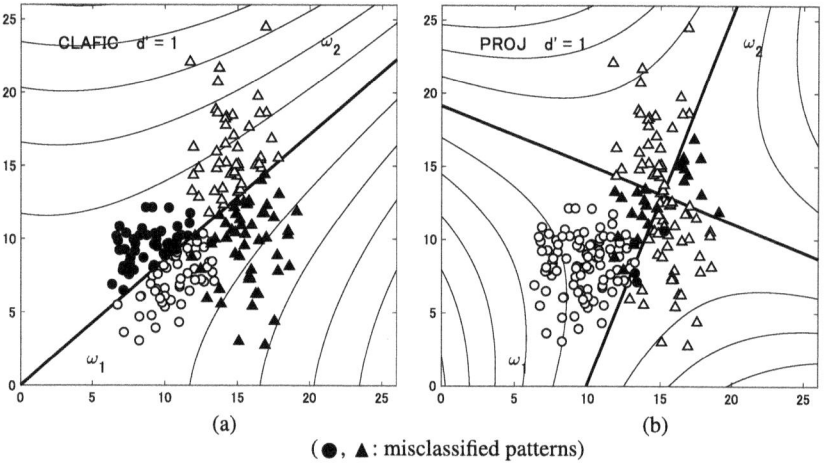

(●, ▲: misclassified patterns)

Fig. 11.11 Linear subspace method (2). (**a**) CLAFIC method. (**b**) Projection distance method

Figure 11.11 illustrates an example of applying linear subspace methods to a two-class ($c = 2$), two-dimensional ($d = 2$) data set shown in Fig. 11.3 after shifting all patterns by (10, 8). In (a), the CLAFIC method is applied, and in (b), the results of the projection distance method are shown. This data set is also linearly nonseparable, but the difficulty of class separation due to the overlapping distributions between the two classes is greater compared to Fig. 8.1. The number of errors (error rate) is 99 patterns (49.50%) for the CLAFIC method and 30 patterns (15.00%) for the projection distance method.

In the two examples mentioned above, the projection distance method shows a higher classification accuracy than the CLAFIC method, but it is not possible to draw a definitive conclusion about the performance difference between the two. Subspace methods are inherently designed with the assumption that patterns are distributed in relatively high dimensionalities and utilize the differences in subspaces characterizing the distributions of each class for classification. They are not well-suited for low-dimensional patterns. Additionally, regarding the CLAFIC method, Satoshi Watanabe has argued that one should not move the origin since each class's characteristics should appear in the mean patterns of each class (Watanabe 1969).

This linear subspace method has been widely applied in Japan, especially in tasks like character recognition, due to its ability to achieve multi-class classification with low computational complexity and high classification performance. However, it has two major limitations.

Firstly, when data distributions spread along nonlinear axes, the principal component directions determined by linear PCA may not have meaningful interpretations. Therefore, the obtained class subspaces may not necessarily be meaningful in characterizing the classes.

Secondly, when the ratio of the feature space dimensionality to the number of classes, denoted as d/c, is small, there is increased overlap between the subspaces of different classes, which generally leads to reduced classification performance. Increasing the number of features increases the dimensionality of the feature space, but it is not always easy to increase the number of features that are effective for classification. In particular, when applying subspace methods to tasks with a large number of classes to be classified, such as Japanese character recognition, it is necessary to pay attention to the dimensionality of the feature space.

To compensate for the first disadvantage, a classification method has been proposed that combines nonlinear PCA using auto-associative neural networks with the subspace method. This approach can absorb the nonlinearity of the data distribution when constructing class subspaces, potentially leading to higher classification performance compared to linear subspace methods. However, it requires learning by a neural network to construct class subspaces, which comes with the risk of getting stuck in local minima. Additionally, since it employs auto-associative circuits oriented to dimensionality reduction, it cannot overcome the second limitation of linear subspace methods mentioned earlier.

On the other hand, the nonlinear transformations defined through kernel functions used in SVMs are transformations to quite high dimensionalities. Therefore, if we can combine this kernel nonlinear transformation with subspace methods, we can expect to overcome the two limitations mentioned earlier in subspace methods. Furthermore, subspace methods can be formulated using kernel functions, allowing classification without explicitly using ϕ, which is advantageous in terms of computational cost. Subspace methods are indeed an application where kernel methods can maximize their abilities.

The following discusses the application of kernel methods to the broad category of linear subspace methods, resulting in a nonlinear version known as kernel subspace methods.

11.6.2 Kernelization of Subspace Methods

In the following, we will explain the CLAFIC method as an example. As shown in
Eq. (11.95), the inner product matrix on the Φ-space of $\mathbf{X}$ is represented as follows:[5]

$$\mathbf{Q}_{X_\phi} = \mathbf{X}_\phi \mathbf{X}_\phi^t = \mathbf{K}(\mathbf{X}, \mathbf{X}). \tag{11.139}$$

Here, $\mathbf{X}_\phi$ is defined as shown in Eq. (11.92),

$$\mathbf{X}_\phi = (\boldsymbol{\phi}(\boldsymbol{x}_1), \ldots, \boldsymbol{\phi}(\boldsymbol{x}_i), \ldots, \boldsymbol{\phi}(\boldsymbol{x}_n))^t. \tag{11.140}$$

The top d' eigenvalues of the matrix $\mathbf{Q}_{X_\phi}$, denoted as λ_i ($i = 1, \ldots, d'$), correspond
to normalized orthogonal eigenvectors $\mathbf{u}_i$. Similarly, $\mathbf{X}_\phi^t \mathbf{X}_\phi$ also has the same
eigenvalues as $\mathbf{Q}_{X_\phi}$ and corresponding normalized orthogonal eigenvectors $\mathbf{v}_i$.

We define $\mathbf{U}_{d'}$, $\mathbf{V}_{d'}$, and $\boldsymbol{\Lambda}_{d'}^{1/2}$ by $\mathbf{u}_i$, $\mathbf{v}_i$ and λ_i like Eqs. (11.106), (11.107) and
(11.108), as shown in the following equations:

$$\mathbf{U}_{d'} = (\mathbf{u}_1, \ldots, \mathbf{u}_{d'}) \qquad\qquad (\in \mathbb{R}^{n \times d'}), \tag{11.141}$$

$$\mathbf{V}_{d'} = (\mathbf{v}_1, \ldots, \mathbf{v}_{d'}) \qquad\qquad (\in \mathbb{R}^{D \times d'}), \tag{11.142}$$

$$\boldsymbol{\Lambda}_{d'}^{1/2} = \begin{pmatrix} \sqrt{\lambda_1} & & 0 \\ & \ddots & \\ 0 & & \sqrt{\lambda_{d'}} \end{pmatrix} \qquad (\in \mathbb{R}^{d' \times d'}). \tag{11.143}$$

Additionally, instead of Eq. (11.109), the following equation holds:

$$\mathbf{X}_\phi = \mathbf{U}_{d'} \boldsymbol{\Lambda}_{d'}^{1/2} \mathbf{V}_{d'}^t. \tag{11.144}$$

Here, let $P(\mathbf{z})$ denote the length of the projection vector of $\boldsymbol{\phi}(\mathbf{z})$ onto the subspace
spanned by $\mathbf{V}_{d'} = (\mathbf{v}_1, \ldots, \mathbf{v}_{d'})$. As shown in Eq. (11.114), the length of the
projection of the vector $\boldsymbol{\phi}(\mathbf{z})$ onto $\mathbf{v}_i$ is given by $\mathbf{v}_i^t \boldsymbol{\phi}(\mathbf{z})$. Therefore, we can express
$P^2(\mathbf{z})$ as

$$P^2(\mathbf{z}) = \sum_{i=1}^{d'} \left(\mathbf{v}_i^t \boldsymbol{\phi}(\mathbf{z})\right)^2 = \|\mathbf{V}_{d'}^t \boldsymbol{\phi}(\mathbf{z})\|^2. \tag{11.145}$$

From Eq. (11.119), we have

---

[5] If we were to use the projection distance method, the vectors and matrices we would be working
with are not $\mathbf{X}_\phi$ and $\mathbf{Q}_{X_\phi}$, but rather $\mathbf{X}_{\phi M}$ and $\mathbf{Q}_{\mathbf{X}_{\phi M}}$.

$$\mathbf{K}(\mathbf{X}, \mathbf{Z}) = \mathbf{X}_\phi \, \mathbf{Z}_\phi^t. \tag{11.146}$$

Here, the matrices $\mathbf{Z}$ and $\mathbf{Z}_\phi$ in the above equation are represented by the following expressions:

$$\mathbf{Z} = (\mathbf{z}_1, \ldots, \mathbf{z}_m)^t, \tag{11.147}$$

$$\mathbf{Z}_\phi = (\phi(\mathbf{z}_1), \ldots, \phi(\mathbf{z}_m))^t. \tag{11.148}$$

Considering the special case of the above equation where $m = 1$, that is, $\mathbf{Z} = \mathbf{z}^t$ and $\mathbf{Z}_\phi = \phi(\mathbf{z})^t$, then from Eqs. (11.146) and (11.144) we have

$$\mathbf{K}(\mathbf{X}, \mathbf{z}^t) = \mathbf{X}_\phi \, \phi(\mathbf{z}) \tag{11.149}$$

$$= \mathbf{U}_{d'} \mathbf{\Lambda}_{d'}^{1/2} \mathbf{V}_{d'}^t \phi(\mathbf{z}). \tag{11.150}$$

The matrix $\mathbf{X}_\phi$ is represented by Eq. (11.140). Thus, the $\mathbf{K}(\mathbf{X}, \mathbf{z}^t)$ in Eq. (11.149) is an n-dimensional vector with components being the inner products of $\phi(x_i)$ ($i = 1, \ldots, n$) and $\phi(\mathbf{z})$. Also, d' represents the dimensionality of the class subspace.

By multiplying both sides of Eq. (11.150) from the left with $\mathbf{\Lambda}_{d'}^{-1/2}\mathbf{U}_{d'}^t$, we get

$$\mathbf{\Lambda}_{d'}^{-1/2}\mathbf{U}_{d'}^t \mathbf{K}(\mathbf{X}, \mathbf{z}^t) = \mathbf{V}_{d'}^t \phi(\mathbf{z}). \tag{11.151}$$

By substituting the above equation into Eq. (11.145), we can derive the subsequent expression:

$$P^2(\mathbf{z}) = \|\mathbf{\Lambda}_{d'}^{-1/2}\mathbf{U}_{d'}^t \mathbf{K}(\mathbf{X}, \mathbf{z}^t)\|^2. \tag{11.152}$$

To perform the classification, as shown in Fig. 11.9(a), one should calculate $P^2(\mathbf{z})$ for each class and output the class that yields the maximum value as the classification result. As discussed in the previous section, $\mathbf{U}_{d'}$ and $\mathbf{\Lambda}_{d'}^{1/2}$ are constituted by the eigenvalues and eigenvectors of Eq. (11.139), and therefore can be derived from $K(\mathbf{X}, \mathbf{X})$. Consequently, even without knowing the form of ϕ, classification of unknown patterns is feasible provided the definition of the kernel function $k(x, y)$ and the learning patterns are given.

Similar to the linear subspace method, the dimensionality d' of the class subspace in the kernel subspace method affects the classification results. Typically, d' is either kept constant regardless of the class, or its value is set such that the cumulative contribution ratio is consistent across classes. The cumulative contribution ratio is defined as shown in Eq. (6.194) , and here it is given by

$$c_p = \sum_{i=1}^{d'} \lambda_i \, \bigg/ \, \sum_{i=1}^{\min\{D,n\}} \lambda_i. \tag{11.153}$$

The optimal values of d' and c_p vary depending on the task. For details on the classification method of the projection distance method, one should refer to Problem 11.5.

11.6.3 Classification Example Using Kernel Subspace Method

Figures 11.12 and 11.13 show the results of applying the kernel subspace method to the data presented in Fig. 8.1. The former employs the Gaussian kernel, while the latter uses the polynomial kernel. In each figure, (a) and (b) represent the results of applying the CLAFIC method and the projection distance method, respectively, displaying the calculated contours, decision boundaries, and misclassified patterns. The parameters of the Gaussian kernel σ, the polynomial kernel parameter p, and the dimensionality of the subspace d' are indicated in the top left of each figure. By nonlinearizing, curved decision boundaries are obtained. The number of misclassifications is one pattern (error rate of 0.25%) in Fig. 11.12(b) and five patterns (error rate of 1.25%) in Fig. 11.13(b), while in all other cases there are no misclassifications. Misclassified patterns are marked with ● and ▲.

On the other hand, the results of applying the kernel subspace method under the same conditions to the data shown in Fig. 11.3 are depicted in Figs. 11.14 and 11.15. The number of misclassifications are seven patterns (error rate of 3.50%) for both Fig. 11.14(a) and (b), while for Fig. 11.15(a) and (b) they are eight patterns (4.00%) and 17 patterns (8.50%) respectively. The number of misclassifications is reduced compared to the linear subspace method in Fig. 11.11.

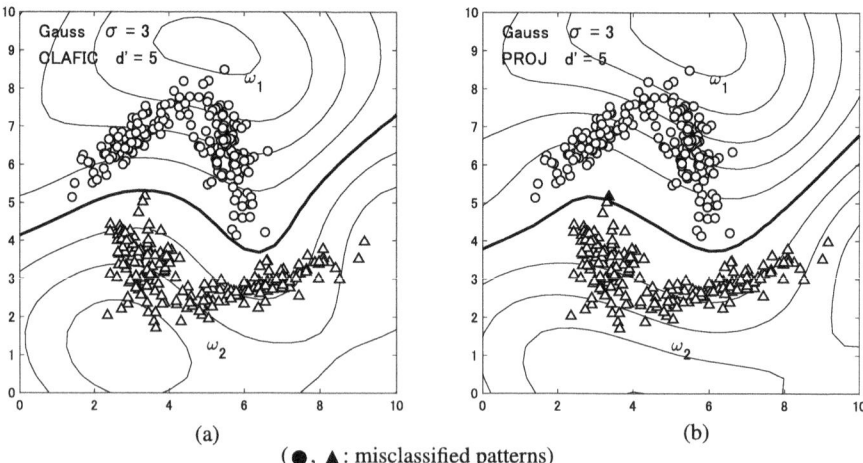

(a)   (b)

(●, ▲: misclassified patterns)

Fig. 11.12 Kernel subspace method using Gaussian kernel (1). (**a**) CLAFIC method. (**b**) Projection distance method

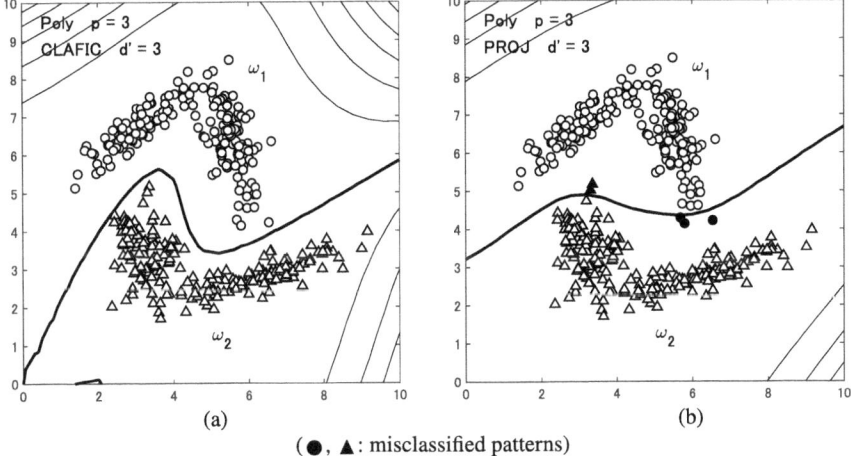

Fig. 11.13 Kernel subspace method using polynomial kernel (1). (**a**) CLAFIC method. (**b**) Projection distance method

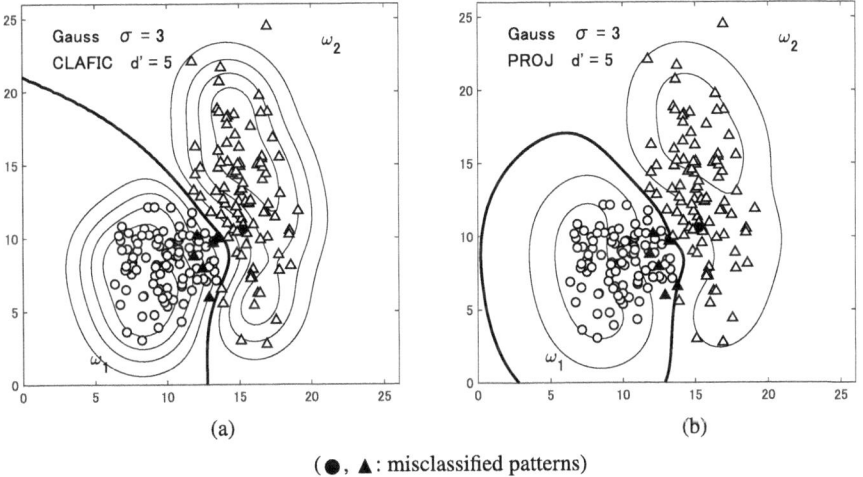

Fig. 11.14 Kernel subspace method using Gaussian kernel (2). (**a**) CLAFIC method. (**b**) Projection distance method

As mentioned at the beginning of this chapter, the strength of the subspace method lies in its suitability for multi-class discrimination. This makes it an apt technique for recognizing characters in languages like Japanese, which has a multitude of character types. Below, we present experimental results of applying the kernel subspace method to multi-class, high-dimensional classification problems. The data set used is the well-known handwritten digit data set, MNIST (refer to Sect. 13.6). The number of classes is 10 ($c = 10$). For learning, there are 1000 patterns for each class, totaling 10,000 patterns. For testing, there are 800 patterns

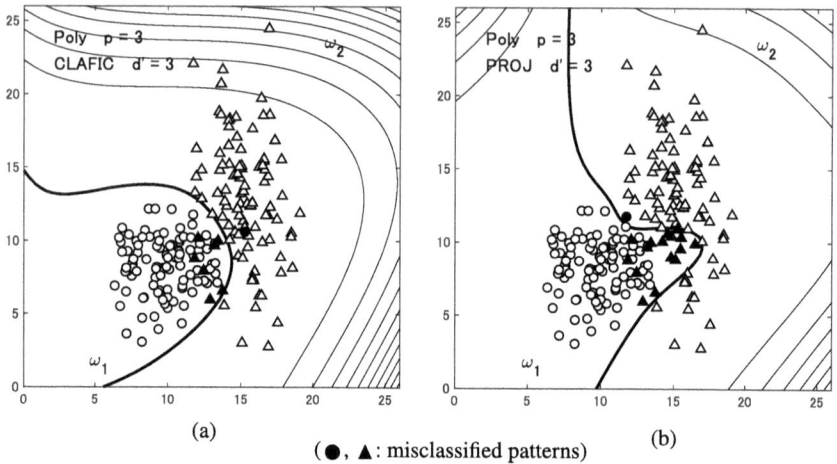

(a)   (●, ▲: misclassified patterns)   (b)

Fig. 11.15 Kernel subspace method using polynomial kernel (2). (**a**) CLAFIC method. (**b**) Projection distance method

Table 11.1 Experimental results of linear subspace method on MNIST data

Exp.: Experimental number,   1-NN: Nearest neighbor rule,
LSS: Linear subspace method,   CLF: CLAFIC method,   PRJ: Projection distance method,
L. pat.: Learning pattern,   T. pat: Test pattern

Exp.	Method	CLF/PRJ	c_p	d' Mean	S.D.	Error rate (%) L. pat.	T. pat.
1	1-NN	–	–	–	–	4.80	5.94
2	LSS	CLF	0.80	10.3	3.7	7.10	8.28
3	LSS	CLF	0.90	28.4	8.2	3.46	5.46
4	LSS	CLF	0.95	59.2	13.6	2.09	5.51
5	LSS	CLF	0.98	119.8	21.8	1.19	7.39
6	LSS	PRJ	0.80	30.0	6.9	2.81	5.13
7	LSS	PRJ	0.90	62.0	11.6	1.79	5.75
8	LSS	PRJ	0.95	107.0	16.9	1.12	7.29
9	LSS	PRJ	0.98	176.8	25.1	0.82	9.07

for each class, adding up to 8000 patterns (see the footnote 1 in Appendix D). The dimensionality d is given by $d = 28 \times 28 = 784$.

The results are presented in Tables 11.1, 11.2, and 11.3. Table 11.1 shows the experimental results of the nearest neighbor decision rule and linear subspace method, which were conducted for comparison purposes. Tables 11.2 and 11.3 display the experimental results of the kernel subspace method. In the tables, the error rates for the learning and test patterns are given in percentages.

Experiment 1 in Table 11.1 presents the results of the nearest neighbor rule. Experiments 2–5 display the results when the linear subspace method (LSS) is combined with the CLAFIC method (CLF). In the table, c_p represents the cumulative contribution ratio, indicating the selection of eigenvectors in descending order of

Table 11.2   Experimental results of kernel subspace method on MNIST data (1)

KSS: Kernel subspace method, σ: Gaussian kernel parameter,

p: Polynomial kernel parameter

				d'		Kernel		Error rate (%)	
Exp.	Method	CLF/PRJ	c_p	Mean	S. D.	σ	p	L. pat.	T. pat
10	KSS	CLF	0.30	96.5	48.4	3		0.45	4.85
11	KSS	CLF	0.50	253.4	97.6	3		0.00	4.38
12	KSS	CLF	0.70	458.5	146.1	3		0.00	4.30
13	KSS	CLF	0.90	740.2	150.0	3		0.00	4.16
14	KSS	CLF	0.95	840.7	122.1	3		0.00	4.16
15	KSS	CLF	0.30	2.1	0.7		3	17.64	13.84
16	KSS	CLF	0.50	9.0	3.8		3	8.75	6.70
17	KSS	CLF	0.70	43.5	18.4		3	4.11	4.64
18	KSS	CLF	0.90	244.2	74.9		3	1.11	3.90
19	KSS	CLF	0.95	409.8	101.8		3	0.53	3.77

Table 11.3   Experimental results of kernel subspace method on MNIST data (2)

				d'		Kernel		Error rate (%)	
Exp.	Method	CLF/PRJ	c_p	Mean	S. D.	σ	p	L. pat.	T. pat
20	KSS	CLF	0.90	895.1	14.4	1		0.00	5.90
21	KSS	CLF	0.90	740.2	150.0	3		0.00	4.16
22	KSS	CLF	0.90	451.8	150.9	5		0.00	3.57
22	KSS	CLF	0.90	191.4	81.5	7		0.09	3.44
24	KSS	CLF	0.90	64.6	29.7	9		0.81	3.67
25	KSS	CLF	0.90	28.2	8.2		1	3.38	5.10
26	KSS	CLF	0.90	244.2	74.9		3	1.11	3.90
27	KSS	CLF	0.90	262.0	65.2		5	2.17	4.63
28	KSS	CLF	0.90	189.4	51.0		7	3.88	5.99
29	KSS	CLF	0.90	118.0	41.7		9	5.50	7.75

their eigenvalues until the specified cumulative contribution ratio is reached. The number of selected eigenvectors determines the dimensionality d' of the subspace, which varies for each class. The table shows the average and standard deviation of d' for various c_p values. Experiments 6–9 in the table present the results of combining the subspace method with the projection distance method (PRJ).

On the other hand, all the experiments shown in Table 11.2 are the results of combining the kernel subspace method (KSS) with the CLAFIC method. In experiments 10–14, a Gaussian kernel with $\sigma = 3$ was used, while in experiments 15–19, a polynomial kernel with $p = 3$ was employed. As with Table 11.1, the table displays the average and standard deviation of d' for various values of c_p. In this study, the values of σ and p were kept constant, and the value of c_p was varied. Conversely, in the results shown in Table 11.3, the value of c_p was kept constant, while the values of σ and p were varied.

In experiments 20–24, with $c_p = 0.90$, the value of σ for the Gaussian kernel was varied as 1, 3, 5, 7, 9. Similarly, in experiments 25 to 29, the value of p for the polynomial kernel was varied in the same manner as 1, 3, 5, 7, 9. Other considerations are the same as in Table 11.2.

For a discussion on the experimental results from Table 11.1, 11.2, and 11.3, refer to Problem 11.10.

11.7 Manifold Learning

The term dimensionality reduction refers to the representation of patterns distributed in a d-dimensional space using a lower-dimensional space. It serves both as a means of information compression, representing patterns with a reduced amount of information, and as a method of feature transformation. With the introduction of kernel methods, efforts have been made to nonlinearize pattern recognition and multivariate analysis methods. The lower-dimensional space representing the pattern distribution is called a *manifold*, and the process of finding such a space, especially using nonlinear methods, is referred to as *dimensional embedding* or *manifold learning*.

In 2000, methods such as *Isomap* (isometric mapping) (Tenenbaum et al. 2000) and *LLE* (locally linear embedding) (Roweis and Saul 2000) were published in the journal Science. They are based on the idea presented in a paper by Lawrence Saul and others, titled "Think Globally, Fit Locally" (Saul et al. 2003). These methods aim to capture the overall main axes while preserving local proximities between patterns. For example, Isomap focuses only on the k nearest neighbors of each pattern, constructs paths based on this, and defines the distance between patterns along these paths.

In Fig. 11.16, the distribution in a three-dimensional space is called *Swiss roll* and is obtained as a result of embedding a distribution from a two-dimensional space into a three-dimensional space. In this example, if k is set to an appropriate value, paths are set only on the sheet of the Swiss roll, and no paths are created between patterns spanning layers, as indicated by the arrows in the figure.The distance along this path is taken as the distance between patterns. In order to perform MDS, Isomap uses the distances obtained in this way. Such dimensionality reduction methods are also discussed in connection with information processing in the human brain (Seung and Lee 2000). For example, various facial images of many people with diverse expressions are stored in the brain, and we can classify individuals and recognize emotions. The facial images of each class can be interpreted as forming a nonlinear manifold in feature space. When information in the high-dimensional space of facial images is sliced along a certain axis, it is represented in a nonlinear lower-dimensional space (Fig. 11.17). Subsequently, various nonlinear dimensionality reduction methods have been proposed as manifold learning methods. In addition to the aforementioned Isomap and LLE, many methods, such as *graph Laplacian*

Fig. 11.16   Swiss roll distribution

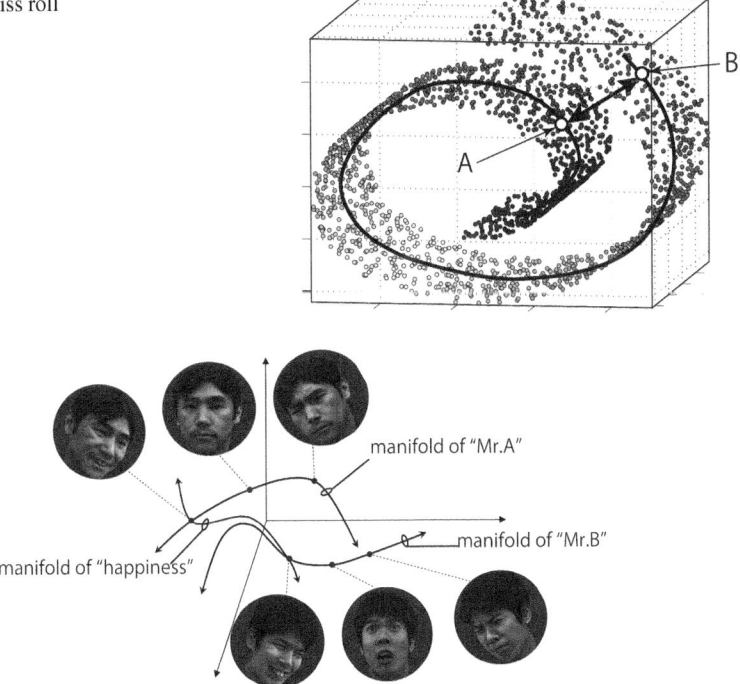

Fig. 11.17   An example of a manifold representing a specific individual or expression in the facial image space

(Belkin and Niyogi 2002), are known to be interpretable as kernel PCA when a special kernel function (gram matrix) is chosen (Ham et al. 2004).

─────────────────────────────   **Coffee Break**   ─────────────────────────────

Kernel Methods for Structured Data

Kernel methods are also used to evaluate the similarity of *structured data* such as strings, trees, and graphs. A typical kernel method for structured data is the *convolution kernel* proposed by Haussler (1999). The convolution kernel is a function that enumerates all the substructures constituting the structured data and evaluates their similarity. The concept of convolution kernels has been applied to various structured data such as strings (Lodhi et al. 2002), tree structures (Collins and Duffy 2001), and graphs (Kashima et al. 2003). It is especially used in fields like natural language processing, which deals with natural sentences represented in tree or graph forms, and bioinformatics that focuses on analyzing the base sequences of a large number of genes or amino acid sequences of proteins. The convolution kernel was one of the early successful examples that expanded its application domain by "embedding" non-vector data, as represented by natural language, into a vector space.

Problems

11.1 For two d-dimensional vectors x and y, there is a function $k(x, y)$. Using n d-dimensional vectors $x_1, \ldots, x_n$, we define a matrix $\mathbf{K}$ where the (i, j) element is $k(x_i, x_j)$ as follows:

$$\mathbf{K} = \begin{pmatrix} k(x_1, x_1) & k(x_1, x_2) & \ldots & k(x_1, x_n) \\ k(x_2, x_1) & k(x_2, x_2) & \ldots & k(x_2, x_n) \\ \vdots & \vdots & \ldots & \vdots \\ k(x_n, x_1) & k(x_n, x_2) & \ldots & k(x_n, x_n) \end{pmatrix}. \tag{11.154}$$

If the function $k(x, y)$ is a kernel function, show that the matrix $\mathbf{K}$ is positive semidefinite.

11.2 Suppose we are given a matrix $\mathbf{K}$ of size $n \times n$. Show that if the matrix $\mathbf{K}$ is positive semidefinite, $\mathbf{K}$ can be expressed as in Eq. (11.154), using n d-dimensional vectors $x_1, \ldots, x_n$ and the kernel function $k(x, y)$.

11.3 Show that the recursive formula of Eq. (11.76) holds for the case $d = 3$, $l = 2$.

11.4 In the fields of pattern recognition and machine learning, it is common to solve the optimization problem of finding $x \, (= x^*)$ that minimizes[6]

$$J = \|\mathbf{A}x - \mathbf{b}\|^2 + \lambda \|x\|^2 \tag{11.155}$$

In the above equation, $\mathbf{A}$ is an $n \times d$ matrix, x is a d-dimensional column vector, and $\mathbf{b}$ is an n-dimensional column vector.

(1) Derive the following equation by partial differentiation of J in the above equation and setting it to $\mathbf{0}$:

$$x^* = (\mathbf{A}^t\mathbf{A} + \lambda\mathbf{I})^{-1}\mathbf{A}^t\mathbf{b}. \tag{11.156}$$

(2) Transform the above solution x^* to the solution,

$$x^* = \mathbf{A}^t(\mathbf{A}\mathbf{A}^t + \lambda\mathbf{I})^{-1}\mathbf{b}, \tag{11.157}$$

using $\mathbf{A}\mathbf{A}^t$ instead of $\mathbf{A}^t\mathbf{A}$ in the equation.

11.5 The projection distance method classifies a pattern x as the class ω_i where the squared distance D^2 between x in Fig. 11.9(b) and P_i is shortest. When applying the kernel method to this projection distance method, show that

---

[6] Refer to Sect. 11.5.

$$D^2(\mathbf{z}) = G_X(\mathbf{z}^t, \mathbf{z}^t) - \|\mathbf{\Lambda}_{d'}^{-1/2} \mathbf{U}_{d'}^t G_X(\mathbf{X}, \mathbf{z}^t)\|^2 \tag{11.158}$$

holds, where, $\mathbf{U}_{d'}$, $\mathbf{V}_{d'}$, and $\mathbf{\Lambda}^{1/2}d'$ are matrices defined in the same manner as in Eqs. (11.141), (11.142), and (11.143). Specifically, for $G_X(\mathbf{X}, \mathbf{X}) = \mathbf{X}\phi_M \mathbf{X}_{\phi_M}^t$, the top d' eigenvalues are λ_i $(i = 1, \ldots, d')$, and the orthonormal eigenvectors of $\mathbf{X}_{\phi_M} \mathbf{X}_{\phi_M}^t$ and $\mathbf{X}_{\phi_M}^t \mathbf{X}_{\phi_M}$ corresponding to eigenvalue λ_i are denoted as $\mathbf{u}_i$ and $\mathbf{v}_i$, respectively.

11.6 Verify that Eq. (11.136) holds.

11.7 Prove the following:

(1) If the distance matrix $\mathbf{D}$ is defined by the Euclidean distance, and $d_{ij}^2 = \|x_i - x_j\|^2$, then $\mathbf{L}$ in Eq. (11.135) can be expressed as $\mathbf{L} = \mathbf{X}_M \mathbf{X}_M^t$ using the mean deviation matrix $\mathbf{X}_M$ of Eq. (11.93), and $\mathbf{L}$ is a positive semidefinite symmetric matrix.

(2) Conversely, if $\mathbf{L}$ is a positive semidefinite symmetric matrix, then $\mathbf{D}$ defined by the Euclidean distance can be determined.

(3) If the distance matrix $\mathbf{D}$ is defined by the Euclidean distance, then multidimensional scaling is equivalent to principal component analysis.

11.8 Show that Eq. (11.138) holds, and that the mean of $y_1, \ldots, y_n$ obtained from the calculated $\mathbf{Y}$ coincides with the origin.

11.9 † Given six comparable cases $x_1, \ldots, x_6$, let's assume the degree of difference (or distance) between x_i and x_j is quantified as d_{ij} (where $i, j = 1, \ldots, 6$). Here, a 6×6 matrix $\mathbf{D}^{(2)}$, with elements d_{ij}^2 as its (i, j) components, is provided as follows:

$$\mathbf{D}^{(2)} = \begin{pmatrix} 0 & 49 & 65 & 16 & 13 & 113 \\ 49 & 0 & 16 & 65 & 34 & 50 \\ 65 & 16 & 0 & 49 & 26 & 10 \\ 16 & 65 & 49 & 0 & 5 & 73 \\ 13 & 34 & 26 & 5 & 0 & 52 \\ 113 & 50 & 10 & 73 & 52 & 0 \end{pmatrix}. \tag{11.159}$$

In this case, set $d' = 2$, find the pattern matrix $\mathbf{Y}$ using the multidimensional scaling method, and plot $y_1, \ldots, y_6$ in a d'-dimensional space.

11.10 Analyze the experimental results regarding the subspace method presented in Tables 11.1, 11.2, and 11.3, and provide the observations for them.

Chapter 12
Neural Networks

Abstract The perceptron is regarded as the pioneering work in machine learning, and it was the forerunner of neural network research. Although the perceptron had a learning function, it had various limitations, and over time it fell into decline. Later, the backpropagation method was proposed as a learning method for multi-layer neural networks. This innovative idea attracted attention as a powerful method for acquiring a nonlinear discriminant function by learning, but it also revealed a number of problems and went into decline. Deep learning, which will be introduced in the next chapter, solved these various problems and emerged as a learning method for multi-layer neural networks that can be fit for practical use. Therefore, in this chapter, as a preparation for the introduction to the next chapter, the backpropagation method and other important techniques supporting neural networks are introduced.

12.1 Dawn of the Neural Boom

Since the advent of computers, attempts to realize human intellectual functions in machines have attracted researchers in various fields and the vigorous challenge continues to this day. The pioneering work in this field was the perceptron (1957), introduced in Chap. 2 which marked the beginning of the *first neural boom*. However, although the perceptron had a learning function, its two layer structure, consisting of an input layer and an output layer, allowed only a linear discriminant function, limiting its application. To overcome this limitation, several attempts have been made to realize nonlinear discriminant functions. One of them is to upgrade the two-layered perceptron to a multi-layered one, but it has not been possible to devise a method for learning multi-layered networks.

The piecewise linear discriminant function introduced in Chap. 2 is a discriminant function composed of a combination of linear discriminant functions and is equivalent in the limit to the multi-layer neural network, as already mentioned

---

Supplementary Information The online version contains supplementary material available at https://doi.org/10.1007/978-981-95-1478-6_12.

K. Ishii et al., *Pattern Recognition and Machine Learning for Self-Study I*, Springer Asia Pacific Mathematics Series 1, https://doi.org/10.1007/978-981-95-1478-6_12

(Nilsson 1965). The fact that there is no effective learning method for piecewise linear discriminant functions is similar to that of the multi-layer neural networks at the time.

The generalized linear discriminant function method and the potential function method introduced in Chaps. 8 and 9, respectively, realize nonlinear discriminant functions, and they are also equipped with learnability. The generalized linear discriminant function method maps original patterns to the Φ space by a nonlinear transformation, converts them to a linearly separable state, and then applies the perceptron learning rule. In this case, the decision of the Φ function remains arbitrary, and there is no guarantee that the result of mapping to Φ space is linearly separable. Also, the potential function method is equivalent to the generalized linear discriminant function method, as already mentioned, so it has the same problem. Thus, both methods were based on the perceptron learning rule and did not provide a new framework for learning methods. This led to the end of the boom and the beginning of a long winter period.

Nearly 30 years later, in 1986, David E. Rumelhart proposed the backpropagation method as a learning method for multi-layer neural networks. Neural networks were once again in the limelight, and the *second neural boom* was ushered in. Because of the long winter period, expectations for neural networks were high, and many researchers actively engaged in research activities.

However, fatal problems such as the vanishing gradient problem, over-fitting, etc., which will be described later, were revealed in this neural network research. As a result, the research entered a winter period again. The solution to these problems had to wait for the advent of deep learning. Deep learning was the technology that sparked the *third neural boom*. Although often overshadowed by the topic of deep learning, the backpropagation method was an epoch-making idea that drove the second neural boom and has played an important role in the subsequent deep learning research.

Therefore, in this chapter, as a preparation for the introduction to deep learning in the next chapter, we first introduce some of the important technologies that supported the second neural boom, such as the backpropagation method. The concepts of the rectified linear function and dropout are techniques that emerged in the process of deep learning research. However, they are not specialized for deep learning, and can be used effectively with conventional neural networks. So, they are taken up in this chapter.

12.2 Backpropagation Method

In the following, we deal with the multi-class problem of classifying an input pattern into one of c classes $\omega_1, \ldots, \omega_c$. A *neural network* has a layered structure. In particular, when emphasizing that many layers are stacked, it is called a *multi-layer neural network*. Figure 12.1 shows a neural network of L layers, which consists of the first layer, i.e., the *input layer*, the last layer, i.e., the *output layer*, and several

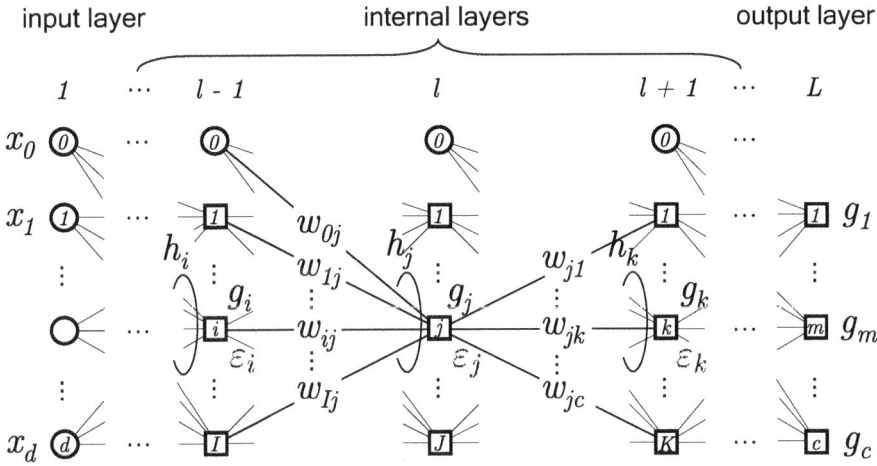

Fig. 12.1 Structure of a neural network (the *l*-th layer is an internal layer)

internal layers[1] in between. The signal flow is unidirectional from the input layer to the output layer. Such a neural network is called a *feedforward neural network*.

Each layer has multiple *units* represented by □ and ○. In the figure, the □ symbol represents the unit performing nonlinear transformation.[2] The ○ symbol represents a unit that has no input from the previous layer and only outputs to the next layer. That is, $(d + 1)$ ○ s in the input layer are units that output the feature values $x_0 (= 1), x_1, \ldots, x_d$ of the input pattern **x** as are, and ○ s in the middle layer are units that always output 1. This configuration corresponds to the fact that x_0 of Eq. (2.7) is set to constant 1 in order to include bias w_0 of Eq. (2.8) as a learning target.

The unit numbers are written in the symbols. However, there is no unit 0 in the output layer, and for a *c*-class classification problem, there are *c* units from unit 1 to unit *c*. Coupling between units exists only between adjacent layers. Each unit in the second and subsequent layers is connected to all units in the previous layer via weights, and this type of coupling is called *fully-connected*.

In the following, the *l*-th layer is discussed. If $2 \leq l \leq L - 1$, the *l*-th layer is an internal layer, and if $l = L$, it is an output layer. The *j*-th unit in the *l*-th layer is called unit j ($= 0, 1, \ldots, J$) and the *i*-th unit in the previous $(l - 1)$-th layer is called unit i ($= 0, 1, \ldots, I$). If the *l*-th layer is an internal layer, there is a $(l + 1)$-th layer, i.e., one more layer later, and its *k*-th unit is called unit k ($= 0, 1, \ldots, K$). The *m*-th unit in the *L*-th layer, i.e., the output layer, is called unit m ($= 1, \ldots, c$).

In Fig. 12.1, the case where the *l*-th layer is the internal layer is shown. Let w_{ij} be the weight of the connection from unit i to unit j and w_{jk} be the weight

---

[1] The internal layer is also called a *hidden layer*.

[2] This unit includes the threshold logic unit introduced in Sect. 3.2.1 and the activation functions described later.

of the connection from unit j to unit k. If the pattern $\mathbf{x} = (x_0, x_1, \ldots, x_d)^t$ is input ($x_0 \equiv 1$) and the output from the output layer is $g_1, \ldots, g_m, \ldots, g_c$, the classification process is based on

$$\max_m\{g_m\} = g_{m^*} \quad \Longrightarrow \quad \mathbf{x} \in \omega_{m^*}. \tag{12.1}$$

The perceptron learning rule can learn only the output layer, and is powerless for learning the internal layers. The *backpropagation method* described in this section, solves this drawback and extends the scope of learning to multi-layer networks. The backpropagation method is abbreviated as below.

In the following, we describe BP using Fig. 12.1. We take the weight w_{ij} between the $(l-1)$-th and l-th layers as the learning target. Given a learning pattern, we denote the input to unit j in the l-th layer as h_j and the output from unit j as g_j. In the same way, h_i and g_i in the $(l-1)$-th layer and h_k and g_k in the $(l+1)$-th layer can be defined.[3] The input h_j is a linear sum of the outputs from all units in the $(l-1)$-th layer that are coupled to unit j, so we can write as follows:[4]

$$h_j = \sum_{i=0}^{I} w_{ij}\, g_i \qquad (j = 1, \ldots, J). \tag{12.2}$$

Furthermore, using the nonlinear function $f(\cdot)$, the output g_j from unit j is expressed as

$$g_j = f(h_j) \qquad (j = 1, \ldots, J). \tag{12.3}$$

This nonlinear function $f(\cdot)$ is called the *activation function*. In a multi-layer neural network, the nonlinear activation function plays a significant role. If the activation function is a linear function, multi-layering only yields a linear discriminant function, and does not provide a highly nonlinear discriminant function. The advantages of multi-layering can only be realized by introducing a nonlinear element such as the activation function.

The above describes the input to the l-th layer and the output from the same layer. If the l-th layer is an internal layer, as in Fig. 12.1, the input h_k and output g_k of the $(l+1)$-th layer can be written as

$$h_k = \sum_{j=0}^{J} w_{jk}\, g_j \qquad (k = 1, \ldots, K) \tag{12.4}$$

---

[3] Strictly speaking, the layers should be indicated by h_j^l, g_j^l, w_{ij}^l, etc. to specify in which layer they are, but we omitted the notation of layers because it would be too complicated. Instead, as shown in Fig. 12.1, each layer of $l-2$, l, $l+1$ is distinguished by the subscripts i, j, k.

[4] Note that i starts at 0 while j starts at 1. Also, $g_0 \equiv 1$. The same is true for j, k in Eqs. (12.4) and (12.5).

$$g_k = f(h_k) \qquad (k = 1, \ldots, K), \tag{12.5}$$

in the same way as in Eqs. (12.2) and (12.3).

To obtain optimal weights through learning, some evaluation method is necessary. For this purpose, we set a teaching signal and repeatedly modify the weights during the learning process so that the output is as close to the teaching signal as possible. The evaluation measure is defined as the degree of dissimilarity between the actual outputs and the teacher signal, and the goal is to minimize the degree of dissimilarity.

The teaching signal for the two-class problem ($c = 2$) is shown in Eq. (2.29). The teaching signal for the multi-class problem ($c > 2$) can be given as a c-dimensional teaching vector. If we use the same teaching vector for patterns belonging to the same class, the teaching vector will be Eq. (3.2). That is, for patterns of class ω_i, the teaching vector $\mathbf{t}_i$ should be set as

$$\mathbf{t}_i = (b_1, \ldots, b_i, \ldots, b_c)^t \quad (b_i > b_j, \quad j \neq i). \tag{12.6}$$

It is simpler to set $\mathbf{t}_i$ as a one-hot vector as follows:

$$\mathbf{t}_i = (0, \ldots, 0, \overset{i}{1}, 0, \ldots, \overset{c}{0}). \tag{12.7}$$

As a teaching signal, instead of 0 and 1, b_m of Eq. (12.6) can be set to

$$\sum_{m=1}^{c} b_m = 1 \qquad (0 \leq b_m \leq 1), \tag{12.8}$$

which is more versatile. For example, for an ambiguous pattern that is difficult to determine as either "0" or "6", the corresponding b_m of the two classes should be 0.5 each, and the other b_ms should be 0. In other words, the teacher judges that there is a fifty-fifty chance of both "0" and "6". Therefore, b_m can be regarded as representing the probability that the pattern belongs to class ω_m.

Given a learning pattern $\mathbf{x}_p$ ($p = 1, \ldots, n$), if we denote the degree of dissimilarity between the output and the teaching signal as J_p, the degree of dissimilarity J_a for all learning patterns can be written as

$$J_a = \sum_{p=1}^{n} J_p. \tag{12.9}$$

In learning, the weights that minimize J_a are obtained by the steepest descent method. When batch learning is applied as the learning method, the weights are modified all at once after all learning patterns are presented, so the weight modification follows the formula below:

$$w'_{ij} = w_{ij} - \rho \frac{1}{n} \cdot \frac{\partial J_a}{\partial w_{ij}} \qquad (12.10)$$

$$= w_{ij} - \rho \frac{1}{n} \sum_{p=1}^{n} \frac{\partial J_p}{\partial w_{ij}} \qquad (i = 0, \ldots, I \quad j = 1, \ldots, J), \quad (12.11)$$

where w'_{ij} is the weight after modifying w_{ij} and $\rho\,(> 0)$ is the learning rate. On the other hand, in online learning, the weights are modified each time a learning pattern is classified, so the weight modification follows the formula below:

$$w'_{ij} = w_{ij} - \rho \frac{\partial J_p}{\partial w_{ij}} \qquad (i = 0, \ldots, I \quad j = 1, \ldots, J). \qquad (12.12)$$

In either learning method, the weight modification is applied repeatedly until convergence is achieved. In the following, we will apply the online learning.

A possible evaluation measure J_p of the degree of dissimilarity is the *square error* between the output and the teaching signal. Suppose that the learning pattern x_p $(p = 1, \ldots, n)$ is input and the output from unit m in the output layer is g_m $(m = 1, \ldots, c)$, the square error $J_p{}^5$ for the pattern x_p is

$$J_p = \frac{1}{2} \sum_{m=1}^{c} (g_m - b_m)^2. \qquad (12.13)$$

In learning, g_m and b_m are compared each time a pattern is classified, and the weights are repeatedly modified to minimize J_p.

Such a learning method is called the minimum square error learning, as described in Sect. 3.1. Another evaluation measure is the cross entropy, which will be discussed later.

The partial derivative $\partial J_p/\partial w_{ij}$ in Eq. (12.12) can be rewritten as follows:

$$\frac{\partial J_p}{\partial w_{ij}} = \frac{\partial J_p}{\partial h_j} \cdot \frac{\partial h_j}{\partial w_{ij}} \qquad (i = 0, \ldots, I \quad j = 1, \ldots, J). \qquad (12.14)$$

The first term $\partial J_p/\partial h_j$ on the right side of the above equation is an important term as described later, so we put it as ε_j as follows:

---

[5] Strictly speaking, g_m in Eq. (12.13) should be written as g_{mp} to indicate that it is the output for the pth pattern x_p, and J_p should be written as

$$J_p = \frac{1}{2} \sum_{m=1}^{c} (g_{mp} - b_m)^2.$$

However, to avoid complications, the subscript p indicating the pattern number is omitted hereafter except for J_p. A constant term $1/2$ in the equation is to simplify the subsequent formulas.

$$\varepsilon_j \overset{\text{def}}{=} \frac{\partial J_p}{\partial h_j}. \tag{12.15}$$

Also, by using Eq. (12.2), the second term $\partial h_j / \partial w_{ij}$ on the right side of Eq. (12.14) is

$$\frac{\partial h_j}{\partial w_{ij}} = g_i. \tag{12.16}$$

From Eqs. (12.14), (12.15), and (12.16), Eq. (12.12) representing the weight modification can be written as follows:

$$w'_{ij} = w_{ij} - \rho \cdot \varepsilon_j \cdot g_i \qquad (i = 0, \ldots, I \quad j = 1, \ldots, J). \tag{12.17}$$

Now, let's take a closer look at ε_j that appears in the above equation. From Eq. (12.15), ε_j can be written as

$$\varepsilon_j = \frac{\partial J_p}{\partial h_j} = \frac{\partial J_p}{\partial g_j} \cdot \frac{\partial g_j}{\partial h_j} \tag{12.18}$$

$$= \frac{\partial J_p}{\partial g_j} f'(h_j). \tag{12.19}$$

We used the fact that $\partial g_j / \partial h_j$ in Eq. (12.18) can be written as

$$\frac{\partial g_j}{\partial h_j} = f'(h_j), \tag{12.20}$$

since it is the derivative by h_j of the activation function $g_j = f(h_j)$ shown in Eq. (12.3). The partial derivative $\partial J_p / \partial g_j$ in Eq. (12.19) is calculated differently depending on whether the l-th layer is the output layer or the internal layer. So, consider two separate cases as follows.

Case 1: The l-th Layer Is the Output Layer ($l = L$)
This is the case where the unit j is in the output layer. As shown in Eq. (12.13), the evaluation measure J_p is defined using the output g_m from the output layer and is directly differentiable by g_m. Thus, by replacing m in Eq. (12.13) with j, $\partial J_p / \partial g_j$ in Eq. (12.19) becomes

$$\frac{\partial J_p}{\partial g_j} = g_j - b_j. \tag{12.21}$$

As a result, the following equation is obtained from Eq. (12.19):

$$\varepsilon_j = (g_j - b_j) f'(h_j) \tag{12.22}$$

Case 2: The l-th Layer Is the Internal Layer ($2 \leq l \leq L - 1$)

This is the case where unit j is in the internal layer. In this case, $\partial J_p / \partial g_j$ is represented using the *chain rule* of the partial derivative including unit k in the $(l + 1)$-th layer. The result is as follows:

$$\frac{\partial J_p}{\partial g_j} = \sum_{k=1}^{K} \frac{\partial J_p}{\partial h_k} \cdot \frac{\partial h_k}{\partial g_j}. \tag{12.23}$$

In the same way as in Eq. (12.15), ε_k can be defined by the following equation:

$$\varepsilon_k \stackrel{\text{def}}{=} \frac{\partial J_p}{\partial h_k}. \tag{12.24}$$

The following equation is also obtained from Eq. (12.4):

$$\frac{\partial h_k}{\partial g_j} = w_{jk}. \tag{12.25}$$

Substituting these into Eq. (12.23) yields

$$\frac{\partial J_p}{\partial g_j} = \sum_{k=1}^{K} \varepsilon_k \, w_{jk}, \tag{12.26}$$

then, from Eq. (12.19), the following equation is obtained:

$$\varepsilon_j = \left(\sum_{k=1}^{K} \varepsilon_k \, w_{jk} \right) f'(h_j) \qquad (j = 1, \dots, J). \tag{12.27}$$

What is noteworthy about the above equation is that ε_j on the left side appears as ε_k on the right side. That is, ε_j in the l-th layer is recursively obtained using ε_k in the next $(l + 1)$-th layer. Thus, ε_i in the $(l - 1)$-th layer can be obtained using ε_j in the l-th layer as

$$\varepsilon_i = \left(\sum_{j=1}^{J} \varepsilon_j \, w_{ij} \right) f'(h_i) \qquad (i = 1, \dots, I). \tag{12.28}$$

So far, we have described the learning method for the neural network. In order to actually apply this learning method, the activation function $f(\cdot)$ contained in Eqs. (12.22) and (12.27) must be defined. The activation function will be described in detail in Sect. 12.3. Here, we leave it as $f(\cdot)$ for the time being, and summarize the learning method of the weight w_{ij} as follows:

$$w'_{ij} = w_{ij} - \rho \, \varepsilon_j \, g_i \tag{12.29}$$

$$\varepsilon_j = (g_j - b_j) f'(h_j) \qquad \text{(the } l\text{-th layer is the output layer)} \tag{12.30}$$

$$\varepsilon_j = \left(\sum_{k=1}^{K} \varepsilon_k \, w_{jk} \right) f'(h_j) \qquad \text{(the } l\text{-th layer is the internal layer)} \tag{12.31}$$

What is noteworthy here is the ε_j obtained for each unit. This is related to the difference between the output of the neural network and the teaching signal, as will become clear in the next section. Therefore, ε_j is called the *error* of unit j in the l-th layer.

The weight modification procedure of the BP method is shown below. The one-hot vector of Eq. (12.7) is used as the teaching signal.

Backpropagation Method

Step 1   Prepare learning patterns together with their teaching signals.
Step 2   Set initial values of weights.
Step 3   Select one of the learning patterns and classify it by the neural network.
Step 4   Compare the output g_m from each unit in the output layer with the teaching signal b_m ($m = 1, \ldots, c$), and if they match,[6] return to Step 3, otherwise go to Step 5. If g_m and b_m match for all learning patterns, the process terminates.
Step 5   Using the output g_m and the teaching signal b_m, calculate the error ε_m in the output layer by Eq. (12.30) (replace j in Eq. (12.30) with m).
Step 6   Modify the weights between the $(L-1)$-th and L-th layers by Eq. (12.29).
Step 7   By Eq. (12.31), calculate the error ε_j in the $(L-1)$-th layer using the error ε_m obtained above and the modified weights.
Step 8   Modify the weights between the $(L-2)$-th and $(L-1)$-th layers by Eq. (12.29) in the same way as above.
Step 9   For each layer prior to the $(L-2)$-th layer, the weights are modified by Eq. (12.29) calculating the error recursively by Eq. (12.31).
Step 10   After completing the weight modification between the 2nd and 1st layers, return to Step 3 and repeat the same process.

The reason why this method is called the backpropagation method is that the weights are modified by propagating the error ε_j from the output layer to the input layer in the reverse direction. On the other hand, the signal of the classification

---

[6] Usually, the output g_m does not exactly match the teaching signal b_m. Therefore, when judging convergence, for example, g_m is regarded as 0 if $0 \leq g_m \leq 0.1$ and g_m is regarded as 1 if $0.9 \leq g_m \leq 1$.

process is propagated from the input layer to the output layer, and this signal flow is called the *forward propagation*.[7]

Comparing Eq. (12.29) of the BP method with Eq. (3.30) of the Widrow–Hoff learning rule for two-layer networks, namely the delta rule, the former is extended to a more generalized multi-layer network. In this sense, the former is sometimes referred to as *generalized delta rule* (Rumelhart and McClelland 1986).

The BP method was explained above using online learning as an example. For batch learning, simply replace Eq. (3.24) with Eq. (3.23).

12.3   Activation Function and Evaluation Measure

The above described the processing details of the BP method. To actually use this method, the activation function $f(\cdot)$ that appears in Eqs. (12.30) and (12.31) must be defined. As the evaluation measure J_p, we used the square error defined by Eq. (12.13), but it is not limited to this. In the following, specific examples of activation functions and evaluation measures other than the square error are described.

Various activation functions have been proposed. The sigmoid function, the rectified linear function, and the softmax function are discussed below.

12.3.1   Sigmoid Function

The most well-known activation function is the following *sigmoid function*, shown by the thick solid line in Fig. 12.2:

$$f(u) = \frac{1}{1 + \exp(-u)}. \tag{12.32}$$

Originally, $f(u)$ should be the threshold function indicated by the thick dotted line in the figure, but this function is not differentiable. Therefore, the sigmoid function is used as a differentiable nonlinear function to approximate the threshold function. The sigmoid function has the following important properties:

$$f'(u) = \frac{\exp(-u)}{(1 + \exp(-u))^2} \tag{12.33}$$

$$= f(u)(1 - f(u)). \tag{12.34}$$

---

[7] Forward propagation, changes the first index of the weights as shown in Eq. (12.2). BP, on the other hand, changes the second index, as shown in Eq. (12.27).

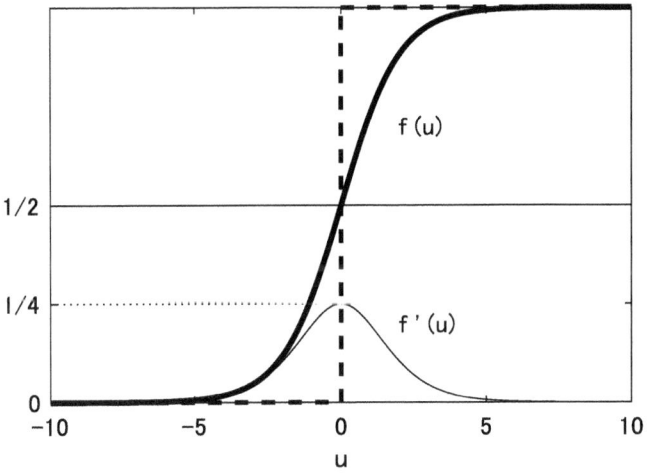

Fig. 12.2 Sigmoid function $f(u)$

If $f(\cdot)$ of Eq. (12.3) is the sigmoid function, the following equation holds from Eq. (12.34):

$$f'(h_j) = f(h_j)\left(1 - f(h_j)\right) \tag{12.35}$$

$$= g_j(1 - g_j). \tag{12.36}$$

Furthermore, using the square error shown in Eq. (12.13) as the evaluation measure J_p, Eqs. (12.30) and (12.31) can be written as follows, respectively:

$$\varepsilon_j = (g_j - b_j)g_j(1 - g_j) \qquad \text{(the } l\text{-th layer is the output layer)}, \tag{12.37}$$

$$\varepsilon_j = \left(\sum_{k=1}^{K} \varepsilon_k\, w_{jk}\right) g_j(1 - g_j) \quad \text{(the } l\text{-th layer is the internal layer)}. \tag{12.38}$$

With the above ε_j, the weights are modified by Eq. (12.29).

In Eqs. (12.37) and (12.38), $0 < g_j(1 - g_j) < 1$. As is clear from Eq. (12.29), the modification of the weights is the largest when the output value g_j is 0.5, and the modification becomes smaller as g_j approaches 0 or 1. The characteristic that the modification becomes larger when the output g_j deviates from 0 and 1 seems to contribute to the stabilization of the resulting neural network.

Here, we mention an important problem posed by the sigmoid function. Using ε_j defined in the l-th layer, let us find ε_i in the previous $(l - 1)$-th layer. The error ε_j of unit j in the l-th layer is represented by Eq. (12.31). Similarly, from Eq. (12.28), the error ε_i of unit i in the previous $(l - 1)$-th layer can be written as

$$\varepsilon_i = f'(h_i) \sum_{j=1}^{J} \varepsilon_j w_{ij} \tag{12.39}$$

$$= f'(h_i) \sum_{j=1}^{J} \left(f'(h_j) \sum_{k=1}^{K} \varepsilon_k w_{jk} \right) w_{ij} \tag{12.40}$$

$$= \sum_{j=1}^{J} \sum_{k=1}^{K} f'(h_i) f'(h_j) \varepsilon_k w_{ij} w_{jk} \qquad (i = 1, \ldots, I). \tag{12.41}$$

It is clear from Eq. (12.32) and Fig. 12.2 that the range of $f(u)$ is

$$0 < f(u) < 1. \tag{12.42}$$

Also, from the relation between $f(u)$ and $f'(u)$ shown in Eq. (12.34), the following formula is valid:

$$0 < f'(u) \le 1/4. \tag{12.43}$$

That is, the value of $f'(u)$ is positive and remains at most $1/4$. This can be confirmed by the graph (thin line in Fig. 12.2) plotting $f'(u)$ against u using Eq. (12.33).

Let us look at Eq. (12.41). In determining the error ε_j of the l-th layer, $f'(h_j)$ is computed. Next, $f'(h_j)$ is multiplied by $f'(h_i)$ when recursively calculating the error ε_i in the previous $(l-1)$-th layer. As a result, Eq. (12.41) contains the term $f'(h_i) f'(h_j)$. Each time this operation is repeated toward the input layer, $f'(\cdot)$ is multiplied, and this term becomes $f'(\cdot)^r$ after going back r layers. If the neural network has about 3 layers, this term is not so small, but when it has multiple layers, even if $f'(\cdot)$ has the maximum value of $1/4$, it is extremely small as $(1/4)^5 \approx 0.000977$, $(1/4)^7 \approx 0.000061$, $(1/4)^{10} \approx 0.000001$. As a result, the error term of Eq. (12.29) becomes $\varepsilon_j \approx 0$ in the layer close to the input layer, and the weight modification hardly works.

Although a highly advanced classifier can be realized by increasing the number of layers in a neural network, this phenomenon is a major obstacle in realizing the learning method for such a neural network. This phenomenon is called the *vanishing gradient problem*, and it caused the second neural boom to slow down. One of the biggest problems with using a sigmoid function as the activation function is the vanishing gradient problem. This problem has been solved by new methods of deep learning, such as *pretraining* (Goodfellow et al. 2016) and the improvement of the activation function described below.

12.3.2 Rectified Linear Function (ReLU Function)

Several methods have been proposed to avoid the gradient vanishing problem, one of which is to use an activation function other than the sigmoid function. One of the candidates is the *rectified linear function*, which was proposed in the process of the deep learning research. The form of the function is expressed in the following equation and its shape is shown in Fig. 12.3:

$$f(u) = \begin{cases} u & (u \geq 0) \\ 0 & (u < 0). \end{cases} \tag{12.44}$$

Fig. 12.3 ReLU function

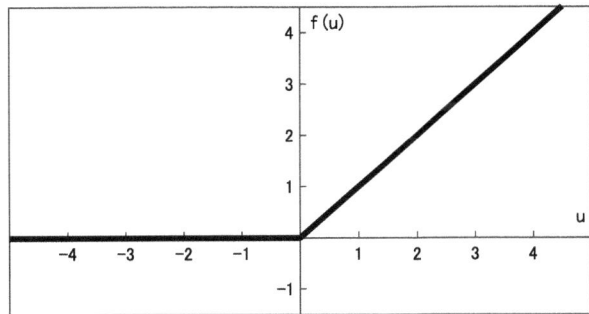

A unit with such a property is called the *rectified linear unit*, and the rectified linear function is sometimes called the *ReLU function*. If $f(u) = u$ over the whole domain, $f(u)$ is just an identity map and a linear function, but $f(u)$ is a nonlinear function since it becomes 0 for $u < 0$.

We describe the weight modification when the rectified linear function is used as the activation function. In the case of the rectified linear function, Eqs. (12.30) and (12.31) are still applicable. Assuming that the square error of Eq. (12.13) is used as the evaluation measure J_p as before, then Eq. (12.21) holds.

On the other hand, from Eq. (12.44), the differential term $f'(h_j)$ is expressed as

$$f'(h_j) = \begin{cases} 1 & (h_j \geq 0) \\ 0 & (h_j < 0), \end{cases} \tag{12.45}$$

so using the rectified linear function, Eqs. (12.30) and (12.31) can be written as the following equations respectively:

$$\varepsilon_j = \begin{cases} g_j - b_j & (h_j \geq 0) \\ 0 & (h_j < 0), \end{cases} \qquad \text{(the } l\text{-th layer is the output layer)} \tag{12.46}$$

$$\varepsilon_j = \begin{cases} \displaystyle\sum_{k=1}^{K} \varepsilon_k w_{jk} & (h_j \geq 0) \\ 0 & (h_j < 0). \end{cases} \quad \text{(the } l\text{-th layer is the internal layer)} \quad (12.47)$$

For weight modification, Eq. (12.29) can be used here again.

If the activation function $f(h_j)$ is the rectified linear function, then Eq. (12.45) holds, so the error term does not become extremely small even after multilayering, and the gradient vanishing problem does not occur.

12.3.3   Softmax Function

The sigmoid function and the rectified linear function introduced so far as activation functions can be used for both internal and output layers. However, the softmax function introduced here is, in principle, used only in the output layer and is particularly suitable as an activation function when dealing with multi-class problems.

Using the softmax function, if h_m is the input to unit m in the output layer (the L-th layer), the output g_m from the same unit is represented by

$$g_m = f(h_1, \ldots, h_m, \ldots, h_c)$$

$$= \frac{\exp(h_m)}{\displaystyle\sum_{r=1}^{c} \exp(h_r)} \quad (m = 1, \ldots, c). \quad (12.48)$$

This $f(h_1, \ldots, h_m, \ldots, h_c)$ is called the *softmax function*. The activation functions introduced so far are univariate functions, as shown in Eq. (12.3). In contrast, the softmax function is a function of all unit inputs $h_1, \ldots, h_c$, as shown in Eq. (12.48). As is clear from the equation,

$$\sum_{m=1}^{c} g_m = 1 \quad (12.49)$$

holds, so the output g_m can be regarded as the probability that the input pattern belongs to the class ω_m. The evaluation measure J_p when the softmax function is used as the activation function of the output layer is described below.

In the learning process, the weights are modified so that the output g_m is as close as possible to the teaching signal b_m. Let us now use b_m of Eq. (12.8) as the teaching signal. The one-hot vector of Eq. (12.7) is included as a special case of Eq. (12.8). Since the output g_m satisfies Eq. (12.49) and b_m satisfies Eq. (12.8), both can be treated as probabilities. Let us express the teaching signal and the output by vectors as follows:

$$\mathbf{t} = (b_1, \ldots, b_c)^t, \tag{12.50}$$

$$\boldsymbol{g}(\mathbf{w}) = (g_1, \ldots, g_c)^t, \tag{12.51}$$

where $\mathbf{t}$ in Eq. (12.50) is the teaching vector. In Eq. (12.51), $\mathbf{w}$ represents the total weights of the neural network, and $\boldsymbol{g}(\mathbf{w})$ indicates that the output is a function of $\mathbf{w}$.[8] Equation (12.51) shows that the weight $\mathbf{w}$ is adjusted by learning, and that $g_1, \ldots, g_c$ are determined as a result.

Suppose that a pattern x is repeatedly classified N times. The teaching signal b_m is considered to be the normalized value of the decision frequency of each class in the range of 0 to 1. Therefore, if the classifier can accurately realize the teaching signal as it is, the number of times x is classified as ω_m should be Nb_m out of N $(m = 1, \ldots, c)$.

Let us consider the case where the classifier outputting equation Eq. (12.51) classifies the pattern x N times. In this case, the probability $P(\mathbf{t}; \boldsymbol{g}(\mathbf{w}))$ of classifying x as ω_m $(m = 1, \ldots, c)$ is as follows:

$$P(\mathbf{t}; \boldsymbol{g}(\mathbf{w})) = \prod_{m=1}^{c} g_m^{Nb_m}. \tag{12.52}$$

The above expression $P(\mathbf{t}; \boldsymbol{g}(\mathbf{w}))$ is a function of $\mathbf{w}$ indicating how closely the output of this classifier reflects the teaching signal. In other words, $P(\mathbf{t}; \boldsymbol{g}(\mathbf{w}))$ represents the likelihood[9] introduced in Sect. 4.2. In learning, the likelihood $P(\mathbf{t}; \boldsymbol{g}(\mathbf{w}))$ is used as the evaluation measure, and the weights are modified so that the output $\boldsymbol{g}(\mathbf{w})$ maximizes the likelihood (see Sect. 4.2). Taking the logarithm of Eq. (12.52) yields[10]

$$\log P(\mathbf{t}; \boldsymbol{g}(\mathbf{w})) = N \sum_{m=1}^{c} b_m \log g_m. \tag{12.53}$$

Dividing the right hand side of the above equation by N and writing the term with the negative sign as $H(\mathbf{t}, \boldsymbol{g}(\mathbf{w}))$, we obtain the following equation:

$$H(\mathbf{t}, \boldsymbol{g}(\mathbf{w})) \stackrel{\text{def}}{=} -\sum_{m=1}^{c} b_m \log g_m. \tag{12.54}$$

The above equation is known as the *cross entropy*.

---

[8] Strictly speaking, Eq. (12.51) should be written as $\boldsymbol{g}(\mathbf{w}) = (g_1(\mathbf{w}), \ldots, g_c(\mathbf{w}))^t$, but $\mathbf{w}$ is omitted on the right side because it would be too complicated.

[9] Since the likelihood $P(\mathbf{t}; \boldsymbol{g}(\mathbf{w}))$ does not represent the conditional probability, it is not appropriate to write $P(\mathbf{t}| \boldsymbol{g}(\mathbf{w}))$.

[10] The left-hand side of Eq. (12.53) is called the *log likelihood*.

Maximizing the likelihood $P(\mathbf{t}; \ g(\mathbf{w}))$ is equivalent to minimizing the cross entropy $H(\mathbf{t}, \ g(\mathbf{w}))$. The vector $g(\mathbf{w})$ that minimizes the cross entropy is $g(\mathbf{w}) = \mathbf{t}$ (see Problem 12.1). From the above, the cross entropy can be defined as the evaluation measure J_p as follows:

$$J_p = H(\mathbf{t}, \ g(\mathbf{w})) = -\sum_{m=1}^{c} b_m \log g_m. \tag{12.55}$$

When the softmax function is used as the activation function of the output layer and the cross entropy is used as the evaluation measure, a simple calculation confirms that the error ε_m in the output layer is expressed as follows (see Problem 12.2):

$$\varepsilon_m = g_m - b_m. \tag{12.56}$$

It can be seen that the above equation has a simpler form compared to Eq. (12.37), which is the result of using the sigmoid function and the square error.

The error in the layer before the output layer is represented by Eq. (12.38) if the activation function is the sigmoid function, and represented by Eq. (12.47) if the activation function is the rectified linear function. The modification of the weights is performed as before by Eq. (12.29).

There is a point to note when using softmax functions. The softmax function, Eq. (12.48), does not change if we multiply the denominator and the numerator by the same constant. Therefore, if a is a constant, multiplying the denominator and the numerator by $\exp(a)$ yields

$$g_m = \frac{\exp(a)\exp(h_m)}{\exp(a)\sum_{r=1}^{c}\exp(h_r)} = \frac{\exp(h_m + a)}{\sum_{r=1}^{c}\exp(h_r + a)}. \tag{12.57}$$

The above equation shows that instead of the input h_m ($m = 1, \ldots, c$) to each unit of the output layer, the same output can be obtained by using $h_m + a$, which is obtained by uniformly adding a constant a. As a result, the weights are not uniquely determined, which causes a problem that the learning efficiency decreases. To eliminate such redundancy, some constraint is necessary. One way to do this is to force one of the inputs $h_i (i = 1, \ldots, c)$ to zero, for example by always setting the constant a to $a = -h_i$.

As examples of activation functions, we introduced the sigmoid function, the rectified linear function, and the softmax function. We also introduced not only the square error of Eq. (12.13) but also the cross entropy of Eq. (12.55) as the evaluation measure J_p. Here, looking at Eqs. (12.30), (12.46), and (12.56) of ε_j and ε_m derived from each activation function, all contain $(g_m - b_m)$ corresponding to the difference between the output and the teaching signal. This is the reason why ε_j and ε_m are called errors.

Coffee Break

Light and Shadow Brought by the Sigmoid Function

Needless to say, the second neural boom was sparked by the BP method proposed by Rumelhart. The perceptron, which was highlighted in the first neural boom, is a two-layer neural network, and it was known that it could be advanced by increasing the number of layers. However, it was not possible to devise a learning method for a multi-layer neural network, and it entered a long winter period. The activation function used at that time was a threshold function, which hindered the development of learning methods.

The sigmoid function proposed by Rumelhart as the activation function is truly a breakthrough. The sigmoid function approximates the threshold function, is globally differentiable, and has other properties suitable for learning processes, such as those found in Eq. (12.34). The sigmoid function was undoubtedly the core technology that supported the BP method.

However, as already mentioned, the sigmoid function had a very serious problem of the vanishing gradient problem. It was quite ironic that this breakthrough technology brought another winter to the neural boom.

12.4   Over-Fitting and Dropout

As already mentioned, one of the reasons why the second neural boom entered the winter period was the vanishing gradient problem. Another cause that must be mentioned is *over-fitting*. Over-fitting tends to occur in the following cases:

(1) There are few learning patterns.
(2) The dimensionality of the feature vectors is large.
(3) There are many parameters to be adjusted by learning.

In the second neural boom, many researchers aimed at high-dimensional features and multi-layer neural networks, using the BP method as a powerful tool. These efforts inevitably led to an increase in the number of parameters to be adjusted. Computational power at that time was poor compared to today, and was not equipped to process the huge number of learning patterns corresponding to the number of parameters. As a result, the decision boundaries for classifying a small number of learning patterns were approximated with high precision using a large number of parameters. The resulting classifier lacked the generalization ability, leading to the so-called over-fitting problem. In other words, it can be said that the efforts of the second neural boom have all of the above (1), (2), and (3). Various methods have been devised to prevent over-fitting, such as *early stopping*, regularization, *ensemble learning*, etc.

Dropout (Srivastava et al. 2014) introduced in this section is a technique which suppresses over-fitting. Dropout was devised in the course of research on deep learning. However, this is not a technique specific to the deep neural networks, and is applicable to the fully-connected neural networks discussed in this chapter. Therefore, in the following, dropout is explained using a three-layer neural network consisting of an input, an internal and an output layer, as shown in Fig. 12.4(a). In the figure, the number of units in the input, internal and output layers are set to 4, 4 and 3 respectively.

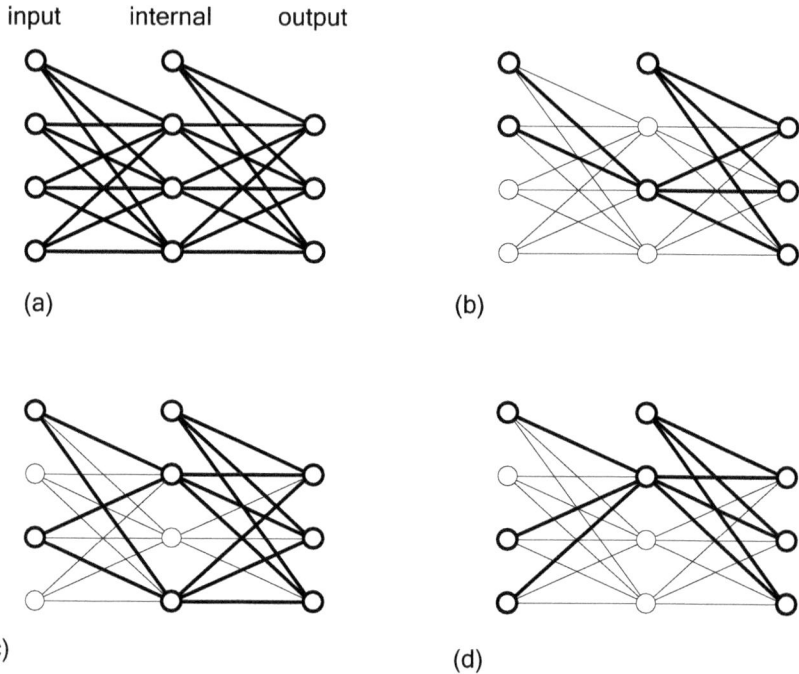

Fig. 12.4 Original network and partial networks obtained by dropout. (**a**) Original neural network. (**b**) Number of dropout units = (2, 2). (**c**) Number of dropout units = (2, 1). (**d**) Number of dropout units = (1, 2)

At each epoch, among the units in the input and internal layers, the units actually used are randomly selected with probability p_r. In other words, each unit is dropped out with probability $(1 - p_r)$. The first units in the input and internal layers are not subject to dropout. The probability p_r can be set to a different value for each layer, but here we use the same value for all layers for simplicity. This allows us to construct a partial network consisting of some units and some weight parameters extracted from the original network.

At a certain epoch, learning is performed on this partial network to modify the weight parameters. At the next epoch, the same process is repeated on a new partial network obtained by randomly re-selecting the units at each layer. Figure 12.4(b)–(d) show examples of partial networks resulting from the dropout. In the figures, the units constituting the partial network and the links joining them are shown in thick lines, while the dropped-out units and links are shown in thin lines. In (b), (c) and (d) of the figure, the captions indicate how many units were dropped out in the input and internal layers, respectively. For example, (2, 1) in (c) indicates that 2 and 1 units were dropped out in the input and internal layers, respectively.

When classification is performed using the weight parameters obtained by learning, all units are used. Therefore, during classification, the output from each unit (except the first unit) in the input and internal layers is multiplied by p_r in order to compensate for differences from the network structure used during learning. The results of the dropout experiments are presented by Fig. 13.20 in Sect. 13.5.5.

Ensemble learning, mentioned at the beginning of this section, is a technique where multiple classifiers are prepared, learned individually, and classification is performed by averaging the outputs from multiple classifiers. Dropout is closely related to the ensemble learning for the following reasons.

Dropout uses neural networks with different structures for each epoch, so it can be regarded as learning multiple classifiers. Also, since the outputs of all units are multiplied by p_r during classification, it can be regarded as averaging the outputs of multiple classifiers. In ensemble learning, multiple classifiers are required for both learning and classification, resulting in a huge amount of computation. In contrast, dropout achieves processing equivalent to ensemble learning by preparing only one type of neural network, and it can be said to be efficient.

─────────────────────── **Coffee Break** ───────────────────────

Generalization $\neq$ Extraction of General Laws

The concept of *generalization* is important in learning theory. In general, generalization is essentially the extraction of general rules that lie behind individual cases. In learning theory, however, generalization is used to express the degree to which a learning machine can produce correct outputs for unknown patterns. The deviation of the actual output from the true output is defined as the expected value of the square error between the two. This is called the *generalization error*. The smaller this error is, the higher the generalization ability. Therefore, whether the generalization ability is high or low is an argument about the size of the generalization error, and has nothing to do with such grandiose matters as the extraction of general laws.

The learning process is an approximation of the function between the input and output, in other words, it is a fitting problem. As an approximation method, perceptrons use a linear approximation, which is essentially the same method as the multiple regression analysis. Neural networks, on the other hand, are nonlinear approximations and can therefore approximate complex functions with high accuracy. However, this can be a disadvantage in terms of the generalization ability. That is, if a small number of learning patterns are approximated by a complex function, the output for unknown patterns will be unreliable. This is discussed again in Sect. 4.4.

Thus, it should be noted that the ability to approximate a function does not necessarily correspond to the ability to generalize, except when there are a large number of learning patterns.

12.5 Experiments on Three-Layer Neural Networks

In this section, in order to investigate the processing mechanism of neural networks in more detail, we take a neural network with a simple configuration and perform experiments applying it to small-scale learning data. The learning patterns are the following six patterns[11] distributed in a two-dimensional feature space that are shown in Eq. (3.66) of Problem 3.2:

$$\left.\begin{array}{lll} x_1 = (0, 5)^t, & x_2 = (1, 1)^t, & x_3 = (5, 0)^t \\ x_4 = (6, 2)^t, & x_5 = (2, 6)^t, & x_6 = (2, 2)^t \end{array}\right\}. \tag{12.58}$$

Of these, x_1, x_2, x_3 belong to class ω_1 and x_4, x_5, x_6 belong to class ω_2. These patterns are linearly nonseparable, as shown in the solution of Problem 3.2.

The structure of the neural network used in the experiment is shown in Fig. 12.5. The neural network is three-layered, with one internal layer between the input and output layers. The number of units in the input, internal and output layers is 3, 3 and 1 respectively, but the first units in the input and internal layers uniformly output 1, so the actual number of units is 2, 2 and 1 respectively. The sigmoid function is used as the activation function.

In the case of two classes ($c = 2$), as in this experiment, the number of units in the output layer is set to one instead of two, as shown in Fig. 12.5. The classification is performed by the following rule using the output g:[12]

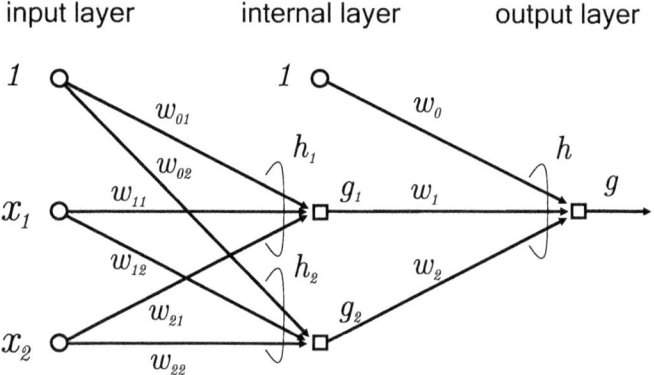

Fig. 12.5 Structure of three-layer neural network used in the experiment

---

[11] In the actual classifier design, when the number of learning patterns is small like this, it is not necessary to introduce a neural network. This example is only intended to illustrate the processing mechanism of a neural network.

[12] For a more advanced classification, the decision should be set strictly, for example

$x \in \omega_1$ if $g > 0.9$,
$x \in \omega_2$ if $g < 0.1$,

otherwise, it should be rejected .

$$\left.\begin{aligned}
g > 0.5 &\implies x \in \omega_1 \\
g < 0.5 &\implies x \in \omega_2 \\
g = 0.5 &\implies \text{reject.}
\end{aligned}\right\} \tag{12.59}$$

In this case, the weights are modified by setting the teaching signal as

$$b_p = \begin{cases} 1 & (x_p \in \omega_1) \\ 0 & (x_p \in \omega_2) \end{cases} \quad (p = 1, \ldots, n). \tag{12.60}$$

The classification process by this neural network is explained with reference to Fig. 12.5. First, for the input pattern $x = (x_1, x_2)^t$, the weights w_{ij} ($i = 0, 1, 2$, $j = 1, 2$) set between the input and internal layers are used to calculate h_1, h_2 as shown in the following formula:

$$\left.\begin{aligned}
h_1 &= w_{01} + w_{11}x_1 + w_{21}x_2 \\
h_2 &= w_{02} + w_{12}x_1 + w_{22}x_2.
\end{aligned}\right\} \tag{12.61}$$

As shown in the figure, these h_1 and h_2 are the inputs to the two units in the internal layer. As the two units in the internal layer are both threshold logic units, their outputs g_1, g_2 are obtained using the sigmoid function $f(u)$ of Eq. (12.32) as follows:

$$\left.\begin{aligned}
g_1 &= f(h_1) \\
g_2 &= f(h_2).
\end{aligned}\right\} \tag{12.62}$$

The input h to the unit in the output layer can be written using the weights w_0, w_1 and w_2 between the intermediate and output layers as

$$h = w_0 + w_1 g_1 + w_2 g_2, \tag{12.63}$$

and, in the same way as Eq. (12.62), the final output g is

$$g = f(h). \tag{12.64}$$

In learning, a total of nine weights, w_{ij} ($i = 0, 1, 2$, $j = 1, 2$) and w_k ($k = 0, 1, 2$), must be determined, as shown in the figure. The initial values of these weights were set by random numbers and online learning was applied with the learning rate of Eq. (12.29) as $\rho = 0.5$. Convergence was judged when the output was greater than 0.9 for all patterns of class ω_1 and less than 0.1 for all patterns of class ω_2, resulting in convergence at epoch 369. The process leading to convergence is shown in Fig. 12.6.

In the figures, the upper graph shows the transition of the square error J_a indicated by Eq. (12.9) and the lower graph shows the transition of the unit output g for the six learning patterns. The horizontal axis of both graphs shows the number

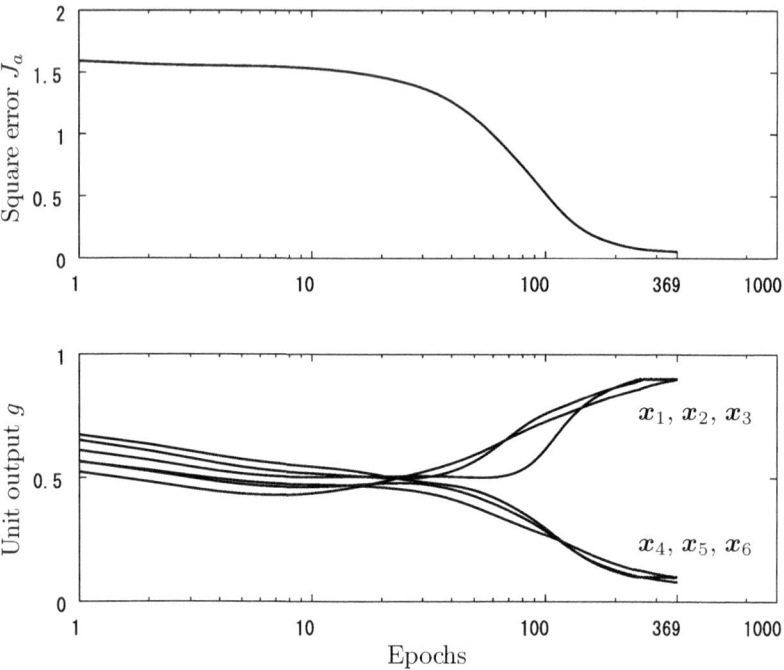

Fig. 12.6 Convergence process of three-layer neural network

of epochs on a logarithmic scale. The square error J_a decreases as the learning progresses with increasing number of epochs. The unit output g does not differ significantly between classes in the initial stage of learning. However, as the learning progresses, they approach 1 or 0, and at the convergence, they are clearly separated into x_1, x_2, and x_3 of class ω_1 and x_4, x_5, and x_6 of class ω_2.

The decision boundary obtained by this neural network is shown in Fig. 12.7, along with the learning patterns. The learning patterns for the classes ω_1 and ω_2 are indicated by black and white circles, respectively. From Eq. (12.59), the decision boundary is the trajectory of $x = (x_1, x_2)$ satisfying $g = 0.5$, which is indicated by the thick line in the figure. As can be seen from the figure, the decision boundary correctly separates the two linearly nonseparable classes.

For reference, the decision boundary obtained by applying the linear discriminant function to the same data is indicated by the dotted line in the figure. This decision boundary is the result obtained by the minimum square error learning described in Sect. 3.1 (Problem 3.2), which fails to separate the classes. Comparing the two, the utility of the neural network is obvious.

Fig. 12.7 Decision boundary
obtained by three-layer neural
network

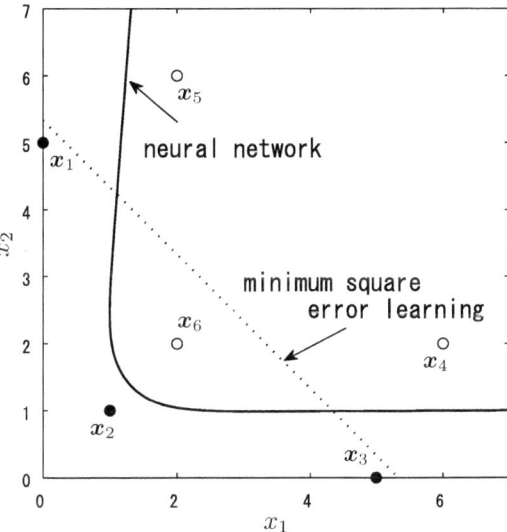

12.6   Role of Internal Layers

As can be seen from the experiments in the previous section, the BP of the neural
network can be used to obtain the decision boundary with no misclassification
even for linearly nonseparable distributions. In the following, we show that this
functionality is provided by the internal layers.

First, a three-layer neural network and then a multi-layer neural network are
discussed, and the role of the internal layers is verified through experiments.

12.6.1   Three-Layer Neural Networks

Let us focus on the internal and output layers shown in Fig. 12.5. As shown in the
figure, the outputs from the internal layer are 1, g_1 and g_2. The input h to the unit
in the output layer is obtained by a product-sum operation using $(1, g_1, g_2)$ and
weights (w_0, w_1, w_2), as shown in Eq. (12.63). The output g from the output
layer is obtained by thresholding h as shown in Eq. (12.64). If we consider the
outputs from the internal layer as new features, the operation performed between
the internal and output layers is nothing but a perceptron processing. This is evident
by comparing Eq. (12.63) with Eq. (2.4). Therefore, if learning patterns that were
linearly nonseparable in the original feature space are all correctly classified, as
in the example in the previous section, the distribution of new features from the
internal layer should be linearly separable. From the above, we can say that the
neural network applies the perceptron procedure after performing more advanced
feature extraction in the internal layer.

Let us confirm the above-mentioned points with the experimental example in the previous section. Figure 12.8(a) is a plot of (h_1, h_2) obtained by Eq. (12.61) on a two-dimensional space for the six learning patterns. This figure shows how each pattern is transformed and input to the internal layer, and the position of each pattern in (h_1, h_2) space is shown in the figure. As is clear from Eq. (12.61), since h_1, h_2 are just linear sums of the input x_1, x_2, any distributions that are linearly nonseparable in the original space (x_1, x_2) are also linearly nonseparable in the space (h_1, h_2). This can also be confirmed by Fig. 12.8(a).

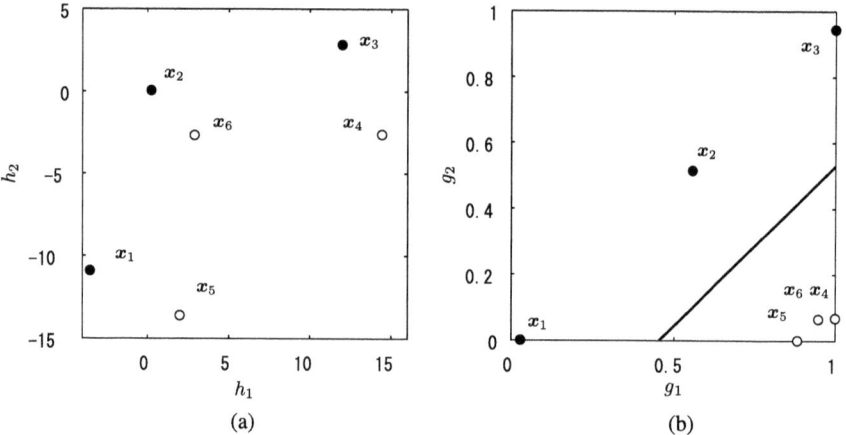

Fig. 12.8 Distribution of patterns in the internal layer (h_1, h_2). (**a**) Input to the internal layer (h_1, h_2). (**b**) Output from the internal layer (g_1, g_2)

Figure 12.8(b) is a similar plot of (g_1, g_2) obtained by Eq. (12.62). This figure shows how each pattern is represented as an output from the internal layer. Unlike Fig. 12.8(a), the distribution of the patterns is found to be linearly separable. Therefore, in this (g_1, g_2) space, both classes can be correctly separated by a linear discriminant function as follows.

That is, the decision boundary satisfies the following equation from Eqs. (12.59) and (12.64):

$$g = f(h) = 0.5. \tag{12.65}$$

From the form of the sigmoid function, $h = 0$ can be derived from the above equation. Therefore, from Eq. (12.63),

$$h = w_0 + w_1 g_1 + w_2 g_2 = 0 \tag{12.66}$$

is obtained as the decision boundary. The above equation represents a straight line in (g_1, g_2) space, corresponding to a linear discriminant function. The weights obtained as a result of the learning in this experiment are

$$(w_0, w_1, w_2) = (2.345, -5.165, 5.317). \tag{12.67}$$

The decision boundary obtained by substituting this value into Eq. (12.66) is indicated by the line in Fig. 12.8(b). The figure confirms that the two classes are correctly separated by this decision boundary. This decision boundary is mapped to the original (x_1, x_2) space, which is the nonlinear decision boundary shown in Fig. 12.7.

As already mentioned in Sect. 12.2, stacking multiple linear operations only achieves a linear operation and does not lead to improved classification ability. It is the nonlinear operations by the activation functions in the internal layers that improve the classification ability.

12.6.2 Multi-Layer Neural Networks

In the above experiment, the function and role of the internal layer were explained using a three-layer neural network as an example. In the experiment, the number of units in the internal layer was set to 3 (effectively 2), but more complex decision boundaries can be set if the number of units is increased. A similar effect can be achieved by increasing the number of layers. In the following, we discuss multi-layer neural networks with more layers.

What has been described so far for three-layer neural networks also applies to multi-layer neural networks. For example, consider an L-layer ($L > 3$) neural network. The processing between the output layer and the previous $(L-1)$-th layer is equivalent to that of the perceptron. As the processing progresses from the 2nd layer to the 3rd layer and to the 4th layer, the features extracted at each internal layer become more and more advanced. Since a perceptron is eventually applied, the learning of multi-layer neural networks aims at getting as close to a linearly separable distribution as possible at the $(L-1)$-th layer. Here, let us confirm by experiment how more advanced features are extracted as the layer gets deeper.

In the experiment, we used MSH784 introduced in Appendix D for learning, which consists of a total of 10,000 character patterns with $c = 10$ and $d = 784$. The neural network consists of five layers, and the number of units in each of the 2nd, 3rd, and 4th internal layers is 81. Therefore, the number of units from the input layer to the output layer is 784, 81, 81, 81, and 10, respectively. Note that these numbers exclude the first units that always output 1. The sigmoid function was used as the activation function. For learning, the learning rate was set to $\rho = 0.1$. Outputs above 0.9 and below 0.1 were considered to be 1 and 0, respectively, and were compared with teaching signals.

Under the above conditions, the weights of the neural network were determined by BP. The iteration was terminated at epoch 2000.[13] The classification rate for

---

[13] The learning can be repeated until all patterns match the teaching signals, but learning may be terminated when the square error becomes less than a certain value, or when the square error is no longer reduced.

the learning patterns was 99.9% (10 errors). (Incidentally, the classification rate for the test patterns MSH784-T consisting of 8000 character patterns was 93.1% (554 errors).)

As already mentioned, 81 dimensional features can be considered to be extracted from the input patterns at each of the three internal layers. What is needed is a method to evaluate the effectiveness of these features. In this experiment, the estimate of the Bayes error was used. For more details, refer to Sect. 5.6.2. The evaluation results are shown in Fig. 12.9. On the horizontal axis of the graph, 1, . . . , 5, correspond to the 1st to 5th layers of the neural network, and the vertical axis indicates the estimated Bayes error in % for the distribution at each layer. Note that the values for the 1st and 5th layers are calculated as 784-dimensional and 10-dimensional feature vectors, respectively, and these two values are for reference only.

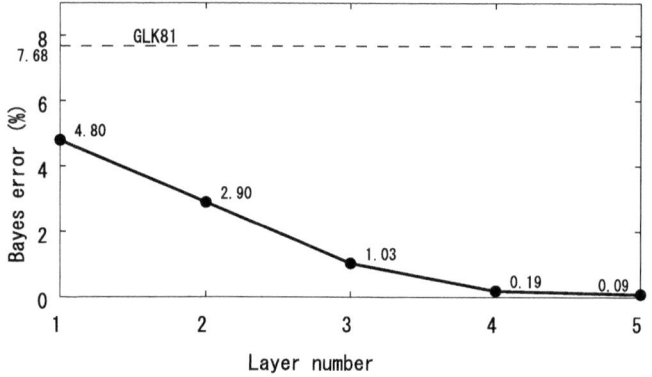

Fig. 12.9 Features that become more advanced with deepening layers

The graph shows that the Bayes error decreases to 2.90, 1.03, and 0.19% as the layers become deeper, such as the 2nd, 3rd, and 4th layers. It can be confirmed that the deeper the layer, the more advanced features are extracted. In the figure, the Bayes error (7.68%) obtained by applying Glucksman's 81-dimensional feature (GLK81) calculated from the learning patterns is also shown for comparison. The features extracted in the 2nd to 4th layers of the neural network are also 81-dimensional, but it is confirmed that they are far superior to GLK81. This result is reasonable because Glucksman's feature is originally considered to be good for printed character recognition but not for handwritten character recognition. For more information on Glucksman's feature, refer to Appendix C.

The role of internal layers of the neural network has been described above. It is known that even with only one internal layer, it is possible to set sufficiently complex decision boundaries by increasing the number of units. On the other hand, for some problems, it may be more efficient to increase the number of layers than to increase the number of units. Increasing the number of units or applying multilayering will increase the number of weights that can be adjusted. As a result, it is expected that neural networks can be made more advanced. On the other hand, however, these attempts cause various problems. For example, the amount of computation required for learning increases, there is a possibility of falling into a *locally optimal solution*,[14] the problem of over-fitting, the vanishing gradient problem, etc.

───────────────────── **Coffee Break** ─────────────────────

Nonlinear Transformation of Feature Space: Hourglass-Type Neural Network

As already mentioned, the KL expansion described in Sect. 6.3 is a method of data compression based on linear transformation, and is equivalent to the principal component analysis. On the other hand, as data compression based on nonlinear transformation, a method using a multi-layer neural network is known. This is, so to speak, a nonlinear version of the PCA.

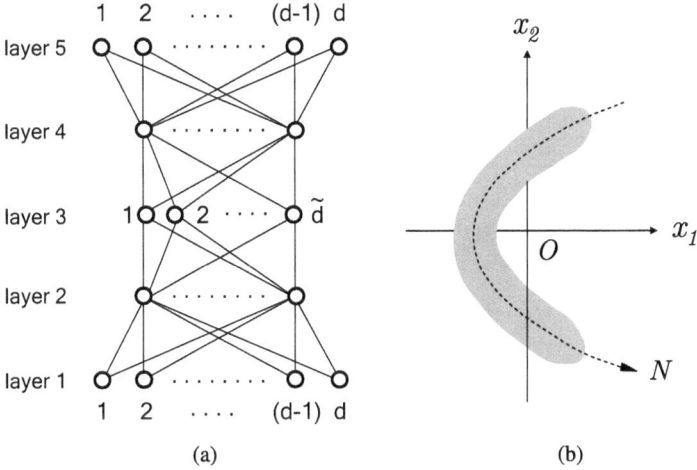

$$(a) \qquad\qquad (b)$$

Fig. 12.10 Examples of nonlinear spatial transformations. (**a**) Hourglass-type five-layer neural network. (**b**) Nonlinear subspace N

This is illustrated by the *hourglass-type neural network* in Fig. 12.10(a). This example shows a neural network consisting of five layers from the input layer (layer 1) to the output layer (layer 5). The input and output layers each have the same number of d units, and one of the internal layers (layer 3 in the figure) has $\tilde{d}(< d)$ units. This neural network is used to learn the identity mapping that produces the same output values as the input. Then, the outputs of this internal layer units after learning are represented as a $\tilde{d}$-dimensional vector, which can be regarded as the result of dimensionality reduction of the original d-dimensional feature vector. In this case, when the identity mapping is realized using a three-layer neural network, it has been proved that the ability to approximate the subspace obtained in the internal layer is inferior compared to the KL expansion (Cottrell and Munro 1988), so experiments are usually conducted using multi-layer neural networks with five or more layers. Using this nonlinear method, it is possible to determine the skewed axis

─────────────────────────

[14] For more information on multi-layer neural networks and locally optimal solutions, refer to the following coffee break.

N for feature vectors with a skewed distribution such as in Fig. 12.10(b). For exactly the same reason that the KL expansion is not necessarily effective for classification, the space transformed by an hourglass-type neural network is not guaranteed to be effective for classification.

The hourglass-type neural network is known as an *autoencoder*, and has become widely used in deep learning. In deep learning, a huge number of weights in a multi-layer neural network must be determined by learning. Conventionally, the initial values of the weights are set using random numbers, which causes the problem of time-consuming convergence. The method of setting the initial values of the weights by an autoencoder instead of random numbers has been introduced, and the time required for learning has been greatly reduced. Before obtaining the weights by learning, the initial values are learned in advance by the autoencoder, which is a method of pretraining. Since autoencoders and pretraining are not discussed in this book, refer to Goodfellow et al. (2016) for details.

Problems

12.1 Show that the cross-entropy $H(\mathbf{t}, \mathbf{g})$ of Eq. (12.54) is minimum at $\mathbf{g} = \mathbf{t}$.

12.2 Using the softmax function $f(h_1, \ldots, h_m, \ldots, h_c)$ of Eq. (12.48) as the activation function of the output layer and the cross-entropy $H(\mathbf{t}, \mathbf{g})$ of Eq. (12.55) as the evaluation measure J_p, show that ε_m in the output layer is

$$\varepsilon_m = g_m - b_m.$$

Chapter 13
Convolutional Neural Networks

Abstract As mentioned in the previous chapter, the second neural boom, like the first neural boom, also entered the winter period. The trigger to emerge from that long winter period was deep learning, which evoked another neural boom. What makes this boom different from previous ones is that high expectations are placed on its high practicality. Deep learning has solved various problems caused by increasing the number of layers in a neural network, and neural network research has made unprecedented progress. Among them, convolutional neural networks are the core technology of deep learning. Therefore, in this chapter, we focus on convolutional neural networks and discuss them emphatically.

13.1 Passage to Deep Learning

At the time of the second neural boom, although there were a considerable number of papers published, none of the proposals reached a level of performance that would allow them to be put to practical use. As a result, neural network research at the time was evaluated as "research at the paper level with little practical utility".

However, we must not forget that there were researchers who continued to work steadily in the face of such adversity. One of them is Geoffrey Hinton and another is Yann LeCun. The ILSVRC (ImageNet Large Scale Visual Recognition Challenge), which has been held since 2010 is a well-known international competition in image recognition. This competition is an opportunity to compete in image recognition methods using the huge image database ImageNet. In the 2012 competition, the neural network AlexNet (Krizhevsky et al. 2012) proposed by Hinton's group won with outstanding results. The *deep learning* technology he adopted suddenly came into the limelight and triggered the third neural boom.

By taking advantage of *big data* and dramatically increased computational power, Hinton found effective solutions to various problems that had been faced

Supplementary Information The online version contains supplementary material available at https://doi.org/10.1007/978-981-95-1478-6_13.

by multi-layer neural networks. While these contributions are of course significant, the greatest factor in the success of AlexNet is its network structure. The multi-layer neural network he proposed is based on *convolutional neural network*. Deep learning is a generic term for learning methods applied to multi-layer neural networks (four or more layers), and since the advent of AlexNet, most of the methods advocating deep learning have been designed based on convolutional neural networks. The multi-layer neural network used in deep learning is called a *deep neural network*.

This convolutional neural network was named LeNet by LeCun in 1989, 20 years before Hinton's AlexNet, and achieved a high classification rate when applied to handwritten character recognition (LeCun et al. 1989). However, going back another 10 years to LeNet, the basic idea of convolutional neural network has been developed by Kunihiko Fukushima, who proposed *neocognitron* (Fukushima 1980). LeCun also clearly states this in his aforementioned paper. In his own book (LeCun 2019), LeCun often cites Fukushima's work and highly praises his foresight, showing LeCun's deep respect for his efforts.

Deep learning has been applied not only to speech and image recognition, natural language understanding, and machine translation, but also to games such as Go and chess, and its practical value is being confirmed. Deep learning is still developing, and new methods and applications continue to be proposed and reported.

In such a situation, it is impossible to introduce the whole picture of deep learning in a limited number of pages. In this book, therefore, we focus on convolutional neural networks, which are the core technology of deep learning, and discuss them emphatically with experimental results. For more information on deep learning in general, refer to Goodfellow et al. (2016).

Coffee Break

The Year of the Deep Learning Revolution, 2012

It was already proved in the 1980s that a three-layer neural network consisting of an input layer, an internal layer, and an output layer could realize arbitrary mappings with arbitrary accuracy (Funahashi 1989). It was also expected that multi-layer neural networks would show higher potential. However, multi-layer neural networks did not always achieve the expected performance due to their huge number of parameters and the large amount of computation required for learning. For this reason, the number of researchers working with neural networks has decreased. Under such circumstances, it was Hinton of the University of Toronto who not only used multi-layer neural networks as tools, but also pursued their mathematical analysis and experimental verification based on their characteristics.

This triggered a great deal of attention to multi-layer neural networks based on deep learning, or deep neural networks (Hinton et al. 2012). A similar trend was occurring in parallel in the field of speech recognition. Researchers at Microsoft Research (MSR) published results on speech recognition using deep learning in 2011 (Seide et al. 2011), and in 2012, four research groups from the University of Toronto, MSR, Google, and IBM jointly wrote a review paper, which clarified the flow of the trend (Hinton et al. 2012).

What they all had in common was a dramatic improvement in classification performance, which had been improved little by little every year by many researchers. This impact quickly spread around the world and had a great impact beyond the boundaries of academic fields. In this sense, 2012 can be called the year of the deep learning revolution (Sejnowski 2018; LeCun 2019).

13.2   Structure of Convolutional Neural Network

An example of a convolutional neural network is shown in Fig. 13.1. This figure is a typical configuration example, and a number of variants exist in practice. As in the previous chapter, we deal with a multi-class problem with c classes.

A convolutional neural network consists of an input layer, *convolution layers*, *pooling layers*, and an output layer. The convolution and pooling layers are paired in this order, and there are multiple pairs between the input and output layers. Since these pairs are usually counted as a single layer, the figure shows an example of a five-layer convolutional neural network including the input and output layers. The signal flow is in one direction from the input layer to the output layer.

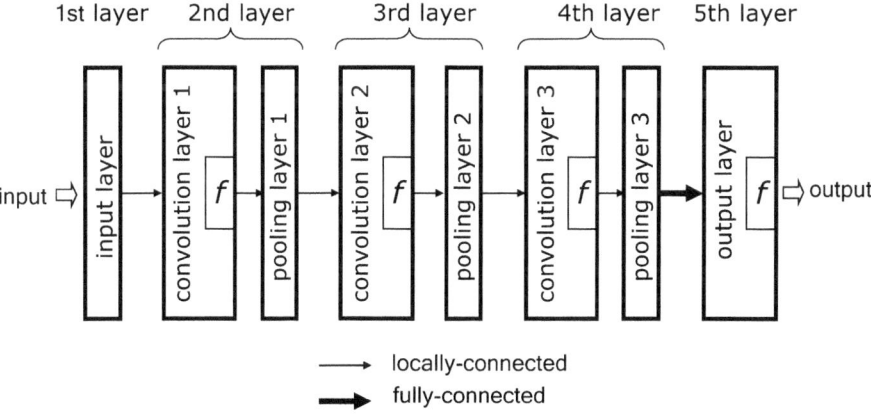

Fig. 13.1 Configuration example of a convolutional neural network

A pair of the convolution layer and the pooling layer corresponds to the internal layer in an ordinary neural network, but there is a significant difference. In conventional neural networks, the connections between adjacent layers are fully-connected. That is, each unit in a layer is connected to all units in the previous layer via weights. Convolutional neural networks, on the other hand, adopt *locally-connected* structure where each unit in a layer is only connected to some units in the previous layer. However, the connection between the output layer and the previous layer is fully-connected. In the figure, fully-connected and locally-connected connections are indicated by thick and thin lines, respectively. The output from the convolution layer is passed through a nonlinear activation function $f(\cdot)$ before being input to the pooling layer.

The weights between the layers are obtained by learning. However, the weights between the convolution layer and the pooling layer are fixed and are not subject to learning. Convolutional neural networks are particularly powerful for image recognition. Therefore, we will use a two-dimensional image as the input pattern.

13.3   Processing of Convolutional Neural Networks

Although the convolutional neural network is a new technology that supports deep learning, there is a long history leading up to its conception, as described in Sect. 13.1. In this section, we will again describe the historical background of this technology, going back to neurophysiological discoveries, followed by a concrete description of the processing of convolutional neural networks.

13.3.1   Insights from Neurophysiology

We mentioned that the convolution and pooling layers are the components that characterize the structure of convolutional neural networks. This basic structure is based on neurophysiological discoveries by David H. Hubel and Torsten N. Wiesel. They studied the visual cortex of cats and found that there are two basic cell types that respond selectively only when straight edges with specific orientations are presented in the visual field. One is the *simple cell* and the other is the *complex cell*. The former responds only when a pattern is presented at a specific position in the visual field, whereas the latter responds even if the pattern is slightly misaligned. In other words, it can be interpreted that the simple cell performs local feature extraction of the pattern, while the complex cell blurs the results to ensure robustness against misalignment.

Inspired by the neurophysiological findings of Heubel and Weisel, Fukushima proposed the neocognitron. The neocognitron has a multi-layered structure with a layer of simple cells paired with a layer of complex cells, and has a structure that can be called the prototype of the convolutional neural network. The layer of simple cells corresponds to the convolution layer, and the layer of complex cells corresponds to the pooling layer.

LeCun's LeNet used the backpropagation method to learn and optimize the weights of the hierarchical model proposed by the neocognitron. LeCun applied LeNet to handwritten character recognition and achieved a high classification performance. While following this basic structure, Hinton solved various problems that conventional neural networks had, such as the vanishing gradient problem and over-fitting, and proposed AlexNet, which recognizes general images.

13.3.2   Convolution Operation

The convolution and pooling layers perform *convolution* and *pooling* operations, respectively. The following describe these two operations in detail, starting with the convolution.

Let us consider the case where grayscale images are used for classification, and the original pattern is directly input for classification without any human-designed feature extraction method. In conventional neural networks, an input pattern is converted to a vector whose elements are the gray level values of each pixel. However, this method does not fully utilize the two-dimensional structure of the pattern. In general, there is a high correlation between neighboring pixels in an image, and their pixel values are similar to each other. Representing such an image as a vector loses the information about the adjacency relationship. To solve this drawback, convolutional neural networks perform their processing while preserving the two-dimensional structure of the image. The specific processing method for this is the convolution operation.

Suppose we have an image x of size $A \times A$. Traditionally, the image is represented as a d-dimensional vector

$$x = (x_1, \ldots, x_d)^t, \tag{13.1}$$

where $d = A \times A$. Note that x_i is the i-th pixel value.

On the other hand, in a convolutional neural network, an image is represented as a two-dimensional $\mathbf{X} = (x_{ij})$ with pixel values x_{ij}, as in Fig. 13.2. The figure shows an example of $A = 8$. For that, we prepare a *filter* $\mathbf{V} = (v_{kl})$ of size $F \times F$, which is smaller than the image. In the figure, v_{kl} is the weight assigned to the (k, l) component of the filter. The figure shows a filter with $F = 3$. To simplify the calculation, F is assumed to be an odd number.

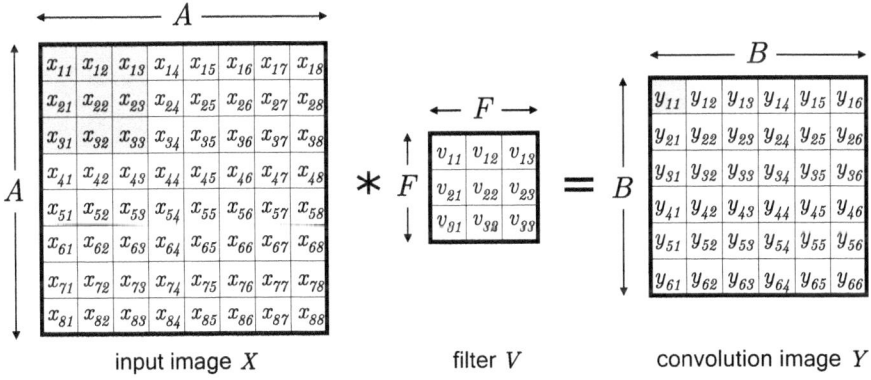

input image X filter V convolution image Y

Fig. 13.2 Operation of convolution

Convolution is a product-sum operation defined between the input image $\mathbf{X}$ and the filter $\mathbf{V}$, resulting in a convolution image $\mathbf{Y} = (y_{ij})$ of size $B \times B$, as follows:

$$y_{ij} = \sum_{k=1}^{F} \sum_{l=1}^{F} x_{i+k-1,\, j+l-1} \cdot v_{kl} \qquad (i, j = 1, \ldots, B). \tag{13.2}$$

That is, the image $\mathbf{Y}$ is obtained by overlaying the filter $\mathbf{V}$ on the image $\mathbf{X}$, scanning the image while shifting it by one pixel, and performing the operation Eq. (13.2) over the entire image. Figure 13.2 shows that the gray pixel y_{11} in the convolution image $\mathbf{Y}$ is obtained by applying Eq. (13.2) between the gray region in the input image $\mathbf{X}$ and the filter $\mathbf{V}$.

Convolution is a term originally used in the field of signal processing, and is used as the name of the above operation. Although the above formula differs from the original definition of convolution (see Eq. (13.37)), the operation of Eq. (13.2) is called convolution hereafter. This point is discussed in detail in Sect. 13.5.2.

Denoting the convolution operation by the symbol $*$, we can write Eq. (13.2) as

$$\mathbf{Y} = \mathbf{X} * \mathbf{V}. \tag{13.3}$$

The size B of the image $\mathbf{Y}$ is

$$B = A - F + 1, \tag{13.4}$$

which is always less than or equal to the size of the image $\mathbf{X}$. This is an unavoidable result as long as the filter is scanned within the image. The figure shows that $B = 8 - 3 + 1 = 6$, yielding an image $\mathbf{Y}$ of 6×6. In order to avoid changing the image size by convolution, we can add a P-pixel-wide band around the image to increase the image size to $(A + 2P) \times (A + 2P)$. This operation is called *padding* of width P. The value of P for which the size B of the convolution image is $B = A$ can be obtained by replacing A by $A + 2P$ in Eq. (13.4) as

$$B = A + 2P - F + 1 = A, \tag{13.5}$$

then

$$P = (F - 1)/2. \tag{13.6}$$

If no padding is performed, $P = 0$, and Eq. (13.4) holds. It is simplest to set the pixel value of the newly added P-pixel-wide band to 0, but this is not always optimal. There is no clear guideline as to whether or not padding should be applied (see Problem 13.1).

In the convolution described above, the filter was moved one pixel at a time. This amount of movement is called a *stride* . The stride need not be limited to 1, but can be set to 2, 3, etc. In this book, the stride in the convolution layer is set to 1 for simplicity.

In the convolutional layer, after the operation Eq. (13.2), a bias v_0 is added to each pixel, then the image is transformed by a nonlinear activation function $f(\cdot)$, and sent to the next pooling layer. The image $\mathbf{Y}' = (y'_{ij})$ obtained in this way has the same size as the image $\mathbf{Y}$ and can be written as follows:

$$y'_{ij} = f(y_{ij} + v_0). \tag{13.7}$$

The bias v_0 in the above equation is the same for all i, j. Now define an image $\mathbf{Y}_0$ of the same size as the image $\mathbf{Y}$ and all pixel values are v_0. The operation to obtain $\mathbf{Y}'$ from $\mathbf{Y}$ according to Eq. (13.7) is abbreviated as follows:

$$\mathbf{Y}' = f(\mathbf{Y} + \mathbf{Y}_0). \tag{13.8}$$

In the above equation, $f(\cdot)$ represents the operation of applying the activation function f to each pixel value of the image $(\mathbf{Y} + \mathbf{Y}_0)$ to obtain the corresponding pixel value of $\mathbf{Y}'$.

Often used as the activation function, $f(\cdot)$ is the sigmoid function. However, when the number of layers is increased, the vanishing gradient problem becomes more serious, as described in the previous chapter, so other activation functions, such as rectified linear functions, are often used in multi-layer neural networks. Each pixel in the convolution layer is not connected to all pixels $(A \times A)$ in the previous layer, but only locally-connected to some pixels $(F \times F)$. Furthermore, the pixels involved in local connectivity occupy a small area close to each other on the two-dimensional image, as is evident from Eq. (13.2). Therefore, the convolution operation shown in Eq. (13.2) can be regarded as a local feature extraction process. This is nothing but the function of the simple cell described earlier.

13.3.3 Pooling Operation

Next, the pooling operation is described. An image $\mathbf{Y}'$ with a size of $B \times B$ is sent to the pooling layer. In pooling, small regions of $H \times H$ in the image $\mathbf{Y}'$ are merged into a single pixel and the image $\mathbf{Y}'$ is transformed into an image $\mathbf{Z} = (z_{ij})$ of $G \times G$. Assume that $G = B/H$ and B is divisible by H. The pooling operation is shown in Fig. 13.3. The figure shows an example for $H = 2$, and the resulting image $\mathbf{Z}$ has size 3×3. In pooling operations, these small regions are usually set so that they do not overlap each other. That is, the stride is H equal to the size of the small regions. The figure shows that the gray region in the image $\mathbf{Y}'$ is merged and converted to the gray pixel z_{11} in the pooling image $\mathbf{Z}$.

The pixel value z_{ij} of the pooling image $\mathbf{Z}$ can be regarded as the representative value of the small region of $H \times H$, and there are several methods for setting it. The most commonly used method is to set the maximum value among $H \times H$ pixels as the representative value, which is called *max pooling*. If the small region $H \times H$ corresponding to the pixel value z_{ij} is represented by R_{ij}, z_{ij} can be written as

$$z_{ij} = \max_{(k,l) \in R_{ij}} y'_{kl}. \tag{13.9}$$

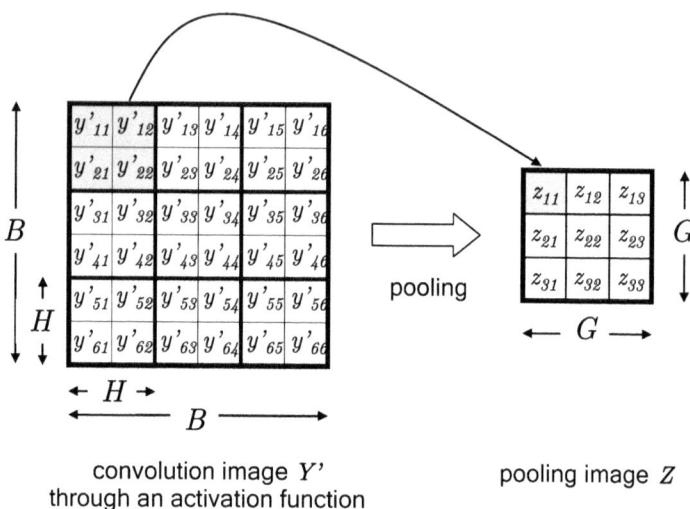

convolution image Y'
through an activation function

pooling image Z

Fig. 13.3 Pooling operation

There is another method called *average pooling* that uses the average value of $H \times H$ pixels as the representative value, which is expressed by the following equation:

$$z_{ij} = \frac{1}{H^2} \sum_{(k,l) \in R_{ij}} y'_{kl}. \tag{13.10}$$

In this book, we will use max pooling.

Here, we describe the function of the pooling. Let us take a handwritten character pattern as an example. A character consists of a combination of character strokes of specific lengths and slopes. For example, the character "Y" consists of diagonal strokes in the upper left and upper right, and a vertical stroke below them. Since the position of the strokes varies depending on writers, strictness is not required, as long as the rough position is correct.

The convolution operation can detect the presence of a specific feature at a specific position. However, a stable classifier cannot be realized if it is too sensitive to this position information. It can be said that the pooling operation reduces the sensitivity of the feature to positional fluctuations, thereby achieving robustness against positional fluctuations. This is the function of the complex cell mentioned earlier.

The connection between the convolutional and pooling layers can also be thought of as locally-connected. However, whether max pooling or average pooling, the weights of the connection are fixed and are not updated by learning.

As described above, the combination of the convolution and pooling operations is indeed an engineering realization of the neurophysiological findings by Hubel and Wiesel. Let us clarify from various perspectives that the basic functions of the convolutional neural networks are superior to those of conventional neural networks. Therefore, in order to facilitate comparison, we will use the shallowest three-layer neural networks below for both the convolutional neural network and the conventional neural network. The convolutional neural network used in the experiments is shown in Fig. 13.4 and will be called *CNN3*.[1] The conventional neural network used for comparison is shown in Fig. 13.5 and will be called *NNW3*. In Fig. 13.5, all connections between layers are fully-connected, and are indicated by thick lines.

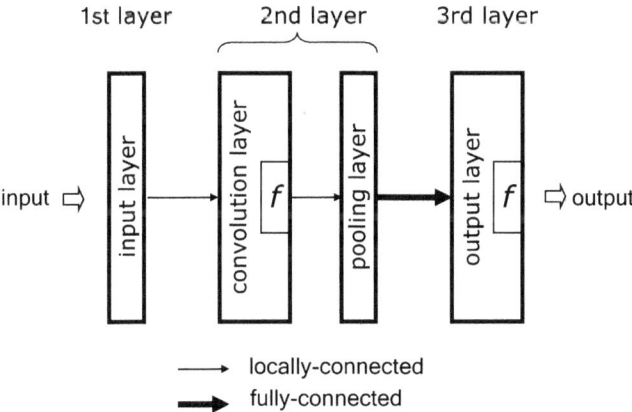

Fig. 13.4 Three-layer convolutional neural network (CNN3)

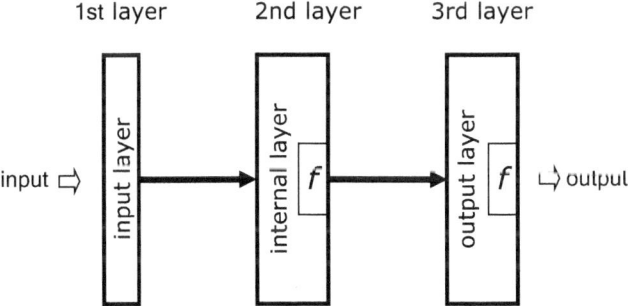

Fig. 13.5 Conventional three-layer neural network (NNW3)

Images are generally *multi-channel*. If the number of channels of the input image is expressed by K, $K = 3$ for a color image since three channels of R, G and B are required. The image to be treated below is a simple grayscale image with $K = 1$.

---

[1] The convolutional neural network is usually abbreviated as CNN, but LeCun himself did not like this name, and he used ConvNet instead (LeCun 2019).

13.3.4   Classification Operation of Convolutional Neural Network

Conventionally, the input to classifiers such as neural networks has been patterns after feature extraction, and feature extraction has been designed based on human intuition and experience. On the other hand, the input to the convolutional neural network is the original pattern itself, and an optimal convolution filter is obtained through learning. Since the convolutional operation can be considered as a feature extraction process using a filter, it can be said that convolutional neural networks can also learn feature extraction methods.

Since only one feature can be extracted by a single filter, multiple features can be extracted by using multiple filters to achieve a more advanced classifier. In the actual convolutional neural network, the M filters $\mathbf{V}_1, \ldots, \mathbf{V}_M$ are used to obtain M convolution images $\mathbf{Y}_1, \ldots, \mathbf{Y}_M$. They are transformed into images $\mathbf{Y}'_1, \ldots, \mathbf{Y}'_M$ by the activation function, and then M pooling images $\mathbf{Z}_1, \ldots, \mathbf{Z}_M$ are obtained.

It is clear from what has already been explained that the following relation holds. First, in the same way as in Eq. (13.3), we can write

$$\mathbf{Y}_m = \mathbf{X} * \mathbf{V}_m \qquad (m = 1, \ldots, M). \tag{13.11}$$

The process of obtaining the image $\mathbf{Y}'_m$ from the image $\mathbf{Y}_m$ is shown in Eqs. (13.7) and (13.8). That is, if $\mathbf{Y}_m = (y_{ijm})$ and $\mathbf{Y}'_m = (y'_{ijm})$, then as in Eq. (13.7),

$$y'_{ijm} = f(y_{ijm} + v_{0m}) \qquad (m = 1, \ldots, M). \tag{13.12}$$

Defining an image $\mathbf{Y}_{0m}$ with all pixel values v_{0m} and using the shorthand notation of Eq. (13.8), the above equation can be written as

$$\mathbf{Y}'_m = f(\mathbf{Y}_m + \mathbf{Y}_{0m}) \qquad (m = 1, \ldots, M). \tag{13.13}$$

In Fig. 13.6, the process up to the classification is shown using the example of CNN3 shown in Fig. 13.4. This figure shows an example where the number of filters is M and the input pattern's belonging class is determined to be one of the c classes $\omega_1, \ldots, \omega_c$.

The pooling images $\mathbf{Z}_1, \ldots, \mathbf{Z}_M$ obtained in the pooling layer are all expanded to one dimension and they are fully-connected to the units in the output layer. Pixel values in the pooling image are output directly from the pooling layer without passing through any activation function. If the number of outputs from the pooling layer is Q, then $Q = G^2 M$, and the number of units in the output layer is equal to the number of classes c. Let $g_1, \ldots, g_Q$ be the outputs from the pooling layer, and add g_0, which takes the value of 1 identically. Let $h_1, \ldots, h_c$ be the inputs to each unit of the output layer. Since the connection between the pooling layer and the output layer is fully-connected, using the weight w_{jk}, the following equation holds:

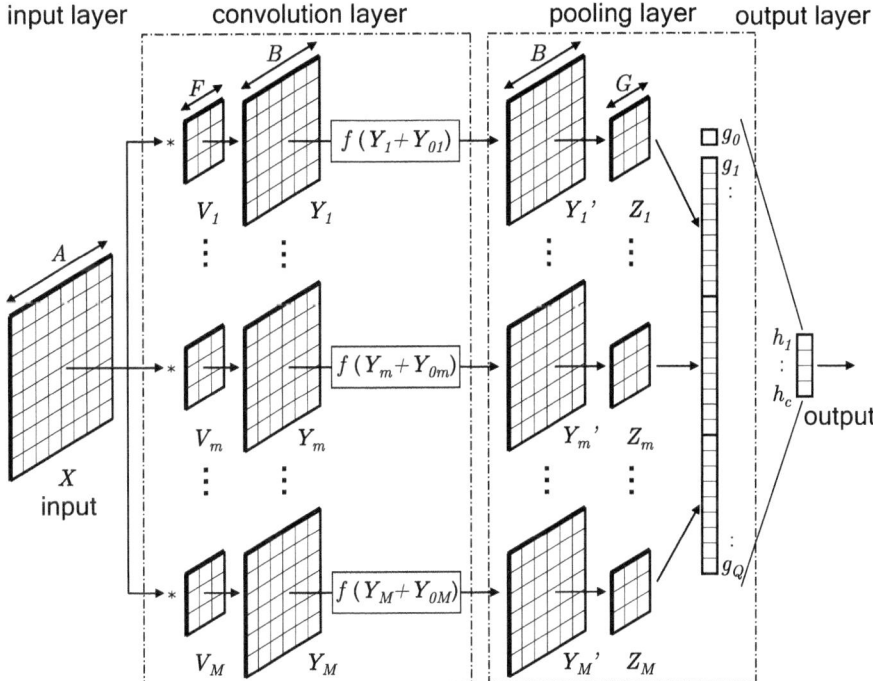

Fig. 13.6 Classification processing by CNN3

$$h_k = \sum_{j=0}^{Q} w_{jk} g_j \tag{13.14}$$

$$= w_{0k} + w_{1k} g_1 + \cdots + w_{Qk} g_Q \qquad (k = 1, \ldots, c). \tag{13.15}$$

In the equation, w_{0k} is the bias term. The classification follows the next equation:

$$\max_{k=1,\ldots,c} \{h_k\} = h_i \quad \Longrightarrow \quad \mathbf{X} \in \omega_i. \tag{13.16}$$

As with conventional neural networks, when dealing with multi-class problems, the softmax function is often used in the output layer during learning.

In the above, the classification process by the convolutional neural network was explained step by step. Here, the various symbols used so far are organized and summarized in Table 13.1. In addition, Eqs. (13.17) to (13.19) show the relational expressions among these symbols.

$$B = A + 2P - F + 1, \tag{13.17}$$

$$G = B/H, \tag{13.18}$$

$$Q = G^2 M. \tag{13.19}$$

Table 13.1 Symbols used

Symbol	Content
$A \times A$	Size of original image $\mathbf{X}$
$B \times B$	Size of convolution image $\mathbf{Y}$
$F \times F$	Size of convolution filter
$G \times G$	Size of pooling image $\mathbf{Z}$
$H \times H$	Size of small region for pooling
K	Number of channels of input image (here $K = 1$)
M	Number of convolutional filters
P	Padding width
Q	Number of outputs from pooling layer

Let us now examine the number of weights to be learned for the CNN3. The input layer and the convolutional layer of this network are locally-connected. The locally-connected weights are v_{kl} $(k, l = 1, \ldots, F)$ in Eq. (13.2) and v_0 in Eq. (13.7), which are shared in the process of computing all pixel values in the image $\mathbf{Y}$. This is called *weight sharing*. That is, the number of weights to be learned is $(F^2 + 1)$ per filter, so for M filters, the number is

$$N_1 = M(F^2 + 1). \tag{13.20}$$

The weights between the convolution and pooling layers are fixed and are not subject to learning.

On the other hand, the connection between the pooling layer and the output layer is fully-connected. As is clear from Fig. 13.6 and Eq. (13.15), the number N_2 of weights w_{jk} that should be learned is given by the following equation:

$$N_2 = (Q + 1)c \tag{13.21}$$

Next, the number of weights to be learned for NNW3 of Fig. 13.5 prepared for comparison is determined. The problem is how to set the number of units in the internal layer for a fair comparison. Since the internal layer in Fig. 13.5 corresponds to the "convolution layer + pooling layer" in Fig. 13.4, it seems reasonable to regard the number of outputs Q from the pooling layer as the number of units in the internal layer. In this case, the number of weights N_1' between the input layer and the internal layer is

$$N_1' = (A^2 + 1)Q, \tag{13.22}$$

and these weights are the learning target. The reason why " $+ 1$" is included in the above equation is due to the bias term.

Since the connection between the internal and output layers of NNW3 is also fully-connected, the number of weights that exist between these layers is N_2, which is the same as Eq. (13.21).

Comparing N_1 in Eq. (13.20) and $N_1{}'$ in Eq. (13.22), $N_1 \ll N_1{}'$, we can see that the number of weights to be learned for CNN3 is overwhelmingly small, which is advantageous. This tendency is especially noticeable when the image size A increases, because $N_1{}'$ increases in proportion to A^2, while N_1 is completely independent of A and remains constant.

Coffee Break

From Plane to Bird

"The human dream of flying in the sky was fulfilled not by imitating the flight mechanism of birds, but by propeller-driven airplanes. It is reasonable to approach pattern recognition research in the same way."

Many pattern recognition researchers have long relied on this attitude. In other words, the advanced pattern recognition ability of humans is regarded as a black box, and as long as the relationship between inputs and outputs is valid, the content of the recognition system does not matter. However, the third neural boom blossomed rather in the opposite direction to this attitude.

The second neural boom was driven by researchers who set their research goals on *PDP: parallel distributed processing*. This group includes D. Rumelhart, J. McClelland, G. Hinton, T. Sejnowski, and other distinguished members who supported the subsequent development of neural networks. Moreover, their specialties were very diverse, including physics, mathematics, computer science, neuroscience, molecular biology, and psychology.

Reading the book by one of the members, Sejnowski (2018), we can see that discussions with researchers in neuroscience, cognitive psychology, and other fields that study human beings were the driving force behind the birth of deep learning. Sejnowski himself is a researcher with a background in physics and neurobiology. It is interesting that the book vividly describes the history of neural network research from perceptron to deep learning, along with some unknown episodes.

The current boom in deep learning research was sparked by the discoveries of D. Hubel and T. Wiesel. Since then, there has been a growing interest in elucidating the functions and thinking mechanism of the human brain and in attempting to apply them to pattern recognition. It is interesting to note that researchers seem to be shifting their attention from airplanes to birds.

13.4 Learning Method of Convolutional Neural Network

The backpropagation method described in the previous chapter as a learning method for conventional neural networks can also be applied to convolutional neural networks. First, the backpropagation method can be directly applied to modify weights between layers that are fully-connected in a convolutional neural network. On the other hand, there are some points to be noted for the convolution and pooling layers. Let's take CNN3 shown in Fig. 13.6 as an example.

The images $\mathbf{Z}_1, \ldots, \mathbf{Z}_M$ obtained by pooling are expanded into $(Q + 1)$ scalar data $g_0, g_1, \ldots, g_Q$, which are directly output from the pooling layer without passing through any activation function. The input to the c units of the output layer is $h_1, \ldots, h_c$, which are fully-connected with the output from the pooling layer as already shown in Eq. (13.14), so the backpropagation method described so far is directly applicable.

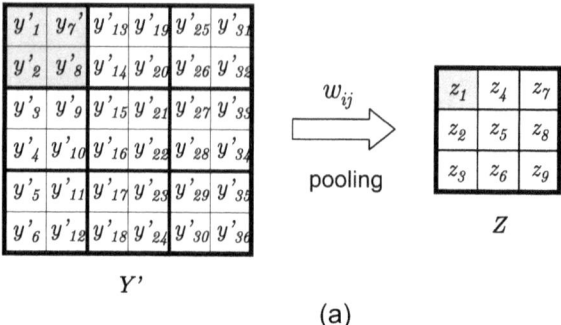

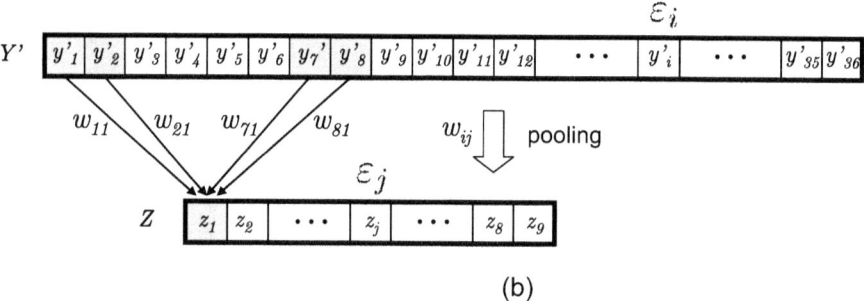

Fig. 13.7 Backpropagation process in the pooling layer. (**a**) Pooling operation (two-dimensional notation). (**b**) Pooling operation (vector notation)

The image $\mathbf{Y}'_m$ is transformed to the pooling image $\mathbf{Z}_m$ by means of locally-connected weights, which are fixed and are not modified in the learning process. However, backpropagation of errors is necessary. In the following, we take Fig. 13.3 as an example to illustrate the backpropagation process in the pooling layer. In Fig. 13.7(a), the components of the image $\mathbf{Y}'$ are rewritten from $y'_{11}, \ldots, y'_{66}$ to $y'_1, \ldots, y'_{36}$, and the components of the image $\mathbf{Z}$ are rewritten from $z_{11}, \ldots, z_{33}$ to $z_1, \ldots, z_9$, while keeping their two-dimensional structures. Furthermore, in Fig. 13.7(b), they are represented as vectors. As a result, the images $\mathbf{Y}'$ and $\mathbf{Z}$ can be represented as 36-dimensional and 9-dimensional vectors, respectively, as shown in the following equations:

$$\mathbf{Y}' = (y'_1, \ldots, y'_i, \ldots, y'_{36}), \tag{13.23}$$

$$\mathbf{Z} = (z_1, \ldots, z_j, \ldots, z_9). \tag{13.24}$$

Here, we consider that the ith component y'_i of $\mathbf{Y}'$ and the jth component z_j of $\mathbf{Z}$ are connected via the weight w_{ij}. Since they are in a locally-connected relationship, z_j is connected only to a part of $y'_1, \ldots, y'_{36}$. For example, z_1 is connected only to y'_1, y'_2, y'_7, y'_8, and the connected components are shown in gray and connections are denoted by straight lines in the figure. Now, let us consider y_i' connected to z_j and denote the set of such i as R_j. For example, R_1 and R_2 are as follows:

$$R_1 = \{1, 2, 7, 8\}, \tag{13.25}$$

$$R_2 = \{3, 4, 9, 10\}. \tag{13.26}$$

Using this symbol, the weight w_{ij} for max pooling is represented as

$$w_{ij} = \begin{cases} 1 & (i = \underset{k \in R_j}{\mathrm{argmax}} \ y_k') \\ 0 & (\text{otherwise}). \end{cases} \tag{13.27}$$

In the above example, let us assume that $j = 1$ and

$$\max_{k \in R_1} y_k' = \max\{y_1', \ y_2', \ y_7', \ y_8'\} = y_7'. \tag{13.28}$$

Then, since

$$i = \underset{k \in R_1}{\mathrm{argmax}} \ y_k' = 7, \tag{13.29}$$

the following equation is obtained from Eq. (13.27):

$$\begin{cases} w_{71} = 1 \\ w_{i1} = 0 & (i \neq 7). \end{cases} \tag{13.30}$$

Let ε_i be the error at the ith component y_i' of the image $\mathbf{Y}'$ and ε_j be the error at the jth component z_j of the image $\mathbf{Z}$. Since y_i' is a value obtained without passing through the activation function, from Eq. (12.28) the backpropagation formula is

$$\varepsilon_i = \sum_j \varepsilon_j \, w_{ij}. \tag{13.31}$$

In the above example, the error ε_1 at z_1 should be propagated as the error ε_7 at y_7'. Therefore, when max pooling is used, it is necessary to remember which component of $\mathbf{Y}'$ was chosen as the maximum value at the classification. With the above points in mind, the conventional backpropagation method can be applied as it is. In the case of average pooling, the following formula for the weights can be used instead of Eq. (13.27):

$$w_{ij} = \begin{cases} \dfrac{1}{H^2} & (i \in R_j) \\ 0 & (\text{otherwise}). \end{cases} \tag{13.32}$$

The basic processing of backpropagation in the convolution layer is the same as that of conventional neural networks. Let us explain using the three-layer convolutional neural network CNN3 that we have discussed so far.

The convolution operation shown in Eq. (13.11) is a product-sum operation of the original image and the convolution filter. The output from the convolution layer is obtained by adding the bias to the product-sum result and applying a nonlinear transformation using an activation function. This process is similar to the forward propagation of conventional neural networks. However, unlike conventional neural networks, the operation to obtain the product-sum result $\mathbf{Y}_m$ from the input image $\mathbf{X}$ by Eq. (13.11) assumes the local connectivity. Therefore, the connection between the images $\mathbf{X}$ and $\mathbf{Y}_m$ is sparse and a few weights are shared. Consequently, it is necessary to prepare a table that shows which unit of $\mathbf{X}$ is connected to which unit of $\mathbf{Y}_m$. In the backpropagation process, this table can be used to modify only the weights between locally-connected units. The process itself is the same as that of conventional neural networks.

--- **Coffee Break** ---

Drosophila in Machine Learning Research

The well-known database of handwritten digits, MNIST, was constructed and released by the National Institute of Standards and Technology (NIST) and is also used in the experiments in this book. Deep learning pioneer G. Hinton mentioned that "MNIST is the *Drosophila* of machine learning" (Goodfellow et al. 2016). *Drosophila* is the scientific name for a fruit fly, a well-known model organism[2] in embryology and genetics. MNIST was the ideal material to verify and evaluate machine learning algorithms in a laboratory environment. This expression by Hinton is taken from "Chess as the Drosophila of AI" by J. McCarthy[3] (McCarthy 1990).

Similar databases corresponding to *Drosophila* also exist in the field of image processing. The factors that have contributed to the significant progress of artificial intelligence research in this field since 2010 are (1) Hinton's research progress in deep learning, (2) the development and popularization of the GPU by NVIDIA, and (3) the construction and release of ImageNet. These three factors are often pointed out. In other words, it is important that the three elements of software (algorithms), hardware, and data have all come together. ImageNet, mentioned in (3) above, is truly the *Drosophila* of the image processing and machine learning fields. ImageNet is a large-scale labeled image database constructed by Professor Fei-Fei Li of Stanford University. It was used in the ILSVRC competition introduced in Sect. 13.1, and has greatly contributed to subsequent image recognition research. The aim and background are described in her TED talk "How we teach computers to understand pictures."

---

[2] A model organism is an animal species that is widely used for a specific research problem in biology. Mice, nematodes, and zebrafish are well-known model organisms. *Drosophila* has been widely used as a material for genetics and embryology because of its low cost of rearing and fast generation change.

[3] An artificial intelligence researcher who hosted the Dartmouth Conference, where the term "artificial intelligence" was used for the first time in human history.

13.5   Experiments Using Artificial Images

The convolution operation is a feature extraction process using filters, and its procedure has already been described. In the following, we will verify the effect of the convolution operation on images by simple experiments. Experiments were performed in two domains, the spatial domain and the spatial frequency domain, and a comparison is made between the two.

13.5.1   Feature Extraction Function of Convolution Filter

The filters used in the experiments are shown in Fig. 13.8. The size of the filter is 5×5 ($F = 5$), and the numbers in the figure show the values of the filter weights v_{kl} ($k, l = 1, \ldots, 5$) in Fig. 13.2. For all filters, the sum of v_{kl} is zero in the range of 5×5, so that in the region where the gray level is constant, the result of the product-sum operation for the convolution shown in Eq. (13.2) is zero. On the other hand, in the region where the gray level changes rapidly, the convolution value appears as a pair of positive and negative peak values.[4] Therefore, the convolution operation using the filters in Fig. 13.8 can detect the edges that exist in the image. Therefore, the convolution operation using the filters shown in Fig. 13.8 can detect the edges existing in the image.

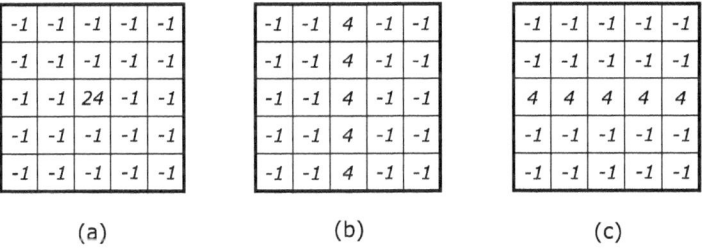

(a)   (b)   (c)

Fig. 13.8 Filters for edge detection (5×5)

The results of the convolution operation are shown in Fig. 13.9. Figure (a) shows a grayscale image with 384×384 pixels ($A = 384$) and 256 gray levels. Here, white is 0 and black is 255. Using filters (a), (b), and (c) of Fig. 13.8, the convolution is performed to this original image, and the results are shown in (b), (c), and (d) of Fig. 13.9, respectively. They are represented as grayscale images with the same number of gray levels as the original image. Although the resulting convolution images are affected by noise to some extent, it can be confirmed that the results are

---

[4] In the following experiments, only positive values were used.

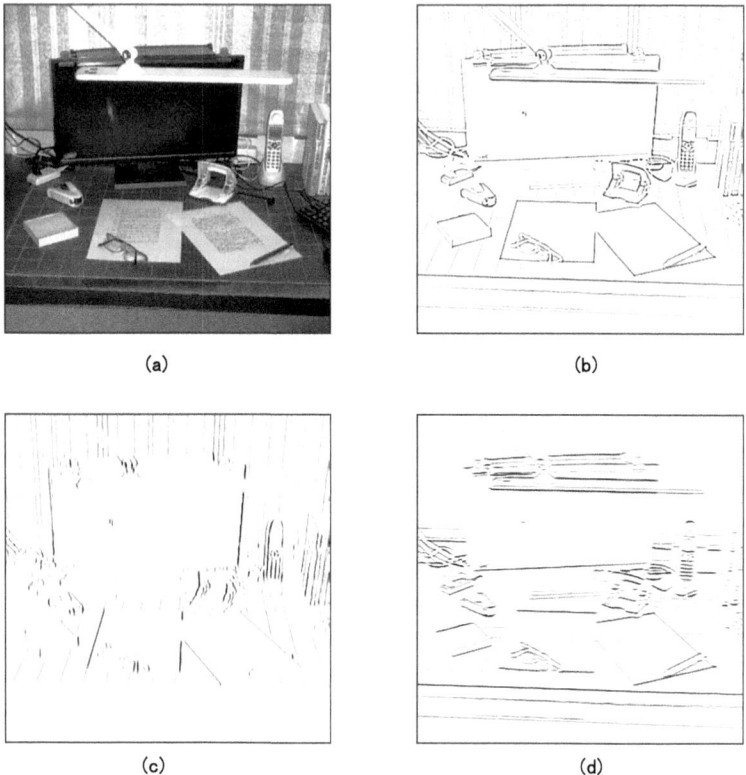

Fig. 13.9 Convolution operation with 5×5 filters. (**a**) Original image. (**b**) Edge detection for all directions. (**c**) Vertical edge detection. (**d**) Horizontal edge detection

as expected. That is, in (b), edges in all directions are detected, in (c) and (d), edges in the vertical and horizontal directions are detected, respectively.

From the above experiments, it can be seen that the convolution operation has a feature extraction function. Although we have taken edges as examples of features, the convolution operation can be applied to extract a variety of features by changing the filter settings (Rosenfeld and Kak 1982).

13.5.2   Convolution Operations in the Spatial Frequency Domain

The effects of applying the convolution operation to the original image are described in Sect. 13.5.1. Such processing in real space is called the processing in the *spatial domain*. In contrast, in the following, we consider what kind of function the convolution operation has in the *spatial frequency domain*.[5]

Convolution operations play a very important role in signal processing. The convolution $h(t) \overset{\text{def}}{=} f(t) * g(t)$ defined for two signals $f(t)$ and $g(t)$ with time t as a variable is expressed as follows:

$$h(t) = f(t) * g(t) = \int_{-\infty}^{\infty} f(\tau)g(t - \tau)d\tau. \tag{13.33}$$

Here, if the *Fourier transform* of the function $f(t)$ is expressed by $\mathcal{F}(f(t))$, the following equation holds:

$$\mathcal{F}(h(t)) = \mathcal{F}(f(t)) \cdot \mathcal{F}(g(t)). \tag{13.34}$$

The above equation is the most powerful tool in the field of signal processing, known as the *convolution theorem*. The equation indicates that the convolution of two signals is expressed as the product of the two in the frequency domain (see Problem 13.2).

This theorem can be extended to two-dimensional image signals with gray levels $f(x, y)$ and $g(x, y)$ at position (x, y). The convolution image $h(x, y) \overset{\text{def}}{=} f(x, y) * g(x, y)$ is represented by

$$h(x, y) = f(x, y) * g(x, y) = \iint_{-\infty}^{\infty} f(\alpha, \beta)g(x - \alpha, y - \beta)d\alpha d\beta. \tag{13.35}$$

Here, if the two-dimensional Fourier transform of the function $f(x, y)$ is expressed as $\mathcal{F}(f(x, y))$, the convolution theorem for two-dimensional images is expressed by the following equation:

$$\mathcal{F}(h(x, y)) = \mathcal{F}(f(x, y)) \cdot \mathcal{F}(g(x, y)). \tag{13.36}$$

Although Eqs. (13.33) to (13.36) deal with continuous functions, they can be used for discrete functions as well. The image $\mathbf{X}$, the filter $\mathbf{V}$, and the convolution image $\mathbf{Y}$ which have been discussed so far are discrete functions. We can interpret that $\mathbf{X}$, $\mathbf{V}$, and $\mathbf{Y}$ correspond to $f(x, y)$, $g(x, y)$, and $h(x, y)$ in Eq. (13.35), respectively. However, a note of caution is necessary here. In the integral operation of Eq. (13.35), the variable of f is (α, β), while the variable of g is $(-\alpha, -\beta)$, i.e., the sign is reversed. This operation is equivalent to rotating the filter g by $180°$.

---

[5] We are dealing here with two-dimensional still images, which are spatially varying signals. To distinguish it from "frequency", which is used for time-varying signals, we use the term *spatial frequency*. Spatial frequency is used to measure the number of regular repetitions of grayscale changes per unit length.

Considering the above, the discrete convolution operation corresponding to Eq. (13.35) is described as

$$y_{ij} = \sum_{k=1}^{F} \sum_{l=1}^{F} x_{i+k-1,\,j+l-1} \cdot v_{F-k+1,\,F-l+1} \qquad (i, j = 1, \ldots, B), \quad (13.37)$$

which is different from the definition given in Eq. (13.2). Taking the 3×3 ($F = 3$) filter shown in Fig. 13.2 as an example, Eq. (13.37) shows that the filter **V** is rotated $180°$ as shown in Fig. 13.10 and then superimposed on the image **X** and the product-sum operation is repeated, shifting the image by one pixel.

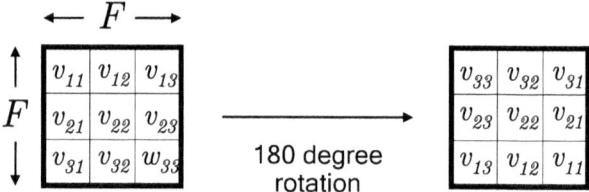

Fig. 13.10 Filter for convolution

Defining the convolution operation by Eq. (13.37), the convolution theorem corresponding to Eq. (13.36) is expressed as

$$\mathcal{F}(\mathbf{Y}) = \mathcal{F}(\mathbf{X}) \cdot \mathcal{F}(\mathbf{V}). \qquad (13.38)$$

Note that $\mathcal{F}$ in the above equation represents a discrete two-dimensional Fourier transform. For more details on signal processing, refer to Lathi (1968) and Rosenfeld and Kak (1982).

13.5.3   Image Data and its Power Spectrum

As shown in Eq. (13.38), the convolution operation can be expressed as the product of two signals in the spatial frequency domain. Using this property, complex processing in the spatial domain becomes simple in the spatial frequency domain. From this, convolutional neural networks are most effective for images that can be processed effectively in the spatial frequency domain. Therefore, the following artificial image data is prepared. Generated images contain stripe patterns of size 16×16 pixels in an area of 32×32 pixels. The background excluding the striped pattern is occupied by noise. Examples are shown in Fig. 13.11. The pixel values of the stripes are binary of -1 and 1, and the pixel values of the background noise are continuous between -1 and 1 generated by uniform random numbers. In the figure, black corresponds to 1 and white to -1. The stripes are two-pixels wide for both

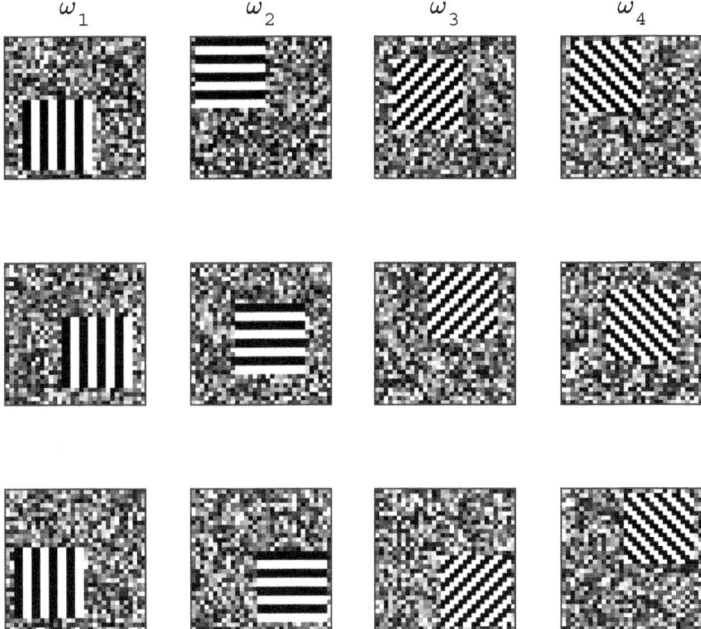

Fig. 13.11 Stripe patterns in noise

black and white, and have four different directions: vertical, horizontal, diagonal to the right, and diagonal to the left. They correspond to the four classes ω_1, ω_2, ω_3, and ω_4, respectively, depending on which of them they contain. Two data sets were prepared, one for learning and the other for testing, and the number of patterns generated was 1000 per class for a total of 4000 patterns, for both learning and testing. In generating the patterns, the stripe patterns are randomly placed within the range of the image, and uniform random numbers are applied to determine their positions. The background noise is different for each pattern. The figure shows three patterns for each class as examples.

Figure 13.12 shows the results of observing the above patterns in the spatial frequency domain. Figure (a) shows the original images extracted one pattern from each class. Figure (b) shows the results of applying the two-dimensional Fourier transform to the patterns of each class and obtaining their *power spectrum,*[6] with darker values indicating larger values. The number of pixels is 32×32, the same as the original images. The spatial frequency domain is represented by (u, v), whereas the original image domain is represented by (x, y), and their axes are shown in the figure. The u, v represent the spatial frequency in the x, y directions of the original image, respectively. The point $(u, v) = (0, 0)$ where the two axes intersect

---

[6] In general, the signal obtained by Fourier transform is a complex number with real and imaginary parts. The power spectrum is obtained as the square of its absolute value $|\mathcal{F}(\mathbf{X})|^2$.

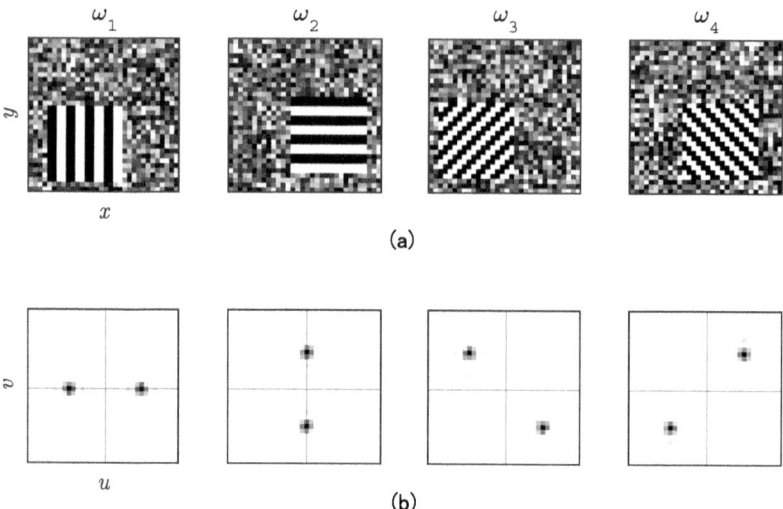

Fig. 13.12 Patterns in the spatial frequency domain. (**a**) Original image. (**b**) Power spectrum

corresponds to the DC component (the average value of the signal) of the original images.

In the power spectrum, peaks corresponding to periodic signals of the stripe patterns can be observed. Moreover, the positions of the peaks are different depending on the class, indicating that this information can be used for classification between classes.

Let us take, for example, the image of class ω_1. The pattern of vertical stripes in this image is a periodic signal that varies along the x-axis, and the power spectrum has a pair of large peaks at the corresponding position on the u-axis. In the power spectrum, peaks can be observed in each direction from $(u, v) = (0, 0)$ orthogonal to the stripes. The same is true for horizontal and diagonal stripes. The position of the peak in the power spectrum does not change even if the position of the stripe pattern changes. This characteristic of the power spectrum, which is not affected by positional fluctuations, is advantageous for processing in the spatial frequency domain and can be utilized to a great extent.

13.5.4 Band-Pass Filter Settings and Convolution Operation

As mentioned above, it was shown that the artificial images containing stripe patterns exhibit noticeable peaks in the spatial frequency domain. In the following, we show that a convolution filter for extracting the stripe pattern can be obtained by setting a band-pass filter in the spatial frequency domain.

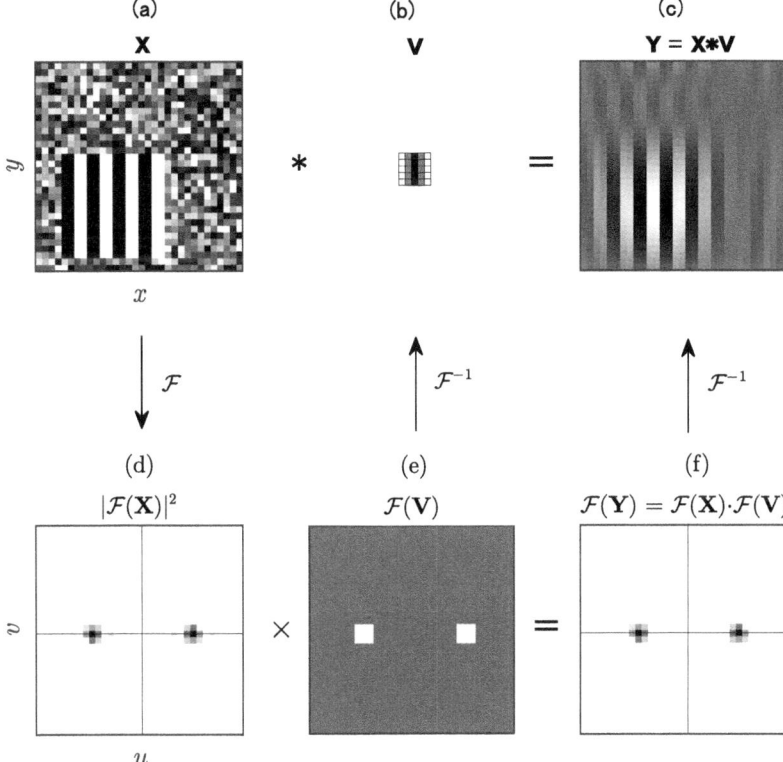

Fig. 13.13 Processing in the spatial domain and in the spatial frequency domain. (**a**) Original image. (**b**) Convolution filter. (**c**) Convolution image. (**d**) Power spectrum. (**e**) Band-pass filter. (**f**) After filtering

Figure 13.13 shows the relationship between processing in the spatial and spatial frequency domains, respectively, taking the convolution operation for the image of class ω_1 as an example. Images (a) and (d) in the figure are the same as the leftmost images in Fig. 13.12, and are the original image $\mathbf{X}$ and its power spectrum $|\mathcal{F}(\mathbf{X})|^2$, respectively. The symbol $\mathcal{F}$ indicates the two-dimensional Fourier transform.

Next, we prepare the *band-pass filter* for extracting the regions that show peaks in the power spectrum. The band-pass filter (e) in the figure shows that only the spatial frequencies corresponding to the white areas are transmitted, while others are blocked. In other words, (e) can be thought of as an image with pixel values of 1 for white areas and 0 for other areas. The size of the transparent area is 3×3 pixels. As the figure shows, the peaks of the power spectrum of the ω_1 image exist at two points on the u-axis. We prepared a filter (e) that transmits only these spatial frequency regions. Then, multiplying (d) by (e) yields the image shown in (f).[7] The image (c) was obtained by applying the *inverse Fourier transform* to (f). The symbol $\mathcal{F}^{-1}$ in the figure indicates the inverse Fourier transform operation. In the image (c), the stripe pattern of the original image $\mathbf{X}$ is emphasized, and it can be confirmed that the stripe pattern is extracted by the band-pass filter.

The lower part of Fig. 13.13 represents the operation of the product in the spatial frequency domain, corresponding to the operation of Eq. (13.38). According to the convolution theorem, this process corresponds to a convolution operation in the spatial domain, represented by Eq. (13.3), i.e., Eq. (13.37). Here, the filter $\mathbf{V}$ required for convolution can be obtained by performing the inverse Fourier transform to the band-pass filter (e), and the result is shown in (b). The filter $\mathbf{V}$ in this figure is the result of cutting out 5×5 region including the central part from the 32×32 image obtained by the inverse Fourier transform. From now on, unless otherwise noted, the size of the filter $\mathbf{V}$ is 5×5. The filter weight v_{kl} includes both positive and negative values. Darker pixels indicate larger values, and the same applies hereafter.

The upper part of Fig. 13.13 represents the operation in the spatial domain, corresponding to the convolution operation Eq. (13.3). In fact, when the convolution operation of Eq. (13.2) was performed using the above filter $\mathbf{V}$, it was confirmed that the obtained image coincided with the image (c) obtained by the inverse Fourier transform of (f). In the convolution, the padding of width 2 was applied, so that the size of the convolution image was 32×32, which was the same as the original image. Since the filter $\mathbf{V}$ in (b) does not change even if the image is rotated $180°$, the result is the same even if the convolution operation is performed by Eq. (13.2). Thus, it was experimentally confirmed that the convolution of two images in the spatial domain (Eq. (13.2)) can be expressed as the product of two images in the spatial frequency domain (Eq. (13.38)).

The above example is for an image of class ω_1 containing a vertical stripe pattern, but the same process can be applied to other classes. The band-pass filters corresponding to the classes ω_1 to ω_4 are shown in (a) to (d) of Fig. 13.14, respectively. The convolution filters $\mathbf{V}_1$ to $\mathbf{V}_4$ obtained by the inverse Fourier transform are shown above each band-pass filter. The filter $\mathbf{V}_1$ in this figure is the same as $\mathbf{V}$ in Fig. 13.13(b). Using the obtained convolution filters $\mathbf{V}_2$ to $\mathbf{V}_4$ and applying convolution operations to the images of class ω_2 to ω_4, horizontal and diagonal stripe patterns can be extracted as (c) of Fig. 13.13.

In the above explanation, the four filters shown in Fig. 13.14 are selected for each class and the convolution is performed. However, the same result can be obtained using only one filter. This is shown in Fig. 13.15.

As is clear from Fig. 13.12, there are two spatial frequency regions where the power spectrum shows peaks for each class, for a total of eight regions for all classes. Therefore, as shown in Fig. 13.14, different band-pass filters may be prepared for

---

[7] When obtaining the product of (d) and (e), we actually use $\mathcal{F}(\mathbf{X})$ as (d) instead of the power spectrum $|\mathcal{F}(\mathbf{X})|^2$ shown in the figure.

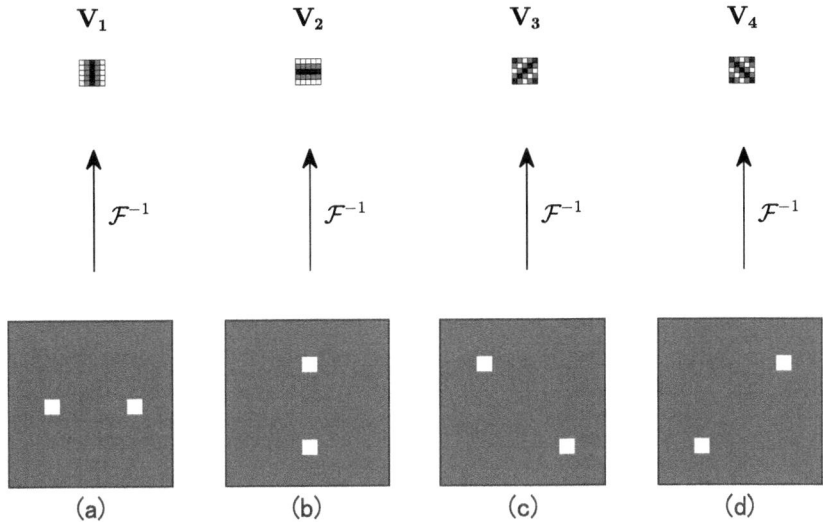

Fig. 13.14 Band-pass filters and corresponding convolution filters

each class, but a single band-pass filter that transmits all eight spatial frequency regions can be expected to have almost the same effect. Such a band-pass filter is shown at the bottom of the center column of Fig. 13.15, and $\mathbf{V}_5$ above it is the convolution filter obtained by the inverse Fourier transform of the band-pass filter. The result of the convolution is shown in the rightmost column of Fig. 13.15. It can be seen that the stripe patterns are extracted by the convolution for all classes, although they are somewhat less clear due to the background noise. The above convolution filters $\mathbf{V}_1$ to $\mathbf{V}_5$ of 5×5 are shown in Fig. 13.16 by the enlarged form.

From the above experiments, we can see that the convolution filter has a feature extraction function that extracts the features of the stripe patterns. Although we used a simple image as an example in this experiment, feature extraction for more complex images would be possible by increasing the number of filters to extract various features. The convolution operation is a local process and this alone does not lead to final classification. The classification can be achieved only after the global process of grasping where and how the extracted features are located in the image. This role is played by the fully-connected network described later. Thus, the combination of local feature extraction and global processing that integrates them can be said to be the most important factor in realizing the convolutional neural network as an advanced classification system.

13.5.5 Learning and Classification Experiments

Taking up CNN3 as a convolutional neural network, experiments on learning and classification are conducted using the artificial images introduced so far. In the

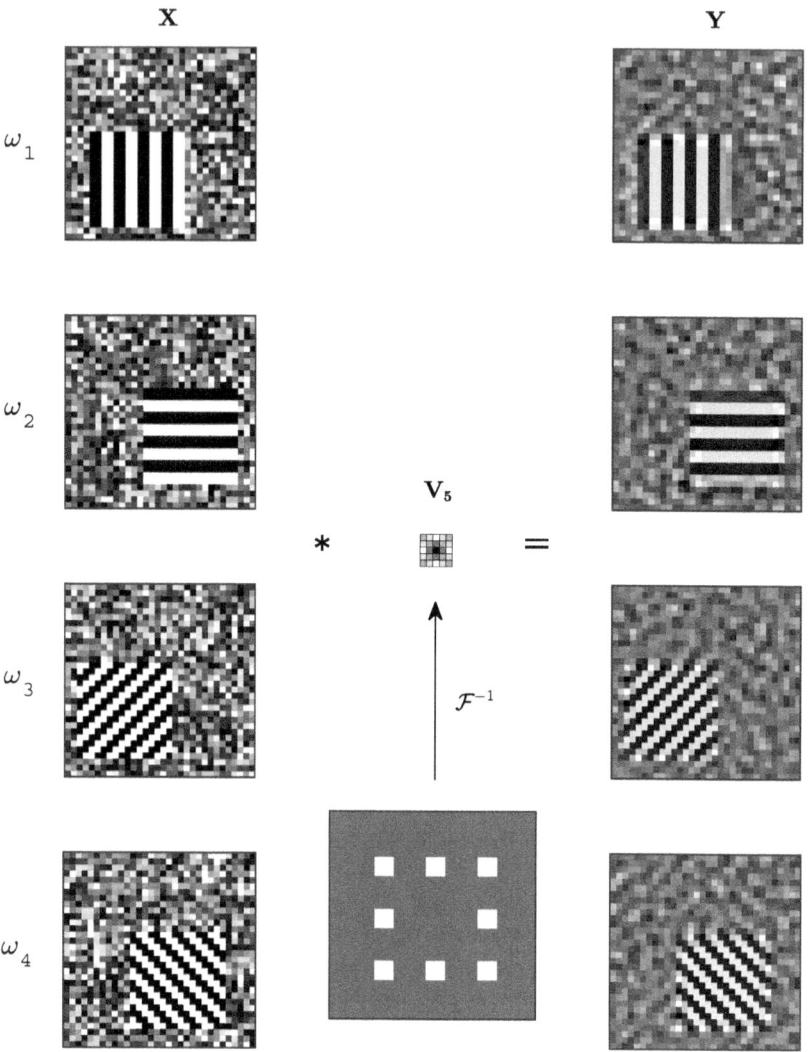

Fig. 13.15 Results of applying convolution filter $\mathbf{V}_5$

experiments, we will compare with the conventional neural network NNW3, and clarify the functions and advantages of the convolutional neural network.

In the experiment, the size of the filter used in CNN3 was set to $F = 5$. The padding width was set to $P = 2$ so that the convolution image size B is the same as the original image size A. That is, $B = A = 32$ from Eq. (13.17). For the pooling, $H = 2$, so $G = B/H = 16$ from Eq. (13.18), and the pooling image is 16×16 in size. If the number of filters used is M, the number of output Q from the pooling layer is $Q = G^2 M = 256M$ from Eq. (13.19), which corresponds to the number of units in the internal layer of NNW3. The number of filters M in the convolutional neural network CNN3 was varied in five ways: 1, 2, 4, 8, 16.

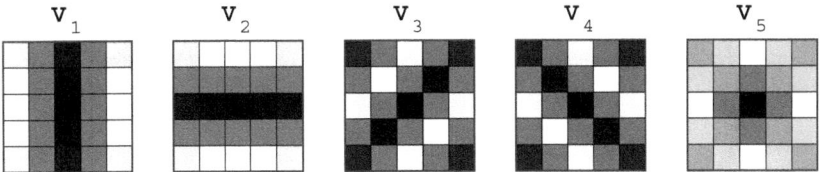

Fig. 13.16 Convolution filters corresponding to band-pass filters

In NNW3 for comparison, the number of units in the internal layer Q was set to 256, 512, 1024, 2048, 4096, corresponding to CNN3. For the activation function of CNN3, we used a sigmoid function for the convolution layer and a softmax function for the output layer. Similarly, for NNW3, we used a sigmoid function in the internal layer and a softmax function in the output layer.

For both neural networks, learning was performed using a total of 4000 learning patterns, 1000 patterns for each class. Both neural networks converged in a state where all learning patterns were correctly classified. The result of classifying the same number of test patterns using the neural network obtained by learning is shown in Fig. 13.17. The horizontal axis of the figure indicates the number of filters M for CNN3 and the number of units Q in the internal layer for NNW3 in parentheses. The vertical axis indicates the error rate as a percentage.

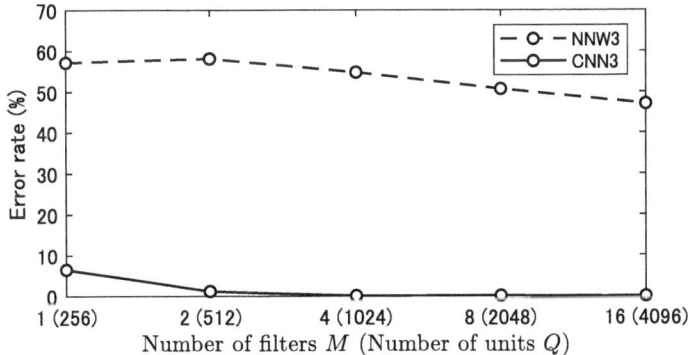

Fig. 13.17 Classification experiments using artificial images

The figure shows that the error rates of CNN3 and NNW3 decrease with the number of filters M and with the number of units Q in the internal layer, respectively, which are reasonable results. However, there is a large difference between the error rates of the two, with CNN3 performing much better than NNW3.

As shown in Fig. 13.11, it is the direction of the stripe patterns occupying part of the images that distinguishes the classes. Noise in the background is not useful for classification and must be ignored in the learning process. Experimental results show that the convolutional neural network CNN3 is able to ignore the background noise and the learning which focuses only on the stripe patterns can be realized thanks to its frequency filtering function. In contrast, NNW3 attempts to classify

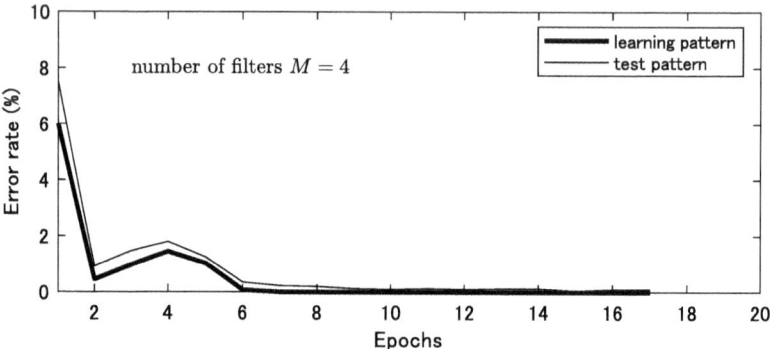

Fig. 13.18 Learning curve for CNN3

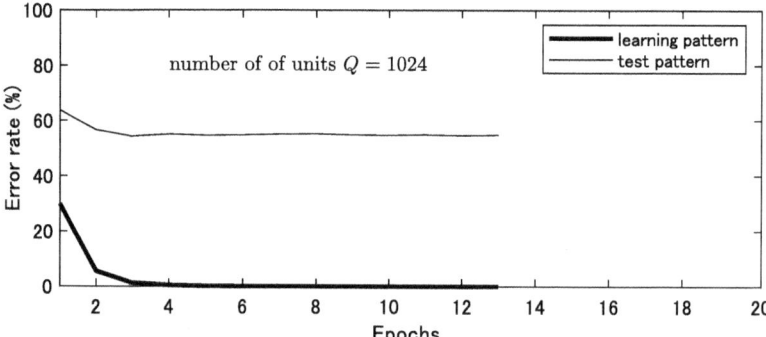

Fig. 13.19 Learning curve for NNW3

patterns by including even background noise, and therefore achieves only a low classification rate for the test patterns. Experiments confirmed that NNW3 can achieve the same classification performance as CNN3 when the background noise is removed or when the background noise is made common to all patterns.

Let us compare between CNN3 and NNW3, how the error rates for learning and test patterns change in the process of learning. In the experiment, the number of convolution filters was $M = 4$ in CNN3, and the number of units in the internal layer was $Q = 1024$ in NNW3 to correspond to CNN3. As before, the number of learning and test patterns were 4000 patterns each.

First, the result for CNN3 is shown in Fig. 13.18. The horizontal axis of the figure shows the number of epochs, and the vertical axis shows the error rate as a percentage. The thick and thin lines in the figure correspond to the learning and test patterns, respectively. Such a graph is called a *learning curve* . Learning is completed in 17 epochs. We can see that the error rate for the learning patterns is decreasing and the error rate for the test patterns is also decreasing. At convergence, the error rates are 0% and 0.10% for the learning and the test patterns, respectively.

Similarly, the result for NNW3 is shown in Fig. 13.19. Note that the horizontal axis is the same as Fig. 13.18, but the scale of the vertical axis is different. As can be seen from this figure, although the learning process is completed in 13 epochs and an error rate of 0% is achieved for the learning patterns, the error rate of the test patterns hardly decreases. The error rate at convergence is 54.80% for the test patterns. The large discrepancy in error rate between the learning and test patterns seen in NNW3 example is due to the attempt to separate a small number of learning patterns with a complex decision boundary. This phenomenon is called over-fitting, as described in Sect. 12.4.

The simplest way to prevent over-fitting is to increase the number of learning patterns, and in fact, experiments have confirmed that increasing the number of learning patterns improves the classification performance of NNW3. However, this requires a huge number of learning patterns and is not efficient. As an efficient method to prevent over-fitting, we introduced the dropout method in Sect. 12.4. In the following, we confirm the effect of dropout under the same conditions as Fig. 13.19.

Figure 13.20 shows the learning curve for NNW3 after introducing dropout. In this experiment, $p_r = 0.5$ was set as the probability of non-dropout. Note that the scale of the horizontal axis is different from that of Fig. 13.19. To avoid complexity, the error rate was calculated every 10 epochs instead of every epoch. For both the learning and test patterns, the error rates fluctuate but show a steady decrease. Learning converged after 330 epochs, and the error rates at convergence were 0 and 0.03% for the learning and test patterns, respectively. Although the number of epochs required for convergence increases, it is confirmed that over-fitting is avoided.

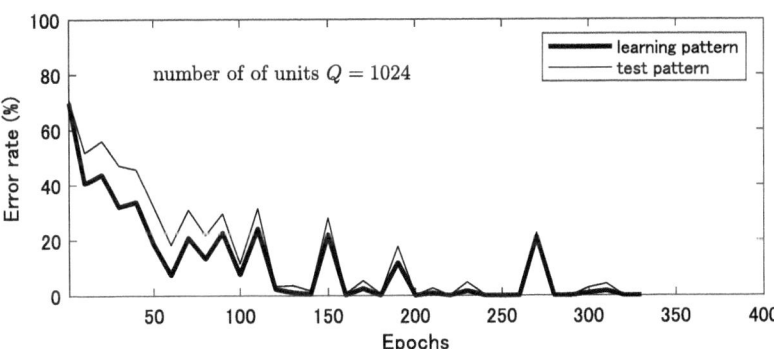

Fig. 13.20 Learning curve for NNW3 with dropout introduced

As is clear from what has been described so far, the convolutional neural network can accurately extract local features without being affected by background noise, even with a small number of learning patterns and a small number of weights. As a result, it can achieve higher classification performance than conventional three-layer neural networks. Also, CNN3 has the advantage of learning overwhelmingly small number of weights.

Let's examine what kind of convolution filters are obtained by learning of CNN3 in an experiment where the number of filters is set to $M = 4$. The upper part of Fig. 13.21 shows the initial filters of CNN3, i.e., the convolution filters set at the beginning of the learning. The weights v_{kl} of the initial filters were determined by uniform random numbers that take values between -1 and 1. On the other hand, the lower part of the figure shows the convolution filters $\mathbf{F}_1$ to $\mathbf{F}_4$ obtained by learning. The arrows in the figure indicate from which initial filter each filter was obtained.

In the learning of a convolutional neural network for the purpose of class separation, such as in this experiment, the convolution filter, whose main purpose is feature extraction, should be adjusted so that it can have the function of class separation. The convolutional filters shown in the lower part of Fig. 13.21 are the results of such adjustments. In this figure, however, the adjustment details are not necessarily clear, nor is the inheritance from the initial filters.

initial filters set by random numbers

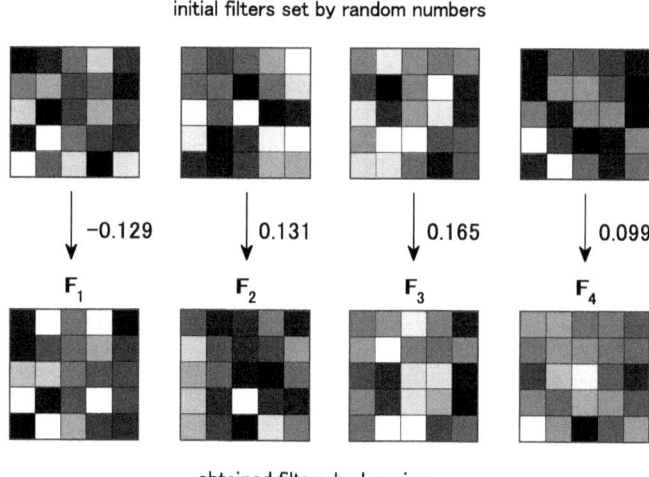

Fig. 13.21 Convolution filters obtained by CNN3 with $M = 4$ (initial filters set by random numbers)

In order to quantitatively check the degree of inheritance from the initial filter, we try to calculate the similarity between the obtained filter and the initial filter. The similarity of two column vectors $\mathbf{x}$ and $\mathbf{y}$ is calculated by *cosine similarity* s defined by the following formula:

$$s = \frac{\mathbf{x}^t \mathbf{y}}{\|\mathbf{x}\| \, \|\mathbf{y}\|}. \tag{13.39}$$

The cosine similarity s takes the value $-1 \leq s \leq 1$. The larger the value, the more similar $\mathbf{x}$ and $\mathbf{y}$ are. The maximum value of 1 is obtained when $\mathbf{y} = a\,\mathbf{x}$, where a is a positive constant. The 5×5 filters shown in Fig. 13.21 were transformed into 25-dimensional vectors, and the similarities between filters $\mathbf{F}_i$ ($i = 1, \ldots 4$) and the corresponding initial filters were calculated. They are indicated on the right of the arrows. The numerical values show that the similarity of each filter to the initial filter is small, at most 0.165 for filter $\mathbf{F}_3$.

Then, the following experiment was performed. Filters $\mathbf{V}_1$ to $\mathbf{V}_4$ of Fig. 13.16 were used as the initial filters instead of filters determined by random numbers. As already mentioned, these filters are not intended for classification, but are effective for detecting four types of stripes with different orientations. Experimental results show that learning was completed in 16 epochs, and the error rate at convergence was 0% and 0.05% for the learning and test patterns, respectively. In the case where the initial filters with random numbers were used, these were 17 epochs, 0%, and 0.10%, respectively, so using $\mathbf{V}_1$ to $\mathbf{V}_4$ as the initial filters is slightly better. It is considered that such results were obtained because the initial filters with feature extraction function were used, so the learning process requires only adjustments to acquire the classification function.

The convolution filters $\mathbf{F}_{V1}$ to $\mathbf{F}_{V4}$ obtained as a result of learning are shown in the lower part of Fig. 13.22, and the corresponding initial filters are shown in the upper part. The similarities to the initial filters are indicated on the right side of the arrows. The view of the figure is the same as Fig. 13.21. The figure shows that filters $\mathbf{F}_{V1}$ to $\mathbf{F}_{V4}$ are quite similar to the initial filters $\mathbf{V}_1$ to $\mathbf{V}_4$, respectively. For example, the similarity between filter $\mathbf{V}_3$ and filter $\mathbf{F}_{V3}$ is 0.951, which is close to 1. The other filter pairs are also close to 1, indicating that they are similar to each other. Compared with the results of Fig. 13.21, the high similarities in Fig. 13.22 are remarkable.

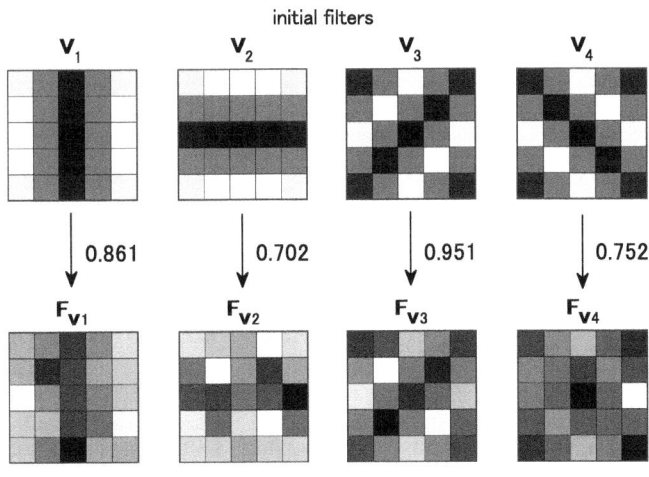

Fig. 13.22 Convolution filters obtained by CNN3 with $M = 4$ (initial filters set to $\mathbf{V}_1$ to $\mathbf{V}_4$)

Filters $\mathbf{F}_{V1}$ to $\mathbf{F}_{V4}$ are similar to filters $\mathbf{V}_1$ to $\mathbf{V}_4$ because the former maintains and utilizes the feature extraction function of the latter. On the other hand, they do not completely match and differ because the former incorporates not only the feature extraction function but also the class separation function. In general, it is said that each layer has a separate role to play: local feature extraction in the convolution layer and class separation in the pooling and output layers. However, the convolution filter acquires not only the feature extraction function but also the class separation function through learning, so the network as a whole can achieve high classification ability.

Coffee Break

Double Descent

Many aspects of the characteristics and behavior of deep neural networks have not yet been elucidated. The biggest mystery is why it shows high generalization ability despite its huge number of parameters. During the second neural boom in the 1980s, the biggest drawback of neural networks was the existence of locally optimal solutions caused by the huge number of parameters. However, numerous experiments have demonstrated that deep neural networks can outperform existing methods despite having hundreds of millions of parameters. This fact suggests that most of the solutions that exist in deep neural networks are suboptimal. There are still many aspects of the characteristics and behavior of deep neural networks that have yet to be elucidated. The greatest mystery is why they exhibit high generalization ability despite having a huge number of parameters.

In general, the relationship between the number of parameters (model's degrees of freedom) and the misclassification rate can be explained by the "Bias and Variance Dilemma" introduced in the coffee break in Sect. 5.5.1. A large-scale neural network may show good performance, but if the scale is too large, the misclassification rate will increase. Although this phenomenon has been described in many textbooks, it has not been clarified why deep neural networks can exhibit high performance.

Regarding this, the phenomenon of *double descent* is being experimentally discovered (Belkin et al. 2019). As has been previously reported, as the number of parameters is increased, the misclassification rate, which had been decreasing, begins to increase at a certain point. However, as the number of parameters is further increased, then the misclassification rate begins to decrease again.

Suboptimal local solutions and double descent are probably phenomena that occur in high-dimensional spaces. The high-dimensional world is still full of mysteries, and we look forward to future research developments.

13.6 Classification Experiments Using Handwritten Digit Patterns

The data used in the experiments of the previous section were artificially created patterns. In this section, we present the results of experiments in which we applied the convolutional neural network to actual handwritten digit patterns. The handwritten digit patterns were obtained from the data set *MNIST: Mixed National Institute of Standards and Technology database*. For the 10 classes of handwritten digits 0–9, 1000 patterns for each class, 10,000 patterns in total, were used as learning patterns, and 800 patterns for each class, 8000 patterns in total, as test patterns.[8] Each pattern is a grayscale image with 28×28 pixels and each pixel has a value between 0 and 1.

In the experiments, as in Sect. 13.5, we use a three-layer convolutional neural network CNN3 shown in Fig. 13.6. Patterns of 28×28 ($A = 28$) are input to the input layer, and then convolution operations are performed by filters of 5×5 ($F = 5$). Since padding of width 2 ($P = 2$) is applied, $B = 28$ is obtained from Eq. (13.17), and the size of the convolution image is 28×28, the same as the input image. If the number of filters used is M, M convolution images of 28×28 are obtained. These are the convolution layer data. Nonlinear transformations are applied to these convolution images using a sigmoid function as the activation function. A pooling operation is applied to each image after nonlinear transformation. Since a filter of 2×2 ($H = 2$) is applied to the pooling, $G = 14$ is obtained from Eq. (13.18), and the size of the pooling images is 14×14. These images are the pooling layer data. The data in the pooling layer is transferred to the output layer via fully-connected weights, without passing through the activation function. In the output layer, a softmax function is applied as the activation function, and the belonging class of the input pattern is output as probabilities.

The input layer handles $28 \times 28 = 784$-dimensional vectors, the convolution layer handles $784 \times M$-dimensional vectors, the pooling layer handles $14 \times 14 \times M$-dimensional vectors, and the output layer handles 10-dimensional vectors. Each layer requires the same number of units.

As in the case of Sect. 13.5, experiments were performed on a three-layer convolutional neural network CNN3 and on a conventional three-layer neural network NNW3 for comparison. As before, for CNN3, the number of convolution filters M was varied in five ways: 1, 2, 4, 8, 16, and for NNW3, the number of units Q in the internal layer was set to 196, 392, 784, 1568, 3136 by Eq. (13.19) to correspond to CNN3. The results are shown in Fig. 13.23. The view of the figure is the same as Fig. 13.17. The solid and dashed lines represent CNN3 and NNW3, respectively.

Although there are some exceptions, the error rates of CNN3 and NNW3 decrease as the number of filters M and the number of units Q increase, respectively, which is a reasonable result. Except for $M = 1$ ($Q = 196$), CNN3 shows lower error rates than NNW3, confirming the superiority of CNN3. However, the difference is not so large as Fig. 13.17. The reason for this may be that, in this experiment, there is no background noise and no extreme over-fitting occurred, as is observed in the Sect. 13.5 experiment.

Next, we introduce an experiment to examine the robustness of the classifier against misalignment of patterns. For this experiment, we prepare five sets of test patterns with misalignment applied to the learning patterns. The number of patterns in each set is 10,000 in total, the same as the number of learning patterns, and

---

[8] Originally, MNIST provides 60,000 patterns for learning and 10,000 patterns for testing. In this experiment, we used 10,000 and 8000 patterns extracted from them, respectively. See Appendix D for details of this data.

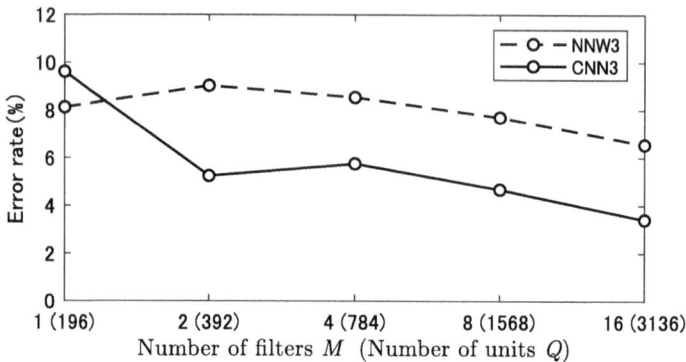

Fig. 13.23 Classification experiments using handwritten digit patterns

examples are shown in Fig. 13.24. The direction of misalignment was set randomly, and the magnitude of misalignment was changed in five steps from one to five pixels. A total of 10,000 patterns were prepared at each step, making one set. The patterns in each row in the figure are examples of the same magnitude of misalignment, and the leftmost numbers indicate the magnitude of misalignment (pixel). The patterns

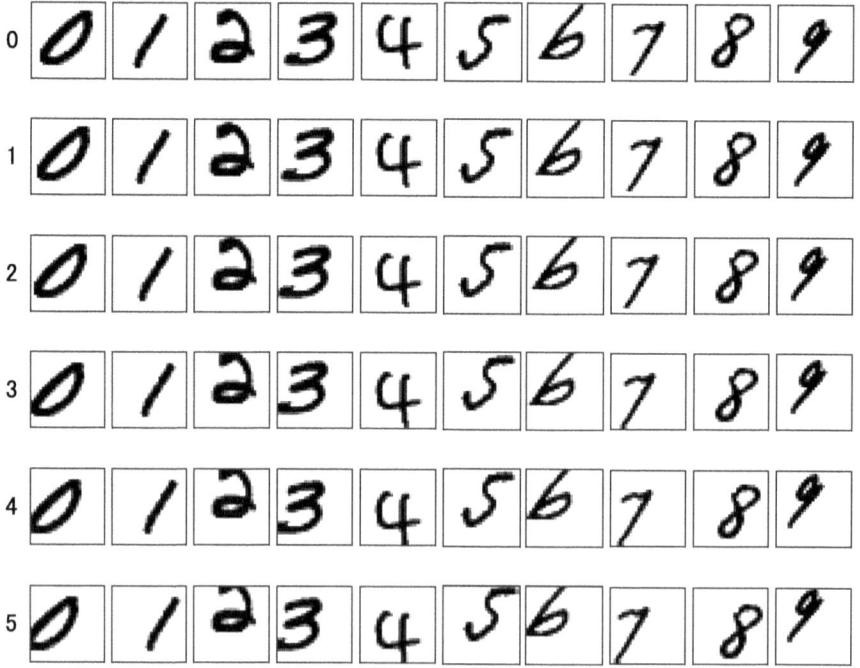

Fig. 13.24 Digit patterns with misalignment

in the top row are examples of zero misalignment, i.e., the original learning patterns. Again, we attempted to compare CNN3 and NNW3. Learning was performed with $M = 4$ for CNN3 and $Q = 784$ for NNW3, and both converged with zero errors. For classification, 10,000 misalignment patterns of each set shown in Fig. 13.24 were used as test patterns. The experimental results are shown in Fig. 13.25. The horizontal axis represents the magnitude of the misalignment, and the vertical axis represents the error rate as a percentage. As in Fig. 13.17, the solid and dashed lines represent CNN3 and NNW3, respectively. The error rate at zero misalignment on the horizontal axis is 0% for both neural networks because it is the result of classifying the learning patterns themselves. For both neural networks, the error rate increases as the magnitude of the misalignment increases. However, CNN3 has a slower growth curve than NNW3 and it can be said that CNN3 is more robust against misalignment.

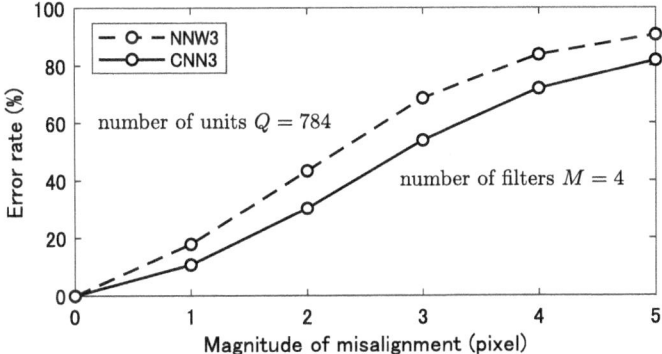

Fig. 13.25   Evaluation of robustness against misalignment

13.7   Multi-Channel and Multi-Layer Structures

So far, the convolutional neural networks used in the experiments have been limited to a three-layer structure in order to compare them with conventional neural networks. To realize a more advanced and practical convolutional neural network, it is essential to increase the number of layers. The structure of a multi-layer convolutional neural network is shown as a schematic diagram in Fig. 13.1, which shows a five-layer example. The concrete processing details of the convolution and pooling layers are shown in Fig. 13.6. Multi-layering is possible by stacking multiple layers according to Fig. 13.1. To make it a more advanced convolutional neural network, it is necessary to expand its functions further. One of them is to increase the number of channels as described below.

In the convolutional neural networks discussed so far, as shown in Fig. 13.6, the input to the convolution layer is only one image, i.e., the number of channels $K = 1$.

In general, a convolutional neural network operation to process K (> 1) channel images is as shown in Fig. 13.26.

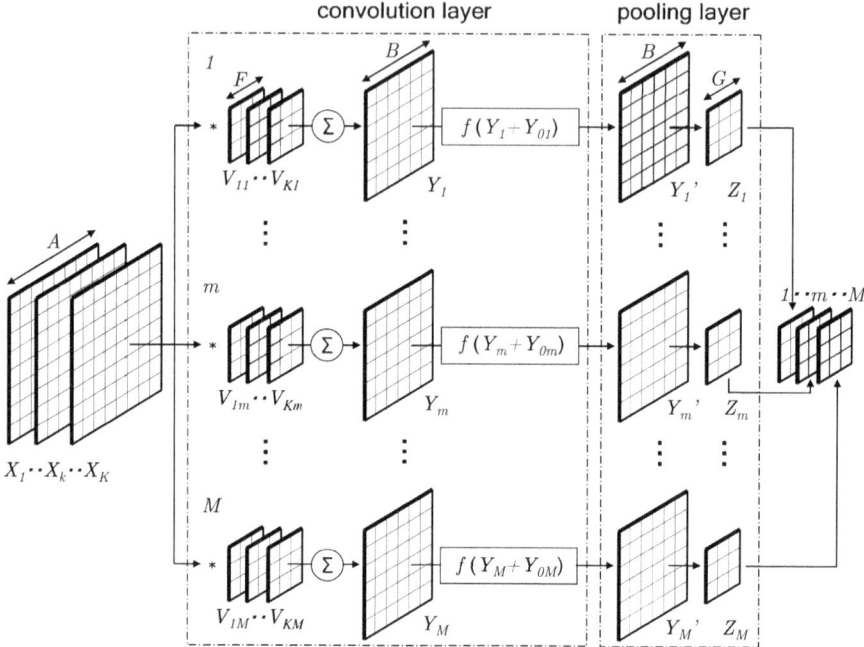

Fig. 13.26 Multi-channel and multi-layer structure for convolutional neural network

Since the input image has K channels, they are $\mathbf{X}_1, \ldots, \mathbf{X}_K$ instead of $\mathbf{X}$. The M types of convolution operations yield convolution images $\mathbf{Y}_1, \ldots, \mathbf{Y}_M$ as before, but the calculation method is slightly different. As shown in the figure, the m-th ($m = 1, \ldots, M$) convolution operation prepares the same number of convolution filters as the number of input channel, $\mathbf{V}_{1m}, \ldots, \mathbf{V}_{Km}$. The convolution image $\mathbf{Y}_m$ is obtained by the following equation instead of Eq. (13.11):

$$\mathbf{Y}_m = \sum_{k=1}^{K} \mathbf{X}_k * \mathbf{V}_{km} \qquad (m = 1, \ldots, M). \qquad (13.40)$$

The computation of the image $\mathbf{Y}'_m$ to be sent to the pooling layer is the same as Eq. (13.13), and this yields M kinds of pooling images $\mathbf{Z}_1, \ldots, \mathbf{Z}_M$. That is, the input images of K channels, each of which has a size of $A \times A$, are sent through the convolution and pooling layers to the next layer as the output images of M channels, each of which has a size of $G \times G$. The next layer regards these images as the M-channel input images and repeats the same process.

―――――――――――――――――――― **Coffee Break** ――――――――――

Beyond Modality Barriers

There used to be a high barrier between the three modalities of image, speech, and natural language, especially between image and natural language. They were treated as distinctly different research fields, and researchers in each field rarely interacted.

However, as introduced in the coffee break in Sect. 10.5, the situation changed drastically after the work by Joachims (1998) in 1998, which applied support vector machines to document processing. Statistical machine learning technologies have been actively used in the field of natural language processing since around 2000, and the barriers between fields and between modalities have gradually been lowered. The pioneering research in the fusion of images and languages can be seen in image captioning and text-enhanced image search.

Deep learning, on the other hand, has played the same role in removing modality barriers. As a result, with the development of deep learning since 2010, the barriers between image, speech, and natural language have become extremely low. Convolutional neural networks, which have been studied in the field of image, are now also used in the field of speech, and technologies such as the transformer (Vaswani et al. 2017) developed in machine translation research are beginning to be used in image processing as well.

Support vector machines, kernel methods, and deep learning are epoch-making technologies in the history of pattern recognition and machine learning. One of their greatest contributions have been the removal of the modality barriers, thereby facilitating the activation of multimodal research fields.

Problems

13.1 [†] Three images of size 7×7 are shown in Fig. 13.27(a)–(c) below. The pixel value of each pixel is set to 1 for black and 0 for white. Convolution operations are performed on these three images. The filter used is of size 3×3, as shown in (d), and the weight values are as shown in the figure.

Show the results of the convolution operation without padding and with padding, and compare them. When padding is applied, the padding width should be set so that the convolution image has the same size as the original image, and the value of the added pixels should be set to 0.

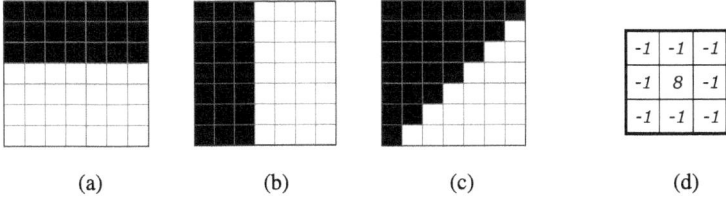

(a)   (b)   (c)   (d)

Fig. 13.27 Three images and the filter for convolution

13.2 Prove the convolution theorem given in Eq. (13.34).

Part III
Bayesian Unified Framework

Chapter 14
Generalization of Learning Algorithm

Abstract The previous chapters have described various learning algorithms for pattern recognition. This chapter introduces the framework of expected loss minimization and organizes the learning algorithms described in the previous chapters in a more general and unified perspective. The square error, zero-one loss, and continuous loss are explained as loss functions. In the framework of expected loss minimization, the so-called least squares method is derived in the case of square error loss. In the case of zero-one loss, a decision function is derived from achieving the minimum error probability (Bayes error). In the case of continuous loss, a classifier with better generalization performance is obtained. Moreover, we explain the probabilistic descent method, which is a general learning method for achieving expected loss minimization. The basic idea and validity of the probabilistic descent method are summarized in the Robbins–Monro algorithm. In this chapter, the above will be explained in more detail. Sect. 14.1 describes the stochastic descent method, which is a learning method for achieving expected loss minimization. The discussion in this chapter is a preparation for clarifying the interrelationships among the learning algorithms described so far and their relationship with the Bayes decision rule in the next chapter.

14.1 Minimum Expected Loss Learning

Now, consider the pattern classification problem with c classes $(\omega_1, \omega_2, \ldots, \omega_c)$. Let $l(\omega_j | \omega_i)$ be the loss when the input pattern of class ω_i is classified as class ω_j.[1] The average loss for a given x, when x is classified as ω_j, is the weighted average

Supplementary Information The online version contains supplementary material available at https://doi.org/10.1007/978-981-95-1478-6_14.

---

[1] By using loss $l(\omega_j | \omega_i)$, we can, for example, change the loss between misclassifying character 1 and misclassifying character 5, or $l(\omega_j | \omega_i) \neq l(\omega_i | \omega_j)$. However, in most practical applications, the losses are assumed to be equal for simplicity. See the coffee break in Sect. 14.2.3 for more information on losses.

of $l(\omega_j|\omega_i)$ with the posterior probability of its class. That is,

$$L(\omega_j|\pmb{x}) = \underset{\omega_i|\pmb{x}}{\mathrm{E}}\ \{l(\omega_j|\omega_i)|\pmb{x}\} \tag{14.1}$$

$$= \sum_{i=1}^{c} l(\omega_j|\omega_i)P(\omega_i|\pmb{x}). \tag{14.2}$$

Here, $\underset{\omega_i|\pmb{x}}{\mathrm{E}}\ \{l|\pmb{x}\}$ denotes the conditional expectation of l with respect to its class ω_i given $\pmb{x}$.

Let $\Psi(\pmb{x})$ denote the *decision function*[2] for an input $\pmb{x}$. That is, for a c-class classification problem, $\Psi(\pmb{x})$ outputs a certain class symbol among $\{\omega_1, \ldots, \omega_c\}$ for the input $\pmb{x}$.[3] Then Eq. (14.2) can be rewritten as

$$L(\Psi(\pmb{x})|\pmb{x}) = \underset{\omega_i|\pmb{x}}{\mathrm{E}}\ \{l(\Psi(\pmb{x})|\omega_i)|\pmb{x}\} \tag{14.3}$$

$$= \sum_{i=1}^{c} l(\Psi(\pmb{x})|\omega_i)P(\omega_i|\pmb{x}). \tag{14.4}$$

Thus, the loss $L(\Psi)$ for all possible inputs $\pmb{x}$ is

$$L(\Psi) = \underset{\pmb{x}}{\mathrm{E}}\{L(\Psi(\pmb{x})|\pmb{x})\} = \underset{\pmb{x},\omega_i}{\mathrm{E}}\ \{l(\Psi(\pmb{x})|\omega_i)\} \tag{14.5}$$

$$= \int L(\Psi(\pmb{x})|\pmb{x})p(\pmb{x})d\pmb{x} \tag{14.6}$$

$$= \sum_{i=1}^{c} \int l(\Psi(\pmb{x})|\omega_i)P(\omega_i|\pmb{x})p(\pmb{x})d\pmb{x} \tag{14.7}$$

$$= \sum_{i=1}^{c} P(\omega_i) \int l(\Psi(\pmb{x})|\omega_i)p(\pmb{x}|\omega_i)d\pmb{x}. \tag{14.8}$$

In Eq. (14.5), $\underset{\pmb{x},\omega_i}{\mathrm{E}}$ denotes expectation with respect to $\pmb{x}$ and ω_i. Note that Bayes' theorem is used in the transformation from Eqs. (14.7) to (14.8). $L(\Psi)$ is called *expected loss*, and the process for finding the decision function that minimizes $L(\Psi)$ from the learning patterns is called *minimum expected loss learning*. In the

---

[2] The decision function should not be confused with the discriminant function.

[3] If the decision function uniquely determines the decision result of classification, it outputs a class symbol indicating the class $\omega_i \in \{\omega_1, \ldots, \omega_c\}$ in the case of a c-class problem. On the other hand, note that, as described later, depending on the loss function used, it may output a c-dimensional vector if it does not uniquely determine the decision result and indicates the possibility of classification across c classes.

following subsections, we describe in detail the specific examples of losses, the learning patterns, and the actual decision functions.

14.2   Various Loss Functions

In this section, we describe various loss functions: the square error loss function, the zero-one loss function, and the continuous loss function.

14.2.1   Square Error

In the case of the square error loss, the decision function Ψ is set to be a c-dimensional vector-valued function for x.

$$y = \Psi(x) = (y_1, \ldots, y_i, \ldots, y_c)^t. \tag{14.9}$$

Then, if

$$y_k > y_j \quad (\forall j \neq k), \tag{14.10}$$

we classify x as class ω_k. Therefore, in supervised learning, where input pattern x is given with a c-dimensional teaching vector $\mathbf{t}_i$ (see Sect. 3.1.1) indicating its belonging class, Ψ is determined so that $y(= \Psi(x))$, the classification result, matches $\mathbf{t}_i$ as closely as possible. That is, by using the square error as a loss function

$$l(\Psi(x)|\omega_i) = \|\Psi(x) - \mathbf{t}_i\|^2, \tag{14.11}$$

Eq. (14.8) becomes

$$L(\Psi) = \sum_{i=1}^{c} P(\omega_i) \int \|\Psi(x) - \mathbf{t}_i\|^2 p(x|\omega_i) dx. \tag{14.12}$$

The above equation gives the expected value of the square error, i.e., the mean square error. Therefore, the decision function Ψ that minimizes Eq. (14.12) is called a decision based on the *least mean square error criterion*, or simply a decision based on the *least squares method*. The latter name will be used hereafter. If the decision function Ψ is an arbitrary nonlinear function, decisions based on the least-squares method are closely related to the Bayes decision rule. This is discussed in detail in the next chapter. As a teaching vector for the class ω_i satisfying Eq. (14.11), a c-dimensional one-hot vector as shown in Eq. (3.3) is used.

14.2.2 Zero-One Loss

The simplest and most natural loss function is $l(\omega_j|\omega_i)$ as shown in the following equation:

$$l(\omega_j|\omega_i) = \begin{cases} 0 & \text{if } j = i, \\ 1 & \text{otherwise.} \end{cases} \tag{14.13}$$

That is, it will give loss 1 when it misclassifies a pattern of class ω_i and loss 0 otherwise.[4] In this case, Eq. (14.2) is as follows:

$$L(\omega_j|x) = \sum_{i \neq j} P(\omega_i|x) = 1 - P(\omega_j|x). \tag{14.14}$$

Since minimizing $L(\Psi)$ is equivalent to minimizing $L(\Psi(x)|x)$, we obtain the following equation:

$$\Psi(x) = \omega_k \quad \text{if} \quad P(\omega_k|x) = \max_i\{P(\omega_i|x)\}. \tag{14.15}$$

This is called the decision rule by the *zero-one loss criterion* and is nothing but a decision rule to achieve the minimum misclassification rate (Bayes error) described in Eq. (5.18), i.e., Bayes decision rule.

In other words, when the zero-one loss criterion is used, we can confirm that the relation "expected loss minimization $\equiv$ posterior probability maximization" is established. The loss obtained in this case is then denoted as Bayes risk.

The zero-one loss criterion leads to the Bayes decision rule only when the decision function Ψ outputs the optimal classification in terms of the minimum expected loss. It should be noted that the classifier does not necessarily implement the Bayes decision rule, even if the class labels assigned to the learning patterns are correct. This is because when the distributions of each class overlap, the class obtained by the Bayes decision rule does not necessarily match the class label of the learning pattern near the decision boundary. A more practical loss function that alleviates this problem is the continuous loss criterion described in the next section.

──────────── **Coffee Break** ────────────

Too Much Can Be as Bad as Too Little

When designing a classifier by learning with given learning patterns, if the classifier is thoroughly designed to match the teaching signals associated with the learning patterns, the classifier may actually perform worse on the test pattern than if the learning is terminated early. This degradation of classification performance is due to the fact that exhaustive learning creates decision boundaries that are considerably more complex than

[4] The binary error evaluation described in Sect. 3.2.1 can be regarded as learning with a zero-one loss criterion.

those obtained by the Bayes decision rule. This problem is known as over-fitting and is generally more pronounced the higher the degrees of freedom of the classifier model, the smaller the number of learning patterns, and the higher the dimensionality of the feature vectors, as in the case of neural networks.

A simple solution to over-fitting is a simplified method called early stopping, in which a portion of the learning patterns are removed as test patterns, the classification results of the test patterns are evaluated during learning, and learning is stopped when the classification performance begins to deteriorate. The essence of learning is to estimate the probability structure behind the data. Learning methods that somehow take into account not only the learning patterns at hand but also unknown patterns are extremely important for practical use.

14.2.3 Continuous Loss

In the zero-one loss criterion, the classification result was judged binary, "correct" or "incorrect", but Amari (1967) proposed a *continuous loss criterion* in the framework of the discriminant function method, which shows not only the classification result but also the degree of misclassification (misclassification measure).

Let $g_i(x; \theta)$ be a discriminant function for the class ω_i. Here, θ is a parameter that specifies the discriminant function. The decision function by the discriminant function method is expressed by the following equation using a c-dimensional vector:[5]

$$\Psi(x; \theta) = (g_1(x; \theta), g_2(x; \theta), \ldots, g_c(x; \theta)), \tag{14.16}$$

$$\max_i \{g_i(x; \theta)\} = g_k(x; \theta) \quad \Longrightarrow \quad x \in \omega_k. \tag{14.17}$$

Assuming $g \geq 0$, Amari proposed

$$d_i(x) = \sum_{j \in S_i} \frac{1}{m_i}(g_j(x; \theta) - g_i(x; \theta)) \tag{14.18}$$

as the misclassification measure for patterns in $x \in \omega_i$, where m_i denotes the number of elements of S_i. Here, S_i is the set of class numbers of discriminant functions greater than the value of the discriminant function of class ω_i, i.e.,

$$S_i = \{j \mid g_j(x; \theta) > g_i(x; \theta)\}. \tag{14.19}$$

---

[5] In this case, the decision function is a c-dimensional vector whose elements are the discriminant functions of each class.

In order for $x(\in \omega_i)$ to be correctly classified, we must have $g_i(x; \theta) > g_j(x; \theta)$ $(\forall j \neq i)$. Therefore, Eq. (14.18) shows that when $d_i(x) \leq 0$, x is correctly classified to the degree of $|d_i(x)|$ and when $d_i(x) > 0$, x is misclassified to the degree of $d_i(x)$.

Since Eq. (14.18) is not guaranteed to be continuous with respect to the parameter θ, it is not compatible with gradient-type algorithms as a minimization technique. In contrast, as shown in the following equation, a continuous misclassification measure for θ was proposed, and a solution for $g < 0$ was also presented (Juang and Katagiri 1992).

$$d_i(x) = -g_i(x; \theta) + \left[\frac{1}{c-1} \sum_{j \neq i} g_j(x; \theta)^\eta \right]^{1/\eta}, \qquad (14.20)$$

where η is a positive constant. As this value increases, the largest value of $g_j(x; \theta), \forall j \neq i$ becomes dominant in the second term on the right-hand side of Eq. (14.20). In its limit $\eta \to \infty$, Eq. (14.20) becomes

$$d_i(x; \theta) = -g_i(x; \theta) + g_k(x; \theta). \qquad (14.21)$$

Here, $g_k(x; \theta) = \max_{j \neq i}\{g_j(x; \theta)\}$.

By introducing a misclassification measure, we can obtain the degree of good or bad classification of x, as described above, and we can reflect this degree in the loss. For example, the following function was proposed for the loss (Juang and Katagiri 1992):

$$l(\Psi(x)|\omega_i) = \frac{1}{1 + \exp(-\xi d_i)}, \qquad (14.22)$$

where ξ is a positive constant. The above loss function asymptotically approaches zero, as $d_i(x)$ gets smaller, and conversely, the loss approaches one, as $d_i(x)$ increases. Around $d_i(x) = 0$, the same amount of loss is given regardless of whether the classification result is correct or incorrect. This means that learning patterns that are located near decision boundaries and whose class labels differ from Bayes decisions are given appropriate losses, resulting in smoother decision boundaries compared to zero-one losses. The degree of smoothness depends on the parameters ξ and η. Needless to say, the degree of smoothness of the decision boundaries must be set appropriately for the problem. This setting problem is of great practical importance in relation to the discriminative performance for unknown patterns, through the problem of determining the model parameters, as already mentioned in Sect. 4.5.

Coffee Break

How to Avoid Poisoned Apples

Snow White once fell into a deep sleep after consuming a poisoned apple sent by the queen's messenger. If Snow White had possessed the skill of pattern recognition, her fate might have been different. Now let's consider two methods of judgment:

- Judgment A: Treats all apples as normal apples.
- Judgment B: Utilizes a pattern recognition technique to extract apple features and determine if it's a poisoned apple. This judgment correctly identifies 99% of normal apples as normal and 99% of poisoned apples as poisonous.

Imagine a pile of 10,000 apples containing 1% (100) poisoned apples. In this scenario, which judgment should be used? First, let's compare them in terms of error rate. Judgment A misclassifies 100 poisoned apples as normal but correctly judges the remaining 9900 apples as normal. Therefore, the error rate is 1%. On the other hand, judgment B misjudges one of the 100 poisoned apples as a normal apple, and 1% (99) of the remaining 9900 normal apples are misjudged as poisoned apples. So, 1 + 99 = 100 apples are misjudged in judgment B, resulting in a 1% error rate as well. In terms of the error rate for this pile of apples, there is no difference between judgments A and B.

Next, let's evaluate their effectiveness in preventing deaths from poisoned apples. If 10,000 people are each given an apple and told to eat it if judged normal, or discard it if judged poisonous, how many will die with judgments A and B? In judgment A, all 100 people who received the poisoned apple will die as they will eat it, believing it's normal. In contrast, with judgment B, only one of the 100 people who received the poisoned apple will die due to consuming the poison. Instead, 99 people would not be able to eat the apple despite having received a normal one. Needless to say, judgment B is superior. This is because the loss resulting from eating a poisoned apple and dying is infinitely greater than the loss from throwing away a normal apple.

The evaluation by error rate can be interpreted as assuming that the loss associated with a misjudgment is the same regardless of the type of misjudgment. For example, in the above example, the same loss of 1 is given when judging a poisoned apple as an normal apple and when judging an normal apple as a poisoned apple. In other words, the error rate is the expected loss when the zero-one loss criterion (Eq. (14.13)) is adopted. Note that the square error criterion (Eq. (14.11)) and continuous loss criterion (Eq. (14.22)) give different losses for different classes. However, they are similar to the zero-one loss criterion in that they are not loss functions that take subjectivity into account. In other words, the square error criterion and the continuous loss criterion can be interpreted as advanced versions of the zero-one loss criterion in that they consider the degree of misjudgment based on the output value of the discriminant function.

Although we have invoked a fairy tale here, the question of how losses should be set is a practical one that is often faced in practice as well. For example, a medical diagnosis from X-ray images is obvious that the loss to the patient is greater if an abnormality is detected but judged to be normal than vice versa. In addition, in character recognition, if the number represents money, it should also be obvious that the degree of loss is different when 1 is mistaken for 2 and when 1 is mistaken for 9. Even more troublesome is the fact that the degree of loss depends on whose perspective is being measured. For example, in the case of the medical diagnosis mentioned above, it is desirable for both patients and physicians to have fewer misdiagnoses. However, they may have very different views of the loss caused by misdiagnosis. Thus, it is difficult to define a loss function in general. For practical use, therefore, it is usual to construct a classifier using the objective loss criteria described in this section, and then modify it to reflect subjective losses. Taking medical diagnosis as an example, this method is equivalent to shifting the decision boundary so that any suspicious item is judged to be abnormal.

Snow White, having learned these things, has adopted a new judgment C, which judges all apples to be poisoned apples, and decides not to eat any apples in order to protect her own safety. In other words, she considered the loss of misjudging a poisoned apple as a normal apple to be infinite. However, Snow White was faced with a new problem of pattern recognition: how to distinguish an apple from a multitude of other foods. The last resort, then, is to decide not to eat any food at all. However, it is difficult to evaluate whether this decision is correct or not.

This is because it is not so easy to compare the expected loss of starvation from not eating food with the expected loss of death from poisoning by poisoned apples from eating food.

14.3 Probabilistic Decent Method

In the previous section, we discussed in detail minimum expected loss learning, and in this section, when Ψ is expressed as $\Psi(x; \theta)$ with parameter θ, we will discuss how to design Ψ to minimize the expected loss, i.e., how to estimate θ.

Following the previous notation, the loss should be written as $l(\Psi(x; \theta)|\omega_i)$, but for simplicity, it will be written as $l_i(x; \theta)$ in the following. In this case, L in Eqs. (14.5)–(14.8) can be written as a function of θ:

$$L(\theta) = \mathop{\mathrm{E}}_{x, \omega_i} \{l_i(x; \theta)\}, \tag{14.23}$$

$$= \sum_{i=1}^{c} \int l_i(x; \theta) P(\omega_i|x) p(x) dx. \tag{14.24}$$

Thus, the optimal θ is obtained as the solution of $\partial L / \partial \theta = 0$. However, in actual applications where only n patterns are given, since $p(x)$ and $P(\omega_i|x)$ are unknown, it is not possible to calculate $\partial L / \partial \theta$ directly. So, as shown below, instead of L, we consider the minimization of the *empirical loss* defined by n patterns. Specifically, if $p(x)$ in Eq. (14.24) is approximated by an *empirical distribution function*[6] representing the distribution of n given patterns

$$p(x) = \frac{1}{n} \sum_{k=1}^{n} \delta(x - x_k), \tag{14.25}$$

and further, let $P(\omega_i|x)$ in Eq. (14.24) be the following equation based on the given class labels,

$$P(\omega_i|x) = \begin{cases} 1 & \text{if } x \in \omega_i, \\ 0 & \text{otherwise.} \end{cases} \tag{14.26}$$

---

[6] This is a delta function set at n pattern positions, divided by n so that the sum is 1.

Then, the empirical loss $L_e(\theta)$ is from Eq. (14.24)

$$L_e(\theta) = \frac{1}{n} \sum_{i=1}^{c} \sum_{k=1}^{n} \int l_i(x; \theta) v(x \in \omega_i) \delta(x - x_k) dx$$

$$= \frac{1}{n} \sum_{k=1}^{n} \sum_{i=1}^{c} l_i(x_k; \theta) v(x_k \in \omega_i). \tag{14.27}$$

Here $v(x \in \omega_i)$ is a function below

$$v(x \in \omega_i) = \begin{cases} 1 & \text{if } x \in \omega_i, \\ 0 & \text{otherwise.} \end{cases} \tag{14.28}$$

Assuming L_i is differentiable, the derivative of L_e with respect to θ is

$$\frac{\partial L_e}{\partial \theta} = \frac{1}{n} \sum_{k=1}^{n} \sum_{i=1}^{c} \frac{\partial l_i(x_k; \theta)}{\partial \theta} \cdot v(x_k \in \omega_i). \tag{14.29}$$

So the θ that minimizes $L_e(\theta)$ can be estimated sequentially by the steepest descent method, as follows, even if $\partial L_e / \partial \theta = 0$ cannot be solved analytically:

$$\theta(t + 1) = \theta(t) - \rho(t) \frac{\partial L_e}{\partial \theta}$$

$$= \theta(t) - \rho(t) \frac{1}{n} \sum_{k=1}^{n} \sum_{i=1}^{c} \nabla l_i(x_k; \theta(t)) v(x_k \in \omega_i). \tag{14.30}$$

Here, t is the index of the tth iteration, and $\rho(t)$ is the learning rate and is positive:

$$\nabla l_i(x_k; \theta(t)) \stackrel{\text{def}}{=} \left. \frac{\partial l_i(x_k; \theta)}{\partial \theta} \right|_{\theta = \theta(t)} \tag{14.31}$$

A careful look at Eq. (14.30) shows that all learning patterns are used simultaneously to modify θ, and this formula represents the batch learning. In contrast, online learning, where only one pattern is presented at a time and θ is modified each time, i.e., adaptive learning under which patterns are given sequentially, is also possible. As a specific algorithm, the *probabilistic descent method* described below is available (Amari 1967).

In the probabilistic descent method, the parameter modification $\delta\theta$ is not necessarily modified in the decreasing direction of L_e, but rather in the direction of decreasing the expected value $E\{L_e\}$ for L_e. That is, for a given x, θ may be modified in the direction of increasing L_e, but for the entire x presented at a given time, θ is modified in the direction of increasing L_e since $E\{\delta L_e\} < 0$. This is exactly a probabilistic descent and is just like a drunk person attempting to go down a slope. That is, at a certain point in time, a drunk may be going up the slope, but over a certain time span, he/she will be going down the slope.

Let $\boldsymbol{\theta}(t)$ be the estimated value of $\boldsymbol{\theta}$ at the tth iteration. Suppose that $\boldsymbol{x}(t)$ is revised by $\delta\boldsymbol{\theta}$ at the $(t+1)$th iteration when $\boldsymbol{x}(t)$ is presented. That is,

$$\boldsymbol{\theta}(t+1) = \boldsymbol{\theta}(t) + \delta\boldsymbol{\theta}(t). \tag{14.32}$$

Assuming $\delta\boldsymbol{\theta}(t)$ is infinitesimal, the change in L_e associated with $\delta\boldsymbol{\theta}(t)$ is given by Taylor expansion as

$$\begin{aligned}
\delta L_e(t) &= L_e(\boldsymbol{\theta}(t) + \delta\boldsymbol{\theta}(t)) - L_e(\boldsymbol{\theta}(t)) \\
&\approx L_e(\boldsymbol{\theta}(t)) + \delta\boldsymbol{\theta}(t)^t \, \nabla L_e(\boldsymbol{\theta}(t)) + O(|\delta\boldsymbol{\theta}(t)|^2) - L_e(\boldsymbol{\theta}(t)) \\
&\approx \delta\boldsymbol{\theta}(t)^t \, \nabla L_e(\boldsymbol{\theta}(t)). \tag{14.33}
\end{aligned}$$

In the above equation, $\delta\boldsymbol{\theta}(t)^t$ represents the transpose of the vector $\delta\boldsymbol{\theta}(t)$ and O denotes the order of the computation, and $\nabla L_e(\boldsymbol{\theta}(t))$ represents

$$\nabla L_e(\boldsymbol{\theta}(t)) \overset{\text{def}}{=} \left. \frac{\partial L_e}{\partial \boldsymbol{\theta}} \right|_{\boldsymbol{\theta}=\boldsymbol{\theta}(t)}. \tag{14.34}$$

Note that $\nabla L_e(\boldsymbol{\theta}(t))$ does not depend on $\boldsymbol{x}$, and taking the expected values for $\boldsymbol{x}$ and ω_i for Eq. (14.33), we obtain

$$\underset{\boldsymbol{x},\omega_i}{\mathrm{E}} \{\delta L_e(t)\} = \underset{\boldsymbol{x},\omega_i}{\mathrm{E}} \{\delta\boldsymbol{\theta}(t)\}^t \, \nabla L_e(\boldsymbol{\theta}(t)). \tag{14.35}$$

The subscripts $\boldsymbol{x}$, ω_i are omitted hereafter for brevity. To achieve probabilistic descent, we need only $\mathrm{E}\{\delta L_e(t)\} < 0$. To do so, it is necessary that $\mathrm{E}\{\delta\boldsymbol{\theta}(t)\}$ in Eq. (14.35) can be written as

$$\mathrm{E}\{\delta\boldsymbol{\theta}(t)\} = -\rho(t) \, C \, \nabla L_e(\boldsymbol{\theta}(t)), \tag{14.36}$$

using any positive definite matrix C (but assumed to be a real symmetric matrix).[7] In fact, at this time, we can confirm that

$$\begin{aligned}
\mathrm{E}\{\delta L_e(t)\} &= \mathrm{E}\{\delta\boldsymbol{\theta}(t)\}^t \, \nabla L_e(\boldsymbol{\theta}(t)) \\
&= -\rho(t) \, \nabla L_e(\boldsymbol{\theta}(t))^t \, C \, \nabla L_e(\boldsymbol{\theta}(t)) < 0. \tag{14.37}
\end{aligned}$$

---

[7] A matrix A is said to be positive definite if the quadratic form is positive for any $\boldsymbol{x}$, i.e. $\boldsymbol{x}^t A \boldsymbol{x} > 0$. Here, as the simplest positive definite matrix, the identity matrix can be used. As for the positive definiteness is already explained in Sect. 11.2.2.

The last inequality in Eq. (14.37) is obviously due to the definition of a positive
definite matrix. Also, substituting the following relation:

$$\nabla L_e(\boldsymbol{\theta}(t)) = E\{\nabla l_i(\boldsymbol{x}(t); \boldsymbol{\theta}(t))\}, \tag{14.38}$$

from Eq. (14.24) into Eq. (14.36), we obtain

$$E\{\delta\boldsymbol{\theta}(t)\} = -\rho(t)\, C\; E\{\nabla l_i(\boldsymbol{x}(t); \boldsymbol{\theta}(t))\}. \tag{14.39}$$

Thus, for $\boldsymbol{x}(t) \in \omega_i$, we modify it as

$$\delta\boldsymbol{\theta}(t) = -\rho(t)C\nabla l_i(\boldsymbol{x}(t); \boldsymbol{\theta}(t)). \tag{14.40}$$

The validity of this modification is based on the *stochastic approximation method*
described below. From Eqs. (14.32) and (14.40), the algorithm for successive
estimation of $\boldsymbol{\theta}$ by the probabilistic descent method is as follows.

Probabilistic Descent Method

Step 1 Initialize $\boldsymbol{\theta}(0)$ arbitrarily. $t \leftarrow 0$ (Initialization)
Step 2 The following is iterated until the appropriate convergence conditions[8] are
met:

$$\boldsymbol{\theta}(t+1) = \boldsymbol{\theta}(t) - \rho(t)C \sum_{i=1}^{c} \nabla l_i(\boldsymbol{x}(t); \boldsymbol{\theta}(t))\, v(\boldsymbol{x}(t) \in \omega_i), \tag{14.41}$$

$$t \leftarrow t + 1.$$

When $\rho(t)$ satisfies the following conditions, $\boldsymbol{\theta}$ is theoretically guaranteed to
converge to $\boldsymbol{\theta}$ that gives the local minimum of L_e.

$$\sum_{t=0}^{\infty} \rho(t) = \infty \quad \text{and} \quad \sum_{t=0}^{\infty} \rho(t)^2 < \infty. \tag{14.42}$$

As one of the candidates for $\rho(t)$ satisfying the above equation, we can consider the
following:

$$\rho(t) = \frac{1}{t}. \tag{14.43}$$

---

[8] For example, $\|\boldsymbol{\theta}(t+1) - \boldsymbol{\theta}(t)\|/\|\boldsymbol{\theta}(t)\| < \text{Thd}$, Thd is the threshold.

From Eq. (14.42), we can see that for every single pattern presented in the probabilistic descent method, θ modification is made. Comparing Eqs. (14.30) and (14.42), we can see that the difference between the two is only a difference between batch type and online type. In fact, in order to apply Eq. (14.42) to the batch type, rewrite $x(t)$ as x_k in Eq. (14.42) and add $\frac{1}{n}\sum_{k=1}^{n}$, and set C as the identity matrix, then we get exactly Eq. (14.30).

The stochastic descent method can also be interpreted as a formulation of the stochastic approximation method[9] This is briefly explained below.[10]

The basic idea of the stochastic approximation method follows the *Robbins–Monroe algorithm*. Suppose now that we have functions $f(w)$ and $h(w)$ of w, and we wish to find the root of $f(w) = 0$. Given a set of pairs of $(w, h(w))$ and

$$E\{h(w)\} = f(w) \tag{14.44}$$

is assumed to hold. Suppose that the value of $h(w)$ is obtained, but the value of $f(w)$ is unknown. If a large set of pairs of $(w, h(w))$ is given at once, then we can model $f(w)$ to estimate the roots of $f(w) = 0$. But here, to deal with the case where the pattern is given sequentially, we use $(w, h(w))$ data at a time. Equation (14.44) is equivalent to the following equation with noise:

$$\begin{cases} h(w) = f(w) + \text{noise}, \\[2mm] E\{\text{noise}\} = 0. \end{cases} \tag{14.45}$$

Here, $f(w)$ is called the *regression function* of $h(w)$.

Let $\hat{w}$ be the root of $f(w) = 0$ and assume the following equation for $f(w)$:

$$\begin{cases} f(w) > 0 & \text{(for } w > \hat{w}), \\ f(w) < 0 & \text{(for } w < \hat{w}). \end{cases} \tag{14.46}$$

This assumption does not lose its generality because, for $f(w)$ showing the opposite trend, we can replace $-f(w)$ by $f(w)$ to satisfy the above equation.

Following the Robbins–Monroe algorithm, the root of $f(w) = 0$ can be estimated as follows:

$$w(t+1) = w(t) - \rho(t) \cdot h(w(t)). \tag{14.47}$$

[9] The name seems to suggest a method to estimate some probability distribution but as described in the text, it has nothing to do with estimating probability distributions. The name is indeed confusing.

[10] The basics of the stochastic approximation method and its application to the design of linear discriminant functions are described in detail in Chapter 6 of *Pattern Recognition Principles* by Tou and Gonzalez (1974).

The convergence of the algorithm is guaranteed if Eq. (14.42) is satisfied. For proof of convergence, see Fukunaga (1990). In the Robbins–Monroe algorithm, even if the value of $f(w)$ is not known, we can find the root of $f(w) = 0$ using Eq. (14.47), as long as the value of $h(w)$ in the relationship Eq. (14.44) is known.

On the other hand, the probabilistic descent method cannot compute the value of ∇L_e, but it can compute the value of ∇L_i in the relationship Eq. (14.38). That is, Eq. (14.38) corresponds to considering h as ∇L_i and f as ∇L_e in Eq. (14.44), so the Robbins–Monroe algorithm can be applied. The result is Eq. (14.42). However, h and f are single variable functions, whereas ∇L_i and ∇L_e are multivariable functions with each element of the vector $\boldsymbol{\theta}$ as a variable. Therefore, we need to have an extension of the Robbins–Monroe algorithm for multi-variables to obtain Eq. (14.42).[11]

In any case, it can be seen that the probabilistic descent method shares its basic idea with that of the stochastic approximation method.

For experiments with the Robbins–Monroe algorithm, see Problem 14.1. Also, see Sect. 3.1.3 that described Widrow–Hoff learning rule can also be derived as a result of applying the Robbins–Monroe algorithm (Problem 14.2).

Problems

14.1 Consider the function $h(w)$ and its regression function $f(w)$ defined by the Eq. (14.45), where w is a scalar and $f(w) = \cos(w)$ $(\pi \leq w \leq 2\pi)$. The noise in Eq. (14.45) is represented by a uniform random number that takes the value $-0.1 \leq$ noise ≤ 0.1. When the Robbins–Monroe algorithm Eq. (14.47) is applied iteratively, verify by experiment that $w(t)$ comes as close as possible to the root $w = 3\pi/2$ of $f(w) = 0$. Note that the initial value is set to $w(1) = 3.5$, and the learning rate $\rho(t)$ is calculated using Eq. (14.43).

14.2 * Consider the least squares approximation of a Bayes discriminant function by a linear discriminant function. Show that Widrow–Hoff learning rule can be derived as a result of applying the Robbins–Monroe algorithm to this process.

---

[11] See page 382 of Fukunaga (1990) for details.

Chapter 15
Learning Algorithms and Bayes Decision Rule

Abstract In the previous chapter, we described the learning algorithm in a more general and unified perspective in the framework of expected loss minimization and explained square error loss, zero-one loss, and continuous loss as loss functions. Furthermore, in the framework of minimum expected loss learning, we showed that the least squares method is derived for square error loss. In this chapter, learning based on the least squares method is described in detail. Specifically, we first derive the least squares solution of the decision function for the linear model and the nonlinear model. Next, the relationship between the least squares method and linear and nonlinear discriminant methods will be discussed. We also explain the relationship between the least squares method and the Bayes decision rule, and its relationship with Widrow–Hoff learning rule and the backpropagation method.

15.1 Learning by Least Squares Method

In this section, we clarify the relationship between learning by the least squares method and the discriminant method, as well as its relationship to the Bayes decision rule.

15.1.1 Least Squares Solution

Learning by the least squares method that is shown in Sect. 14.2.1, the learning method is to find the decision function Ψ that minimizes the following equation:

$$L(\Psi) = \mathop{\mathrm{E}}_{x|\omega_i} \{||\Psi(x) - t_i||^2\}, \tag{15.1}$$

Supplementary Information The online version contains supplementary material available at https://doi.org/10.1007/978-981-95-1478-6_15.

$$= \sum_{i=1}^{c} P(\omega_i) \int \|\Psi(x) - \mathbf{t}_i\|^2 p(x|\omega_i)dx. \tag{15.2}$$

As a specific algorithm for finding Ψ from the learning pattern, the probabilistic descent method was described in Sect. 14.3, but in fact, the analytical solution of Ψ that minimizes Eq. (15.2) has already been revealed (Otsu 1981).[1] In this section, we derive the analytical solution for each of the linear and nonlinear models as Ψ.

(a) Linear Models

For simplicity, consider linear models for two-class case.[2] Originally, a discriminant function is defined for each class, but in the case of a two-class problem, as mentioned in Sect. 2.3, one discriminant function $g(\mathbf{x})$ can be defined as

$$g(x) = g_1(x) - g_2(x) = \mathbf{w}^t \mathbf{x}. \tag{15.3}$$

The above equation corresponds to the case where

$$\Psi(x) = \mathbf{w}^t \mathbf{x} \tag{15.4}$$

in Eq. (14.9).[3] Therefore, Eq. (15.2) is written as

$$L(\Psi) = L(\mathbf{w}) \tag{15.5}$$

$$= P(\omega_1) \operatorname*{E}_{x \in \omega_1} \{(\mathbf{w}^t \mathbf{x} - b_1)^2 | \omega_1\} + P(\omega_2) \operatorname*{E}_{x \in \omega_2} \{(\mathbf{w}^t \mathbf{x} - b_2)^2 | \omega_2\}. \tag{15.6}$$

---

[1] In the nonlinear case, the analytical solution cannot be actually computed because it contains unknowns such as the true distribution. Be careful not to jump to the conclusion that if the analytical solution is known, there is no need to use algorithms such as probabilistic descent. In other words, the derivation of the analytical solution is a preparation for the discussion of mathematical properties such as the relationship between the least squares method, the Bayes decision rule, and discriminant analysis.

[2] In the multi-class case, the linear mapping defined by $\mathbf{A} = [\mathbf{w}_1, \mathbf{w}_2, \ldots, \mathbf{w}_{\tilde{d}}]$ is as follows:

$$\Psi(x) = \mathbf{A}^t \mathbf{x} = (\Psi_1, \Psi_2, \ldots, \Psi_{\tilde{d}})^t \qquad (\Psi_i = \mathbf{w}_i^t \mathbf{x}, \quad i = 1, 2, \ldots, \tilde{d}).$$

In this case, the decision function becomes a $\tilde{d}$-dimensional vector whose ith component outputs a real value. The optimal solution, in this case, can be derived in the same way as in the two-class case. For details, refer to Otsu (1981).

[3] See footnote 3 in Chap. 14. If $\Psi(x)$ is regarded as decision rule, we have

$$\Psi(x) = \begin{cases} \omega_1 & (g(x) > 0) \\ \omega_2 & (g(x) < 0). \end{cases}$$

Here, b_1 and b_2 are teaching signals for class ω_1 and ω_2, respectively. The notion $\underset{x \in \omega_1}{E} \{(w^t x - b_1)^2 | \omega_1\}$ is the expectation of $(w^t x - b_1)^2$ with respect to $x \in \omega_1$.

Further calculations show that

$$L(w) = P(\omega_1) \underset{x \in \omega_1}{E} \{w^t xx^t w - 2w^t x b_1 + b_1^2 | \omega_1\}$$

$$+ P(\omega_2) \underset{x \in \omega_2}{E} \{w^t xx^t w - 2w^t x b_2 + b_2^2 | \omega_2\}$$

$$= w^t R_0 w - 2w^t r + \text{const.} \tag{15.7}$$

Here, const is a term independent of w, and R_0 is

$$R_0 \overset{\text{def}}{=} \underset{x}{E}\{xx^t\} \tag{15.8}$$

$$= \underset{x}{E} \left\{ \begin{pmatrix} 1 & x^t \\ x & xx^t \end{pmatrix} \right\} = \begin{pmatrix} 1 & m^t \\ m & R \end{pmatrix} = \begin{pmatrix} 1 & m^t \\ m & \Sigma_T + mm^t \end{pmatrix}. \tag{15.9}$$

In the above equation, $R = E\{xx^t\}$ is the autocorrelation matrix and we used Eq. (6.56). In Eq. (15.7), r is

$$r = P(\omega_1) b_1 \underset{x \in \omega_1}{E} \{x | \omega_1\} + P(\omega_2) b_2 \underset{x \in \omega_2}{E} \{x | \omega_2\}$$

$$= P(\omega_1) b_1 \underset{x \in \omega_1}{E} \left\{ \begin{pmatrix} 1 \\ x \end{pmatrix} | \omega_1 \right\} + P(\omega_2) b_2 \underset{x \in \omega_2}{E} \left\{ \begin{pmatrix} 1 \\ x \end{pmatrix} | \omega_2 \right\}$$

$$= \begin{pmatrix} P(\omega_1)b_1 + P(\omega_2)b_2 \\ P(\omega_1)b_1 m_1 + P(\omega_2)b_2 m_2 \end{pmatrix}. \tag{15.10}$$

Here, setting the partial derivative with respect to w to 0, we obtain

$$\frac{\partial L(w)}{\partial w} = 2R_0 w - 2r = 0. \tag{15.11}$$

That is,

$$R_0 w = R_0 \begin{pmatrix} w_0 \\ w \end{pmatrix} = r. \tag{15.12}$$

By substituting Eqs. (15.9) and (15.10) into Eq. (15.12), we obtain

$$\begin{pmatrix} m^t w + w_0 \\ \Sigma_T w + m(m^t w + w_0) \end{pmatrix} = \begin{pmatrix} P(\omega_1)b_1 + P(\omega_2)b_2 \\ P(\omega_1)b_1 m_1 + P(\omega_2)b_2 m_2 \end{pmatrix}. \tag{15.13}$$

Using Eq. (15.13) and the relation $\mathbf{m} = P(\omega_1)\mathbf{m}_1 + P(\omega_2)\mathbf{m}_2$, we can derive for $\boldsymbol{w}$ as

$$
\begin{aligned}
\boldsymbol{\Sigma}_T \boldsymbol{w} &= -(\mathbf{m}^t \boldsymbol{w} + w_0)\mathbf{m} + P(\omega_1)b_1\mathbf{m}_1 + P(\omega_2)b_2\mathbf{m}_2 \\
&= -(P(\omega_1)b_1 + P(\omega_2)b_2)\mathbf{m} + P(\omega_1)b_1\mathbf{m}_1 + P(\omega_2)b_2\mathbf{m}_2 \\
&= k_1\mathbf{m}_1 + k_2\mathbf{m}_2.
\end{aligned}
\tag{15.14}
$$

Here, we set

$$
\left.
\begin{aligned}
k_1 &= -P(\omega_1)^2 b_1 - P(\omega_1)P(\omega_2)b_2 + P(\omega_1)b_1 \\
k_2 &= -P(\omega_2)^2 b_2 - P(\omega_1)P(\omega_2)b_1 + P(\omega_2)b_2.
\end{aligned}
\right\}
\tag{15.15}
$$

Using $P(\omega_1) + P(\omega_2) = 1$, we have

$$
\left.
\begin{aligned}
k_1 &= P(\omega_1)P(\omega_2)(b_1 - b_2) \\
k_2 &= -P(\omega_1)P(\omega_2)(b_1 - b_2).
\end{aligned}
\right\}
\tag{15.16}
$$

By substituting those for Eq. (15.14) and solving for $\boldsymbol{w}$, we obtain

$$
\boldsymbol{w} = P(\omega_1)P(\omega_2)(b_1 - b_2)\boldsymbol{\Sigma}_T^{-1}(\mathbf{m}_1 - \mathbf{m}_2).
\tag{15.17}
$$

In addition, from Eq. (15.13) we obtain

$$
\begin{aligned}
w_0 &= -\mathbf{m}^t \boldsymbol{w} + P(\omega_1)b_1 + P(\omega_2)b_2 \\
&= -P(\omega_1)P(\omega_2)(b_1 - b_2)\mathbf{m}^t \boldsymbol{\Sigma}_T^{-1}(\mathbf{m}_1 - \mathbf{m}_2) \\
&\quad + P(\omega_1)b_1 + P(\omega_2)b_2.
\end{aligned}
\tag{15.18}
$$

From the above, the decision function $\Psi(\boldsymbol{x})$ is[4]

$$
\Psi(\boldsymbol{x}) = \boldsymbol{w}^t \boldsymbol{x} + w_0.
\tag{15.19}
$$

Since

$$
\boldsymbol{w} \propto \boldsymbol{\Sigma}_T^{-1}(\mathbf{m}_1 - \mathbf{m}_2),
\tag{15.20}
$$

the direction of $\boldsymbol{w}$ does not depend on how b_1 and b_2 are taken, but note that w_0 depends on b_1 and b_2. That is, the position of the decision boundary changes depending on the way the teaching vector is taken. The teaching vector, as mentioned in Sect. 14.2.1, is usually set to a unit vector. Since we are dealing with

---

[4] As with Eq. (15.4), see footnote 3 in Chap. 14.

two-class problem, let the teaching signal for each class be $b_1 = +1$ and $b_2 = -1$ according to Eq. (3.36), and from the above results, we have

$$w = 2P(\omega_1)P(\omega_2)\Sigma_T^{-1}(\mathbf{m}_1 - \mathbf{m}_2), \tag{15.21}$$

$$w_0 = -2P(\omega_1)P(\omega_2)\mathbf{m}^t \Sigma_T^{-1}(\mathbf{m}_1 - \mathbf{m}_2) + P(\omega_1) - P(\omega_2). \tag{15.22}$$

The optimization of the linear model for the two-class problem discussed here is the process of minimizing the expected loss $L(\mathbf{w})$ defined in Eq. (15.6). Finally, we obtain Eqs. (15.21) and (15.22). Similar results are derived in other chapters. For example, the minimization of the square error $J(\mathbf{w})$ defined as Eq. (3.38) in Sect. 3.1 is the same as the attempt described in this section, and its solution is shown in Eq. (3.41) as the optimal solution of the minimum square error learning. It is easy to verify that Eq. (3.41) coincides with Eqs. (15.21) and (15.22) (Problem 15.1).

In Sect. 4.3, we introduced how to obtain the optimal linear discriminant function using an evaluation function $J = J\left(\tilde{m}_1, \tilde{m}_2, \tilde{\sigma}_1^2, \tilde{\sigma}_2^2\right)$ defined by Eq. (4.28). As an example of J, we introduced the square error in Problem 4.4. One can see that its optimal solution is also consistent with Eqs. (15.21) and (15.22) (Problem 15.2).

Furthermore, we can see that w, which is obtained by Fisher's method introduced in Sect. 6.4, also has the same form as in Eq. (15.21).

Thus, it can be seen that the expressions Eqs. (15.21) and (15.22) are comprehensive expressions for the optimal linear discriminant function for the two-class problem.

(b) Nonlinear Models

Extending the decision function Ψ to a nonlinear model, the optimal solution Ψ that minimizes Eq. (15.2) can be easily derived using calculus of variations (Otsu 1981). That is, the minimization of Eq. (15.2) is a variational problem with respect to variable function Ψ as the functional $L(\Psi)$.[5]

Now, setting

$$F(\mathbf{x}, \Psi(\mathbf{x})) \overset{\text{def}}{=} \sum_{i=1}^{c} P(\omega_i)\|\Psi(\mathbf{x}) - \mathbf{t}_i\|^2 p(\mathbf{x}|\omega_i), \tag{15.23}$$

we can see that L is the simplest form of the functional as

$$L(\Psi) = \int F(\mathbf{x}, \Psi(\mathbf{x}))d\mathbf{x}. \tag{15.24}$$

---

[5] When a certain number y corresponds to x in a domain, y is called a function of the variable x, whereas when a certain number v corresponds to a function $u(x)$ in a family of functions, $u(x)$ is called a variable function and v is called a functional depending on the variable function $u(x)$, and v is denoted by $v = v[u(x)]$.

Hence, its stationary solution must satisfy the following Euler equation:[6]

$$\frac{\partial}{\partial \Psi} F(x, \Psi(x)) = \mathbf{0}. \tag{15.25}$$

The calculation yields

$$2 \sum_{i=1}^{c} P(\omega_i)(\Psi(x) - \mathbf{t}_i) p(x|\omega_i) = \mathbf{0}. \tag{15.26}$$

Solving this for Ψ yields the following optimal solution $\Psi^*(x)$:

$$\Psi^*(x) = \sum_{i=1}^{c} \frac{P(\omega_i) p(x|\omega_i)}{p(x)} \mathbf{t}_i$$

$$= \sum_{i=1}^{c} P(\omega_i|x) \mathbf{t}_i. \tag{15.27}$$

Here, we used Bayes' theorem

$$\frac{P(\omega_i) \, p(x|\omega_i)}{p(x)} = P(\omega_i|x). \tag{15.28}$$

Thus, we see that the optimal solution of the nonlinear model under the least squares method is represented by a linear combination of the teaching vector $\mathbf{t}_i$ with the Bayes posterior probability $P(\omega_i|x)$ as the weight coefficient.

15.1.2   Least Squares Method and Discriminant Method

Intuitively, minimizing $L(\Psi)$ means minimizing the mean square error of mapping each class pattern x by $y = \Psi(x)$ around teaching vectors $\mathbf{t}_i$ in the *discriminant space*, as shown in Fig. 15.1. This means that y for $\upsilon x \in \omega_i$ are concentrated around $\mathbf{t}_i$ for easy classification. The mapping from the feature space $\mathcal{F}(\subset \mathbb{R}^d)$ to a new discriminant space of smaller dimension effective for classification is discriminant mapping. Ψ is exactly the mapping from the feature space to the discriminant space,

$$\Psi : \mathcal{F} \to \mathcal{D}.$$

When Ψ is restricted to a linear model, it is *linear discriminant mapping*, and extending Ψ to a nonlinear model, it means *nonlinear discriminant mapping*. There

---

[6] See textbooks on variational calculus.

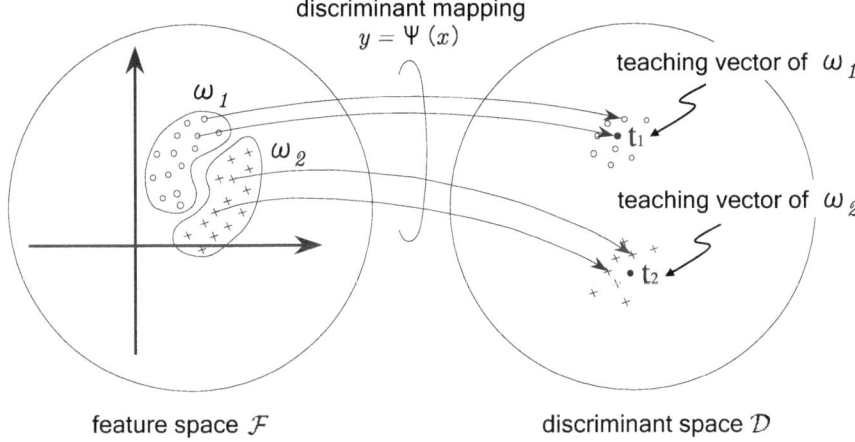

Fig. 15.1 Discriminant mapping

is a close relationship between the least squares solution obtained in the previous
section and this discrimination method. This is described in detail below.

15.1.2.1 Least Squares Method and Linear Discriminant Method

As already mentioned, the optimal solution in the linear model with $b_1 = +1$ and
$b_2 = -1$ as the teaching signal is represented by Eq. (15.19) using w in Eq. (15.21)
and w_0 in Eq. (15.22). Using Eq. (6.127), one can see that

$$w \propto \Sigma_W^{-1}(m_1 - m_2) \tag{15.29}$$

holds. This shows that in the two-class case, w obtained by the least squares method
is identical to the projection axis obtained by Fisher's method. However, it differs
from Fisher's method in that it even obtains a classification rule (decision boundary):
if $w^t x + w_0 > 0$, then x belongs to class ω_1 and if $w^t x + w_0 < 0$, then x belongs
to class ω_2.

15.1.2.2 Least Squares Method and Nonlinear Discriminant Method

In the *nonlinear discriminant method*, simultaneous optimization of maximization
of between-class variance and minimization of within-class variance is required, and

the optimal solution is actually similar to the least squares method, as shown in the following equation:[7]

$$\Psi^*(x) = \sum_{i=1}^{c} P(\omega_i | x)\, \hat{\mathbf{t}}_i. \qquad (15.30)$$

Here, the vector $\hat{\mathbf{t}}_i$ is obtained from eigenvalue problem of a confusion matrix $S = (s_{ij})$ defined by

$$s_{ij} = \int P(\omega_j | x)\, p(x | \omega_i)\, dx = P(\omega_j | \omega_i). \qquad (15.31)$$

As can be seen from the definition, s_{ij} represents the probability that a given pattern x of class ω_i is classified as class ω_j.

Comparing the optimal solution Eq. (15.30) with the optimal solution in the nonlinear model Eq. (15.27), we can see that they have the same form except for the teaching vector part. In the least squares method, as mentioned earlier, the teaching vector $\mathbf{t}_i$ $(i = 1, \ldots, c)$ is fixed in advance, and the square error $\sum_{p,i} \| y_p - \mathbf{t}_i \|^2$ between the mapping point $y_p \in \omega_i$ $(i = 1, \ldots, c)$ and its teaching vector is minimized. Here, fixing $\mathbf{t}_i$ means that between-class variance is fixed in advance, and minimizing $\sum_{p,i} \| y_p - \mathbf{t}_i \|^2$ means that variation around $\mathbf{t}_i$, that is, within-class variance is to be small. Thus, the optimal solution by the nonlinear model, in the language of discriminant analysis, can be interpreted as the minimization of the within-class variance under the pre-fixing of the between-class variance. In other words, nonlinear discriminant mapping by the least squares method corresponds to a special case of nonlinear discriminant analysis in that the between-class variance is fixed in advance. If the teaching vector is $\hat{\mathbf{t}}_i$, the nonlinear discriminant mapping obtained by the least squares method is in perfect agreement with the nonlinear discriminant method.

15.1.3 Least Squares Method and Bayes Decision Rule

(a) Linear Models

We investigate the relationship between the linear discriminant function obtained by learning by the least squares method and Bayes decision rule. In Eq. (15.6), setting $b_1 = +1$ and $b_2 = -1$, we obtain

$$L(\mathbf{w}) = P(\omega_1) \int (\mathbf{w}^t x - 1)^2 p(x | \omega_1) dx + P(\omega_2) \int (\mathbf{w}^t x + 1)^2 p(x | \omega_2) dx.$$

---

[7] Since nonlinear discriminant analysis itself is not directly related to the main focus of this chapter, i.e., minimum expected loss learning, the derivation of the optimal solution is omitted. For details, refer to the literature (Otsu 1981).

$$(15.32)$$

On the other hand, since the Bayes discriminant function $g_0(x)$ is written as

$$g_0(x) = P(\omega_1|x) - P(\omega_2|x)$$
$$= \frac{p(x|\omega_1)P(\omega_1) - p(x|\omega_2)P(\omega_2)}{p(x)} \qquad (15.33)$$

by the Bayes rule, $L(\mathbf{w})$ is as follows:

$$L(\mathbf{w}) = \int \left(\mathbf{w}^t\mathbf{x} - 1\right)^2 P(\omega_1)p(x|\omega_1)dx + \int \left(\mathbf{w}^t\mathbf{x} + 1\right)^2 P(\omega_2)p(x|\omega_2)dx$$
$$= \int \left(\mathbf{w}^t\mathbf{x}\right)^2 p(x)dx - 2 \int \mathbf{w}^t\mathbf{x}g_0(x)p(x)dx + 1$$
$$= \int \left(\mathbf{w}^t\mathbf{x} - g_0(x)\right)^2 p(x)dx + \left(1 - \int g_0^2(x)p(x)dx\right). \qquad (15.34)$$

Since the second term in Eq. (15.34) does not depend on $\mathbf{w}$, $\mathbf{w}$ that minimizes $L(\mathbf{w})$ is the $\mathbf{w}$ that minimizes the first term. Thus, the linear discriminant function, $g(x) = \mathbf{w}^t\mathbf{x}$, obtained in this way is a linear discriminant function that approximates the Bayes discriminant function $g_0(x)$ in the least squares sense.

───────────────── **Coffee Break** ─────────────────

What is the Best Linear Discriminant Function?

As described in Sect. 5.3, Bayes discriminant functions are discriminant functions that minimize misclassification, That is, they lead to Bayes errors and therefore can be regarded as ideal discriminant functions. As described in this section, learning by the least squares method produces a linear discriminant function that is a least squares approximation of the Bayes discriminant function. Therefore, the linear discriminant function obtained by the least squares method appears to be a linear discriminant function that minimizes the misclassification rate.

Fig. 15.2 Bayes discriminant function and linear discriminant function obtained by learning by the least squares method

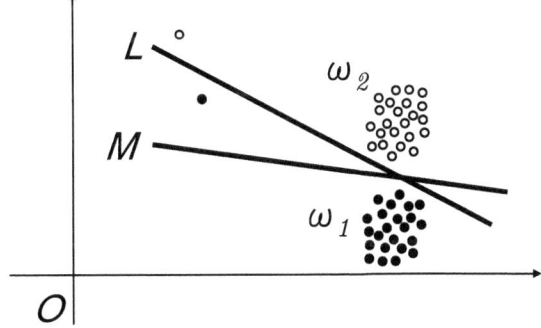

As shown in Fig. 15.2, however, this is not necessarily true. An example of a decision boundary for a linear discriminant function that minimizes the error probability in the figure is L. However, the decision boundary for a linear discriminant function that approximates the Bayes discriminant function as the least squares method is M, and the error probability is not zero. This is due to the fact that, using the criterion of minimum square error, the contribution is larger where the number of patterns is large, i.e., where $p(x)$ is large. Therefore, the linear discriminant function, which is a least squares approximation of the Bayes discriminant function, is not necessarily the best in terms of error probability. Note that as shown in Fig. 15.2, the situation is the same even for linearly separable case. The same is described in Sect. 3.2.2.

(b) Nonlinear Models

The optimal solution of a nonlinear model Eq. (15.27)

$$y^* = \Psi^*(x) = \sum_{i=1}^{c} P(\omega_i | x) t_i$$

is an extremely important relationship that links the least squares method and the Bayes decision rule. Note that

$$P(\omega_i | x) \geq 0 \quad \text{and} \quad \sum_{i=1}^{c} P(\omega_i | x) = 1. \tag{15.35}$$

It can be seen that each pattern x is geometrically transferred by the optimal map $\Psi^*(x)$ to an interior point that divides the representative point t_i of each class by the ratio of its Bayes posterior probabilities.

The space spanned by y^* is a $(c - 1)$-dimensional hyperplane, passing through c representative points t_i ($i = 1, \ldots, c$) of c classes in c-dimensional space. Then, without any loss of generality, as the teaching vector t_i for class ω_i, we can choose a c-dimensional unit vector such that the ith component is one and all other components are zero.[8] That is,

$$t_i = (0, \overset{1}{\ldots}, 0, \overset{i}{1}, 0, \ldots, \overset{c}{0}) \quad (i = 1, \ldots, c). \tag{15.36}$$

In this case, the pattern x is mapped by the optimal mapping $y^* = \Psi^*(x)$ to the following Bayes probability vector:

$$\Psi^*(x) = (P(\omega_1 | x), \ldots, P(\omega_c | x))^t \overset{\text{def}}{=} \Psi_B(x), \tag{15.37}$$

---

[8] In this case, each teaching vector is the vertex of a $(c - 1)$-dimensional simplex. This also confirms that for $c = 2$, the discriminant function needs only one dimension.

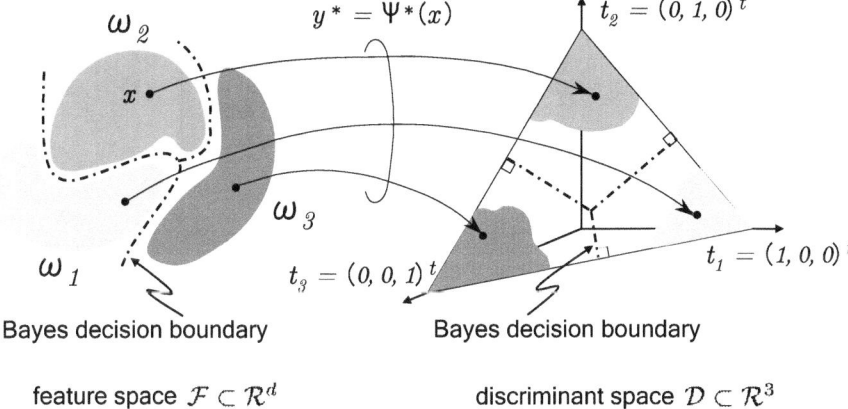

Fig. 15.3 Nonlinear mapping by the least squares method

where the ith component is the Bayes posterior probability of the class ω_i. That is, as is clear from Eqs. (14.9) and (14.10), the decision rule determined by the optimal map $y^* = \Psi_B(x)$ is perfectly consistent with the Bayes decision rule.

Let us consider how the *Bayes decision boundary* (the boundary by the Bayes decision rule) in the feature space is in the discriminant space by the optimal discriminant mapping Ψ_B. The geometric interpretation of the points mapped by nonlinear discriminant mapping based on the least squares method is as follows. For example, the three classes of patterns $x \in \mathcal{R}^d$ distributed in the d-dimensional feature space $\mathcal{F}$ as shown in Fig. 15.3 are mapped onto or within the perimeter of a triangle with $\mathbf{t}_1$, $\mathbf{t}_2$ and $\mathbf{t}_3$ as vertices. So, calculating the square distance between $y^*(= \Psi^*(x))$ and $\mathbf{t}_i$ on the discriminant plane, we obtain[9]

$$D_i^2 = \|y^* - \mathbf{t}_i\|^2$$
$$= \|y^*\|^2 - 2(y^*)^t \mathbf{t}_i + 1$$
$$= \|y^*\|^2 - 2P(\omega_i|x) + 1. \tag{15.38}$$

Thus, the minimization of the square distance D_i^2 with respect to i is equivalent to the maximization of the posterior probability. This means that in the feature space $\mathcal{F}$, the Bayes decision rule implies the class selection that maximizes the posterior probability. On the other hand, in the discriminant space $\mathcal{D}$, the class selection minimizes the square distance between y and $\mathbf{t}_i$. Therefore, as shown in Fig. 15.3, the Bayes decision boundary in the feature space $\mathcal{F}$ is a simple center-of-gravity boundary in the $(c-1)$-dimensional simplex, and the complex boundary in

---

[9] Note that $(y^*)^t \mathbf{t}_i = (P(\omega_1|x), \ldots, P(\omega_c|x))(0, \ldots, 0, 1, 0, \ldots, 0)^t = P(\omega_i|x)$.

feature space $\mathcal{F}$

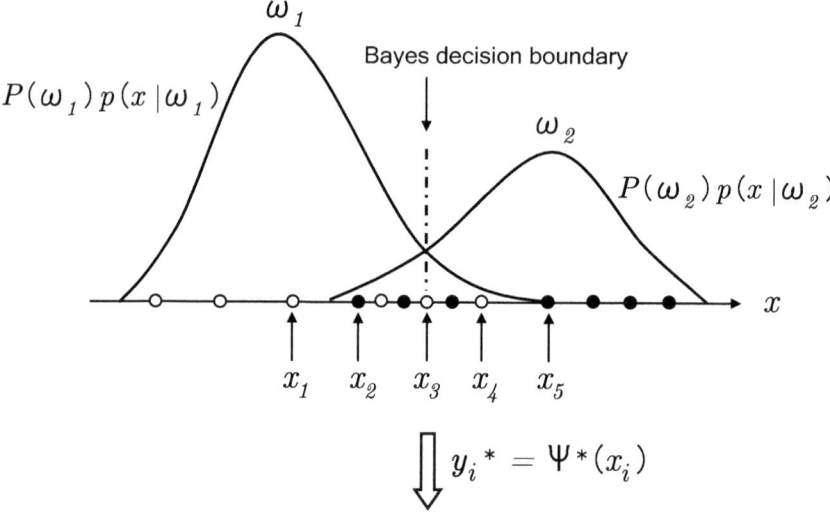

$$y_i{}^* = \Psi^*(x_i)$$

discriminant space $\mathcal{D}$

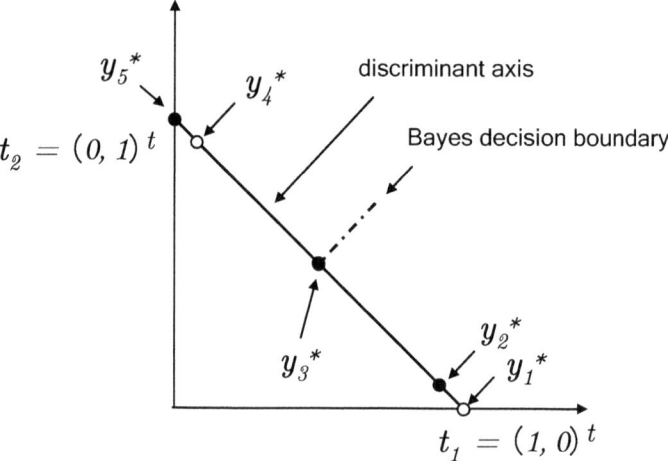

Fig. 15.4 Example of nonlinear discriminant mapping by the least squares method ($d = 1, c = 2$)

$\mathcal{F}$ is a simple linear discriminant boundary in $\mathcal{D}$. In the figure, the Bayes decision boundary is indicated by a thick single-dashed line.

Examples of nonlinear discriminant mapping by the least squares method are shown in Figs. 15.4 and 15.5 for the two-class case with one-dimensional feature and the three-class case with one-dimensional feature, respectively. As in Fig. 15.3, the Bayes decision boundary is indicated by a thick single-dashed line. First, because of

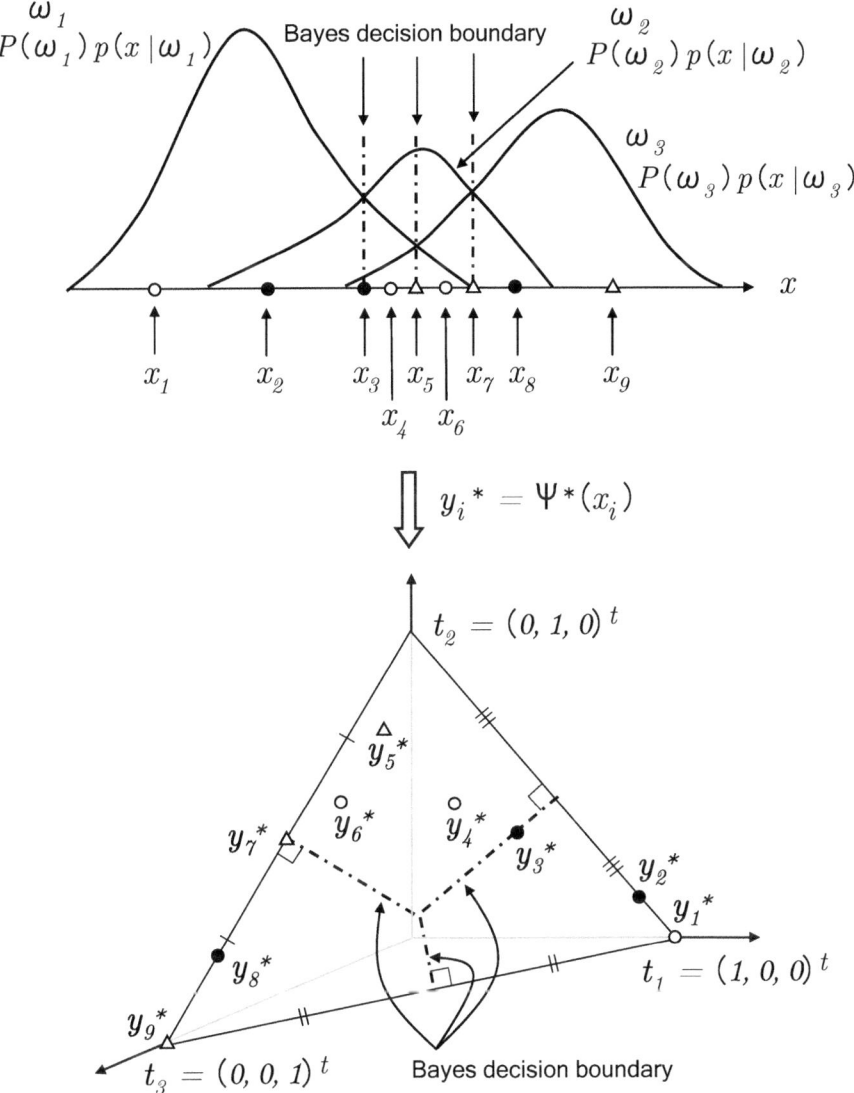

Fig. 15.5 Example of nonlinear discriminant mapping by the least squares method ($d = 1, c = 3$)

the two-class problem in Fig. 15.4, $x_i \in \mathbb{R}$ is moved to the line segment connecting two points $\mathbf{t}_1$ and $\mathbf{t}_2$ by the optimal discriminant mapping $\mathbf{y}^* = \Psi^*(x_i)$. Then, x_3 on the Bayes decision boundary in feature space is indeed transferred to the Bayes decision boundary on the discriminant axis (the midpoint of $\mathbf{t}_1$ and $\mathbf{t}_2$). Although x_2

is originally a pattern of class ω_2, in the feature space since $P(\omega_1|x_2) > P(\omega_2|x_2)$,[10] x_2 is misclassified as class ω_1 according to the Bayes decision rule. On the other hand, since $\|\mathbf{t}_1 - \mathbf{y}_2\| < \|\mathbf{t}_2 - \mathbf{y}_2\|$ holds , $\mathbf{y}_2$ is certainly misclassified as class ω_1 also in the discriminant space.

Next, since Fig. 15.5 is a three-class problem, the pattern x_i is mapped onto or within the perimeter of a triangle with three vertices $\mathbf{t}_1$, $\mathbf{t}_2$ and $\mathbf{t}_3$ by the optimal discriminant mapping. Since $P(\omega_1|x_5) = P(\omega_3|x_5) < P(\omega_2|x_5)$ in the figure, the point x_5 on the Bayes decision boundary of the classes ω_1 and ω_3 in the feature space is mapped to a point satisfying $\|\mathbf{t}_1 - \mathbf{y}_5^*\| = \|\mathbf{t}_3 - \mathbf{y}_5^*\| > \|\mathbf{t}_2 - \mathbf{y}_5^*\|$ inside the above triangle. For x_1 and x_9, $P(\omega_1|x_1) = 1$ and $P(\omega_3|x_9) = 1$, so they are mapped to $\mathbf{y}_1^* = \mathbf{t}_1$ and $\mathbf{y}_9^* = \mathbf{t}_3$ respectively.

15.2 The Least Squares Method and Various Learning Methods

This section describes the relationship between the least squares method and Widrow–Hoff learning rule, and the relationship between the least squares method and the backpropagation method.

15.2.1 Least Squares Method and Widrow–Hoff Learning Rule

In pattern classification by linear discriminant function, the discriminant function g_i of class ω_i is defined as

$$g_i(\mathbf{x}) = \mathbf{w}_i^t \mathbf{x}, \tag{15.39}$$

and $\mathbf{x}$ and $\mathbf{w}_i$ are set to the following equations:

$$\mathbf{x} = \begin{pmatrix} 1 \\ x \end{pmatrix}, \quad \mathbf{w}_i = \begin{pmatrix} w_{i0} \\ \mathbf{w}_i \end{pmatrix}. \tag{15.40}$$

In learning, for a pattern x of class ω_i, the parameters $\mathbf{w}_i$ $(i = 1, \ldots, c)$ are determined so that

$$g_i(\mathbf{x}) > g_j(\mathbf{x}) \quad \forall j \neq i. \tag{15.41}$$

The perceptron learning rule tries to faithfully realize Eq. (15.41) for each class of patterns x, but if the distribution of each class is not linearly separable, it cannot completely realize Eq. (15.41) and the perceptron learning rule will not converge.

---

[10] Bayes' theorem $P(\omega)p(x|\omega) = P(\omega|x)p(x)$ is used.

To address this problem, Widrow and Hoff proposed a learning method that minimizes the square error between the actual value of the discriminant function and the value of the teaching signal which is predefined by a desired output value for each input learning pattern. Widrow–Hoff learning rule is, as shown in Eq. (3.18), derived by minimizing

$$J(\mathbf{w}_1, \mathbf{w}_2, \ldots, \mathbf{w}_c) = \frac{1}{2} \sum_{i=1}^{c} \|\mathbf{X}\mathbf{w}_i - \mathbf{b}_i\|^2 \tag{15.42}$$

by the steepest descent method. Here,

$$\begin{cases} \mathbf{X} = (\mathbf{x}_1, \mathbf{x}_2, \ldots, \mathbf{x}_n)^t \\ \mathbf{b}_i = (b_{i1}, b_{i2}, \ldots, b_{in})^t \end{cases} \quad (i = 1, 2, \ldots, c). \tag{15.43}$$

In the above equation, n is the total number of patterns. Here, setting

$$\Psi(\mathbf{x}) = (\mathbf{w}_1^t \mathbf{x}, \ \mathbf{w}_2^t \mathbf{x}, \ \ldots, \ \mathbf{w}_c^t \mathbf{x})^t, \tag{15.44}$$

and let $\mathbf{t}_i$ be a c-dimensional unit vector shown in Eq. (15.36), then Eq. (15.42) can be rewritten as follows with a slight modification of the equation (Problem 15.3):

$$J(\Psi) = \frac{1}{2} \sum_{p=1}^{n} \sum_{i=1}^{c} \|\Psi(\mathbf{x}_p) - \mathbf{t}_i\|^2 \cdot v(\mathbf{x}_p \in \omega_i). \tag{15.45}$$

Here, $v(\cdot)$ is a function defined by Eq. (14.28).

On the other hand, in the empirical loss in Eq. (14.27), let the loss $l_i(\mathbf{x}_p; \boldsymbol{\theta})$ be

$$l_i(\mathbf{x}_p; \boldsymbol{\theta}) = \|\Psi(\mathbf{x}_p) - \mathbf{t}_i\|^2, \tag{15.46}$$

then Eq. (15.45) agrees with Eq. (14.27) except for a constant factor that is irrelevant to the design of the classifier. In other words, Eq. (15.45) is an approximation of the expected loss with the square error as the loss function by the empirical loss based on the learning pattern. From the above, it can be concluded that Widrow–Hoff learning rule is a rule for realizing the linear discriminant mapping based on the least squares method.

15.2.2   Least Squares Method and Backpropagation Method

When a multi-layer neural network is used for a c-class pattern classification problem, the output of the neural network for the input vector $\mathbf{x}$ is a nonlinear vector-valued function as

$$\mathbf{y} = \mathbf{f}(\mathbf{x}, \mathbf{v}). \tag{15.47}$$

Here, $\mathbf{v}$ is a parameter vector consisting of all the weights and $\mathbf{y}$ is a c-dimensional vector. In learning a neural network based on the backpropagation method, the weights are modified to minimize the square error between the c-dimensional unit vector $\mathbf{t}_i$ and $\mathbf{f}(\mathbf{x}, \mathbf{v})$. Here, for $\mathbf{x} \in \omega_i$, the ith component of the unit vector is one and the others are zero. This makes the decision function

$$\Psi(\mathbf{x}) = \mathbf{f}(\mathbf{x}, \mathbf{v}), \tag{15.48}$$

which is nothing more than learning a nonlinear discriminant function based on the least squares method. Therefore, from the discussion of Sect. 15.1, it can be seen that a neural network can best approximate the Bayes discriminant function within the complexity expressed by its network structure. In other words, neural networks are nonlinear discriminant functions that can approximate Bayes discriminant functions more accurately than linear discriminant functions. However, it should be noted that if the number of internal units is increased more than necessary to improve the approximation accuracy of the neural network, the classification performance will decrease. This is due to the fact that as the number of internal units increases, the degree of freedom of the neural network increases, and the output of the neural network becomes very sensitive to fluctuations in the learning pattern and initial values of parameters (variance increases). Nonlinear models such as neural networks can only reach their potential performance when these characteristics are well understood and used effectively.

As a practical method to reduce the variance of a neural network and to stabilize it appropriately for a task, *weight decay parameter* and ensemble learning are proposed. The former is a method that tries to decrease variance by smoothing the function estimated by a neural network, a kind of *regularization* method. The weight decay parameter controls the degree of stabilization and is also called the *regularization parameter*. Specifically, the new objective function is the result of adding the square of the norm[11] of the neural network weights $||\mathbf{v}||$ times the constant λ to the usual objective function in the backpropagation method, as shown in the following equation:

$$J(\mathbf{v}) = \sum_{i=1}^{c} \sum_{x \in \omega_i} ||\mathbf{f}(\mathbf{x}, \mathbf{v}) - \mathbf{t}_i||^2 + \lambda \, ||\mathbf{v}||^2. \tag{15.49}$$

As can be seen from the above equation, the second term on the right-hand side is the regularization term, where λ is the weight decay parameter. In other words, the second term on the right-hand side plays the role of a penalty term to make the norm of the weights as small as possible. The larger the value of λ, the fewer degrees of freedom the neural network model has, resulting in smoother class decision boundaries. It is important to note that the norm of the weights is used as

---

[11] The absolute norm is sometimes used.

the regularization term, instead of the curvature of the function used in the usual regularization. Although this method appears to be limited, a theoretical analysis of its validity has been done. For more details, refer to Chaps. 9 and 10 of Bishop (1995). The value of λ is obtained by applying the hyperparameter determination method of Sect. 4.5.

The latter, ensemble learning, is a learning method in which M neural networks[12] $\mathbf{f}_1(\mathbf{x}, \mathbf{v}), \ldots, \mathbf{f}_M(\mathbf{x}, \mathbf{v})$ are learnt independently on the same task using learning patterns, and then the average (generally, weighted average) value of the outputs of these neural networks for a certain input is used as the output for the input. That is, in ensemble learning, the output $\mathbf{f}_{ens}$ for $\mathbf{x}$ is calculated using weights[13] $\alpha_m (m = 1, \ldots, M)$ as

$$\mathbf{f}_{ens}(\mathbf{x}, \mathbf{v}) = \sum_{m=1}^{M} \alpha_m \mathbf{f}_m(\mathbf{x}, \mathbf{v}). \tag{15.50}$$

For regression problems (fitting a function $y = f(x; \theta)$ from a real-valued input-output pairs $(x_i, y_i)(i = 1, \ldots, N)$), the analysis of the generalization error by ensemble learning is discussed in Ueda and Nakano (1996). For the classification problems, Breiman (1997) provides more details. Readers who are using or considering using neural networks as classifiers should be familiar with these topics.

Problems

15.1 Show that the optimal solution for the least squares error shown in Eq. (3.41),

$$\mathbf{w} = (\mathbf{X}^t \mathbf{X})^{-1} \mathbf{X}^t \mathbf{b},$$

agrees with Eqs. (15.21) and (15.22).

15.2 In Problem 4.4 in Chap. 4, as the weights $\mathbf{w}$ and w_0 of the optimal linear discriminant function for two-class case, we obtained

$$\mathbf{w} = a \, \Sigma_W^{-1} (\mathbf{m}_1 - \mathbf{m}_2),$$
$$w_0 = -\mathbf{m}^t \mathbf{w} + P(\omega_1) - P(\omega_2).$$

Show that the above equation agrees with Eqs. (15.21) and (15.22) by using Eq. (6.134).

15.3 Derive Eq. (15.45) from Eq. (15.42). Equation (3.11) can be used.

---

[12] They need not be the same model.
[13] There are several types of ensembles.

Appendix A
Proof of the Perceptron Convergence Theorem

In the following, we deal with the two-class case.

A total of n learning patterns $\mathbf{x}_1$, $\mathbf{x}_2$, ..., $\mathbf{x}_n$ are prepared. Note that $\mathbf{x}_p$ ($p = 1, \ldots, n$) is an augmented feature vector. Each pattern belongs to one of the classes ω_1 or ω_2, and they are linearly separable.

Using the augmented weight vector $\mathbf{w}$, the linear discriminant function $g(\mathbf{x})$ is represented by

$$g(\mathbf{x}) = \mathbf{w}^t \mathbf{x}, \tag{A.1}$$

and $\mathbf{w}$ is determined so that

$$\begin{cases} g(\mathbf{x}) = \mathbf{w}^t \mathbf{x} > 0 & \text{(for all } \mathbf{x} \in \omega_1) \\ g(\mathbf{x}) = \mathbf{w}^t \mathbf{x} < 0 & \text{(for all } \mathbf{x} \in \omega_2). \end{cases} \tag{A.2}$$

By replacing all patterns of $\mathbf{x} \in \omega_2$ by their negatives, Eq. (A.2) can be summarized as follows:

$$g(\mathbf{x}) = \mathbf{w}^t \mathbf{x} > 0 \qquad \text{(for all } \mathbf{x}). \tag{A.3}$$

The learning patterns that cannot be correctly classified during the error-correction process of the perceptron are registered in order as

$$\mathbf{x}^1, \mathbf{x}^2, \ldots, \mathbf{x}^k, \ldots, \tag{A.4}$$

where $\mathbf{x}^k$ represents the kth pattern that was not correctly classified and

$$\mathbf{x}^k \in \{\mathbf{x}_1, \mathbf{x}_2, \ldots, \mathbf{x}_n\} \qquad (k = 1, 2, \ldots). \tag{A.5}$$

© The Author(s), under exclusive license to Springer Nature Singapore Pte Ltd. 2026
K. Ishii et al., *Pattern Recognition and Machine Learning for Self-Study I*, Springer
Asia Pacific Mathematics Series 1, https://doi.org/10.1007/978-981-95-1478-6

An example with 4 patterns ($n = 4$) is shown in Fig. A.1. Even if the pattern is the same, when it is not correctly classified across multiple epochs, the number on the right shoulder of $\mathbf{x}$ is changed each time it is registered. For example, in Fig. A.1, both $\mathbf{x}^3$ and $\mathbf{x}^5$ refer to the pattern $\mathbf{x}_4$.

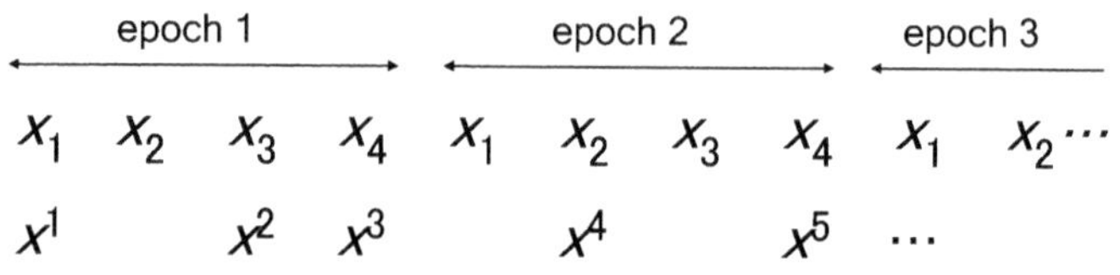

Fig. A.1 Sequence of patterns in the learning process

Since the learning rate $\rho (> 0)$ can be set arbitrarily, let $\rho = 1$ for simplicity.

When the initial value $\mathbf{w}^1$ of the weight vector is set arbitrarily and a pattern $\mathbf{x}^k$ that cannot be correctly classified occurs, the weight vector $\mathbf{w}^k$ is modified to $\mathbf{w}^{k+1}$ according to the perceptron learning rule as follows:

$$\mathbf{w}^{k+1} = \mathbf{w}^k + \mathbf{x}^k \qquad (k \geq 1). \tag{A.6}$$

Assuming that $\hat{\mathbf{w}}$ is one of the solution weight vectors, then

$$\hat{\mathbf{w}}^t \mathbf{x}_p > 0 \qquad (p = 1, 2, \ldots, n). \tag{A.7}$$

Here, using the constant $\alpha \, (> 0)$, the next formula holds from Eq. (A.7):

$$\alpha \hat{\mathbf{w}}^t \mathbf{x}_p > 0 \qquad (p = 1, 2, \ldots, n). \tag{A.8}$$

So, the weight vector $\alpha \hat{\mathbf{w}}$ is also a solution.

In the following, we show that the weight approaches infinitely close to $\alpha \hat{\mathbf{w}}$ by iteration[1] (Fig. A.2).

Subtracting $\alpha \hat{\mathbf{w}}$ from both sides of Eq. (A.6) gives the following equation:

$$\left(\mathbf{w}^{k+1} - \alpha \hat{\mathbf{w}} \right) = \left(\mathbf{w}^k - \alpha \hat{\mathbf{w}} \right) + \mathbf{x}^k \qquad (k \geq 1). \tag{A.9}$$

Taking the norm of both sides and squaring gives

$$\left\| \mathbf{w}^{k+1} - \alpha \hat{\mathbf{w}} \right\|^2$$

---

[1] If the norm of $\hat{\mathbf{w}}$ is small and $\hat{\mathbf{w}}$ is located near the origin with respect to the initial value $\mathbf{w}^1$, the weight moves away from $\hat{\mathbf{w}}$ with each iteration, as shown in Fig. A.2. For any $\hat{\mathbf{w}}$ in the solution domain, there exists some α such that the weight always converges to $\alpha \hat{\mathbf{w}}$. The value of such α is shown in Eq. (A.17).

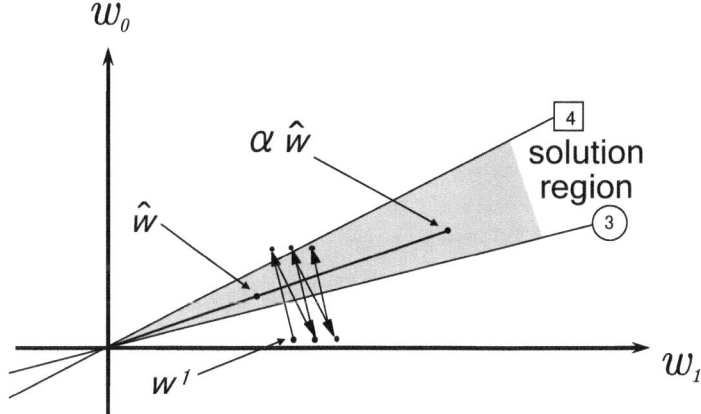

Fig. A.2 Moving weight vector by learning

$$= \left\| \mathbf{w}^k - \alpha \hat{\mathbf{w}} \right\|^2 + 2 \left(\mathbf{w}^k - \alpha \hat{\mathbf{w}} \right)^t \mathbf{x}^k + \left\| \mathbf{x}^k \right\|^2 \tag{A.10}$$

$$= \left\| \mathbf{w}^k - \alpha \hat{\mathbf{w}} \right\|^2 + 2 \left(\mathbf{w}^k \right)^t \mathbf{x}^k - 2\alpha \hat{\mathbf{w}}^t \mathbf{x}^k + \left\| \mathbf{x}^k \right\|^2. \tag{A.11}$$

As shown in Eq. (A.4), $\mathbf{x}^k$ is a pattern that was not correctly classified, so from Eq. (A.3),

$$g(\mathbf{x}^k) = \left(\mathbf{w}^k \right)^t \mathbf{x}^k \leq 0. \tag{A.12}$$

Therefore, Eq. (A.11) is further transformed to obtain

$$\left\| \mathbf{w}^{k+1} - \alpha \hat{\mathbf{w}} \right\|^2 \leq \left\| \mathbf{w}^k - \alpha \hat{\mathbf{w}} \right\|^2 - 2\alpha \hat{\mathbf{w}}^t \mathbf{x}^k + \left\| \mathbf{x}^k \right\|^2. \tag{A.13}$$

Let us define β and γ as follows:

$$\beta \overset{\text{def}}{=} \max_{p=1, \ldots, n} \| \mathbf{x}_p \|, \tag{A.14}$$

$$\gamma \overset{\text{def}}{=} \min_{p=1, \ldots, n} \hat{\mathbf{w}}^t \mathbf{x}_p > 0. \tag{A.15}$$

From Eq. (A.13), the following inequality holds:

$$\left\| \mathbf{w}^{k+1} - \alpha \hat{\mathbf{w}} \right\|^2 \leq \left\| \mathbf{w}^k - \alpha \hat{\mathbf{w}} \right\|^2 - 2\alpha \gamma + \beta^2. \tag{A.16}$$

If α is set to

$$\alpha = \frac{\beta^2}{\gamma},$$ (A.17)

then the following relationship holds:

$$\left\| \mathbf{w}^{k+1} - \alpha\hat{\mathbf{w}} \right\|^2 \leq \left\| \mathbf{w}^k - \alpha\hat{\mathbf{w}} \right\|^2 - \beta^2$$ (A.18)

$$\leq \left\| \mathbf{w}^{k-1} - \alpha\hat{\mathbf{w}} \right\|^2 - 2\beta^2$$ (A.19)

$$\cdots$$

$$\leq \left\| \mathbf{w}^1 - \alpha\hat{\mathbf{w}} \right\|^2 - k\beta^2.$$ (A.20)

As a result, the following inequality is finally obtained:

$$0 \leq \left\| \mathbf{w}^{k+1} - \alpha\hat{\mathbf{w}} \right\|^2 \leq \left\| \mathbf{w}^1 - \alpha\hat{\mathbf{w}} \right\|^2 - k\beta^2.$$ (A.21)

Here, k_0 is set as follows:

$$k_0 = \frac{\left\| \mathbf{w}^1 - \alpha\hat{\mathbf{w}} \right\|^2}{\beta^2}.$$ (A.22)

Since $\left\| \mathbf{w}^{k+1} - \alpha\hat{\mathbf{w}} \right\|^2$ cannot be negative, when k is increased, the process always converges at a number of modifications less than k_0.

Appendix B
Derivatives of Vectors and Matrices

In the following, formulas for derivatives with respect to vectors and matrices are summarized. The formulas discussed below are primarily those used in this book. For other formulas or for more details, refer to the appendices of Fukunaga (1990), Duda et al. (2001), and Bishop (2006).

First, the notations used here are shown below. The notation $\mathbf{A} = (a_{ij})$ indicates that the (i, j) component of the matrix $\mathbf{A}$ is a_{ij}. Also, a *scalar function* is a function whose output is a scalar, and a *vector function* is a function whose output is a vector.

$\boldsymbol{x} = (x_1, x_2, \ldots, x_d)^t$	: d-dimensional column vector
$\boldsymbol{y} = (y_1, y_2, \ldots, y_d)^t$	: d-dimensional column vector
$\mathbf{a} = (a_1, a_2, \ldots, a_d)^t$	: d-dimensional column vector
$f(\boldsymbol{x})$	: scalar function of vector $\boldsymbol{x}$ as a variable
$f_i(\boldsymbol{x}) \quad (i = 1, \ldots, m)$	: scalar function of vector $\boldsymbol{x}$ as a variable
$\mathbf{f}(\boldsymbol{x}) = (f_1(\boldsymbol{x}), \ldots, f_m(\boldsymbol{x}))^t$	: m-dimensional vector function (note the boldface)
$\mathbf{A} = (a_{ij})$	: square matrix of $d \times d$
$\mathbf{Y} = (y_{ij})$	: square matrix of $d \times d$
$\mathbf{S} = (s_{ij})$	: symmetric matrix of $d \times d$
$\mathbf{B} = (b_{ij})$	: rectangular matrix of $m \times d$
$\mathbf{C} = (c_{ij})$	: rectangular matrix of $d \times m$
$\mathbf{X} = (x_{ij})$	: rectangular matrix of $d \times m$
$|\mathbf{A}|, \quad \det(\mathbf{A})$	: *determinant* of $\mathbf{A}$
$\operatorname{tr}(\mathbf{A}) = \sum_{i=1}^{d} a_{ii}$	: *trace* of $\mathbf{A}$ (the sum of the diagonal elements)

© The Author(s), under exclusive license to Springer Nature Singapore Pte Ltd. 2026   411
K. Ishii et al., *Pattern Recognition and Machine Learning for Self-Study I*, Springer
Asia Pacific Mathematics Series 1, https://doi.org/10.1007/978-981-95-1478-6

First, the derivatives with respect to vectors and matrices are defined as follows:

$$\frac{\partial f}{\partial \boldsymbol{x}} \stackrel{\text{def}}{=} \left(\frac{\partial f}{\partial x_1}, \ldots, \frac{\partial f}{\partial x_d} \right)^t, \tag{B.1}$$

$$\frac{\partial \mathbf{f}}{\partial \boldsymbol{x}} \stackrel{\text{def}}{=} \left(\frac{\partial f_i}{\partial x_j} \right) = \begin{pmatrix} \dfrac{\partial f_1}{\partial x_1} & \cdots & \dfrac{\partial f_1}{\partial x_d} \\ \vdots & \ddots & \vdots \\ \dfrac{\partial f_m}{\partial x_1} & \cdots & \dfrac{\partial f_m}{\partial x_d} \end{pmatrix}, \tag{B.2}$$

$$\frac{\partial f}{\partial \mathbf{X}} \stackrel{\text{def}}{=} \left(\frac{\partial f}{\partial x_{ij}} \right) = \begin{pmatrix} \dfrac{\partial f}{\partial x_{11}} & \cdots & \dfrac{\partial f}{\partial x_{1m}} \\ \vdots & \ddots & \vdots \\ \dfrac{\partial f}{\partial x_{d1}} & \cdots & \dfrac{\partial f}{\partial x_{dm}} \end{pmatrix}. \tag{B.3}$$

Equation (B.1) is the derivative of a scalar function with respect to a vector. Equation (B.2) is the derivative of a vector function with respect to a vector, so that the resulting $\partial \mathbf{f}/\partial \boldsymbol{x}$ is a matrix of $m \times d$. Equation (B.3) is the derivative of a scalar function with respect to a matrix.

The following are examples of the derivative of a scalar function with respect to a vector:

$$\frac{\partial}{\partial \boldsymbol{x}}(\mathbf{a}^t \boldsymbol{x}) = \frac{\partial}{\partial \boldsymbol{x}}(\boldsymbol{x}^t \mathbf{a}) = \mathbf{a}, \tag{B.4}$$

$$\frac{\partial}{\partial \boldsymbol{x}}(\boldsymbol{x}^t \mathbf{A} \boldsymbol{x}) = (\mathbf{A} + \mathbf{A}^t)\boldsymbol{x}. \tag{B.5}$$

The following is an example of the derivative of a vector function with respect to a vector:

$$\frac{\partial}{\partial \boldsymbol{x}}(\mathbf{B}\boldsymbol{x}) = \mathbf{B}. \tag{B.6}$$

The following are examples of derivatives of a scalar function with respect to a matrix:

$$\frac{\partial}{\partial \mathbf{X}}\text{tr}(\mathbf{X}\mathbf{B}) = \frac{\partial}{\partial \mathbf{X}}\text{tr}(\mathbf{B}\mathbf{X}) = \mathbf{B}^t, \tag{B.7}$$

$$\frac{\partial}{\partial \mathbf{X}} \mathrm{tr}(\mathbf{X}^t \mathbf{C}) = \frac{\partial}{\partial \mathbf{X}} \mathrm{tr}(\mathbf{C} \mathbf{X}^t) = \mathbf{C}, \tag{B.8}$$

$$\frac{\partial}{\partial \mathbf{X}} \mathrm{tr}(\mathbf{X}^t \mathbf{A} \mathbf{X}) = (\mathbf{A} + \mathbf{A}^t)\mathbf{X}, \tag{B.9}$$

$$\frac{\partial}{\partial \mathbf{A}} \log |\mathbf{A}| = \left(\mathbf{A}^{-1}\right)^t, \tag{B.10}$$

$$\frac{\partial}{\partial \mathbf{Y}} \mathrm{tr}(\mathbf{Y}^{-1} \mathbf{A}) = -\mathbf{Y}^{-1} \mathbf{A} \mathbf{Y}^{-1}. \tag{B.11}$$

Only the results are described above, but the validity of these formulas can be easily proved by comparing both sides of the equations element by element of a vector or a matrix.

Although not a differential operation, the following equations are also often used:

$$\mathbf{x}^t \mathbf{y} = \mathrm{tr}(\mathbf{x} \mathbf{y}^t) = \mathrm{tr}(\mathbf{y} \mathbf{x}^t), \tag{B.12}$$

$$\left|\mathbf{A}^{-1}\right| = |\mathbf{A}|^{-1}, \tag{B.13}$$

$$|a\mathbf{A}| = a^d |\mathbf{A}|, \tag{B.14}$$

where a is an arbitrary constant.[1] Using Eq. (B.12), we obtain the following equation:

$$\mathbf{x}^t \mathbf{S} \mathbf{y} = \mathrm{tr}(\mathbf{x} \mathbf{y}^t \mathbf{S}). \tag{B.15}$$

Equations (B.12) and (B.15) are formulas that convert the inner product of vectors to a trace of a matrix, and are useful when transforming formulas.

---

[1] Care must be taken not to set $|a\mathbf{A}| = a|\mathbf{A}|$.

Appendix C
Glucksman's Feature

This appendix describes *Glucksman's feature* (Glucksman 1967). This feature was invented by Herbert A. Glucksman for character recognition. This feature extraction method focuses on the parts other than character strokes, i.e., background parts. First, the character pattern is binarized, and then the circumscribed rectangle of the pattern is set. Using a handwritten character "5" as an example, the result of binarization and setting the circumscribed rectangle is shown in Fig. C.1. The method is explained below using this figure.

Let A be an arbitrary point in the background area. As shown in the figure, straight lines are extended from point A in the vertical and horizontal directions, and the number of times they intersect with the character strokes is counted. In other words, for each direction, 0 is assigned if the line does not intersect with any character strokes, 1 is assigned if it intersects once, and 2 is assigned if it intersects twice or more times. Thus, for point A in the figure, the code "1210" is assigned in the order of top, bottom, left, and right. All points including point A in the gray area of the figure are assigned the code "1210". Similarly, the gray area including point B is assigned the code "0101". Since either 0, 1, or 2 is assigned to each direction, each point is assigned one of $3^4 = 81$ different codes.[1]

By applying the above process to all points in the background area, the background area is divided into multiple areas with the same code. In the figure, the boundaries of these regions are indicated by thin lines. The area of each region is calculated and represented as a vector, which is Glucksman's feature.

As mentioned above, there are 81 kinds of codes, so Glucksman's feature x for one character pattern is represented as an 81-dimensional ($d = 81$) vector

$$x = (x_1, x_2, \ldots, x_{81})^t, \tag{C.1}$$

---

[1] There are 81 types of codes from "0000" to "2222", but within the circumscribed rectangle, the lines always intersect the character strokes in one of the directions, so the region with the code "0000" does not occur.

K. Ishii et al., *Pattern Recognition and Machine Learning for Self-Study I*, Springer Asia Pacific Mathematics Series 1, https://doi.org/10.1007/978-981-95-1478-6

Fig. C.1 Glucksman's
feature

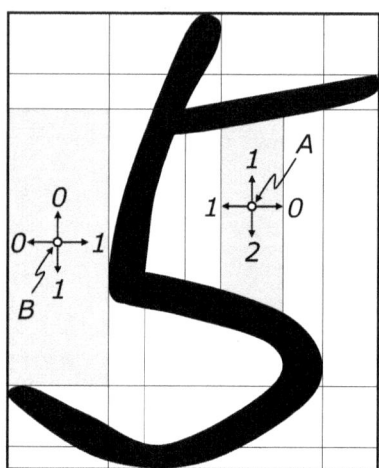

where each element x_j $(j = 1, \ldots, 81)$ of the vector is the area of the region with
each corresponding code.

The above is the original Glucksman's feature. In this book, it has been partially
modified to include the following features.

One of them is that the number of intersections is limited to two types, 0 and
1, instead of three types, 0, 1, and 2. In other words, the feature is a vector that
represents the result of observing only the presence or absence of character strokes
in each direction. In this case, the dimension d of the feature vector is $d = 2^4 = 16$.

The other is a feature with the number of intersections extended to 0, 1, 2, and 3,
in which case the dimension of the feature vector d is $d = 4^4 = 256$.

The elements x_j $(j = 1, \ldots, d)$ of these feature vectors all represent the area
of the corresponding code region. In this book, x_j is normalized by the area of
the circumscribed rectangle, so $0 < x_j < 1$. In this book, the 16-, 81-, and 256-
dimensional features described above are all referred to as Glucksman's features.
Originally, Glucksman's feature was designed for printed character recognition and
is not necessarily suitable for recognizing characters with large deformations such
as handwritten characters. However, the novel approach of this method, which
focuses on the background rather than the character strokes, has had a significant
impact on subsequent character recognition research. It is also a great advantage of
Glucksman's feature that it can be realized with simple processing without requiring
any complicated preprocessing.

Appendix D
Data for Experiments

This appendix summarizes the data used in the experiments in this book. The data is the MNIST (Mixed National Institute of Standards and Technology database) , a data set of handwritten digit patterns often used in the field of pattern recognition and machine learning. This data set contains 10 classes of numerals 0–9, 60,000 patterns for learning and 10,000 patterns for testing.[1] From the above learning patterns, 1000 patterns per class were selected, and a total of 10,000 patterns were used as the learning patterns in the experiments. In addition, 800 patterns per class were selected from the above test patterns, and a total of 8000 patterns were used as the test patterns.

Each pattern is a grayscale image with a size of 28×28 pixels, and each pixel has a gray level value from 0 to 255. The gray level value of each pixel in this image was normalized to a value between 0 and 1 by being divided by 255, and used as the original pattern. A part of the MNIST patterns are shown in Fig. D.1. If each original pattern is regarded as a feature vector as it is, its dimensionality is $d = 28 \times 28 = 784$. The experimental learning and test patterns obtained in this way are called MSH784 and MSH784-T, respectively.

In addition to the above, three more types of data were prepared as experimental feature vectors. That is, after binarizing the original patterns, three types of feature vectors with dimensionality $d = 16$, 81 and 256 were created by Glucksman's feature extraction method (Appendix C). For binarization, the method proposed by Otsu (1979) was used. These three experimental data are called GLK16, GLK81 and GLK256, respectively.

---

[1] The number of patterns per class for both learning and test patterns is not identical across classes, but differs somewhat.

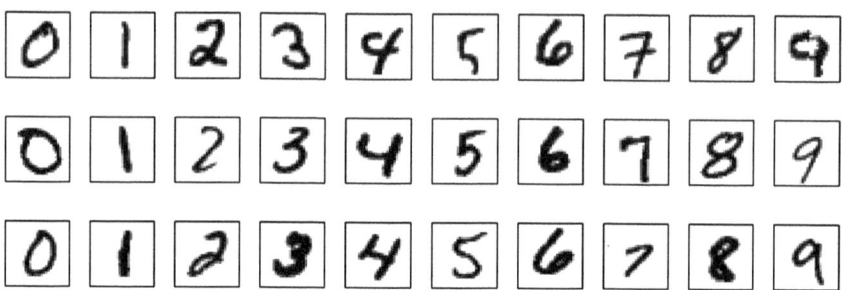

Fig. D.1 Examples of the MNIST digits

Appendix E
Parzen Windows

A method based on the Parzen window is known as a technique of estimating the probability density function from given data. Here, an outline of the method and experimental results using the Parzen window are introduced.

E.1 Estimation of Probability Density Functions

As shown in Eq. (8.12), if we know the probability density function $p(x|\omega_i)$ for each class, we can design the Bayes discriminant function . Here, the following cases are assumed.

(1) The probability density function is completely known.
(2) The form of the probability density function is known, but its parameters are unknown.
(3) The probability density function is completely unknown.

  In (1) above, it is easy to design the discriminant function. As an example of (2), the distribution is known to be a multivariate normal distribution, but its mean vector, and covariance matrix, are unknown. In this case, the probability density function can be obtained by estimating the unknown parameters from the observed data, i.e., the learning patterns. This method of estimating the parameters of the probability density function from the learning patterns is the parametric learning and was described in Sect. 4.1.

  However, the above (1) and (2) cannot be expected in the actual pattern recognition and machine learning problems. Therefore, the actual problem to be faced can be considered as (3) above.

  In the following, we assume that the probability density function is completely unknown, and describe a method for estimating it. This method does not assume the form of the probability density function in advance, but estimates it based only on the learning patterns. This method, which does not involve parameter estimation, is the nonparametric learning, and was also described in Sect. 4.1.

Suppose that n patterns, $x_1, \ldots, x_n$, distributed according to the probability density function $p(x)$, are observed in a d-dimensional space. In nonparametric learning, the probability density function $p(x)$ is estimated from these n patterns. The probability P that a pattern x exists in a region $\mathcal{R}$ is expressed as

$$P = \int_{\mathcal{R}} p(x')dx'. \tag{E.1}$$

Suppose that m out of the n patterns exist in the region $\mathcal{R}$. If n is sufficiently large, we may consider

$$P \approx \frac{m}{n}. \tag{E.2}$$

In Eq. (E.1), if $\mathcal{R}$ is small enough and $p(x)$ can be considered almost constant within this region, we can write

$$P = p(x) \int_{\mathcal{R}} dx' = p(x)V, \tag{E.3}$$

where x is the pattern in the region $\mathcal{R}$ and V is the volume of $\mathcal{R}$. From Eqs. (E.2) and (E.3), we can obtain

$$p(x) \approx \frac{m}{nV}. \tag{E.4}$$

The probability density function $p(x)$ expressed in the above equation is the average value obtained within the region $\mathcal{R}$, and is considered constant within $\mathcal{R}$. Therefore, the following two conditions must be satisfied in estimating $p(x)$ by Eq. (E.4).

Condition 1: To improve the estimation accuracy of the probability density function $p(x)$, the number of patterns m in the region $\mathcal{R}$ must be as large as possible.

Condition 2: To reveal the fine structure of the probability density function $p(x)$, $p(x)$ must be obtained with high resolution with respect to x, and $\mathcal{R}$ must be set as small as possible.

Since the number of patterns used for estimation is finite and n, in order to satisfy condition 2, setting $V \to 0$ necessarily leads to $m \to 0$ which violates condition 1. The above two conditions are in conflict with each other, and the value of V must be a compromise between the two conditions.

To facilitate further discussion, we define a function $\varphi(\mathbf{u})$ whose variable is a d-dimensional vector

$$\mathbf{u} = (u_1, \ldots, u_d)^t \tag{E.5}$$

as follows:

$$\varphi(\mathbf{u}) = \begin{cases} 1 & (|u_j| \leq 1/2, \quad \text{for all } j) \\ 0 & \text{(otherwise)}. \end{cases} \tag{E.6}$$

The $\varphi(\mathbf{u})$ in the above equation is a function that assumes a *hypercube* centered at the origin with a side length of 1, and takes the value 1 if $\mathbf{u}$ is inside it and 0 otherwise. For the function $\varphi(\mathbf{u})$,

$$\varphi(\mathbf{u}) \geq 0, \tag{E.7}$$

$$\int \varphi(\mathbf{u}) \, d\mathbf{u} = 1 \tag{E.8}$$

hold, so $\varphi(\mathbf{u})$ has the property of a probability density function.

Let $\mathcal{R}_x$ denote the region defined by the hypercube whose center is the pattern x in the d-dimensional space and whose side length is h. It is clear that the following formula holds for any pattern x_k:

$$\varphi\left(\frac{x_k - x}{h}\right) = \begin{cases} 1 & (x_k \in \mathcal{R}_x) \\ 0 & \text{(otherwise)}. \end{cases} \tag{E.9}$$

The above expression $\varphi\left((x_k - x)/h\right)$ is a function that takes 1 only when the pattern x_k is within the region $\mathcal{R}_x$. Therefore, using Eq. (E.9), among n patterns $x_1, \ldots, x_n$, the number of patterns m contained within the region $\mathcal{R}_x$ can be written as follows:

$$m = \sum_{k=1}^{n} \varphi\left(\frac{x_k - x}{h}\right). \tag{E.10}$$

Since the volume V of the region $\mathcal{R}_x$ is

$$V = h^d, \tag{E.11}$$

substituting Eqs. (E.10) and (E.11) into Eq. (E.4) yields

$$\hat{p}(x) = \frac{1}{nh^d} \sum_{k=1}^{n} \varphi\left(\frac{x_k - x}{h}\right) \tag{E.12}$$

as the estimated expression $\hat{p}(x)$ for $p(x)$. From the above equation,

$$\int \hat{p}(x) \, dx = \frac{1}{nh^d} \sum_{k=1}^{n} \int \varphi\left(\frac{x_k - x}{h}\right) dx \tag{E.13}$$

holds and the integral on the right side of the above equation is the volume of the region $\mathcal{R}_x$. Therefore, using Eq. (E.11), the above equation becomes

$$\int \hat{p}(x)\, dx = \frac{1}{nh^d} \sum_{k=1}^{n} h^d = 1, \tag{E.14}$$

and the estimated expression $\hat{p}(x)$ satisfies the condition as a probability density function.

There are other possible functions that satisfy Eqs. (E.7) and (E.8). One of them is the Gaussian function as follows:

$$\varphi(\mathbf{u}) = (2\pi)^{-d/2} \exp\left[-\frac{1}{2}\|\mathbf{u}\|^2\right]. \tag{E.15}$$

These functions are called *window functions*. In particular, when they are used to estimate probability density functions, they are called Parzen windows.

E.2   Experiments Using Parzen Windows

In the following, estimation experiments of the probability density function are performed using Parzen windows. As a simple example, let us take a one-dimensional probability density function ($d = 1$). The probability density function to be estimated is the mixed beta distribution

$$p(x) = \frac{2}{5} \operatorname{Be}(x; 3, 9) + \frac{3}{5} \operatorname{Be}(x; 12, 6), \tag{E.16}$$

where $\operatorname{Be}(x; \alpha, \beta)$ is the *beta distribution* with α, β as parameters and is expressed by the following equation:

$$\operatorname{Be}(x; \alpha, \beta) = \frac{x^{\alpha-1}(1-x)^{\beta-1}}{B(\alpha, \beta)} \qquad (0 \le x \le 1). \tag{E.17}$$

The denominator $B(\alpha, \beta)$ of the above equation is called the *beta function*. This is the normalization term required for the integration result of the beta distribution $\operatorname{Be}(x; \alpha, \beta)$ to be 1, and is expressed by the following equation:

$$B(\alpha, \beta) = \int_0^1 v^{\alpha-1} (1-v)^{\beta-1}\, dv \qquad (\alpha > 0,\ \beta > 0). \tag{E.18}$$

The beta distribution is defined on the interval $[0, 1]$. The mixed beta distribution of Eq. (E.16) is shown in Fig. E.1. In the following, we show the results of estimating the mixed beta distribution of Eqs. (E.16) by (E.12). The patterns used for the

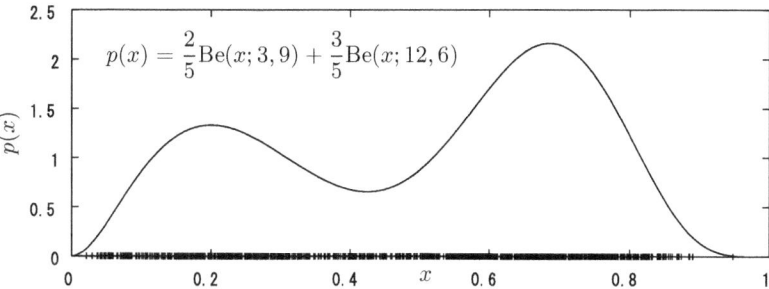

Fig. E.1 Mixed beta distribution

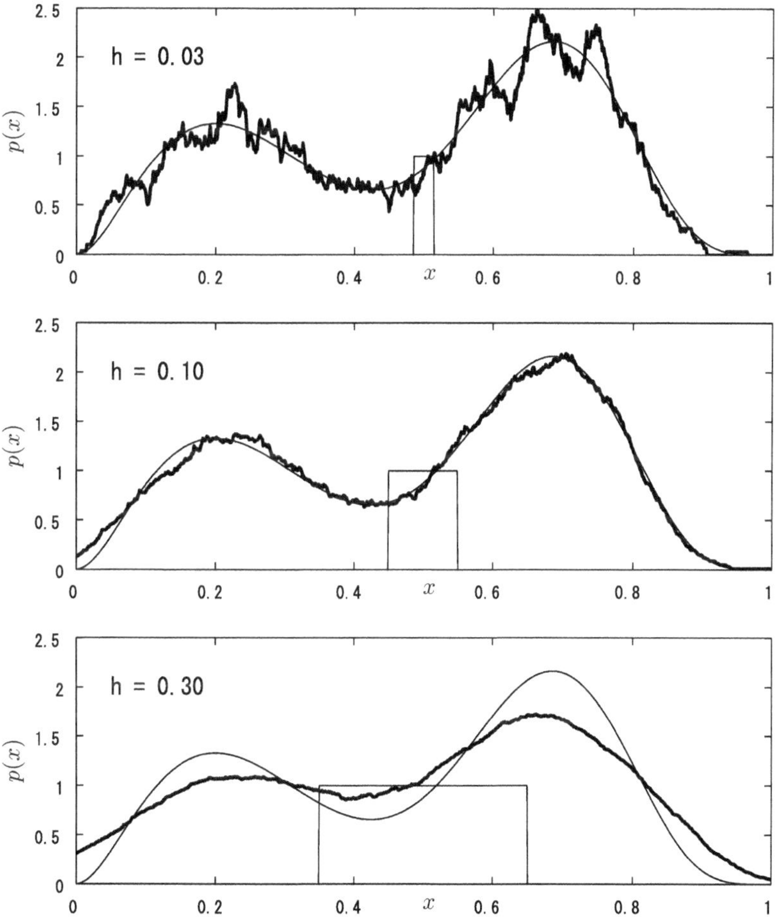

Fig. E.2 Estimation of mixed beta distribution (square wave)

estimation are the 1000 patterns ($n = 1000$) generated according to Eq. (E.16), and they are plotted on the x axis of Fig. E.1.

The results of using $\varphi(u)$ in Eq. (E.6) for the estimation are shown in Fig. E.2. In the figure, the probability density function $p(x)$ to be estimated is shown as a thin line, and the probability density function $\hat{p}(x)$ obtained by the estimation is shown as a thick line. In order to confirm the effect of h, which represents the spread of the window function, we have plotted three graphs with different values of $h = 0.03$, 0.10, and 0.30. The shape of the function $\varphi(u/h)$ used is indicated by the gray line in the center of each graph. In this figure, when $h = 0.03$, the structure of the probability density function cannot be reproduced correctly due to insufficient smoothing. When $h = 0.30$, the fine structure cannot be reproduced due to excessive smoothing. When $h = 0.10$, which is in the middle of both, the original structure is reproduced most faithfully. This is a compromise between conditions 1 and 2 described in Sect. E.1. However, because the function $\varphi(u)$ used has discontinuity points, none of the estimated graphs is smooth. Then, the results of using $\varphi(u)$ of Eq. (E.15) as a function that does not include such discontinuity points are shown in Fig. E.3. In this figure, plots are shown for three different values, $h = 0.01, 0.05$, and 0.10. As in the previous figure, the shape of the function $\varphi(u/h)$ used is indicated by a gray line in the center of each graph.

Also in this figure, when $h = 0.01$, the graph is insufficiently smoothed and when $h = 0.10$, the graph does not follow the changes sufficiently due to excessive smoothing. When $h = 0.05$, which is in the middle of both, a good estimation result is obtained, and a trend similar to Fig. E.2 can be observed. The graph of Fig. E.3 is smoother than that of Fig. E.2, so the effect of using the Gaussian function Eq. (E.15) can be confirmed.

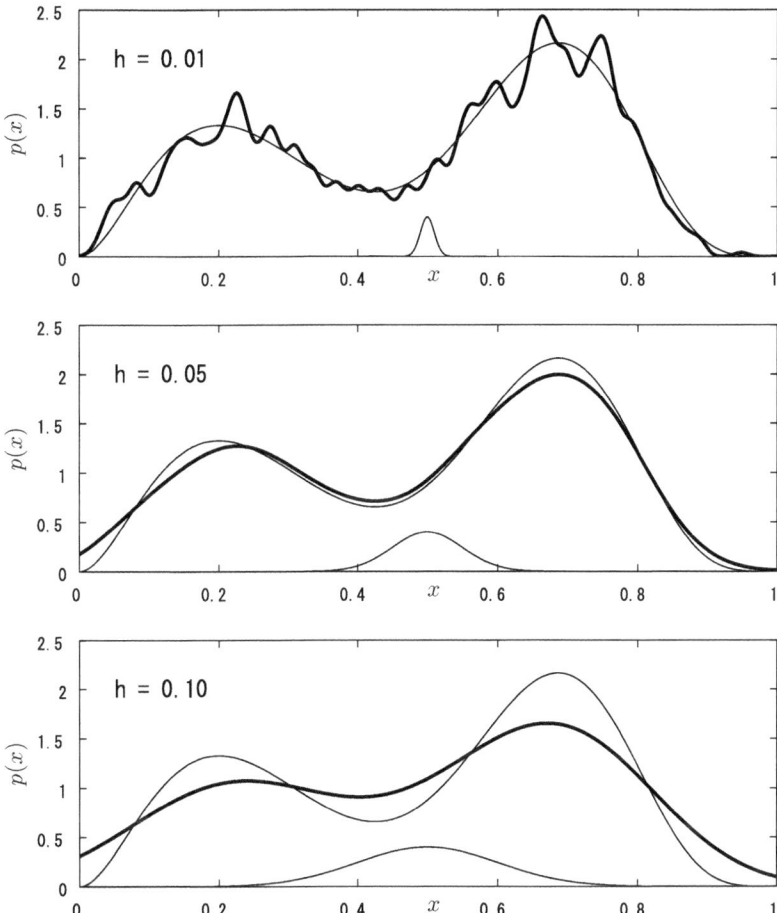

Fig. E.3 Estimation of mixed beta distribution (Gaussian function)

Appendix F
Optimization Problems Under Inequality Constraints

In the following, we address optimization problems under inequality constraints, which are necessary for solving problems of support vector machines. First, let d-dimensional vector x be a variable and $f(x)$ be a function to be optimized. Optimization is to minimize or maximize $f(x)$. Since maximization of $f(x)$ is equivalent to minimization of $-f(x)$, we will assume that optimization means minimization. Next, m functions $g_i(x)$ ($i = 1, \ldots, m$) are defined as functions to set constraints. Assume that both functions $f(x)$ and $g_i(x)$ are differentiable with respect to x. The optimization problem under the inequality constraints can be written as follows.

Optprob. F.1 When the vector x satisfies conditions

$$g_i(x) \leq 0 \quad (i = 1, \ldots, m), \tag{F.1}$$

find $x \; (= x^*)$ that minimizes $f(x)$ and its minimum value $f(x^*)$.

The function $f(x)$ is called the *objective function* and $g_i(x)$ is called the *constraint function*. The optimization problem described above is an *linear programming problem* if all of the functions $f(x)$ and $g_i(x)$ are linear. On the other hand, it is a *nonlinear programming problem* if either $f(x)$ or $g_i(x)$ is nonlinear. The set of x that satisfies the constraint Eq. (F.1) is called the *feasible region*.

Here we introduce the *Lagrangian function* $L(x, \lambda)$ as follows:

$$L(x, \lambda) = f(x) + \sum_{i=1}^{m} \lambda_i g_i(x), \tag{F.2}$$

where λ is expressed as

© The Author(s), under exclusive license to Springer Nature Singapore Pte Ltd. 2026
K. Ishii et al., *Pattern Recognition and Machine Learning for Self-Study I*, Springer
Asia Pacific Mathematics Series 1, https://doi.org/10.1007/978-981-95-1478-6

$$\boldsymbol{\lambda} = (\lambda_1, \ldots, \lambda_m)^t. \tag{F.3}$$

A necessary condition for the existence of a solution $\boldsymbol{x}^*$ to the problem listed in Optprob. F.1 is the existence of $\boldsymbol{\lambda} = \boldsymbol{\lambda}^* = (\lambda_1^*, \ldots, \lambda_m^*)^t$ at $\boldsymbol{x} = \boldsymbol{x}^*$ such that the following equations hold:

$$\frac{\partial L}{\partial \boldsymbol{x}} = \boldsymbol{0}, \tag{F.4}$$

$$g_i(\boldsymbol{x}) \leq 0, \tag{F.5}$$

$$\lambda_i \geq 0, \tag{F.6}$$

$$\lambda_i \, g_i(\boldsymbol{x}) = 0 \tag{F.7}$$

$$(i = 1, \ldots, m).$$

The vector derivative $\partial L / \partial \boldsymbol{x}$ of Eq. (F.4) is defined as

$$\frac{\partial L}{\partial \boldsymbol{x}} = \left(\frac{\partial L}{\partial x_1}, \ldots, \frac{\partial L}{\partial x_d} \right)^t. \tag{F.8}$$

In Eq. (F.4), $\boldsymbol{0}$ represents a column vector whose elements are all zeros. For vector derivatives, see Appendix B.

The above conditions are known as the *Karush–Kuhn–Tucker conditions*, or *KKT conditions* for short (Cristianini and Shawe-Taylor 2000). Of these KKT conditions, the most notable is Eq. (F.7) known as the *complementarity condition*. In the following, we will discuss what this formula indicates.

Let us take the ith constraint $g_i(\boldsymbol{x}) \leq 0$ among the m constraints expressed by Eq. (F.5). This constraint can be divided into two cases: $g_i(\boldsymbol{x}^*) < 0$ when the optimal solution $\boldsymbol{x}^*$ exists inside the domain satisfying the constraint, and $g_i(\boldsymbol{x}^*) = 0$ when it exists on the boundary of the domain.

If $g_i(\boldsymbol{x}^*) < 0$, then $\lambda_i^* = 0$ from Eq. (F.7). In that case, from Eq. (F.2), this i-th constraint does not affect the computation of the optimal solution $\boldsymbol{x}^*$ and can be ignored. Such a constraint is called the *inactive constraint*.

If $g_i(\boldsymbol{x}^*) = 0$, then we can set $\lambda_i^* > 0$ from Eq. (F.7), which is reflected in the Lagrangian function of Eq. (F.2), so this constraint is called the *active constraint*. That is, according to the complementarity condition, the optimal solutions $\boldsymbol{x}^*$ and λ_i^* satisfy either of the following equations:

$$\lambda_i^* > 0 \quad \text{and} \quad g_i(\boldsymbol{x}^*) = 0, \tag{F.9}$$

$$\lambda_i^* = 0 \quad \text{and} \quad g_i(\boldsymbol{x}^*) < 0 \tag{F.10}$$

$$(i = 1, \ldots, m).$$

The above holds for the general functions $f(\boldsymbol{x})$ and $g_i(\boldsymbol{x})$. Let us consider the following case. That is, the constraint functions $g_i(\boldsymbol{x})$ $(i = 1, \ldots, m)$ are linear and the objective function $f(x)$ is quadratic. Such an optimization problem is called the *quadratic programming problem*. The quadratic programming problem can be written as follows.

Optprob. F.2 When the vector x satisfies conditions

$$g_i(x) = a_{i0} + \mathbf{a}_i^t x \leq 0 \quad (i = 1, \ldots, m), \tag{F.11}$$

find $x \ (= x^*)$ that minimizes

$$f(x) = \frac{1}{2} x^t \mathbf{Q} x + \mathbf{v}^t x, \tag{F.12}$$

and its minimum value $f(x^*)$, where $\mathbf{a}_i$ and $\mathbf{v}$ are d-dimensional column vectors, a_{i0} is a scalar, and $\mathbf{Q}$ is a $d \times d$ matrix.

In general, there is not always a globally optimal solution for quadratic programming problems. However, if the matrix $\mathbf{Q}$ is positive semidefinite, then $f(x)$ is a convex function and a globally optimal solution exists.[1] Such a quadratic programming problem is called the *convex quadratic programming problem* and the method of solving the problem is called the *convex quadratic programming*. For the definition of the positive semidefinite matrix, see Sect. H.2.

The optimization problem handled by the SVM is nothing but a convex quadratic programming problem where the matrix $\mathbf{Q}$ is positive semidefinite in Optprob. F.2. In other words, the fact that the SVM is attributed to a convex quadratic programming problem guarantees a globally optimal solution, and brings about the advantages of the SVM described in Sect. 10.1.

So, let us take the following example as a convex quadratic programming problem, where the functions $f(x)$ and $g_i(x)$ are specifically set.

Example F.1 When a variable $x = (x_1, \ldots, x_d)^t$ in a d-dimensional space exists in a region satisfying m inequalities

$$g_i(x) = a_{i0} + \sum_{j=1}^{d} a_{ij} x_j \leq 0 \quad (i = 1, \ldots, m), \tag{F.13}$$

find $x = x^*$ that minimizes

---

[1] A necessary and sufficient condition for $f(x)$ to be a convex function is that $d \times d$ *Hessian* whose (i, j) components are $\partial^2 f(x)/\partial x_i \partial x_j$ is positive semidefinite. The Hessian of $f(x)$ represented by Eq. (F.12) is equal to $\mathbf{Q}$. Therefore, a necessary and sufficient condition for $f(x)$ to be a convex function is that the matrix $\mathbf{Q}$ is positive semidefinite. For details, see Ishii and Ueda (2026).

$$f(x) = \frac{1}{2} \sum_{j=1}^{d} (x_j - c_j)^2,$$ (F.14)

and the minimum value $f(x^*)$, where a_{ij} and c_j are constants.

Define d-dimensional column vectors $\mathbf{a}_i$ and $\mathbf{c}$ as follows:

$$\mathbf{a}_i \overset{\text{def}}{=} (a_{i1}, \ldots, a_{id})^t \qquad (i = 1, \ldots, m),$$ (F.15)

$$\mathbf{c} \overset{\text{def}}{=} (c_1, \ldots, c_d)^t.$$ (F.16)

Using these, Eqs. (F.13) and (F.14) can be written respectively as

$$g_i(x) = a_{i0} + \mathbf{a}_i{}^t x \le 0 \qquad (i = 1, \ldots, m),$$ (F.17)

$$f(x) = \frac{1}{2} \|x - \mathbf{c}\|^2 = \frac{1}{2}(x - \mathbf{c})^t (x - \mathbf{c})$$ (F.18)

$$= \frac{1}{2} x^t x - \mathbf{c}^t x + \frac{1}{2} \|\mathbf{c}\|^2.$$ (F.19)

Comparing with the formula of Optprob. F.2, Eq. (F.17) agrees with Eq. (F.11). The constant term $\|\mathbf{c}\|^2/2$ in Eq. (F.19) has no effect on the magnitude of $f(x)$, so we can ignore it. Furthermore, by setting $\mathbf{Q} = \mathbf{I}_d$ and $v = -\mathbf{c}$ in Eqs. (F.12), (F.12) coincides with Eq. (F.19). Since $\mathbf{I}_d$ is a d-dimensional identity matrix, it is positive definite. Therefore, this example is a convex quadratic programming problem.

By substituting Eqs. (F.17) and (F.18) into Eq. (F.2), the Lagrangian function for this example can be written as

$$L(x, \lambda) = \frac{1}{2}(x - \mathbf{c})^t (x - \mathbf{c}) + \sum_{i=1}^{m} \lambda_i (a_{i0} + \mathbf{a}_i{}^t x).$$ (F.20)

From the above equation and Eq. (F.4) we have

$$\frac{\partial L(x, \lambda)}{\partial x} = x - \mathbf{c} + \sum_{i=1}^{m} \lambda_i \mathbf{a}_i = \mathbf{0},$$ (F.21)

and from this we obtain

$$x^* = \mathbf{c} - \sum_{i=1}^{m} \lambda_i \mathbf{a}_i.$$ (F.22)

Substituting the above expression into Eq. (F.20) and expressing the Lagrangian function as $L(\lambda)$ of only λ gives

$$
L(\lambda) = \frac{1}{2} \left(\sum_{i=1}^{m} \lambda_i \mathbf{a}_i{}^t \right) \left(\sum_{j=1}^{m} \lambda_j \mathbf{a}_j \right)
$$

$$
+ \sum_{i=1}^{m} \lambda_i a_{i0} + \left(\sum_{i=1}^{m} \lambda_i \mathbf{a}_i{}^t \right) \left(\mathbf{c} - \sum_{j=1}^{m} \lambda_j \mathbf{a}_j \right)
$$

$$
= \frac{1}{2} \sum_{i=1}^{m} \sum_{j=1}^{m} \lambda_i \lambda_j \mathbf{a}_i{}^t \mathbf{a}_j + \sum_{i=1}^{m} \lambda_i a_{i0} + \sum_{i=1}^{m} \lambda_i \mathbf{a}_i{}^t \mathbf{c} - \sum_{i=1}^{m} \sum_{j=1}^{m} \lambda_i \lambda_j \mathbf{a}_i{}^t \mathbf{a}_j
$$

$$
= -\frac{1}{2} \sum_{i=1}^{m} \sum_{j=1}^{m} \lambda_i \lambda_j \mathbf{a}_i{}^t \mathbf{a}_j + \sum_{i=1}^{m} \lambda_i (a_{i0} + \mathbf{a}_i{}^t \mathbf{c}). \tag{F.23}
$$

Here we put

$$
h_{ij} = \mathbf{a}_i{}^t \mathbf{a}_j \qquad (i, j = 1, \ldots, m), \tag{F.24}
$$

and define an $m \times m$ matrix $\mathbf{H} = (h_{ij})$ whose (i, j) component is h_{ij}. Also define a vector $\mathbf{q}$ whose i-th $(i = 1, \ldots, m)$ component is $a_{i0} + \mathbf{a}_i{}^t \mathbf{c}$. That is,

$$
\mathbf{q} = \begin{pmatrix} a_{10} + \mathbf{a}_1{}^t \mathbf{c} \\ \vdots \\ a_{m0} + \mathbf{a}_m{}^t \mathbf{c} \end{pmatrix}. \tag{F.25}
$$

Using these, Eq. (F.23) can be expressed as follows:

$$
L(\lambda) = -\frac{1}{2} \lambda^t \mathbf{H} \lambda + \lambda^t \mathbf{q}. \tag{F.26}
$$

Thus, the minimization problem of Example F.1 becomes the following maximization problem:

$$
\left.
\begin{aligned}
\text{Maximize} \quad & L(\lambda) = -\frac{1}{2} \lambda^t \mathbf{H} \lambda + \lambda^t \mathbf{q} \\
\text{subject to} \quad & \lambda \geq \mathbf{0}
\end{aligned}
\right\} \tag{F.27}
$$

Substituting the solution $\lambda = \lambda^* = (\lambda_1^*, \ldots, \lambda_m^*)^t$ into Eq. (F.22), we obtain $\mathbf{x}^*$, and from Eq. (F.18) we obtain $f(\mathbf{x}^*)$.

Example F.1 is called the *primal problem* in the convex quadratic programming problem , while Eq. (F.27) is called the *dual problem*. The reason why minimization in the primal problem becomes maximization in the dual problem can be explained as follows.

From the definitions so far, the following formula holds:

$$L(\lambda) = L(x^*, \lambda) = \min_{x} L(x, \lambda) \tag{F.28}$$

$$\leq L(x, \lambda) = f(x) + \sum_{i=1}^{m} \lambda_i g_i(x) \tag{F.29}$$

$$\leq f(x). \tag{F.30}$$

The relationship between the above Eqs. (F.29) and (F.30) is clear from Eqs. (F.5) and (F.6). The above equations indicate the following. That is, the maximum value of $L(\lambda)$ in the dual problem gives a lower bound of $f(x)$, and the minimum value of $f(x)$ in the main problem gives an upper bound of $L(\lambda)$. It is known that a necessary and sufficient condition for the solution of the main problem to be x^* and for the solution of the dual problem to be λ^* is

$$f(x^*) = L(\lambda^*). \tag{F.31}$$

Since the dual problem is constrained only by the condition that the variables are non-negative, its solution is generally simpler than the main problem.

Next, let's solve the following problem, where specific numerical values are given for Example F.1. This example is helpful to understand the complementarity condition shown in Eq. (F.7).

Example F.2 Suppose that $x = (x_1, x_2)^t$ in two-dimensional space exists in the range satisfying the following three inequalities:

$$g_1(x) = x_1 + x_2 - 4 \leq 0, \tag{F.32}$$

$$g_2(x) = -2x_1 + x_2 - 4 \leq 0, \tag{F.33}$$

$$g_3(x) = x_1 - 5x_2 + 2 \leq 0. \tag{F.34}$$

In this case, find $x = x^*$ that minimizes

$$f(x) = \frac{1}{2} \left((x_1 - 2)^2 + (x_2 - 4)^2 \right), \tag{F.35}$$

and the minimum value $f(x^*)$

This example employs $d = 2$ and $m = 3$, which is illustrated in Fig. F.1. The region satisfying the three inequalities Eqs. (F.32)–(F.34), i.e., the feasible region, is shown in gray in the figure. In the figure, the contour lines of the function $f(x)$

Fig. F.1 Optimization under inequality constraints (1)

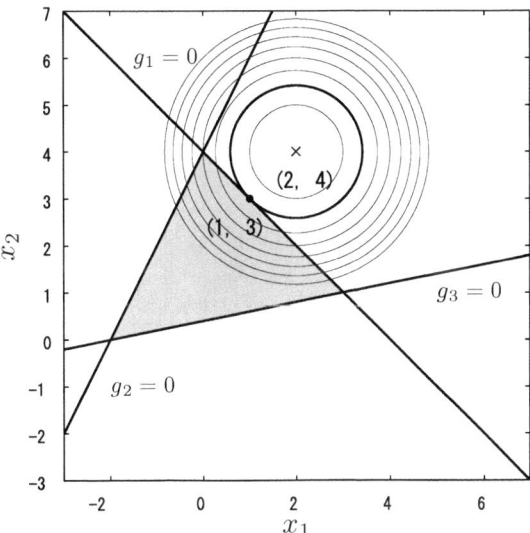

to be minimized are drawn as concentric circles centered at $(2, 4)$, and the inner concentric circles have smaller values of $f(x)$. The center of the concentric circles is marked by $\times$.

The values given in the example are as follows:

$$\left.\begin{array}{l} \mathbf{a}_1 = (1, 1)^t, \ \mathbf{a}_2 = (-2, 1)^t, \ \mathbf{a}_3 = (1, -5)^t \\ a_{10} = -4, \quad a_{20} = -4, \quad a_{30} = 2 \\ \mathbf{c} = (2, 4)^t. \end{array}\right\} \tag{F.36}$$

From these values and Eqs. (F.24) and (F.25), $\mathbf{H}$ and $\mathbf{q}$ are represented as follows:

$$\mathbf{H} = \begin{pmatrix} 2 & -1 & -4 \\ -1 & 5 & -7 \\ -4 & -7 & 26 \end{pmatrix}, \tag{F.37}$$

$$\mathbf{q} = (2, -4, -16)^t. \tag{F.38}$$

By using these, solving the dual problem of Eq. (F.27), gives[2]

$$\boldsymbol{\lambda}^* = (1, 0, 0)^t,$$

$$\text{that is} \quad \lambda_1^* = 1, \quad \lambda_2^* = \lambda_3^* = 0. \tag{F.39}$$

---

[2] Various libraries are available for solving quadratic programming problems. In this book, we used MATLAB's quadprog.

Therefore, from Eq. (F.22),

$$x^* = \mathbf{c} - \lambda_1^* \mathbf{a}_1$$
$$= (2, \, 4)^t - (1, \, 1)^t$$
$$= (1, \, 3)^t \tag{F.40}$$

holds, and the minimum value is obtained as

$$f(x^*) = f(1, 3) = \frac{1}{2}(1^2 + 1^2) = 1. \tag{F.41}$$

Let's check the above result using Fig. F.1. The $x = x^*$ that minimizes the function $f(x)$ is the x on the concentric circle with the smallest radius among the x in the gray area. Therefore, the x^* to be found is the point of contact $(1, 3)$ between the line $g_1(x) = 0$ and the circle indicated by the thick line in the figure, which is shown by the black dot in the figure. This result is consistent with Eq. (F.40).

This result shows that this optimization problem is equivalent to the problem of minimizing the function $f(x)$ under the constraint $g_1(x) = 0$. This is shown in Eq. (F.39). That is, as explained for Eq. (F.7) of the complementarity conditions, one of the KKT conditions, $g_1(x) \leq 0$ is an active constraint, and $g_2(x) \leq 0$ and $g_3(x) \leq 0$ are inactive constraints. Therefore, the constraints on $g_2(x)$ and $g_3(x)$ can be ignored. Next, let us take the following example, which is a partial modification of Example F.2.

Example F.3 If $x = (x_1, x_2)^t$ in two-dimensional space exists in the range satisfying the three inequalities Eqs. (F.32)–Eq. (F.34) shown in Example F.2, find $x = x^*$ that minimizes $f_1(x)$ and $f_2(x)$ in the following equations:

$$f_1(x) = \frac{1}{2}\left((x_1 - 5)^2 + (x_2 - 1)^2\right), \tag{F.42}$$

$$f_2(x) = \frac{1}{2}\left((x_1 - 1)^2 + (x_2 - 1)^2\right). \tag{F.43}$$

Also find the minimum values $f_1(x^*)$ and $f_2(x^*)$, respectively.

The contents of this example are shown in Figs. F.2 and F.3 respectively.

First, we take the minimization of $f_1(x)$. This problem can be transformed into a dual problem of Eq. (F.27) in the same way as before. The matrix $\mathbf{H}$ is the same as Eq. (F.37) and the vector $\mathbf{q}$ is

$$\mathbf{q} = (2, \, -13, \, 2)^t, \tag{F.44}$$

instead of Eq. (F.38). Solving this dual problem yields

Fig. F.2 Optimization under inequality constraints (2)

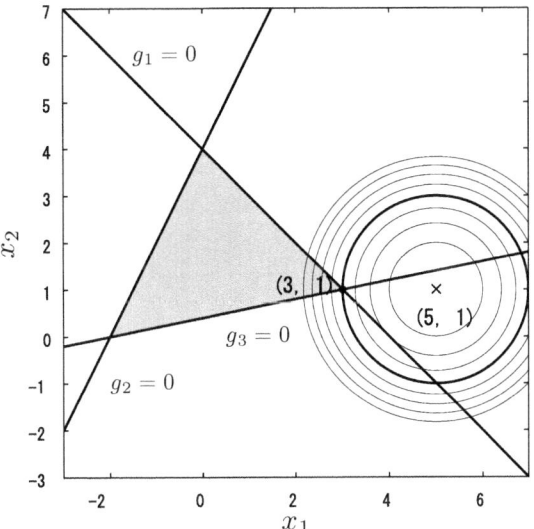

$$\boldsymbol{\lambda}^* = \left(\frac{5}{3},\ 0,\ \frac{1}{3} \right)^t,$$

$$\text{that is,} \quad \lambda_1^* = \frac{5}{3}, \quad \lambda_2^* = 0, \quad \lambda_3^* = \frac{1}{3}. \tag{F.45}$$

By using Eq. (F.22),

Fig. F.3 Optimization under inequality constraints (3)

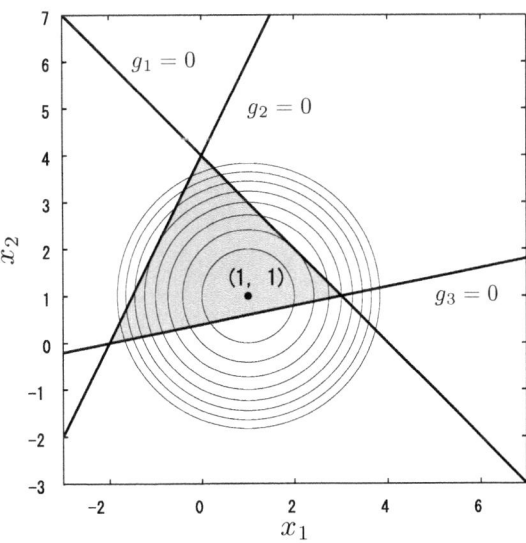

$$x^* = c - \lambda_1^* a_1 - \lambda_3^* a_3$$

$$= (5, \ 1)^t - \frac{5}{3}(1, \ 1)^t - \frac{1}{3}(1, \ -5)^t$$

$$= (3, \ 1)^t, \tag{F.46}$$

and the minimum value is obtained as

$$f(x^*) = f(3, 1) = \frac{1}{2}(2^2 + 0^2) = 2. \tag{F.47}$$

In this case, $g_1(x) \le 0$ and $g_3(x) \le 0$ are active constraints, and $g_2(x) \le 0$ is an inactive constraint. In other words, the constraint on $g_2(x)$ can be ignored. Therefore, the optimal solution x^* is the intersection of $g_1(x) = 0$ and $g_3(x) = 0$ (black points in the figure), and the circle (thick line in the figure) passing through this intersection corresponds to $f_1(x)$ which gives the minimum value.

In a similar way, for the minimization of $f_2(x)$, solving the dual problem using

$$q = (-2, \ -5, \ -2)^t \tag{F.48}$$

yields

$$\lambda^* = (0, \ 0, \ 0)^t,$$

$$\text{that is,} \quad \lambda_1^* = \lambda_2^* = \lambda_3^* = 0, \tag{F.49}$$

and all the given constraints are inactive. Therefore, since the problem is unconstrained, the following results are obtained from Eq. (F.22):

$$x^* = c = (1, \ 1)^t, \tag{F.50}$$

$$f(x^*) = 0. \tag{F.51}$$

Since the optimal solution x^* is inside the gray feasible region, as shown in Fig. F.3, it is confirmed that all the given constraints are inactive and the problem is unconstrained.

Appendix G
Mathematical Expression Using Pattern Matrix

In general, n d-dimensional feature vectors can be represented as a matrix of $n \times d$. Data representation by matrices is convenient in that various statistics and calculations can be organized and described concisely using matrices and vectors. This appendix focuses mainly on those used frequently in this book. The symbols used below are summarized in the list of symbols in this book.

Let $\mathbf{1}_l = (1, \ldots, 1)^t$ be an l-dimensional column vector with all elements 1, then the $l \times m$ matrix $\mathbf{1}_{lm}$ with all elements 1 can be expressed as

$$\mathbf{1}_{lm} = \mathbf{1}_l \mathbf{1}_m^t. \tag{G.1}$$

We have already shown that the matrix

$$\mathbf{X} = (\boldsymbol{x}_1, \ldots, \boldsymbol{x}_n)^t \quad (\in \mathbb{R}^{n \times d}) \tag{G.2}$$

in which n feature vectors

$$\boldsymbol{x}_k = (x_{k1}, \ldots, x_{kd})^t \quad (\in \mathbb{R}^d, \quad k = 1, \ldots, n) \tag{G.3}$$

are arranged is called the pattern matrix.[1]

Consider a subspace spanned by $m(< d)$ linearly independent vectors $\mathbf{a}_1, \ldots, \mathbf{a}_m$ in a d-dimensional space. Let's find a *projection matrix* $\mathbf{P}$ $(\in \mathbb{R}^{d \times d})$ representing the projection onto this subspace. Let $\mathbf{A} = (\mathbf{a}_1, \ldots, \mathbf{a}_m)$ $(\in \mathbb{R}^{d \times m})$ be a matrix whose columns are vectors $\mathbf{a}_1, \ldots, \mathbf{a}_m$. If an arbitrary vector $\mathbf{b}$ on the original d-dimensional space is projected onto this subspace, its projection point x is expressed as follows:

$$x = \mathbf{P}\mathbf{b}. \tag{G.4}$$

---

[1] In Eq. (3.16), $\boldsymbol{x}_k$ is the augmented feature vector, so the size of $\mathbf{X}$ is $n \times (d + 1)$.

© The Author(s), under exclusive license to Springer Nature Singapore Pte Ltd. 2026
K. Ishii et al., *Pattern Recognition and Machine Learning for Self-Study I*, Springer
Asia Pacific Mathematics Series 1, https://doi.org/10.1007/978-981-95-1478-6

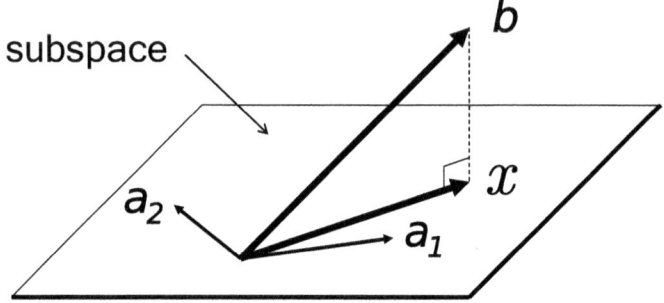

Fig. G.1 Projection onto subspace ($d = 3$, $m = 2$)

Figure G.1 shows an example for $d = 3$, $m = 2$. Since the projection point x is on the subspace, it can be expressed as a linear combination of $\mathbf{a}_1, \ldots, \mathbf{a}_m$ as in the following equation:

$$x = \sum_{j=1}^{m} q_j \mathbf{a}_j = A\hat{x}, \tag{G.5}$$

where $\hat{x}$ in the above equation is defined by

$$\hat{x} = (q_1, \ldots, q_m)^t. \tag{G.6}$$

By the definition of the projection point, the vector $(\mathbf{b} - x)$ is orthogonal to the subspace, so it is orthogonal to the vector $\mathbf{a}_1, \ldots, \mathbf{a}_m$. That is, the following equation holds:

$$A^t(\mathbf{b} - x) = 0. \tag{G.7}$$

Substituting Eq. (G.5) into the above equation yields

$$A^t(\mathbf{b} - A\hat{x}) = 0, \tag{G.8}$$

and from the above equation, the next equation is obtained:

$$\hat{x} = (A^t A)^{-1} A^t \mathbf{b}. \tag{G.9}$$

Substituting the above equation into Eq. (G.5) yields

$$x = A(A^t A)^{-1} A^t \mathbf{b}. \tag{G.10}$$

As shown in Eq. (G.4), the projection matrix $\mathbf{P}$ to be obtained is the transformation matrix from $\mathbf{b}$ to $\mathbf{x}$, so from Eq. (G.10), $\mathbf{P}$ is represented as follows;[2]

$$\mathbf{P} = \mathbf{A}(\mathbf{A}^t\mathbf{A})^{-1}\mathbf{A}^t. \tag{G.11}$$

For the projection matrix $\mathbf{P}$, the following holds:

$$\mathbf{P}^t = \mathbf{A}\left((\mathbf{A}^t\mathbf{A})^{-1}\right)^t \mathbf{A}^t = \mathbf{A}\left((\mathbf{A}^t\mathbf{A})^t\right)^{-1} \mathbf{A}^t = \mathbf{A}(\mathbf{A}^t\mathbf{A})^{-1}\mathbf{A}^t$$

$$= \mathbf{P} \tag{G.12}$$

That is, $\mathbf{P}$ is a symmetric matrix. In addition, the following formula also holds:

$$\mathbf{P}^2 = \mathbf{A}\left((\mathbf{A}^t\mathbf{A})^{-1}\mathbf{A}^t\mathbf{A}\right)(\mathbf{A}^t\mathbf{A})^{-1}\mathbf{A}^t$$

$$= \mathbf{A}\,\mathbf{I}_d\,(\mathbf{A}^t\mathbf{A})^{-1}\mathbf{A}^t$$

$$= \mathbf{P}. \tag{G.13}$$

Let $\mathbf{J}$ ($\in \mathbb{R}^{d\times d}$) be the projection matrix representing the projection onto the orthogonal complement of the subspace spanned by $\mathbf{a}_1, \ldots, \mathbf{a}_m$. Then, from Eq. (G.4)

$$\mathbf{b} - \mathbf{x} = \mathbf{b} - \mathbf{P}\mathbf{b} = (\mathbf{I}_d - \mathbf{P})\mathbf{b} = \mathbf{J}\mathbf{b}, \tag{G.14}$$

so the following holds:

$$\mathbf{J} = \mathbf{I}_d - \mathbf{P}. \tag{G.15}$$

Let $\mathbf{P}_n$ ($\in \mathbb{R}^{n\times n}$) be the projection matrix onto the space spanned by the vector $\mathbf{1}_n$ and let $\mathbf{J}_n$ ($\in \mathbb{R}^{n\times n}$) be the projection matrix onto its orthogonal complement, then the following equation is obtained from Eqs. (G.11) and (G.15):

$$\mathbf{P}_n = \mathbf{1}_n(\mathbf{1}_n^t\mathbf{1}_n)^{-1}\mathbf{1}_n^t = \frac{1}{n}\mathbf{1}_{nn} = \begin{pmatrix} 1/n & 1/n & \cdots & 1/n \\ \vdots & & \ddots & \vdots \\ 1/n & 1/n & \cdots & 1/n \end{pmatrix}, \tag{G.16}$$

$$\mathbf{J}_n = \mathbf{I}_n - \mathbf{P}_n = \mathbf{I}_n - \frac{1}{n}\mathbf{1}_{nn}$$

---

[2] If the vectors $\mathbf{a}_1, \ldots, \mathbf{a}_m$ are orthonormal bases of m-dimensional subspace, then $\mathbf{A}^t\mathbf{A} = \mathbf{I}_m$ holds, and from Eq. (G.11), we obtain $\mathbf{P} = \mathbf{A}\mathbf{A}^t$. See Eqs. (6.67) and (6.68) in Chap. 6.

$$= \begin{pmatrix} 1 - 1/n & -1/n & \cdots & -1/n \\ -1/n & 1 - 1/n & \cdots & -1/n \\ \vdots & & \ddots & \vdots \\ -1/n & -1/n & \cdots & 1 - 1/n \end{pmatrix}. \tag{G.17}$$

Let us describe the basic statistics for the pattern x_k $(k = 1, \ldots, n)$ using the pattern matrix. First, the mean vector $\bar{x}$ of n patterns is represented as

$$\bar{x} = \frac{1}{n} \sum_{k=1}^{n} x_k = \frac{1}{n} X^t 1_n. \tag{G.18}$$

The matrix representing $x_k - \bar{x}$ $(k = 1, \ldots, n)$ as a pattern matrix is called the *mean deviation matrix*. Letting $\mathbf{X}_M$ $(\in \mathbb{R}^{n \times d})$ denote this mean deviation matrix, the following equation is obtained:

$$\mathbf{X}_M = (x_1 - \bar{x}, \ldots, x_n - \bar{x})^t \tag{G.19}$$

$$= \mathbf{X} - \frac{1}{n} 1_{nn} \mathbf{X} \tag{G.20}$$

$$= \mathbf{J}_n \mathbf{X}. \tag{G.21}$$

Since Eqs. (G.12) and (G.13) are valid for any projection matrix, $(\mathbf{J}_n)^t = \mathbf{J}_n$ and $(\mathbf{J}_n)^2 = \mathbf{J}_n$ also hold for $\mathbf{J}_n$. Using this relation, the covariance matrix $\mathbf{C}_X$ $(\in \mathbb{R}^{d \times d})$ calculated from n patterns can be written using $\mathbf{X}$ as follows:

$$\mathbf{C}_X = \frac{1}{n} \sum_{k=1}^{n} (x_k - \bar{x})(x_k - \bar{x})^t \tag{G.22}$$

$$= \frac{1}{n} \mathbf{X}_M^t \mathbf{X}_M \tag{G.23}$$

$$= \frac{1}{n} \mathbf{X}^t (\mathbf{J}_n)^t \mathbf{J}_n \mathbf{X} \tag{G.24}$$

$$= \frac{1}{n} \mathbf{X}^t \mathbf{J}_n \mathbf{X} \tag{G.25}$$

$$= \frac{1}{n} (\mathbf{X}^t \mathbf{X} - \mathbf{X}^t \mathbf{P}_n \mathbf{X}). \tag{G.26}$$

On the other hand, the $n \times n$ matrix whose (i, j) component is the inner product $x_i^t x_j$ of x_i and x_j is called *inner product matrix* of $\mathbf{X}$. Let $\mathbf{Q}_X$ $(\in \mathbb{R}^{n \times n})$ be the inner product matrix of $\mathbf{X}$, then $\mathbf{Q}_X$ is expressed as follows:

$$\mathbf{Q}_X = \mathbf{X} \mathbf{X}^t. \tag{G.27}$$

Furthermore, from the above relation, the inner product matrix $\mathbf{Q}_{X_M}$ for $\mathbf{X}_M$ can be expressed as

$$\mathbf{Q}_{X_M} = \mathbf{X}_M \mathbf{X}_M^t \tag{G.28}$$

$$= (\mathbf{X} - \frac{1}{n}\mathbf{1}_{nn}\mathbf{X})(\mathbf{X} - \frac{1}{n}\mathbf{1}_{nn}\mathbf{X})^t \tag{G.29}$$

$$= (\mathbf{I}_n - \frac{1}{n}\mathbf{1}_{nn})\mathbf{X}\mathbf{X}^t(\mathbf{I}_n - \frac{1}{n}\mathbf{1}_{nn}) \tag{G.30}$$

$$= \mathbf{J}_n \mathbf{Q}_X \mathbf{J}_n. \tag{G.31}$$

Appendix H
Singular Value Decomposition

Singular value decomposition is a method of representing an arbitrary matrix by its singular values and singular vectors, and plays a useful role in pattern recognition and machine learning. In the following, an overview of singular value decomposition, the linear algebra necessary to understand it, and how to derive the formula for singular value decomposition will be described. We also show that spectral decomposition can be applied to symmetric matrices.

H.1 Basic Concepts of Singular Value Decomposition

All matrices treated below are assumed to be real matrices.

Any matrix $\mathbf{X}$ of size $m \times n$ can be represented as the product of three matrices $\mathbf{U}$, $\mathbf{V}$, and $\mathbf{\Lambda}^{1/2}$ as follows:

$$\mathbf{X} = \mathbf{U}\mathbf{\Lambda}^{1/2}\mathbf{V}^t, \tag{H.1}$$

where $\mathbf{U}$ is an $m \times r$ matrix consisting of eigenvectors $\mathbf{u}_1, \ldots, \mathbf{u}_r$ corresponding to non-zero eigenvalues $\lambda_1, \ldots, \lambda_r$ of the matrix $\mathbf{X}\mathbf{X}^t$, $\mathbf{V}$ is an $n \times r$ matrix consisting of eigenvectors $\mathbf{v}_1, \ldots, \mathbf{v}_r$ corresponding to non-zero eigenvalues $\lambda_1, \ldots, \lambda_r$ of the matrix $\mathbf{X}^t\mathbf{X}$. The non-zero eigenvalues of the matrices $\mathbf{X}\mathbf{X}^t$ and $\mathbf{X}^t\mathbf{X}$ are $\lambda_1, \ldots, \lambda_r$, which are identical. Also, $\mathbf{\Lambda}^{1/2}$ is an $r \times r$ diagonal matrix with the square roots of the above r eigenvalues as diagonal components. That is, $\mathbf{U}$, $\mathbf{V}$ and $\mathbf{\Lambda}^{1/2}$ are expressed by the following equations:

$$\mathbf{U} = (\mathbf{u}_1, \ldots, \mathbf{u}_r), \tag{H.2}$$

$$\mathbf{V} = (\mathbf{v}_1, \ldots, \mathbf{v}_r), \tag{H.3}$$

© The Author(s), under exclusive license to Springer Nature Singapore Pte Ltd. 2026
K. Ishii et al., *Pattern Recognition and Machine Learning for Self-Study I*, Springer
Asia Pacific Mathematics Series 1, https://doi.org/10.1007/978-981-95-1478-6

$$\Lambda^{1/2} = \begin{pmatrix} \sqrt{\lambda_1} & & & 0 \\ & \sqrt{\lambda_2} & & \\ & & \ddots & \\ 0 & & & \sqrt{\lambda_r} \end{pmatrix}. \tag{H.4}$$

The representation of a matrix $\mathbf{X}$ as Eq. (H.1) is called *singular value decomposition* and $\sqrt{\lambda_1}, \ldots, \sqrt{\lambda_r}$ are called *singular value*s. Vectors $\mathbf{u}_1, \ldots, \mathbf{u}_r$ and $\mathbf{v}_1, \ldots, \mathbf{v}_r$ are called *singular vectors*.

H.2   Linear Algebra Required

Understanding the singular value decomposition requires some basic concepts and theorems of linear algebra, which are briefly outlined below. For details of the proofs and derivation methods, refer to the references Strang (2023), Boyd and Vandenberghe (2018), and Tsukada et al. (2023).

Theorem H.1 *For the n eigenvectors of a square matrix $\mathbf{A}$ of size $n \times n$ to be an orthonormal basis, it is a necessary and sufficient condition that $\mathbf{A}$ be a symmetric matrix.*

Let λ_i and $\mathbf{u}_i$ be the eigenvalues and eigenvectors of the matrix $\mathbf{A}$, respectively, then

$$\mathbf{A}\mathbf{u}_i = \lambda_i \mathbf{u}_i \qquad (i = 1, \ldots, n). \tag{H.5}$$

If $\mathbf{A}$ is a symmetric matrix, then from Theorem H.1, $\mathbf{u}_1, \ldots, \mathbf{u}_n$ can be the orthonormal basis, and the following equation holds:[1]

$$\mathbf{u}_i^t \mathbf{u}_j = \begin{cases} 1 & (i = j) \\ 0 & (i \neq j). \end{cases} \tag{H.6}$$

---

[1] As is clear from Eq. (H.5), if $\mathbf{u}_i$ is an eigenvector, $a\mathbf{u}_i$ obtained by multiplying it by a constant $a(\neq 0)$ is also an eigenvector. The lower equation of Eq. (H.6) always holds, but to satisfy the upper equation, it is necessary to normalize $\mathbf{u}_i$ so that $\|\mathbf{u}_i\| = 1$.

Theorem H.2 *For all n eigenvalues of a symmetric matrix* $\mathbf{A}$ *of size* $n \times n$ *to be non-negative, it is a necessary and sufficient condition that* $\mathbf{A}$ *be positive semidefinite.*

A matrix $\mathbf{A}$ of size $n \times n$ is said to be positive semidefinite if $\mathbf{A}$ is a symmetric matrix and

$$y^t \mathbf{A} y \geq 0 \tag{H.7}$$

for any n-dimensional column vector $y(\neq \mathbf{0})$, as described in Sect. 11.2.2.

Theorem H.3 *The number of nonzero eigenvalues among the n eigenvalues of the positive semidefinite matrix* $\mathbf{A}$ *of size* $n \times n$ *is equal to the rank of matrix* $\mathbf{A}$.

For any matrix $\mathbf{X}$ of size $m \times n$, the maximum number of linearly independent column vectors in $\mathbf{X}$ is equal to the maximum number of linearly independent row vectors, and these are called the *rank* of $\mathbf{X}$.

If a positive semidefinite matrix $\mathbf{A}$ of size $n \times n$ has rank r, then $(n - r)$ out of n eigenvalues of $\mathbf{A}$ are 0.

H.3   Formula Derivation of Singular Value Decomposition

Given a matrix $\mathbf{X}$ of size $m \times n$, the matrix $\mathbf{XX}^t$ of size $m \times m$ is positive semidefinite. This is because $\mathbf{XX}^t$ is a symmetric matrix and if $\mathbf{A} = \mathbf{XX}^t$ in Eq. (H.7), the next formula holds:

$$y^t \mathbf{XX}^t y = (\mathbf{X}^t y)^t \mathbf{X}^t y$$
$$= \|\mathbf{X}^t y\|^2 \geq 0. \tag{H.8}$$

Let $\lambda_1, \ldots, \lambda_m$ be the eigenvalues of the matrix $\mathbf{XX}^t$ in descending order, and let $\mathbf{u}_1, \ldots, \mathbf{u}_m$ be the corresponding m-dimensional eigen column vectors. Let r be the rank of the matrix $\mathbf{XX}^t$. Since $\mathbf{XX}^t$ is positive semidefinite, from Theorems H.2 and H.3, eigenvalues $\lambda_1, \ldots, \lambda_r$ are positive and all other eigenvalues are 0. Therefore, using nonzero eigenvalues, we can write

$$\mathbf{XX}^t \mathbf{u}_i = \lambda_i \mathbf{u}_i \qquad (i = 1, \ldots, r), \tag{H.9}$$

where $\mathbf{u}_i$ is assumed to be normalized to satisfy Eq. (H.6).

Here, multiplying both sides of Eq. (H.9) by $\mathbf{X}^t$ from the left yields

$$\mathbf{X}^t\mathbf{X}\mathbf{X}^t\mathbf{u}_i = \lambda_i\mathbf{X}^t\mathbf{u}_i \qquad (i = 1, \ldots, r). \tag{H.10}$$

If $\mathbf{v}_i$ is defined as

$$\mathbf{v}_i = \frac{1}{\sqrt{\lambda_i}}\mathbf{X}^t\mathbf{u}_i, \tag{H.11}$$

then Eq. (H.10) becomes

$$\mathbf{X}^t\mathbf{X}\mathbf{v}_i = \lambda_i\mathbf{v}_i \qquad (i = 1, \ldots, r). \tag{H.12}$$

It is easy to verify that $\mathbf{X}^t\mathbf{X}$ in the above equation is an $n \times n$ matrix and, like $\mathbf{X}\mathbf{X}^t$, is a positive semidefinite matrix. It is also clear from the form of Eq. (H.12) that the n-dimensional column vector $\mathbf{v}_i$ is an eigenvector of $\mathbf{X}^t\mathbf{X}$, and the corresponding nonzero eigenvalues $\lambda_1, \ldots, \lambda_r$ are consistent with those of $\mathbf{X}\mathbf{X}^t$. From Eqs. (H.11), (H.9) and (H.6), the following equation holds:

$$\mathbf{v}_i^t\mathbf{v}_j = \frac{\mathbf{u}_i^t\mathbf{X}\mathbf{X}^t\mathbf{u}_j}{\sqrt{\lambda_i\lambda_j}} = \frac{\mathbf{u}_i^t\lambda_j\mathbf{u}_j}{\sqrt{\lambda_i\lambda_j}} = \sqrt{\frac{\lambda_j}{\lambda_i}}\,\mathbf{u}_i^t\mathbf{u}_j$$

$$= \begin{cases} 1 & (i = j) \\ 0 & (i \neq j). \end{cases} \tag{H.13}$$

Consequently, vectors $\mathbf{v}_1, \ldots, \mathbf{v}_r$ also form an orthonormal set.

Rewriting Eq. (H.11) gives the following equation:

$$\mathbf{X}^t\mathbf{u}_i = \sqrt{\lambda_i}\,\mathbf{v}_i \qquad (i = 1, \ldots, r) \tag{H.14}$$

Multiplying both sides of the above equation by $\mathbf{X}$ from the left and using Eq. (H.9) yields

$$\mathbf{X}\mathbf{v}_i = \sqrt{\lambda_i}\,\mathbf{u}_i \qquad (i = 1, \ldots, r). \tag{H.15}$$

It is clear that $\mathbf{X}$ satisfying Eq. (H.15) can be written as

$$\mathbf{X} = \sqrt{\lambda_1}\,\mathbf{u}_1\mathbf{v}_1^t + \cdots + \sqrt{\lambda_r}\,\mathbf{u}_r\mathbf{v}_r^t \tag{H.16}$$

The matrix notation of Eqs. (H.16) yields (H.1). That is, the singular value decomposition of $\mathbf{X}$ can be summarized as

$$\mathbf{X} = \mathbf{U}\mathbf{\Lambda}^{1/2}\mathbf{V}^t = \sum_{i=1}^{r} \sqrt{\lambda_i}\mathbf{u}_i\mathbf{v}_i^t. \tag{H.17}$$

H.4   Spectral Decomposition of Symmetric Matrices

In the singular value decomposition described so far, we considered a general matrix $\mathbf{X}$ of size $m \times n$. In the following, we consider a symmetric matrix $\mathbf{X}$ of size $n \times n$. Let $\sigma_1, \ldots, \sigma_n$ be the n eigenvalues of $\mathbf{X}$ and $\boldsymbol{w}_1, \ldots, \boldsymbol{w}_n$ be the corresponding eigenvectors, then

$$\mathbf{X}\boldsymbol{w}_i = \sigma_i\boldsymbol{w}_i \qquad (i = 1, \ldots, n) \tag{H.18}$$

holds. Assuming that $\boldsymbol{w}_1, \ldots, \boldsymbol{w}_n$ are column vectors, let us define an $n \times n$ matrix $\mathbf{W}$ composed of n eigenvectors as follows:

$$\mathbf{W} \stackrel{\text{def}}{=} (\boldsymbol{w}_1, \ldots, \boldsymbol{w}_n). \tag{H.19}$$

Since the matrix $\mathbf{X}$ is a symmetric matrix, $\boldsymbol{w}_1, \ldots, \boldsymbol{w}_n$ can be an orthonormal basis as in Eq. (H.6) from Theorem H.1. That is,

$$\boldsymbol{w}_i^t\boldsymbol{w}_j = \begin{cases} 1 & (i = j) \\ 0 & (i \neq j) \end{cases} \tag{H.20}$$

holds. Therefore, we can write

$$\mathbf{W}^t\mathbf{W} = \mathbf{I}. \tag{H.21}$$

Since the above equation shows

$$\mathbf{W}^t = \mathbf{W}^{-1}, \tag{H.22}$$

we have

$$\mathbf{W}\mathbf{W}^t = \mathbf{I}, \tag{H.23}$$

in addition to Eq. (H.21). Multiplying $\mathbf{X}$ from the left of Eq. (H.19) and applying Eq. (H.18) yields

$$\mathbf{X}\mathbf{W} = (\mathbf{X}\boldsymbol{w}_1, \ldots, \mathbf{X}\boldsymbol{w}_n) \tag{H.24}$$

$$= (\sigma_1\boldsymbol{w}_1, \ldots, \sigma_n\boldsymbol{w}_n) \tag{H.25}$$

$$= \mathbf{W}\mathbf{\Sigma}, \tag{H.26}$$

where $\mathbf{\Sigma}$ is defined by the following equation:

$$\mathbf{\Sigma} \stackrel{\text{def}}{=} \begin{pmatrix} \sigma_1 & & & 0 \\ & \sigma_2 & & \\ & & \ddots & \\ 0 & & & \sigma_n \end{pmatrix}. \tag{H.27}$$

By multiplying $\mathbf{W}^t$ from the right of Eq. (H.26) and applying Eq. (H.23), we obtain

$$\mathbf{X} = \mathbf{W}\mathbf{\Sigma}\mathbf{W}^t = \sum_{i=1}^{n} \sigma_i \mathbf{w}_i \mathbf{w}_i^t. \tag{H.28}$$

The above equation is called the *spectral decomposition* of the symmetric matrix $\mathbf{X}$.

Let us now consider the relationship between spectral decomposition and singular value decomposition. From the definition of a symmetric matrix,

$$\mathbf{X}^t\mathbf{X} = \mathbf{X}\mathbf{X}^t = \mathbf{X}\mathbf{X} \tag{H.29}$$

holds. Therefore, $\mathbf{X}^t\mathbf{X}$ and $\mathbf{X}\mathbf{X}^t$ have n common eigenvalues and eigenvectors, which coincide with the eigenvalues and eigenvectors of $\mathbf{X}\mathbf{X}$ (including eigenvalues that are zero). The eigenvalues $\lambda_1, \ldots, \lambda_n$ and eigenvectors $\mathbf{u}_1, \ldots, \mathbf{u}_n$ of the matrix $\mathbf{X}\mathbf{X}^t$ are also eigenvalues and eigenvectors of $\mathbf{X}\mathbf{X}$, so the next equation holds:

$$\mathbf{X}\mathbf{X}\mathbf{u}_i = \lambda_i \mathbf{u}_i \qquad (i = 1, \ldots, n). \tag{H.30}$$

Multiplying Eq. (H.18) by $\mathbf{X}$ from the left gives

$$\mathbf{X}\mathbf{X}\mathbf{w}_i = \sigma_i \mathbf{X}\mathbf{w}_i = \sigma_i^2 \mathbf{w}_i \qquad (i = 1, \ldots, n), \tag{H.31}$$

which means that σ_i^2 is the eigenvalue of the matrix $\mathbf{X}\mathbf{X}$, and $\mathbf{w}_i$ is the corresponding eigenvector. Comparison of Eqs. (H.30) and (H.31) yields

$$\mathbf{u}_i = \mathbf{w}_i, \quad \lambda_i = \sigma_i^2. \tag{H.32}$$

Also, since $\mathbf{V} = \mathbf{U}$, Eq. (H.17) becomes

$$\mathbf{X} = \mathbf{U}\mathbf{\Lambda}^{1/2}\mathbf{U}^t = \sum_{i=1}^{n} \sqrt{\lambda_i} \mathbf{u}_i \mathbf{u}_i^t. \tag{H.33}$$

From Eqs. (H.28), (H.33), and (H.32), we see that spectral decomposition is a special case of singular value decomposition.

References

Aizerman, M., Braverman, E., Rozonoer, L.: Theoretical foundations of the potential function method in pattern recognition learning. Autom. Remote Control **25**(6), 821–837 (1964)

Aizerman, M.A., Braverman, E.M., Rozonoer, L.I.: Potential Function Method in the Theory of Machine Learning (in Russian). Nauka, Moscow (1970)

Akaho, S.: A kernel method for canonical correlation analysis. In: Proceedings of the International Meeting of the Psychometric Society (IMPS2001) (2001)

Amari, S.: A theory of adaptive pattern classifiers. IEEE Trans. **EC-16**, 299–307 (1967)

Bach, F.R., Jordan, M.I.: Kernel independent component analysis. J. Mach. Learn. Res. **3**, 1–48 (2003)

Baudat, G., Anouar, F.: Generalized discriminant analysis using a kernel approach. Neural Comput. **12**, 2385–2404 (2000)

Belkin, M., Niyogi, P.: Laplacian eigenmaps for dimensionality reduction and data representation. Neural Computat. **15**(6), 1373–1396 (2002)

Belkin, M., Hsu, D., Ma, S., Mandal, S.: Reconciling modern machine-learning practice and the classical bias–variance trade-off. Proc Natl. Acad. Sci. **116**(32), 15849–15854 (2019)

Bhatia, N., Vandana: Survey of nearest neighbor techniques. Int. J. Comput. Sci. Inf. Secur. **8**(2), 302–305 (2010)

Bishop, C.M.: Neural Networks for Pattern Recognition. Oxford University Press, Oxford (1995)

Bishop, C.M.: Pattern Recognition and Machine Learning. Springer-Verlag, Berlin (2006)

Boser, B.E., Guyon, I.M., Vapnik, V.N.: A training algorithm for optimal margin classifiers. In: Proceedings of the 5th Annual ACM Workshop on Computational Learning Theory, pp. 144 152 (1992)

Boyd, S., Vandenberghe, L.: Introduction to Applied Linear Algebra: Vectors, Matrices, and Least Squares. Cambridge University Press, Cambridge (2018)

Breiman, L.: Bias, variance, and arcing classifiers, vol. 460. Technical Report, Statistics Department, University of California, Berkeley, 1997

Collins, M., Duffy, N.: Convolution kernels for natural language. In: Advances in Neural Information Processing Systems, vol. 14, pp. 625–632. MIT Press, Cambridge (2001)

Cortes, C., Vapnik, V.: Support vector networks. Mach. Learn. **20**, 273–297 (1995)

Cottrell, G.W., Munro, P.: Principal component analysis of images via back propagation. Proc. SPIE 1001. Visual Communications and Image Processing' 88, pp. 1070–1076 (1988)

Cover, T.M.: Classification and generalization capabilities of linear threshold units. Rome Air Development Center Technical Documentary Report, vol. RADC-TDR-64-32, pp. 515–516 (1964)

© The Author(s), under exclusive license to Springer Nature Singapore Pte Ltd. 2026
K. Ishii et al., *Pattern Recognition and Machine Learning for Self-Study I*, Springer
Asia Pacific Mathematics Series 1, https://doi.org/10.1007/978-981-95-1478-6

Cover, T.M., Hart, P.E.: Nearest neighbor pattern classification. IEEE Trans. Inf. Theory **IT-13**(1), 21–27 (1967)

Cristianini, N., Shawe-Taylor, J.: An Introduction to Support Vector Machines and Other Kernel-based Learning Methods. Cambridge University Press, Cambridge (2000)

Dasarathy, B.V., Sanchez, J.S., Townsend, S.: Nearest neighbour editing and condensing tools – synergy exploitation. Pattern Anal. Appl. **3**, 19–30 (2000)

Duda, R.O., Hart, P.E.: Pattern Classification and Scene Analysis. John Wiley & Sons, Inc., Hoboken (1973)

Duda, R.O., Hart, P.E., Stork, D.G.: Pattern Classification, 2nd edn. John Wiley & Sons, Inc., Hoboken (2001)

Efron, B., Tibshirani, R.J.: An Introduction to the Bootstrap. Chapman & Hall, Boca Raton (1993)

Fisher, R.A.: The use of multiple measurements in taxonomic problems. Ann. Eugenics **7**(Part II), 179–188 (1936); Also in: Contributions to Mathematical SyStatistics. John Wiley, New York (1950)

Fukunaga, K.: Bias of nearest neighbor error estimation. IEEE Trans. Pattern Anal. Mach. Intell. **PAMI-9**(1), 103–112 (1987)

Fukunaga, K.: Introduction to Statistical Pattern Recognition, 2nd edn. Academic Press, Inc., Amsterdam (1990)

Fukushima, K.: Neocognitron: a self-organizing neural network model for a mechanism of pattern recognition unaffected by shift in position. Biol. Cybern. **36**, 193–202 (1980)

Funahashi, K.-I.: On the approximate realization of continuous mappings by neural networks. Neural Netw. **2**(3), 183–192 (1989)

Glucksman, H.A.: Classification of mixed-font alphabetics by characteristic loci. In: IEEE Computer Conference, pp. 138–141 (1967)

Goodfellow, I., Bengio, Y., Courville, A.: Deep Learning. MIT Press, Cambridge (2016)

Guyon, I., Boser, B., Vapnik, V.: Automatic capacity tuning of very large vc-dimension classifiers. In: Proceedings of NIPS 1992, pp. 147–155 (1992)

Ham, J., Lee, D.D., Mika, S., Schölkopf, B.: A kernel view of the dimensionality reduction of manifolds. In: Proceedings of the 21st International Conference on Machine Learning (ICML'04), pp. 369–376 (2004)

Hart, P.E.: The condensed nearest neighbor rule. IEEE Trans. Info. Theory **IT-14**, 515–516 (1968)

Haussler, D.H.: Convolution kernels on discrete structures. Technical Report UCS-CRL-99-10, UC Santa Cruz, 1999

Hinton, G., Deng, L., Yu, D., Dahl, G.E., Mohamed, A.-R., Jaitly, N., Senior, A., Vanhoucke, V., Nguyen, P., Sainath, T.N., Kingsbury, B.: Deep neural networks for acoustic modeling in speech recognition: the shared views of four research groups. IEEE Signal Process. Mag. **29**(6), 82–97 (2012)

Hughes, G.F.: On the mean accuracy of statistical pattern recognizers. IEEE Trans. Inf. Theory **IT-14**, 55–63 (1968)

Iijima, T.: A theory of pattern recognition by compound similarity method. In: PRL Technical Report, IEICE of Japan, pp. 929–936, Japan, 1974. IEICE (in Japanese)

Iijima, T., Genchi, H., Mori, K.: A theory of character recognition by pattern matching method. In: Proceedngs of the 1st International Conference on Patter Recognition, pp. 50–56 (1973)

Ikeda, M., Tanaka, H., Motooka, T.: Projection distance method for recognition of hand-written characters. Trans. Inf. Process. Soc. Jpn. **24**(1), 106–112 (1983)

Ishii, K., Ueda, N.: Pattern Recognition and Machine Learning for Self-Study II – Unsupervised Learning. Springer-Verlag, Berlin (2026). (to be published)

Joachims, T.: Making large-scale svm learning practical. LS8-Report 24, Universität Dortmund, LS VIII-Report, 1998

Joachims, T.: Text categorization with support vector machines: learning with many relevant features. In: European Conference on Machine Learning (ECML), pp. 137–142 (1998)

Joachims, T.: Learning to Classify Text Using Support Vector Machines – Methods, Theory, and Algorithms. Kluwer/Springer, Alphen aan den Rijn/Berlin (2002)

Juang, B.H., Katagiri, S.: Discriminant learning for minimum error classification. IEEE Trans. Signal Process. **SP-40**(12), 3043–3054 (1992)

Kashima, H., Tsuda, K., Inokuchi, A.: Marginalized kernels between labeled graphs. In: Proceedings of the 20th International Conference on Machine Learning (ICML'03), pp. 321–328 (2003)

Krizhevsky, A., Sutskever, I., Hinton, G.E.: Imagenet classification with deep convolutional neural networks. In: Proceedings of NIPS 2012, pp. 1097–1105 (2012)

Lai, P.L., Fyfe, C.: Kernel and nonlinear canonical correlation analysis. In: Proceedings of the IEEE-INNS-ENNS International Joint Conference on Neural Networks (IJCNN'00), vol. 4, pp. 614–619 (2000)

Lathi, B.P.: Communication Systems. John Wiley & Sons, Inc., New York (1968)

LeCun, Y.: Quand la machine apprend: La révolution des neurones artificiels et de l'apprentissage profond (in French). Odile Jacob (2019)

LeCun, Y., et al.: Backpropagation applied to handwritten zip code recognition. Neural Comput. **1**, 541–555 (1989)

Lodhi, H., Saunders, C., Shawe-Taylor, J., Cristianini, N., Watkins, C.: Text classification using string kernels. J. Mach. Learn. Res. **2**, 419–444 (2002)

Maeda, E., Murase, H.: Multi-category classification by kernel based nonlinear subspace method. In: Proceedings of the 1999 IEEE International Conference on Acoustics, Speech, and Signal Processing, ICASSP '99, Phoenix, Arizona, USA, March 15–19, 1999, pp. 1025–1028. IEEE Computer Society (1999)

Mardia, K.V., Kent, J.T., Bibby, J.M.: Multivariate Analysis. Academic Press, Amsterdam (1979)

McCarthy, J.: Chess as the drosophila of ai. In: Marsland, T.A., Schaeffer, J. (eds.) Computers, Chess, and Cognition, pp. 227–237. Springer, Berlin (1990)

Mercer, J.: Functions of positive and negative type and their connection with the theory of integral equations. Philos. Trans. Roy. Soc. London, **A209**, 415–446 (1909)

Mika, S., Rätsch, G., Weston, J., Schölkopf, B., Müller, K.: Fisher discriminant analysis with kernels. In: Proceedings of IEEE Neural Networks for Signal Processing Workshop IX (NNSP'99), pp. 41–48 (1999)

Minsky, M., Papert, S.: Perceptrons. MIT Press, Cambridge (1969)

Minsky, M., Papert, S.: Perceptrons, Expanded edn. MIT Press, Cambridge (1988)

Murase, H., Nayar, S.K.: Visual learning and recognition of 3-d objects from appearance. Int. J. Computer Vision **14**, 5–24 (1995)

Nilsson, N.J.: Learning Machines. McGraw-Hill, Columbus (1965)

Oja, E.: Subspace Methods of Pattern Recognition. Research Studies Press Ltd., Baldock (1983)

Otsu, N.: A threshold selection method from gray-level histograms. IEEE Trans. Syst Man Cybern. **SMC-9**(1), 62–66 (1979)

Otsu, N.: Mathematical Studies on Feature Extraction in Pattern Recognition, vol. 818. Researches of the Electrotechnical Laboratory, Tsukuba (1981). In Japanese

Peterson, D.W.: Some convergence properties of a nearest neighbor decision rule. IEEE Trans. Inf. Theory **IT-16**(1), 26–31 (1970)

Rosenfeld, A., Kak, A.: Digital Picture Processing, vols. 1 and 2. Academic Press, Amsterdam (1982)

Rumelhart, D.E., McClelland, J.L.: Parallel Distributed Processing. MIT Press, Cambridge (1986)

Roweis, S.T., Saul, L.K.: Nonlinear dimensionality reduction by locally linear embedding. Science **290**(5500), 2323–2326 (2000)

Saul, L.K., Roweis, S.T., Singer, Y.: Think globally, fit locally: unsupervised learning of low dimensional manifolds. J. Mach. Learn. Res. **4**, 119–155 (2003)

Schölkopf, B.: The kernel trick for distances. In: Proceedings of NIPS 2000, pp. 301–307 (2000)

Schölkopf, B., Smola, A., Müller, K.-R.: Nonlinear component analysis as a kernel eigenvalue problem. Neural Comput. **10**(5), 1299–1319 (1998)

Sebestyen, G.S.: Decision-Making Processes in Pattern Recognition. Macmillan, London (1962)

Seide, F., Li, G., Yu, D.: Conversational speech transcription using context-dependent deep neural networks. In: INTERSPEECH, pp. 437–440. ISCA (2011)

Sejnowski, T.J.: The Deep Learning Revolution. MIT Press, Cambridge (2018)

Seung, S.H., Lee, D.D.: Cognition: the manifold ways of perception. Science **290**(5500), 2268–2269 (2000)

Srivastava, N., Hinton, G.E., Krizhevsky, A., Sutskever, I., Salakhutdinov, R.: Dropout: a simple way to prevent neural networks from overfitting. J. Mach. Learn. Res. **15**, 1929–1958 (2014)

Strang, G.: Introduction to Linear Algebra, 6th edn. Wellesley-Cambridge Press, Wellesley (2023)

Tenenbaum, J.B., de Silva, V., Langford, J.C.: A global geometric framework for nonlinear dimensionality reduction. Science **290**(5500), 2319–2323 (2000)

Tou, J.T., Gonzalez, R.C.: Pattern Recognition Principles. Addison-Wesley Publishing Company, Boston (1974)

Tsuda, K.: Subspace classifier in the hilbert space. Pattern Recogn. Lett. **20**(5), 513–519 (1999)

Tsukada, M., Kobayashi, Y., Kaneko, H., Takahasi, S., Shirayanagi, K., Noguchi, M.: Linear Algebra with Python. Springer-Verlag, Berlin (2023)

Ueda, N., Nakano, R.: Generalization error of ensemble estimators. In: Proceedings of International Conference on Neural Networks (ICNN96), pp. 90–95 (1996)

Vapnik, V.: The Nature of Statistical Learning Theory. Springer, Berlin (1995)

Vaswani, A., Shazeer, N., Parmar, N., Uszkoreit, J., Jones, L., Gomez, A.N., Kaiser, L., Polosukhin, I.: Attention is all you need. In Proceedings of NIPS 2017, pp. 6000–6010 (2017)

Watanabe, S.: Knowing & Guessing — Quantitative Study of Inference and Information. John Wiley & Sons, Inc., New York (1969)

Index

A

Activation function, **310**, 337, 340, 361
Active constraint, **428**, 434, 436
Additive property, **156**
Aizerman, M.A., 214
Akaike Information Criterion (AIC), 88
AlexNet, 335
All-subsets kernel, 271, **272**
Amount of information, 102
ANOVA kernel, **274**, 274
a posteriori probability, **70**
a priori probability, **70**
Asymptotic consistency, **76**
Asymptotic efficiency, **76**
Asymptotic unbiasedness, **76**
Atom, **123**
Augmented feature vector, **20**, 201
Augmented weight vector, **20**
Autocorrelation matrix, **134**, 391
Autoencoder, **334**
Average pooling, **342**, 349

B

Backpropagation method, 46, 308, **310**, 338, 347
Band-pass filter, **357**
Batch learning, **55**, 311, 383
Bayes decision boundary, **399**
Bayes decision rule, 51, **70**, 100, 187, 215, 281, 378, 399
Bayes discriminant function, **102**, 187, 215, 397, 419
Bayes error, 98, **101**, 101, 103, 332

Bayesian estimation, **76**
Bayes posterior probability, 399
Bayes probability vector, 398
Bayes risk, **101**, 378
Bayes' theorem, **70**, 99
Beta distribution, **422**
Beta function, **422**
Between-class covariance matrix, **145**, 151
Between-class scatter matrix, **140**, 152
Between-class variance, **97**, 146, 395, 396
Bias, **20**, 87, 309, 340, 345, 346
Big data, **335**
Bootstrap method, **90**
Border ratio, **193**, 196, 206, 220
Boser, Bernhard, 229, 259
BP, **310**
BP method, 323
BS method, **90**

C

Calculus of variations, 393
Canonical correlation analysis (CCA), **290**
Capacity, **86**
Category, **3**
Chain rule, **314**
Character recognition, **3**, 153, 173, 295, 338, 381, 415
Charged particle, 213
CLAFIC method, **170**, 292, 296
Class, **3**
Class autocorrelation matrix, 172
Classification, **4**
Classification dictionary, **5**, 8

© The Editor(s) (if applicable) and The Author(s), under exclusive license to Springer
Nature Singapore Pte Ltd. 2026
K. Ishii et al., *Pattern Recognition and Machine Learning for Self-Study I*, Springer
Asia Pacific Mathematics Series 1, https://doi.org/10.1007/978-981-95-1478-6